Advances in Geographical and Environmental Sciences

Series Editor

R. B. Singh, University of Delhi, Delhi, India

Advances in Geographical and Environmental Sciences synthesizes series diagnostigation and prognostication of earth environment, incorporating challenging interactive areas within ecological envelope of geosphere, biosphere, hydrosphere, atmosphere and cryosphere. It deals with land use land cover change (LUCC), urbanization, energy flux, land-ocean fluxes, climate, food security, ecohydrology, biodiversity, natural hazards and disasters, human health and their mutual interaction and feedback mechanism in order to contribute towards sustainable future. The geosciences methods range from traditional field techniques and conventional data collection, use of remote sensing and geographical information system, computer aided technique to advance geostatistical and dynamic modeling.

The series integrate past, present and future of geospheric attributes incorporating biophysical and human dimensions in spatio-temporal perspectives. The geosciences, encompassing land-ocean-atmosphere interaction is considered as a vital component in the context of environmental issues, especially in observation and prediction of air and water pollution, global warming and urban heat islands. It is important to communicate the advances in geosciences to increase resilience of society through capacity building for mitigating the impact of natural hazards and disasters. Sustainability of human society depends strongly on the earth environment, and thus the development of geosciences is critical for a better understanding of our living environment, and its sustainable development.

Geoscience also has the responsibility to not confine itself to addressing current problems but it is also developing a framework to address future issues. In order to build a 'Future Earth Model' for understanding and predicting the functioning of the whole climatic system, collaboration of experts in the traditional earth disciplines as well as in ecology, information technology, instrumentation and complex system is essential, through initiatives from human geoscientists. Thus human geosceince is emerging as key policy science for contributing towards sustainability/survivality science together with future earth initiative.

Advances in Geographical and Environmental Sciences series publishes books that contain novel approaches in tackling issues of human geoscience in its broadest sense—books in the series should focus on true progress in a particular area or region. The series includes monographs and edited volumes without any limitations in the page numbers.

More information about this series at http://www.springer.com/series/13113

R. B. Singh · Bathula Srinagesh ·
Subhash Anand
Editors

Urban Health Risk and Resilience in Asian Cities

Springer

Editors
R. B. Singh
Department of Geography
Delhi School of Economics
University of Delhi
Delhi, India

Bathula Srinagesh
Department of Geography
Osmania University
Hyderabad, Telangana, India

Subhash Anand
Department of Geography
Delhi School of Economics
University of Delhi
Delhi, India

ISSN 2198-3542 ISSN 2198-3550 (electronic)
Advances in Geographical and Environmental Sciences
ISBN 978-981-15-1204-9 ISBN 978-981-15-1205-6 (eBook)
https://doi.org/10.1007/978-981-15-1205-6

This Springer imprint is published by the registered company Springer Nature Singapore Pte Ltd.
The registered company address is: 152 Beach Road, #21-01/04 Gateway East, Singapore 189721, Singapore

Preface

Presently, world is facing crucial and grave danger as Coronavirus disease (COVID-19). This is the kind of crisis on human existence and can not be eliminated without understanding the environment, urban development process and health degradation. Different models of cities have been found in the past decades; today, we are more concerned about healthy, sustainable and future cities. Urban areas are often most vulnerable to hazards and disasters. The global urban sustainability and sustainable development goals requires cities to be sustainable by ensuring that they are healthy places to live in and providing opportunities to improve the wellbeing of its inhabitants. Cities are the dominant human habitat, where environmental, social, cultural and economic factors have impacts on human health and wellbeing. The world is becoming more urban day by day. In 2050, the world population will reach about 9.5 billion with around 66% living in cities. The urban population in Asia has increased from 32% in 1990 to 47% in 2019 and years are not very far when Asia will be more than half urbanized.

Global sustainable development challenges concentrate in cities and are closely tied to human health and wellbeing. On 25 September 2015, global community adopted United Nations 17 Sustainable Development Goals (SDGs); out of these, two Sustainable Development Goals address these concerns: Goal 3 is related to "Ensure healthy live and promote wellbeing for all" and Goal 11 discusses to "Make cities inclusive, safe, resilient and sustainable". Highest scientific bodies like International Council for Science (ICSU) started new initiatives: Health and Wellbeing in Changing Urban Environments—A System Analysis Approach. Global sustainability cannot be achieved without local sustainability. Urban health and wellbeing is central to the mission of the International Council for Science (ICS), which is to strengthen international science for the benefit of the society. ICSU's strategic programme: Health and Wellbeing in Changing Urban Environment—A System Analysis Approach was initiated in the above context. The programme recognizes the constraints due to geographical diversity in social status, income level, culture, governance, capacity and most importantly availability of data for implementing systems analysis approach in health management in the region.

Environment, climate and health degradation are big threats to the sustainability of the future earth, particularly in Asia. The factors which influence urban health include urban governance; demographic characteristics; natural and built environment; social and economic conditions; services and health emergency management; and food security. Now, focus is being shifted from Illness to Wellness. The Asia Health and Wellbeing Initiative (AHWIN) promotes vibrant and healthy societies where people can enjoy long and productive lives, and to contribute to the region's sustainable and equitable development and economic growth, it fosters sustainable and self-reliant healthcare system in Asia. By looking at the health problems and issues, an idea came in mind that there should be research volume to provide the analytical overview of health conditions and wellbeing in Asian cities. Interdisciplinary research with biophysical and human geosciences should be promoted and the roles of geographers are crucial to providing the necessary sociocultural linkages by connecting social sciences, humanities and science and technology for the sustainable cities. It is now realized that future research initiatives for sustainability need to be appropriate, indigenous, smart and solution-orientated.

In this volume, conceptual framework of health applied as a natural integrator cutting across different urban sectors attempts to describe the health problems and solutions having case studies. The knowledge gained through the case studies will be very crucial in developing local, regional and global sustainable planning for sustainable health. It provides an overview of environmental, geographical and cultural urban development and related health challenges and explains how good health can be achieved in Asian cities. This volume is an outcome of the valuable contributions made by eminent scientists and academicians who have been striving to develop alternative strategies, solutions and modes for urban sustainability through supplying the conceptual tools as well as evidence-based suggestions to planners. This volume will be useful for academicians, scientists, policymakers, decision makers and various related stakeholders. We are highly thankful to all the contributors for their significant research papers and thoughts.

Delhi, India — Prof. R. B. Singh
Hyderabad, India — Prof. Bathula Srinagesh
Delhi, India — Dr. Subhash Anand
August 2019

Contents

Part I Urban Process, Vulnerability and Risk: Methods and Techniques

1 **Systems Approach for Climate Change Impacts on Urban Health: Conceptual Framework, Modelling and Practice** 3
Yinpeng Li, Peter Urich and Chonghua Yin

2 **"S.I.T.E" (Societal-Institutional-Technical-Economic) Valuation Framework: A Case of Drinking-Water Facilities and Services in Slum Areas of the Hyderabad Region, India** 33
Anil Kumar Palakodeti

3 **Inequalities in Access to Water and Sanitation: A Case of Slums in Selected States of India** 57
Anuradha Banerjee and Bidisha Chattopadhyay

4 **Chennai Floods 2005, 2015: Vulnerability, Risk and Climate Change** 73
T. Vasantha Kumaran, O. M. Murali and S. Rani Senthamarai

5 **An Environmental Study of Solid Waste Management System in Chandrapur City, Maharashtra, India** 101
Y. Y. Dudhapachare

6 **Explaining the links Between Purdah Practice, Women's Autonomy and Health Knowledge in India** 111
Pintu Paul

7 **Development of Urban Heat Island and Its Relation to Heat Waves** 127
Rajashree Borah

Part II Urban Health and Wellbeing

8 Growing Urbanization, Health Infrastructure, and Vector-Borne Diseases: A Study in Khammam Municipal Corporation, Telangana State 143
A. Rambabu, G. S. Srinivasa Gopinath and B. Srinivas

9 Maternal Reproductive Health: A Comparison Between India and Empowered Action Group States 149
Deepali Chanu Sanasam

10 Research Frontiers in Water, Environment, and Human Health 165
Dilek Eren Akyuz

11 Impacts of Resource Consumption and Waste Generation on Environment and Subsequent Effects on Human Health: A Study Based on Ecological Footprint Analysis 179
Tanmoy Roy Tusher, Tonmoy Pondit, Moinul Hasan, Mahmuda Binte Latif and Md. Binyamin

12 The Impact of Climate Change on Human Eyes 205
Murali Krishna V. Iyyanki and Prisilla Jayanthi

13 Health and Well-Being of Ageing Population in India: A Case Study of Kolkata 217
Lakshmi Sivaramakrishnan and Ambika Roy Bardhan

14 Noise Pollution and Its Consequences on Urban Health in Sylhet City 231
Tulshi Kumar Das and Rony Basak

15 Neighbourhood Deprivation and Health; A Study of Low-Income Neighbourhoods in Azamgarh City 253
Saleha Jamal and Uzma Ajmal

16 Quality of Living, Health and Well-Being of Slum Dwelling Women Domestic Workers in Kolkata 271
Sudeshna Roy

17 Urban Growth and Environment and Health Hazards in Kathmandu Valley, Nepal 293
Bandana Pradhan, Puspa Sharma and Pushkar K. Pradhan

Part III Integrated Urban Risk Mitigation and Resilience Measures

18 Impact of Urbanization on Megacities' Lakes Using Remote Sensing Technique: A Case Study of Water Quality Analysis in Ambattur Lake, Chennai, India 327
V. Chandrasekar, Shaik Mahamad and G. Latha

19 Evaluation of Mass Rapid Transit System (MRTS): A Case Study of Delhi 343
Menka Kumari and Anuradha Banerjee

20 The First Survey of Flood Disaster Preparedness in Hanoi, Vietnam 355
Minh Dinh Nguyen

21 Tradition Meets Innovation: Herbal Medicine as a Sustainable Source of Anticancer Agents 367
Shanoo Suroowan and Mohamad Fawzi Mahomoodally

22 Urban Health Infrastructure in Small Cities: Is It Availability or Accessibility? 389
Krishna Udnoor, Priya Narayanan and Ramesh Anguluri

23 An Approach to Social Sustainability in Chennai—Understanding the Dynamics of Public Places 409
Reshmi Panicker

24 Water Scarcity in Delhi: Mapping for Solutions and the Way Forward 421
Rituparna Bhattacharyya and Sanjay Prasad

25 Assessing the Relationship Between Neighbourhood Socio-economic Status and Environmental Quality 445
Saleha Jamal and Uzma Ajmal

Part I
Urban Process, Vulnerability and Risk: Methods and Techniques

Chapter 1
Systems Approach for Climate Change Impacts on Urban Health: Conceptual Framework, Modelling and Practice

Yinpeng Li, Peter Urich and Chonghua Yin

Abstract Climate change has direct and indirect impacts on urban human health and well-being at multiple scales and levels. WHO and other international and national organisations have exerted great effort in understanding and adapting to the add-on or new challenges from climate change. A systems approach is an ideal tool for managing the analysis of such complicated systemic challenges. Systems approaches are accepted in health sector projects and research work in past decades from conceptual frameworks to practical applications. The socio-ecological system (SES) framework is an appropriate approach for such a complex system. Applying SES frameworks starts with the scoping and conceptualisation of risk identification; then forms system dynamics models, or hybrid models, and ends with scenario analysis. This approach can provide different types of support for organisational thinking and decision-making. A system approach software platform UrbanCLIM/RIDS—an integrated decision support system—is briefly described in this chapter. The tool includes mapping, modelling, data and knowledge management functionalities for addressing the integrated challenges of climate change, urban health and well-being. There are many opportunities to push forward in applications of system sciences to analyse climate change and health SES issues. Given the nature of a complex system, there is no simple pathway to get people to understand, and communicate between different sub-systems. However, barriers could be breached with systems approaches, data technologies and economic innovations.

Keywords System approach · Mapping · Modelling · Health · Climate change

Y. Li (✉) · P. Urich · C. Yin
International Global Change Institute and CLIMsystems Ltd, Hamilton 3200, New Zealand
e-mail: yinpengli@climsystems.com

CLIMsystems Ltd, Hamilton 3204, New Zealand

R. B. Singh et al. (eds.), *Urban Health Risk and Resilience in Asian Cities*,
Advances in Geographical and Environmental Sciences,
https://doi.org/10.1007/978-981-15-1205-6_1

1.1 Introduction

Nowadays, cities are the dominant human habitat, where a large number of environmental, social, cultural and economic factors have impacts on human health and well-being. This is a kind of typical complex system (Batty 2013). In addition, climate change has posed extra risks for urban population health and well-being, through the direct impacts on food, nutrition, waste, air quality, transport, infrastructure, housing and indirectly on energy, safety and access to health care, and other urban public services. Currently, there is a limited understanding of the complex processes that shape urban population health and well-being. On the other hand, systems approaches are increasingly being recognised as appropriate tools for examining complex socio-ecological system (SES) issues (ICSU 2011).

Climate change positioned as discrete to the environment and disconnected from health systems is an ineffective approach. A holistic SES perspective that integrates a range of climate change and sustainability-related initiatives and focused on actions (i.e., implementation of solutions) is considered a better pathway. Such an approach typically, explicitly and iteratively connects climate mitigation with cultural attitudes, social and intergenerational justice, human rights, economic transitions and efforts to minimise degradation of human and environment systems (Hall et al. 2017).

SESs are often large complex systems, which contain a significant number of attributes such as nonlinearity, uncertainty, emergence, scale and self-organisation as defined by Bar-Yam (2002). Gunderson et al. (2002) and Ostrom (2009) considered SES systems theory as the amalgamation of natural and social science approaches.

SES analysis frameworks have been developed and applied for different sectors over several decades (Binder et al. 2013; Schlüter et al. 2014).The development of sustainable management strategies and better understanding are new methods based on social and ecological interaction. However, SES frameworks have not been adopted as mainstream tools in population health planning and policymaking (Bai et al. 2012; Maglio et al. 2014). There are several issues hampering its applications, which consist of (a) ranges of uncertainty derived from the use of statistical approaches based on confidence intervals and estimate (b) the difficulty of multidisciplinary collaboration for modelling and simulation, and (c) systems scientists' inability to communicate effectively about the added value of specific tools, as well as (d) limited financial support for population health systems science.

Some of the transdisciplinary potentials of the systems approach has been realised (e.g., Castella et al. 2005). The systems approach has also made headway in organisational development and the analysis of learning organisations. Senge (1990) and a number of other scholars regard systems thinking as key to the understanding of complexity and options for change in organisations and as a mental model to facilitate strategy development. As anthropogenic activities endanger sustainability, research needs to address whether and how systems thinking on human/nature connections can increase the learning capacity of society at large-scale (Clark et al. 2016).

System dynamic models and other new modelling technology can serve as a set of systems science tools to explore social-ecological interactions to examine behaviours and outcomes resulting from these interactions over time (Page et al. 2017; Atkinson et al. 2015). Studied systems are typically very complex. It is, therefore, challenging to take different disciplinary theories and knowledge bases and integrate them. This chapter proposes a systems approach based on a SES framework, and furthermore describes a decision support platform based on the proposed approach, where a set of useful tools are provided to bridge the gaps of research and practice between climate change and human health.

1.2 Human Health and Well-Being

Climate change incessantly impacts human health and well-being by interfering with the environment, social conditions and infrastructure. It is important to understand these interactions so as to better integrate climate change and other drivers of global environmental change like social and economic conditions, habitat loss, land degradation and ecosystem disruptions to address the impact of climate change on human health (UNFCCC 2017). In general, climate change needs to be considered in terms of impact, vulnerability and adaptation in a SES framework (Berry et al. 2017).

We define SES as a composition of four subsystems: institutional, economic, social, and ecological (bio-physical). The key elements, functions and their interdependencies with climate change are briefly described as follows: Figure 1.1 maps out the analysis structure.

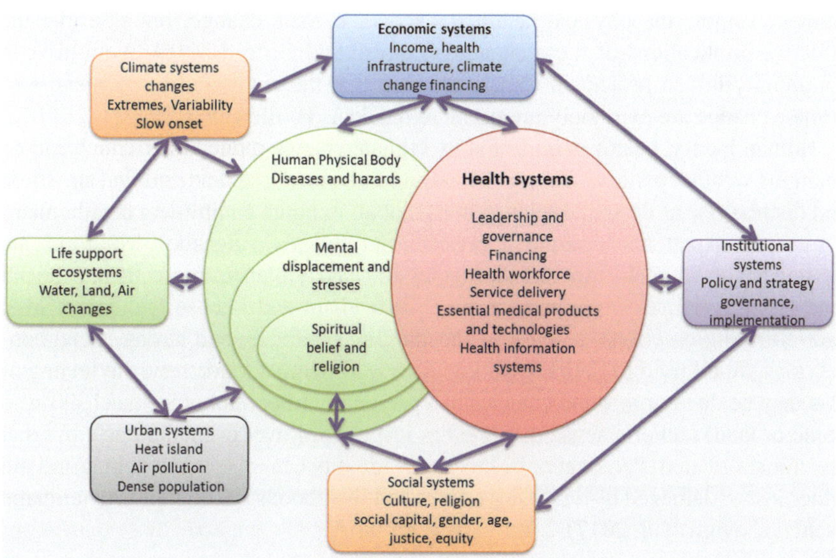

Fig. 1.1 A high-level systemic framework for climate change impacts on urban public health. This framework needs to be further explored for the next level of detail for each system in special case studies *Source* Author created

1.2.1 Physical Body—Diseases

Studies have shown that airborne, vectorborne and waterborne diseases are all sensitive to climate change (Patz et al. 2005; Greer et al. 2008), among which the incidence of mosquito-borne diseases (such as malaria, dengue and viral encephalitis) are most sensitive to climate change (Patz et al. 1996). Brenner et al. (2017) found that airborne infectious diseases spreading dynamics are strongly dependent on the climatic environment of the epidemic outbreak location, and the season during the year when the outbreak is happening. Curriero et al. (2001) found fifty-one percent of waterborne disease outbreaks were preceded by extreme precipitation events that resulted in surface water contamination entering surface and groundwater sources. Forzieri et al. (2017) indicated that two-thirds of the European population will be exposed to climate change related weather distasters. By 2100, and disaster-attributable deaths are expected to increase by roughly 50 times, mainly from heatwaves.

Other categories of human disease related to climate change include asthma, respiratory allergies, airway diseases, cancer, cardiovascular disease and stroke, neurological diseases and disorders (Patz et al. 2005).

1.2.2 Mental Health

The affective dimensions of climate change and changes in the land directly impact emotional health and well-being (Fritze et al. 2008; Shukla 2016). While numerous studies examine the physical health impacts of climate change, few consider the affective implications of these changes (Gifford and Gifford 2016). A multi-year, community-driven project in Canada showed that the emotional consequences of climate change are extremely important to residents (Willox et al. 2013).

Human mental health is impacted by climate change-induced extreme weather, changing weather patterns, damaged food and water resources and polluted air. Stress and distress rise as do social relationships that all impinge on physical health, memory loss, sleep patterns, immune suppression and changes in digestion. Major chronic mental health impacts can include aggression and violence, more mental health emergencies, helplessness, hopelessness, or fatalism and intense feelings of loss. Life-affirming emotions, a sense of meaning and purpose, and strong social connections are all used to define psychological well-being. Conversely, the feeling of loss may be due to profound changes in a personally important place (such as one's home or land) and/or a sense that one has lost control over events in one's life that are climate-related. Personal or occupational identity can arise when belongings are either lost or damaged by a disaster or jobs and livelihoods are disrupted by climatic events (Clayton et al. 2017).

Personal relationships and community interactions can be affected by shifts in climate. Compounded stress from a changing environment, eco-migration, and/or eco-anxiety can affect an entire community's sense of mental well-being. A loss of social identity and cohesion followed by hostility, violence, and interpersonal and intergroup aggression can result.

Climate change can threaten one's ability to process information and make decisions without being disabled by extreme emotional responses. Extreme weather increasingly can be a source of trauma. Such an experience can trigger disabling emotions. More subtle and indirect effects of climate change can add stress to people's lives in varying degrees. Most people are able to cope with a certain amount of stress. However, the accumulated effects can overwhelm a person and flip them from a mentally healthy to a mentally ill condition. Uncertainty can be stressful and a risk factor for psychological problems (Swim et al. 2009). People can become distraught by the news of other's negative experiences, and by fears—founded or unfounded—about their own vulnerability.

1.2.3 Spiritual Well-Being

Spirituality is recognised as a real phenomenon and not merely a "mental illusion" (Rose 2001; Koenig 2015). More detailed assessments to extend our understanding of spiritual wellness are needed, to help diagnose spiritual disease, so well designed spiritual care might be provided to recover spiritual health (Bergmann 2009; Zwingmann et al. 2011). Gomez and Fisher (2003) defined four facets of spiritual well-being: personal; communal; environmental and transcendental spiritual. Rowold (2011) applied a Spiritual Well-Being Questionnaire (SWBQ-G). The SWBQ-G was found to be valid and that each of the four SWBQ-G scales discriminated between mental, physical and emotional well-being. The SWBQ-G also predicted levels of future happiness, psychological well-being (positive relationship) and stress (negative relationship). Such results affirm the validity of the construct of spiritual well-being.

Although religions have considerable reach and act as major social actors and institutions relatively little social science research has addressed the interaction of religious bodies and human-induced climate change (Gerten 2010; Haluza-DeLay 2014). Murphy et al. (2016) found that religious beliefs can have tangible influences on the lived practices of individuals and communities, and vice versa. They can be a critical determinant of the adaptive capacity of a community to climate change (Posas 2007).

1.2.4 Community Health

Besides residents' mental and physical health affecting communities, the impacts of climate on community health can have a particularly strong effect on community fabric and interpersonal relationships (Berry et al. 2008). The fabric of social interaction can be shifted by climate-related events. Relationships between the natural world and individual and the larger society can also change (Black et al. 2011).

1.3 Elements and Structures of SES Subsystems in the Context of Human Health and Climate Changed World

A socio-ecological systems framework emphasises the integrated concept of human beings in a natural environment. The delineation between social systems and ecological (or bio-physical) systems is stressed as being artificial and arbitrary. Social and ecological systems are linked within the SES approach through feedback mechanisms as both display complexity and resilience.

In this section, we describe with regard to SES, the key elements and structures of the subsystems within human health and well-being systems in the context of a climate-changed world.

1.3.1 Ecological (Bio-Physical) Systems

1.3.1.1 Climate System: Climate Variability, Extremes and Slow Onset Changes

Climate change presents, in terms of variability, two broad categories of interpretation: extremes and slow onset. Climate variability includes the seasonality and annual variations. For example, rainfall and temperature vary in patterns year by year. Climate extremes include extreme heat, heat waves, cold, extreme precipitation, drought cyclone (hurricane) and storm surge events. They will definitely change as global mean temperatures increase (Meehl et al. 2000; Stott 2016). Climate extremes have clear impacts on human health (Hashim and Hashim 2016). Slow onset changes include mean temperature change, warm nights, sea ice melting, mean sea level rise and those impacting societies in a slow way (Stocker 2014). Whatever these key elements or variables impact human health directly or indirectly, they need to be considered in a defensible and actionable manner with local historical and projected future climate change data when impacts assessment are carried out.

1.3.1.2 Human Life Support Ecosystem

Although human life is limited, it is a long-term task to protect our earth systems (Cash et al. 2003) as this involves intergenerational equity. The earth system includes all the elements of land, soil, water, food and air each of which is life-supporting element that links directly to human health. They have been studied over history, such as land and soil water degradation (Rosenzweig et al. 2001; Huang et al. 2016), desertification (Xu et al. 2014; Li et al. 2016), soil erosion (Nearing et al. 2004; Panagos et al. 2017), water scarcity (Gosling and Arnell 2016), food production (Deryng et al. 2014; Lesk et al. 2016; Cheeseman 2016) and air pollution (D'Amato et al. 2013). However, the understanding and awareness of protecting this system are still in a developing stage with new evidence being found with time.

Food

In a climate changed world, elevated carbon dioxide stimulates the yields of wheat and other grains; however, it reduces their protein/nitrogen (N) concentration in turn. Other essential nutrients are also subject to change (Broberg et al. 2017). It was found that the staple foods of rice, wheat, barley and potato protein contents decreased by more than 6% with the increase of carbon dioxide. As a consequence, by 2050, assuming today's diets and levels of income inequality stay static, an additional 1.6% or 148.4 million of the world's population may be placed at risk of protein deficiency because of elevated CO_2 (Medek et al. 2017).

Water quality and supply

Climate changes include slow onset changes such as mean temperature increase, and extreme events including, heavy rainfall, cyclones, floods, heatwaves, droughts, extreme cold and wildfires. Changes in water catchments, storage reservoirs, the capacity of water treatment processes or the viability of distribution systems can have far-reaching and potentially harmful impacts on the quality fo drinking water (Khan et al. 2015; Urich et al. 2017). The provision of safe drinking water could be exposed to new risks. This may require the integration of the knowledge gained from examination and recognition of systemic risk relations, in combination with more inclusive collaboration across the water and related sectors (Boholm and Prutzer 2017).

Air quality

The source of emissions, transport process, dilution, chemical transformation and the eventual deposition of air pollutants can be directly impacted by air temperature, humidity, wind speed and direction, and atmospheric mixing layer. Concentrations of other human health-related air contaminants, such as smoke from wildfires, airborne pollens and molds could be exacerbated by climate change. Growing studies focusing on impacts of climate change on air quality show these impacts and the implications for human health may manifest in the near future (Kinney 2008; Hsu et al. 2017; Ebi et al. 2017).

1.3.1.3 Urban System

Urban built-up areas are largely a huge and relative standing out system, which represents support complex and dynamic interactions between societies and natural ecosystems. As the world's population migrates to urban centres, cities will increasingly concentrate and potentially attract greater populations for working opportunities and unbalanced development. Urban populations are characterised by their high population densities and thus can become more vulnerable to climatic change and bacterial and viral disease threats from heat waves, wildfire particulates (PM_{10} and $PM_{2.5}$), air pollution, and for coastal cities the impacts from sea-level sea level rise and extreme still high water events and the impacts stemming from the concatenation of land-based and sea-based flooding (Campbell-Lendrum and Corvalán 2007; Slovic et al. 2016).

Meanwhile, urban areas are consumers of resources and contributors to GHG emissions. This may lead to changes in socio-economic development paths that exacerbate global climate change. The interaction between urban-related energy and transportation and climate change policies, and in relation to health and well-being creates both a need and an opportunity for new understanding and methods to assess complex risks and to support urban planning and development decisions (Hughes et al. 2018).

1.3.2 Social Systems

1.3.2.1 Social Capital

Social capital is defined as "the degree of social cohesion which exists in communities. It refers to the processes between people which establish networks, norms and social trust, and facilitate coordination and co-operation for mutual benefit". With the gradual in depth understanding of the social determinants of health, social capital is increasingly becoming an important component of health research (Kunitz 2004; Harpham 2008; Kim et al. 2008).

Beyond impacts on physical infrastructure, some studies have indicated "Climate change adaptation is a social process inherently" (e.g., Wolf 2011). The strengthening of individuals and communities affected by climate change needs to be on the agenda of adaptation (Adger et al. 2005; Aldrich et al. 2016). Therefore, social capital naturally needs to be addressed in the social system.

1.3.2.2 Justice and Equity

When considering SDG's justice and equity issues are increasingly on the international community's agenda. This is actioned locally through their incorporation in climate risk assessments and concomitant adaptation planning and project implementation. Increasingly climate change and its role in equity and justice amongst the poorest and vulnerable groups are gaining attention at local and sub-national and international levels (Thomas and Twyman 2005).

Guerra et al. (2016) found that health inequalities are linked to social inequities. Imbalances in the distribution of power, prestige and resources can have either a direct or indirect role in determining the health status of population groups. Linking with climate change, Boeckmann and Zeeb (2016) proposed a theoretical perspective focusing on proxy indicators, by measuring the effects of adaptation on determinants of health. This framework was aimed at the measurement of health related inequities, so as to improve climate justice and climate change adaptation evaluation standards.

1.3.2.3 Gender

There are complex and dynamic links between gender and climate change, and they exist at all the dimensions from vulnerability to adaptative capacity and mitigation measures, and at different spatial scales (Terry 2009). Manata and Papazu (2009) described that due to the division of labour between the sexes, which places the primary burden of natural environmental systems, mainly agricultural related work on the women, making women more vulnerable than men to the consequences of climate change, as well as making great demands on women's adaptive capabilities, especially in developing countries (Andersen et al. 2016; Chauhan and Kumar 2016). Studies from Sub-Saharan Africa consistently exhibit high-levels of gender engagement (Bunce and Ford 2015). Natalia (2011) and Bunce (2015) found that women may be affected differently according to their status in physical health, education and agricultural production activities. Other vulnerabilities and adaptive capacities of women were, however, found to be consistent with women in Arctic regions where hunting and collection of marine resources can be important sources of sustenance.

1.3.2.4 Health Literacy and Culture

Inclusive with functional literacy is the notion that increasingly more sophisticated literacy skills are needed to continue to fully participate in society. Therefore, low literacy may have an effect on health and health care (Berkman et al. 2010). Health literacy reflects the knowledge and abilities of persons to reach the demands of health in modern society. Sørensen et al. (2012) developed an integrative conceptual model that focused on four-dimensional matrices of health literacy. These were applied to several health domains: knowledge and motivation; competencies of accessing and understanding; and finally, appraising health care information, disease prevention and health promotion settings.

Regarding climate change related health literacy, interventions and knowledge translation and guidelines for projecting health impacts are lacking (Hess et al. 2014), capacity building is part of the climate change adaptation process (Marinucci et al. 2014; Araos et al. 2016).

1.3.2.5 Belief and Religion

Quality of life and spirituality research is a growing area of interest in the health professions (Panzini et al. 2017). Li et al. (2016) found that frequent attendance at religious services was associated with a significantly lower risk of all-cause, cardiovascular, and cancer mortality among women. When appropriate, physicians have been advised to consider the discussion of religion and spirituality with their patients to gain a more holistic understanding of well-being in relation to their presented health status. Moore (2017) demonstrated that when spirituality, demographic factors, social support and spiritual coping usage were all examined as predictors of mental health, religious and secular forms of spirituality were the only variables that maintained a large predictive strength. The results indicated that living in accordance with one's spiritual values, even when defined in a variety of ways, is characteristic of greater mental health.

The encyclical on Climate Change and Inequality: On Care for Our Common Home (Pope Francis 2015) has had far-reaching influence. The encyclical suggests that sustained exposure to compelling climate messages from trusted sources can increase the performance of activism and ultimately modify human behaviour (Myers et al. 2017).

1.3.3 Economic Systems

1.3.3.1 Economic Development

Many epidemiological transitions are propelled by economic development. For example, differences in GDP per capita explain almost two-thirds of the differences in female obesity among 37 developing countries (Monteiro et al. 2004). Within countries, cardiovascular disease (CVD) is also related to income level. CVD and its risks are concentrated among the lowest socio-economic groups of the more developed (upper-middle and high-income) countries, and among middle and high-income populations of low-middle income countries (Monteiro et al. 2004; Kelly and Fuster 2010).

Health can affect income, work productivity, children's education, savings and investment and demographic structure (Chisholm et al. 2010). Current illness may impact on lifespans and life cycle behaviour. Some studies link health and nutrition in utero, with that of the first few years of life, on adult physical and cognitive development and economic success (World Health Organization 2009; Bloom and Canning 2009). Some health interventions that are relatively inexpensive can have large-scale effects on population health. Such investments are a promising policy tool for growth in developing countries (Deogaonkar et al. 2012).

Climate change has largely been driven by fossil fuel-dependent economic development stemming from the consequences of the industrial revolution and corollary changes in institutional, socio, earth, economic, and technical systems (Fouquet 2016). Eventually, the impacts of climate change on human health became noticeable and costly. To solve problems stemming from climate change, they need to be mainstreamed into economic development processes (Fouquet 2016; Qi et al. 2016), and need to be evaluated and monitored (Peters et al. 2017).

1.3.3.2 Individual Income and Health Expenditure

Individual income inequality has adverse impacts on health and this is manifested by a myriad of differences from the individual to population groups (Lynch et al. 2000; Deaton 2008). Poor households can pay a substantial share of their income for health services and this can push up poverty (World Health Organization 2004). Many households borrow, sell their assets or forgo needed health services and are either incapacitated or are underproductive in the labour market. Households may not be able to escape the trap of ill-health and poverty once they enter it.

Many governments have cut the real per capita budget for health owing to poor overall national economic performance. To keep a resemblance of a public health system operating policymakers have resorted to cost containment and cost recovery strategies and user fees. Households are increasingly paying for necessary health services and often under duress. Public and private resources need to be more equitably allocated to address health related issues (Lynch et al. 2004).

Climate change could bring more problems on income equality in low-income countries (Hanna and Oliva 2016). Normally they are the most vulnerable population to climate change (Lloyd et al. 2016). In low-developing countries, children also encounter air and water pollution, infectious and parasitic diseases and can suffer climate-induced displacement, migration and violence (Hanna and Oliva 2016).

1.3.3.3 Health Infrastructure

The malfunctioning of health care facilities can ensue during and immediately following extreme weather events. This can be devastating for communities during disaster events and hamper sustained recovery efforts (Balbus et al. 2016). Climate change related hazard can have an impact on infrastructure location (service locations, stormwater, site and transportation access issues); infrastructure structure (fixed structural elements, such as roofs and walls); non-structural (utilities, electromechanical systems, communications systems); organisational (supply chain and staff accommodation). The function and accessibility of the infrastructure like the hospital and critical healthcare provision system should be climate resilient (Guenther and Balbus 2014).

To produce more resilient hospital infrastructure, there needs to be better resourced and integrated/collaborative approaches applied to disaster management planning (Chand and Loosemore 2015). Disaster planning decisions would benefit from the more active participation of health facilities managers. Better systems, training, technology and information about health infrastructure performance before, during and after extreme climatic events would strengthen resilience.

1.3.4 Institutional System

1.3.4.1 Governance

Steering and rule-making related functions define governance in the health sector. Governance requires the balancing of competing influences and demands from the different components of the institutions. It can include: defining the strategic direction of policy development and implementation; remapping poorly performing trends; spelling out how health fits with national development priorities; regulation of a complex myriad of actors and agents; and creating and implementing appropriate monitoring and evaluation mechanisms. These are carried out to meet a set of governmental health objectives that are designed to broaden access and improve the overall individual health of the population (Kickbusch and Gleicher 2012). The governance challenges in the health sector are by no means unique (Frenk and Moon 2013), especially linking with climate change governance and city issues (Gupta 2016; Hughes et al. 2018). Systems approaches are needed to solve interdependent complex governance issues.

1.3.4.2 Institutional Arrangement and Capacity

Management of the health effects of climate change and strengthening health systems will require inputs from all sectors of government, civil society, local communities and academic institutes. This is the new way of institutional arrangement and certain capacities to carry out new activities (Costello et al. 2009; Bloland et al. 2012). An integrated and multidisciplinary approach requires multiple levels of action to reduce the adverse health effects of climate change. Institutional capacity building is critical for the implementation of some regions (Furgal and Seguin 2006; McIver et al. 2016).

1.3.4.3 Policy and Strategy

More proactive policy and strategy are required to ensure that development decisions serve the ultimate goal of improving human health (Campbell-Lendrum et al. 2007; World Health Organization 2015). In 2015 the Lancet Commission on Health and

Climate Change mapped out the impacts of climate change, and the necessary policy responses, to attain the standards of health for population worldwide (Watts et al. 2015). Climate change could become a great opportunity for global health (Wang and Horton 2015). Adaptation policy and actions for climate-related health hazards can be both proactive and reactive and they occur at the population, community and individual levels (McMichael 2003).

1.3.4.4 Implementation Capacity

Common barriers to implementation can include: opposition from key stakeholders; a lack of resources (human and financial); poorly defined implementation guidelines, roles and responsibilities; institutional and existing policy conflicts; and, poor coordination and lack of political will to see changes through to fruition (WHO 2017).

Many innovations in the health sector are complex, requiring coordinated use by multiple organisational members to achieve benefits. Often, complex innovations are adopted with great anticipation only to fail during implementation (Haines et al. 2004; Helfrich et al. 2007). This reflects the gap that can often exist between theory and practice. Interventions designed in a research environment fail to achieve the desired end result and desired patient care outcomes when introduced into a diverse array of real world settings. (Damschroder et al. 2009).

1.3.4.5 Healthcare Sector as Systems

Climate change is creating an opening for the health sector, whereby population health can be a foil for improving our ways of living to become more environmentally sensitive and equitable (McMichael et al. 2009). There could be substantial health dividends achieved through restructuring our social practices, technologies, for example in energy production and commercial practices. The current health care system characterised by fragmentation and lack of coherence seems to be limiting interventions and quality health reform outcomes.

A reorientation of goals with an increasing emphasis on patient functioning, social participation and the addition of clinical measures as core outcomes of effective care may be required (World Health Organization 2012). Policies for climate change mitigation and adaptation offer different opportunities for healthcare sector development, especially climate financing mechanisms for innovative solutions (World Health Organization 2017; Junghans et al. 2017).

1.4 Systems Thinking and Modelling for Human Health and Climate Change

Systems thinking requires that one views systems and their sub-components as intimately interrelated and connected to each other. Ultimately, there needs to be the belief that truly understanding how things work requires the interpretation of interactions and relationships within and between systems (Adam and de Savigny 2012; Adam et al. 2012). Such a perspective goes beyond specific events, to recognise behaviour and the underlying systemic interrelationships that define the patterns and their related events (Leischow et al. 2008). When the systems are open they can be complex and adaptive, constantly changing, sometimes resistant, counter-intuitive, non-linear, often the whole is greater than the sum of the parts (Best et al. 2007).

System thinking in health care systems has been reported from South Africa (Gilson et al. 2014), India (Varghese et al. 2014) and China (Zhang et al. 2014). Varghese et al. (2014) guided by a complex of interactive and adaptive systems when they explored changes in vaccination coverage in India. Paxton and Frost (2017) developed a multidisciplinary curriculum that utilised systems thinking to frame and analyse global health policies and practice so as to train future leaders in global health. Caffrey and Munro (2017) applied a systems approach to evaluate health policy.

1.4.1 Existing System Approaches for Health and Well-Being

At their core, systems science methodologies are designed to generate models, or simplified versions of reality. They do this by attempting to capture the real world in important ways while simplifying where ever possible but maintaining the critical aspects relevant to the studied problem. We can, therefore, better understand the structural complexity of real world problems that results from the interaction of specific phenomena and their environments.

System dynamics addresses the dynamic complexity that characterises public health issues (Homer and Hirsch 2006). The approach involves computer simulations that digitise processes of accumulation and feedback. They can be tested systematically to help uncover policies for overcoming policy resistance. Interdisciplinary partnerships have opened up and become stronger linking behavioural–social–ecologic models at various levels with new understanding achieved that all invigorate the debates around public health (Mabry et al. 2008, 2010).

Much like what-if climate change science, Semwanga et al. (2016) presented a system dynamics model of a neonatal health system. The resulting deeper understanding of constraints and opportunities and, the identification of barriers to change led to more inclusive and better adoption of required interventions in the overall system of care. Using 'what-if' scenarios, health practitioners could more easily discuss the consequences and effects of various decisions. Proposed interventions and their impact could be tested through simulation experiments and the results could be fed into policies and interventions with the highest impact for improved healthcare delivery. O'Donnell et al. (2017) and Pitt et al. (2016) found similar results that challenged the traditional use of hierarchies of evidence to support decisions on complex dynamic problems.

1.4.2 Mapping of the Structure and Function of the System

In the system mapping phase, conceptual models of a problem are created, known as causal loop diagrams (CLDs). This is a major component of the systems thinking and modelling approach. The following steps are normally used in causal loop mapping:

(1) Identify main (key) variables (elements).
(2) Draw behaviour charts that include a time element (or reference modes) for the main variables.
(3) Develop causal loop diagrams to show the relationships among the variables.
(4) Discuss the implications of changes in behaviour over time that are implied by the causal loop diagrams.
(5) Identify archetypes that would describe high-level causal patterns.
(6) Identify key leverage points.
(7) Develop intervention strategies.

Although these steps are part of an overarching process, they need to be carefully customised when a specific SES issue is to be investigated. Figure 1.2 is an illustration of what can be considered, but may not be exhaustive when building an SES-based system dynamics model.

Step 1: Risk identification

This step is taken to create a broad list of climate change risks that might affect health in different scales. It is the process that generates a broad list of reasonably foreseeable ways that climate change stressors could hamper an individual or organisation in relation to health. All potential risks should be considered. This is because if risks are not identified in the early stage, they will not be analysed and evaluated in subsequent steps.

Risk identification is often carried out either through survey questionnaires, expert group discussions, brainstorming and literature reviews or the collective application of all of these methods. Expert groups are composed of experienced researchers, managers and officers from different sectors. Their expertise should cover social, economic, ecological, climatological and institutional aspects. Identifying risks is the first step, perhaps the most important step, in health risk management. When risks are not adequately identified other steps in risk management cannot be implemented.

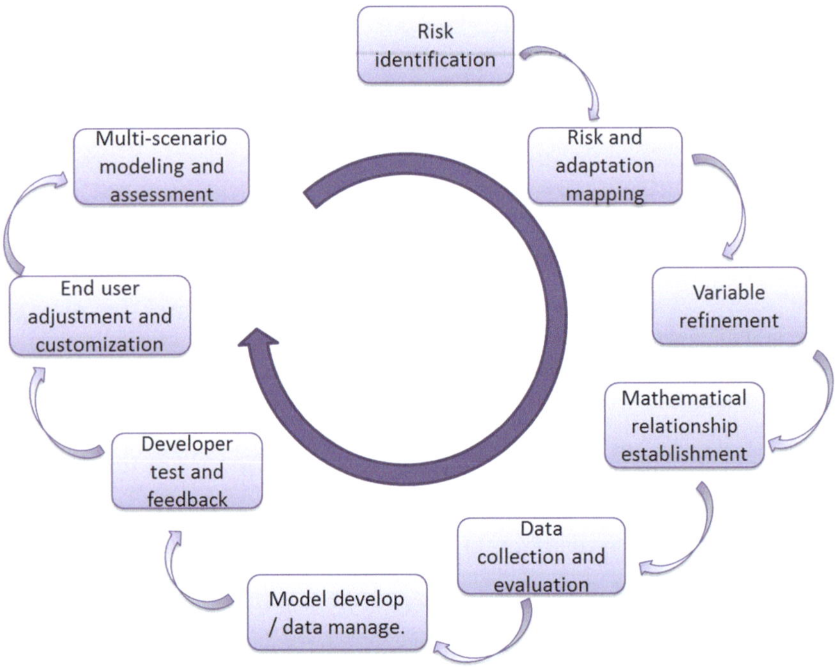

Fig. 1.2 Steps for building system dynamics model in UrbanCLIM/RIDS platform *Source* Reprinted from Li et al. (2016). Copyright Asia-Pacific Network for Global Change Research 2016

It is important to realise that an individual or organisation's exposure to risks may change over time and new and previously unexperienced risks could emerge.

Step 2: Risk and adaptation mapping

The linkages and relationships among the risk (and potential options) factors are mapped to expose the structure of the socio-ecological system. The SES framework is applied at this point. Risks are defined in a SES context, to consider the interdependency of the four subsystems.

Step 3: Variable refinement

Variable refinement is the selection of variables in other words, which depend on data availability. Moreover, only key variables should be selected and utilised in the model building; otherwise, the model may become very complicated and bloat. The key parameter is also selected in this step.

Step 4: Mathematical relationship establishment

The procurement or development of mathematical equations between variables could stem from either a comprehensive review and critical assessment of the literature, or through independent development and the application of existing data. No doubt, mathematical capacity is critical for building a system dynamics model.

Step 5: Data collection and evaluation

According to selected variables and equations, corresponding data need to be collected, wrangled and evaluated. Data can derive from various methods and formats such as statistical or time series, raster or vector in the GIS (Geographical Information System) realm or, from categorical ranking systems.

Step 6: Model development in relation to data management

Available mathematical equations and data are used to build the model. Much thinking is needed. The model should be simple and logical but robust. It involves a designing and programming process. Software manuals can inform the process, however, often the most efficient way is to work with software and programmers familiar with the modeller's needs.

Step 7: Developer test and feedback to designers

This step involves the running of the model and testing its performance. It is an iterative process to tune model parameters in order to get the expected outcome. It is imperative that developers demonstrate the model to their end-user community, and seek their opinions and then modify and improve the model accordingly.

Step 8: End-user adjustment and customization

After the model ran smoothly, end users can start to adjust its parameters in order to acquire better understanding of its functions. The model can also be configured with specific skin colour, graph type, logo, etc. according to user preferences.

Step 9: Multi-scenario simulation applied in assessments

The last step is the creation of real simulations applying scenarios to related stakeholders who could be policymakers or peer-researchers.

As a whole, model building is a loop process of capacity building on risk assessment, risk management, integrated risk governance, evaluation and feedback.

1.4.3 Indicators and Variables in System Modelling

In earlier sections, potential modelling areas were described in the SES elements and structures. In fact, it is only a start for the system modelling indicator and variable identification process. There are many ways to look at indicators of system elements and functions. Moser & Ekstrom (2010) highlighted that in the management of climate change the diagnostic framework's structural elements such as actors and agents, the governance and wider socio-economic context, and the system of concern needed to be carefully considered. In the climate change adaptation process, including understanding, planning and managing (or implementing), each stage has its own working process and barriers. For successful assessment and action to be achieved it needs to be informed by the elements and the structure of the systems.

In modelling practice, the following points could be considered when indicators need to be selected:

- Accessibility and feasibility of each indicator consistently and accurately over time.
- Choosing indicators that help make the case for particular actions or strategies.
- Maintaining flexibility with the results framework, allowing users to add or adjust indicators later for projects and programmes.
- Less is more and a small set of well-tracked indicators are better than several at multiple levels. The complex process could be embedded in the modelling process as intermediate variables.
- Indicators that provide relative data (proportion, percent, ratio) ease tracking of overall progress in the larger context.
- Indicators should be chosen based on information already available or could be derived using data technology.

1.4.4 UrbanCLIM/RIDS Decision Support System

The UrbanCLIM system's architecture was designed to provide robust support for three classes of users—developers, modellers, analysts and policymakers (Li et al. 2016). UrbanCLIM was extended to a Risk Informed Decision Support (RIDS) platform for other sectors and broader applications (International Global Change Institute, New Zealand: www.igci.org.nz/RIDS).

By applying UrbanCLIM/RIDS, it is easy for developers to reach into the deepest software layers to extend existing and/or build a new simulation, modelling and interactive capabilities that integrate seamlessly with other applications. Modellers are able to use a variety of blocks and connectors, user interaction and model aggregation capabilities to create robust models. Analysts and policymakers like to apply simple and powerful analytical tools that smoothly integrate models and other decision-making methods into a support engine for formulating practical approaches to real world challenges. Therefore, its open framework can act as a generic platform for many other areas other than climate change issues by adding outer components.

The UrbanCLIM/RIDS platform was designed to support multi-tiered applications. The interactive layers allow efficient and effective interaction between the model developers and end users. The policy-making tier supports the policy-making processes through the provision of graphs, maps, and technical information. It supports a participatory assessment approach through users' dialogue with urban policymakers and planners from targeted cities.

The features of the UrbanCLIM/RIDS platform include

- Apply a modular design approach including standardised technologies to either reduce or eliminate barriers to the linking of existing and potential future models and related applications;
- An open framework, allowing for multi-scale impact assessments, that can be customised case-by-case for each studied area;
- Integrated analysis tools that enable testing of adaptation and mitigation options against socio-economic drivers, likely impacts and current sustainable development goals (SDGs);
- Climate change uncertainty analysis based on GCM and RCM ensemble approaches and the latest IPCC climate change scenarios;
- GIS compatibility;
- Visualisation and enhanced analysis options for the assessment of results;
- Integration with the Socio-Ecological-System framework.

1.4.5 Model Library Strategies

To realise a publicly accessible and broadly useful library, UrbanCLIM/RIDS has an urban climate change decision support model library that includes impact and risk assessments for major analytical sectors: climate-related hazards such as water, transport and health. Targeted urban policymakers and planners should be part of an integrated approach. Through the UrbanCLIM/RIDS Community of Practice, various tools, data and models can be enhanced and integrated into the platform thus enabling a robust knowledge, technology sharing, transfer and, communication system.

1.4.6 Knowledge Management Tool

The risk-informed decision support system provides a navigator that can act as either a knowledge or project management tool. It can also be used to store and index existing studies, data, models and literature. Each model and study is presented alongside

documentation that explain its genesis, objective, limitations and applicability. A user can create a new study or project, add documentation and models and import and export to help build a community knowledge base.
The knowledge management tool is designed to:

- Empower decision-makers and planners to be more productive, more comprehensive and quicker in responding by providing a friendly browser-like environment.
- Provide a management system for data, models and tools and thus encourage cross-pollination of disparate knowledge bases, the reuse of sound and well tested models and leveraging of analytical tools across a range of analytical domains.

1.4.7 Community of Practice

Community of Practice (CoP) consists of a diverse group of climate change and sectoral modellers, analysts and decision-makers. The development of a CoP serves twofold critical purposes. On one hand, it is the cross-pollination of ideas, techniques and technologies; on the other hand, it is to guide the core development of the UrbanCLIM/RIDS platform.

More specifically:

- Promote science-based climate change and sectoral applications;
- Promote the sharing of climate change model through a community portal that leverages project management (e.g. github.com);
- Provide a method for software and information delivery to its members;
- Invite participation and dialogue between inside and outside perspectives;
- Enable broad software development support for climate change adaptation and mitigation;
- Enable cross-functional collaboration in projects;
- Enhance public awareness of communication through the CoP.

1.5 Challenges and Opportunities

Skill barriers for systems approach applications

Understanding its foundations, concepts and terminology is one of the key barriers to learning and accepting systems thinking (Schneider 2012). Any change that requires how we view the world, our work and how we relate to others can create resistance. Learning takes time and practitioners and organisations generally are committed to their current practices and need to be shown the effectiveness of systems approach.

Ackoff (2007) noted that system thinking must be clearly shown and understood with as little jargon as possible before become widely accepted. This could also include specific guidelines on how to carry out interventions (Small 2005).

The acquisition and application of systems thinking capabilities require both a rethinking of the skills required and their development and implementation for global sustainability studies. The need for 'T-shaped' skills is becoming increasingly important. This is in contrast with the more common I-shaped professionals. Current and, rapidly evolving knowledge and data will require practitioners and researchers that have both deep knowledge in a discipline or system (vertical bar) combined with the capacity to move across disciplines and systems (horizontal bar). This may be even more critical when applying a systems approach with the required climate-related transition to a low or no carbon global system in conjunction with adaptation (Barile et al. 2015). The 'T-shape' skill training needs to be embedded in the education process.

Data and modelling challenges

Climate change and human health/well-being SES is a large and complex system. It could be studied from many different scales, levels and functionalities. A complex system may be composed of many subsystems. Although high modelling performance could be achieved for each subsystem, it still can be very difficult for everyone to take into account the mutual influence of subsystems in order to study the system as a whole. To an extent, some parameterization schemes have to be carried out to simplify the subsystem modelling. Moreover, all of them may require massive data and modelling tools, which potentially limit their usefulness (Schulze et al. 2017).

In recent years, the increase in computer power and informatics tools has enabled the gathering of reliable data quantifying the complexity of socio-technical systems (Vespignani 2012). Data-driven computational models have emerged as appropriate tools to tackle the study of dynamical phenomena as diverse as epidemic outbreaks, information spreading and Internet packet routing. The increasing use of complexity science tools including ABMs, networks, statistical mechanics, which are largely rule-based in social and ecological domains may potentially open up possibilities to include rules that would write new rules as a system at hand restructures.

Looking forward

WHO UNFCCC projects seek to raise awareness of the health impacts of climate change while supporting evidence-based decision-making. This approach should strengthen the climate resilience of health systems while promoting actions that improve health as carbon emissions are rapidly reduced and SDGs are achieved. These efforts provide good opportunities for climate resilient health systems (Wang et al. 2015; Meara et al. 2015; World Health Organization 2017). A WHO action agenda on the Paris agreement has been formed to address the health risks and opportunities, to ensure support for health and climate action; health adaptation to climate change; to ramp up investments in climate change and health (World Health Organization 2016).

These actions need to be taken in a more systematic way. The application of social-ecological systems (SES) is crucial to support the sustainable management of human health and well-being (Mabry et al. 2010). Moreover, successful stories came from cases underscored by participatory approaches in model building. Clear communication, sharing of source code, and tools and strategies for model design and analysis should form the foundation. The UrbanCLIM/RIDS platform and SES framework as described could contribute to the establishment of a new advanced system approach that will allow us to develop a more general theory and practical solutions to human health and well-being in the context of climate change.

References

Ackoff RL (2007) Why few organizations adopt systems thinking. Primer Project. http://www.gperform.com/Ackoff_on_Adoption_of__Systems.pdf

Adam T, de Savigny D (2012) Systems thinking for strengthening health systems in LMICs: need for a paradigm shift. Health Policy Plan 27(suppl_4):iv1–iv3

Adam T, Hsu J, De Savigny D, Lavis JN, Røttingen JA, Bennett S (2012) Evaluating health systems strengthening interventions in low-income and middle-income countries: are we asking the right questions? Health Policy Plan 27(suppl_4):iv9–iv19

Adger WN, Hughes TP, Folke C, Carpenter SR, Rockström J (2005) Social-ecological resilience to coastal disasters. Science 309(5737):1036–1039

Aldrich DP, Page C, Paul CJ (2016) Social capital and climate change adaptation

Andersen LE, Verner D, Wiebelt M (2016) Gender and climate change in Latin America: an analysis of vulnerability, adaptation and resilience based on household surveys. J Int Dev

Araos M, Austin SE, Berrang-Ford L, Ford JD (2016) Public health adaptation to climate change in large cities: a global baseline. Int J Health Serv 46(1):53–78

Atkinson JA, Wells R, Page A, Dominello A, Haines M, Wilson A (2015) Applications of system dynamics modelling to support health policy. Public Health Res Pract 25(3):e2531531

Bai X, Nath I, Capon A, Hasan N, Jaron D (2012) Health and wellbeing in the changing urban environment: complex challenges, scientific responses, and the way forward. Curr Opin Environ Sustain 4(4):465–472

Balbus J, Berry P, Brettle M, Jagnarine-Azan S, Soares A, Ugarte C, Varangu L, Prats EV (2016) Enhancing the sustainability and climate resiliency of health care facilities: a comparison of initiatives and toolkits. Revista Panamericana de Salud Pública 40(3):174–180

Barile S, Saviano M, Polese F, Caputo F (2015) T-shaped people for addressing the global challenge of sustainability. In: Service dominant logic, network and systems theory and service science: integrating three perspectives for a new service agenda. Giannini, Napoli, pp 1–23

Bar-Yam Y (2002) General features of complex systems. In: Encyclopedia of life support systems (EOLSS). UNESCO, EOLSS Publishers, Oxford, UK

Batty M (2013) The new science of cities. MIT Press

Bergmann S (2009) Climate change changes religion: space, spirit, ritual, technology–through a theological lens. Studia Theologica 63(2):98–118

Berkman ND, Davis TC, McCormack L (2010) Health literacy: what is it? J Health Commun 15(S2):9–19

Berry HL, Kelly BJ, Hanigan IC, Coates JH, McMichael AJ, Welsh JA, Kjellstrom T (2008) Rural mental health impacts of climate change. In: Commissioned report for the Garnaut Climate Change Review. The Australian National University, Canberra

Berry P, Sánchez-Arcilla Conejo A, Betts R, Harrison PA (2017) High-end climate change in Europe: impacts, vulnerability and adaptation

Best A, Clark PI, Leischow SJ, Trochim WM (2007) Greater than the sum: systems thinking in tobacco control

Binder C, Hinkel J, Bots P, Pahl-Wostl C (2013) Comparison of frameworks for analyzing social-ecological systems. Ecol Soc 18(4)

Black R, Adger N, Arnell N, Dercon S, Geddes A, Thomas D (2011) Migration and global environmental change: future challenges and opportunities

Bloland P, Simone P, Burkholder B, Slutsker L, De Cock KM (2012) The role of public health institutions in global health system strengthening efforts: the US CDC's perspective. PLoS Med 9(4):e1001199

Bloom DE, Canning D (2009) Population health and economic growth. Health Growth 53

Boeckmann M, Zeeb H (2016) Justice and equity implications of climate change adaptation: a theoretical evaluation framework. Healthcare 4(3):65. Multidisciplinary Digital Publishing Institute

Boholm Å, Prutzer M (2017) Experts' understandings of drinking water risk management in a climate change scenario. Clim Risk Manag 16:133–144

Brenner F, Marwan N, Hoffmann P (2017) Climate impact on spreading of airborne infectious diseases. Eur Phys J Spec Top 226(9):1845–1856

Broberg MC, Högy P, Pleijel H (2017) CO2-induced changes in wheat grain composition: meta-analysis and response functions. Agronomy 7(2):32

Bunce A (2015) Gender and the human dimensions of climate change: global discourse and local perspectives from the Canadian Arctic. Doctoral dissertation, McGill University

Bunce A, Ford J (2015) How is adaptation, resilience, and vulnerability research engaging with gender? Environ Res Lett 10(12):123003

Caffrey L, Munro E (2017) A systems approach to policy evaluation. Evaluation

Campbell-Lendrum D, Corvalán C (2007) Climate change and developing-country cities: implications for environmental health and equity. J Urban Health 84(1):109–117

Campbell-Lendrum D, Corvalán C, Neira M (2007) Global climate change: implications for international public health policy. Bull World Health Organ 85(3):235–237

Cash DW, Clark WC, Alcock F, Dickson NM, Eckley N, Guston DH, Jäger J, Mitchell RB (2003) Knowledge systems for sustainable development. Proc Natl Acad Sci 100(14):8086–8091

Castella JC, Trung TN, Boissau S (2005) Participatory simulation of land-use changes in the northern mountains of Vietnam: the combined use of an agent-based model, a role-playing game, and a geographic information system. Ecol Soc 10(1)

Chand AM, Loosemore M (2015) A socio-ecological analysis of hospital resilience to extreme weather events. Constr Manag Econ 33(11–12):907–920

Chauhan NB, Kumar VH (2016) Gender responsive climate change strategies for sustainable development. Productivity 57(2):182

Cheeseman, J (2016) Food security in the face of salinity, drought, climate change, and population growth. In: Halophytes for food security in dry lands, pp 111–123

Chisholm D, Stanciole AE, Edejer TTT, Evans DB (2010) Economic impact of disease and injury: counting what matters. BMJ 340:c924

Clark WC, van Kerkhoff L, Lebel L, Gallopin GC (2016) Crafting usable knowledge for sustainable development. Proc Natl Acad Sci 113(17):4570–4578

Clayton S, Manning CM, Krygsman K, Speiser M (2017) Mental health and our changing climate: impacts, implications, and guidance. American Psychological Association, and ecoAmerica, Washington, D.C.

Costello A, Abbas M, Allen A, Ball S, Bell S, Bellamy R, Lee M (2009) Managing the health effects of climate change. Lancet 373(9676):1693–1733

Curriero FC, Patz JA, Rose JB, Lele S (2001) The association between extreme precipitation and waterborne disease outbreaks in the United States, 1948–1994. Am J Public Health 91(8):1194–1199
D'Amato G, Baena-Cagnani CE, Cecchi L, Annesi-Maesano I, Nunes C, Ansotegui I, Canonica WG (2013) Climate change, air pollution and extreme events leading to increasing prevalence of allergic respiratory diseases. Multidiscip Respir Med 8(1):12
Damschroder LJ, Aron DC, Keith RE, Kirsh SR, Alexander JA, Lowery JC (2009) Fostering implementation of health services research findings into practice: a consolidated framework for advancing implementation science. Implement Sci 4(1):50
Deaton A (2008) Income, health, and well-being around the world: evidence from the Gallup World Poll. J Econ Perspect 22(2):53–72
Deogaonkar R, Hutubessy R, Van Der Putten I, Evers S, Jit M (2012) Systematic review of studies evaluating the broader economic impact of vaccination in low and middle income countries. BMC Public Health 12(1):878
Deryng D, Conway D, Ramankutty N, Price J, Warren R (2014) Global crop yield response to extreme heat stress under multiple climate change futures. Environ Res Lett 9(3):034011
Ebi KL, Frumkin H, Hess JJ (2017) Protecting and promoting population health in the context of climate and other global environmental changes. Anthropocene 19:1–12
Forzieri G, Cescatti A, e Silva FB, Feyen L (2017) Increasing risk over time of weather-related hazards to the European population: a data-driven prognostic study. Lancet Planet Health 1(5):e200–e208
Fouquet R (2016) Lessons from energy history for climate policy: technological change, demand and economic development. Energy Res Soc Sci 22:79–93
Frenk J, Moon S (2013) Governance challenges in global health. N Engl J Med 368(10):936–942
Fritze JG, Blashki GA, Burke S, Wiseman J (2008) Hope, despair and transformation: climate change and the promotion of mental health and wellbeing. Int J Ment Health Syst 2(1):13
Furgal C, Seguin J (2006) Climate change, health, and vulnerability in Canadian northern Aboriginal communities. Environ Health Perspect 114(12):1964
Gerten D (2010) Religion and dangerous environmental change: transdisciplinary perspectives on the ethics of climate and sustainability, vol 2. LIT Verlag Münster
Gifford E, Gifford R (2016) The largely unacknowledged impact of climate change on mental health. Bull At Sci 72(5):292–297
Gilson L, Elloker S, Olckers P, Lehmann U (2014) Advancing the application of systems thinking in health: South African examples of a leadership of sensemaking for primary health care. Health Res Policy Syst 12(1):30
Gomez R, Fisher JW (2003) Domains of spiritual well-being and development and validation of the Spiritual Well-Being Questionnaire. Personality Individ Differ 35(8):1975–1991
Gosling SN, Arnell NW (2016) A global assessment of the impact of climate change on water scarcity. Clim Change 134(3):371–385
Greer A, Ng V, Fisman D (2008) Climate change and infectious diseases in North America: the road ahead. Can Med Assoc J 178(6):715–722
Guenther R, Balbus J (2014) Primary protection: enhancing health care resilience for a changing climate. US Department of Health and Human Services
Guerra G, Borde E, de Snyder VNS (2016) Measuring health inequities in low and middle income countries for the development of observatories on inequities and social determinants of health. Int J Equity Health 15(1):9
Gunderson LH, Holling CS, Pritchard L, Peterson GD (2002) Resilience of large-scale resource systems. Scope-Sci Comm Probl Environ Int Counc Sci Unions 60:3–20
Gupta J (2016) Climate change governance. Wiley Interdiscip Rev Clim Chang 7
Haines A, Kuruvilla S, Borchert M (2004) Bridging the implementation gap between knowledge and action for health. Bull World Health Organ 82(10):724–731

Hall DM, Feldpausch-Parker A, Peterson TR, Stephens JC, Wilson EJ (2017) Social-ecological system resonance: a theoretical framework for brokering sustainable solutions. Sustain Sci 12(3):381–392
Haluza-DeLay R (2014) Religion and climate change: varieties in viewpoints and practices. Wiley Interdiscip Rev Clim Chang 5(2):261–279
Hanna R, Oliva P (2016) Implications of climate change for children in developing countries. Futur Child 26(1):115–132
Harpham T (2008) The measurement of community social capital through surveys. In: Social capital and health. Springer, New York, pp 51–62
Hashim JH, Hashim Z (2016) Climate change, extreme weather events, and human health implications in the Asia Pacific region. Asia Pac J Public Health 28(2_suppl):8S–14S
Helfrich CD, Weiner BJ, McKinney MM, Minasian L (2007) Determinants of implementation effectiveness: adapting a framework for complex innovations. Med Care Res Rev 64(3):279–303
Hess JJ, Eidson M, Tlumak JE, Raab KK, Luber G (2014) An evidence-based public health approach to climate change adaptation. Environ Health Perspect 122(11):1177
Homer JB, Hirsch GB (2006) System dynamics modeling for public health: background and opportunities. Am J Public Health 96(3):452–458
Hsu WH, Hwang SA, Kinney PL, Lin S (2017) Seasonal and temperature modifications of the association between fine particulate air pollution and cardiovascular hospitalization in New York State. Sci Total Environ 578:626–632
Huang J, Yu H, Guan X, Wang G, Guo R (2016) Accelerated dryland expansion under climate change. Nature Climate Change 6(2):166–171
Hughes S, Chu EK, Mason SG (2018) Introduction. In Climate Change in Cities (pp 1–15). Springer, Cham
ICSU (2011) Report of the ICSU Planning Group on Health and Wellbeing in the Changing Urban Environment: a Systems Analysis Approach. International Council for Science, Paris
Junghans L, Eckstein D, Kreft S, Syberg M, Weischer L (2016) Going to town: How the Green Climate Fund can support a paradigm shift in cities
Kelly BB, Fuster V (Eds.) (2010) Promoting cardiovascular health in the developing world: a critical challenge to achieve global health. National Academies Press
Khan SJ, Deere D, Leusch FD, Humpage A, Jenkins M, Cunliffe D (2015) Extreme weather events: Should drinking water quality management systems adapt to changing risk profiles? Water Res 85:124–136
Kickbusch I, Gleicher D (2012) Governance for health in the 21st century. World Health Organization, Geneva
Kim D, Subramanian SV, Kawachi I (2008) Social capital and physical health. In Social capital and health (pp. 139–190). Springer New York
Kinney PL (2008) Climate change, air quality, and human health. Am J Prev Med 35(5):459–467
Koenig HG (2015) Religion, spirituality, and health: a review and update. Advances in mind-body medicine 29(3):19–26
Kunitz SJ (2004) Social capital and health. Br Med Bull 69(1):61–73
Leischow SJ, Best A, Trochim WM, Clark PI, Gallagher RS, Marcus SE, Matthews E (2008) Systems thinking to improve the public's health. Am J Prev Med 35(2):S196–S203
Lesk C, Rowhani P, Ramankutty N (2016) Influence of extreme weather disasters on global crop production. Nature 529(7584):84–87
Li S, Stampfer MJ, Williams DR, VanderWeele TJ (2016) Association of religious service attendance with mortality among women. JAMA internal medicine 176(6):777–785
Li Y, Wang S, Dan Li, Luo L, Khanh D, Walag FB (2016) Development of an integrated climate change impact assessment tool for urban policy makers (UrbanCLIM). http://www.apn-gcr.org/resources/files/original/a5bb45087352107d0027d7e1cea4db6c.pdf

Lloyd SJ, Kovats RS, Chalabi Z, Brown S, Nicholls RJ (2016) Modelling the influences of climate change-associated sea-level rise and socioeconomic development on future storm surge mortality. Clim Change 134(3):441–455
Lynch JW, Smith GD, Kaplan GA, House JS (2000) Income inequality and mortality: importance to health of individual income, psychosocial environment, or material conditions. BMJ: British Medical Journal, 320(7243), 1200
Lynch J, Smith GD, Harper SA, Hillemeier M, Ross N, Kaplan GA, Wolfson M (2004) Is income inequality a determinant of population health? Part 1. A systematic review. The Milbank Quarterly 82(1):5–99
Mabry PL, Marcus SE, Clark PI, Leischow SJ, Méndez D (2010) Systems science: a revolution in public health policy research
Mabry PL, Olster DH, Morgan GD, Abrams DB (2008) Interdisciplinarity and systems science to improve population health: a view from the NIH Office of Behavioral and Social Sciences Research. Am J Prev Med 35(2):S211–S224
Maglio PP, Sepulveda MJ, Mabry PL (2014) Mainstreaming modeling and simulation to accelerate public health innovation. Am J Public Health 104(7):1181–1186
Manata S, Papazu I (2009) Gendering Climate Change, A Perspective on the Overlooked Gender Dimension of Climate Change in the Developing World
Marinucci GD, Luber G, Uejio CK, Saha S, Hess JJ (2014) Building resilience against climate effects—A novel framework to facilitate climate readiness in public health agencies. International journal of environmental research and public health 11(6):6433–6458
McIver L, Kim R, Woodward A, Hales S, Spickett J, Katscherian D, Naicker J (2016) Health impacts of climate change in Pacific island countries: a regional assessment of vulnerabilities and adaptation priorities. Environ Health Perspect 124(11):1707
McMichael AJ (2003) Global climate change and health: an old story writ large. Climate change and human health: Risks and responses. World Health Organization, Geneva, Switzerland
McMichael AJ, Neira M, Bertollini R, Campbell-Lendrum D, Hales S (2009) Climate change: a time of need and opportunity for the health sector. The Lancet 374(9707):2123
Meara JG, Leather AJ, Hagander L, Alkire BC, Alonso N, Ameh EA, Mérisier ED (2015) Global Surgery 2030: evidence and solutions for achieving health, welfare, and economic development. The Lancet 386(9993):569–624
Medek DE, Schwartz J, Myers SS (2017) Estimated effects of future atmospheric CO 2 concentrations on protein intake and the risk of protein deficiency by country and region. Environ Health Perspect 125(8):087002
Meehl GA, Zwiers F, Evans J, Knutson T, Mearns L, Whetton P (2000) Trends in extreme weather and climate events: issues related to modeling extremes in projections of future climate change. Bull Am Meteor Soc 81(3):427–436
Monteiro CA, Moura EC, Conde WL, Popkin BM (2004) Socioeconomic status and obesity in adult populations of developing countries: a review. Bull World Health Organ 82(12):940–946
Moore JT (2017) Multicultural and idiosyncratic considerations for measuring the relationship between religious and secular forms of spirituality with positive global mental health. Psychology of Religion and Spirituality 9(1):21
Moser SC, Ekstrom JA (2010) A framework to diagnose barriers to climate change adaptation. Proc Natl Acad Sci 107(51):22026–22031
Murphy C, Tembo M, Phiri A, Yerokun O, Grummell B (2016) Adapting to climate change in shifting landscapes of belief. Clim Change 134(1–2):101–114
Myers TA, Roser-Renouf C, Maibach E, Leiserowitz A (2017) Exposure to the Pope's Climate Change Message Activated Convinced Americans to Take Certain Activism Actions. Global Challenges 1(4)
Natalia K (2011) Climate change effects on human health in a gender perspective: some trends in Arctic research. Glob Health Action 4(1):7913
Nearing MA, Pruski FF, O'neal MR (2004) Expected climate change impacts on soil erosion rates: a review. J Soil Water Conserv 59(1):43–50

O'Donnell E, Atkinson JA, Freebairn L, Rychetnik L (2017) Participatory simulation modelling to inform public health policy and practice: Rethinking the evidence hierarchies. J Public Health Policy 38(2):203–215
Ostrom E (2009) A general framework for analyzing sustainability of social-ecological systems. Science 325(5939):419–422
Page A, Atkinson JA, Heffernan M, McDonnell G, Hickie I (2017) A decision-support tool to inform Australian strategies for preventing suicide and suicidal behaviour. Public Health Res Pract 27(2)
Panagos, P, Pasquale Borrelli, Katrin Meusburger, Bofu Yu, Andreas Klik, Kyoung Jae Lim, Jae E. Yang et al. "Global rainfall erosivity assessment based on high-temporal resolution rainfall records." Scientific Reports 7 (2017)
Panzini RG, Mosqueiro BP, Zimpel RR, Bandeira DR, Rocha NS, Fleck MP (2017) Quality-of-life and spirituality. International Review of Psychiatry 29(3):263–282
Patz JA, Campbell-Lendrum D, Holloway T, Foley JA (2005) Impact of regional climate change on human health. Nature 438(7066):310
Patz JA, Epstein PR, Burke TA, Balbus JM (1996) Global climate change and emerging infectious diseases. JAMA 275(3):217–223
Paxton A, Frost LJ (2017) Using Systems Thinking to train future leaders in global health. Global Public Health, 1–9
Peters GP, Andrew RM, Canadell JG, Fuss S, Jackson RB, Korsbakken JI, Nakicenovic N (2017) Key indicators to track current progress and future ambition of the Paris Agreement. Nature Climate Change 7(2):118–122
Pitt M, Monks T, Allen M (2016) Systems modelling for improving health care. BMC Health Services Research 16(101):1
Pope Francis (2015) Encyclical on climate change and inequality: On care for our common home. Melville House
Posas PJ (2007) Roles of religion and ethics in addressing climate change. Ethics Sci Environ Polit 7:31–49
Qi Y, Stern N, Wu T, Lu J, Green F (2016) China's post-coal growth. Nat Geosci 9(8):564–566
Røpke I (2016) Complementary system perspectives in ecological macroeconomics—the example of transition investments during the crisis. Ecol Econ 121:237–245
Rose S (2001) Is the term 'spirituality' a word that everyone uses, but nobody knows what anyone means by it? J Contemp Relig 16(2):193–207
Rosenzweig C, Iglesias A, Yang XB, Epstein PR, Chivian E (2001) Climate change and extreme weather events; implications for food production, plant diseases, and pests. Global change & human health 2(2):90–104
Rowold J (2011) Effects of spiritual well-being on subsequent happiness, psychological well-being, and stress. J Relig Health 50(4):950–963
Schlüter M, Hinkel J, Bots P, Arlinghaus R (2014) Application of the SES framework for model-based analysis of the dynamics of social-ecological systems. Ecol Soc 19(1)
Schneider, SV (2012) Overcoming barriers to understanding systems thinking. Systems Thinking World Journal: Reflection in Action, 1(2)
Schulze J, Müller B, Groeneveld J, Grimm V (2017) Agent-based modelling of social-ecological systems: achievements, challenges, and a way forward. J Artif Soc Soc Simul 20(2):1–8
Semwanga AR, Nakubulwa S, Adam T (2016) Applying a system dynamics modelling approach to explore policy options for improving neonatal health in Uganda. Health Res Policy Syst 14(1):35
Senge PM (1990) Prisoners of the system or prisoners of our own thinking. Fifth Discipl 27–54
Shukla J (2016) Extreme weather events: addressing the mental health challenges. In: Climate change and health. Springer International Publishing, pp 15–27
Slovic AD, de Oliveira MA, Biehl J, Ribeiro H (2016) How can urban policies improve air quality and help mitigate global climate change: a systematic mapping review. J Urban Health 93(1):73–95
Small SA (2005) Bridging research and practice in the family and human sciences. Fam Relat 54:320–334

Sørensen K, Van den Broucke S, Fullam J, Doyle G, Pelikan J, Slonska Z, Brand H (2012) Health literacy and public health: a systematic review and integration of definitions and models. BMC public health 12(1):80
Stocker T (ed) (2014) Climate change 2013: the physical science basis: Working Group I contribution to the Fifth assessment report of the Intergovernmental Panel on Climate Change. Cambridge University Press
Stott P (2016) How climate change affects extreme weather events. Science 352(6293):1517–1518
Swim J, Clayton S, Doherty T, Gifford R, Howard G, Reser J, Weber E (2009) Psychology and global climate change: addressing a multi-faceted phenomenon and set of challenges. A report by the American Psychological Association's task force on the interface between psychology and global climate change. American Psychological Association, Washington
Terry G (2009) Climate change and gender justice. Oxfam, Great Britain
Thomas DS, Twyman C (2005) Equity and justice in climate change adaptation amongst natural-resource-dependent societies. Glob Environ Change 15(2):115–124
UNFCCC (2017) Human health and adaptation: understanding climate impacts on health and opportunities for action synthesis paper by the secretariat. http://unfccc.int/resource/docs/2017/sbsta/eng/02.pdf
Urich PB, Li Y, Burton D (2017) When extreme rainfall disrupts a water supply – the case of Auckland City, New Zealand - March 2017. CLIMsystems Blog. http://www.climsystems.com/blog/post/when-extreme-rainfall-disrupts-a-water-supply-the-case-of-auckland-city-new-zealand. Accessed 3 Oct 2017
Varghese J, Kutty VR, Paina L, Adam T (2014) Advancing the application of systems thinking in health: understanding the growing complexity governing immunization services in Kerala, India. Health Res Policy Syst 12(1):47
Vespignani A (2012) Modelling dynamical processes in complex socio-technical systems. Nat Phys 8(1):32
Wang H, Horton R (2015) Tackling climate change: the greatest opportunity for global health. Lancet 386(10006):1798–1799
Watts N, Adger WN, Agnolucci P, Blackstock J, Byass P, Cai W, Cox PM (2015) Health and climate change: policy responses to protect public health. Lancet 386(10006):1861–1914
Willox AC, Harper SL, Ford JD, Edge VL, Landman K, Houle K, Wolfrey C (2013) Climate change and mental health: an exploratory case study from Rigolet, Nunatsiavut, Canada. Clim Change 121(2):255
Wolf J (2011) Climate change adaptation as a social process. In: Ford JD, Berrang-Ford L (eds) Climate change adaptation in developed nations: from theory to practice. Springer, New York
World Health Organization (2004) The impact of health expenditure on households and options for alternative financing
World Health Organization (2009) WHO guide to identifying the economic consequences of disease and injury
World Health Organization (2012) Modern health care delivery systems, care coordination and the role of hospitals. WHO Regional Office for Europe, Copenhagen
World Health Organization (2015) Climate change and health in the Western Pacific Region: synthesis of evidence, profiles of selected countries and policy direction. WHO Regional Office for the Western Pacific, Manila
World Health Organization (2016) Conference conclusions and Action Agenda, final version to be available online from Tuesday 12 July 2016. http://www.who.int/globalchange/conference-actionagenda-final.pdf
World Health Organization (2017) Building health systems resilience to climate change
Xu D, Li C, Song X, Ren H (2014) The dynamics of desertification in the farming-pastoral region of North China over the past 10 years and their relationship to climate change and human activity. CATENA 123:11–22

Zhang X, Bloom G, Xu X, Chen L, Liang X, Wolcott SJ (2014) Advancing the application of systems thinking in health: managing rural China health system development in complex and dynamic contexts. Health Res Policy Syst 12(1):44

Zwingmann C, Klein C, Büssing A (2011) Measuring religiosity/spirituality: theoretical differentiations and categorization of instruments. Religions 2(3):345–357

Chapter 2
"S.I.T.E" (Societal-Institutional-Technical-Economic) Valuation Framework: A Case of Drinking-Water Facilities and Services in Slum Areas of the Hyderabad Region, India

Anil Kumar Palakodeti

Abstract This paper provides the original research contribution, titled "S.I.T.E valuation of Infrastructure facilities and Services" (Societal-Institutional-Technical-Economic valuation—A systematic and integrated valuation framework for decision-making processes to provide safe and adequate infrastructure facilities and services to urban poor, in a cost-effective manner). In India, there are various policies, programmes, schemes, and projects to improve the infrastructure conditions and many are still ongoing to meet the Millennium Development goals. Nevertheless, the infrastructure associated challenges still persevere, particularly in slum areas, leading to adversative consequences in terms of deprived health, loss of productivity—making the living of the poor population even more miserable. Despite having urban planning instruments that are steering the policies and projects, specifically designed for improving the infrastructure for slum population, the decision-making processes for developing policies and projects were amalgamation of many apprehensions, leading the government agencies unable to meet their set objectives and targets of sustainability, equity, effectiveness, efficiency and good governance in the decisions made, and in providing the requisite infrastructure facilities and services. As a portion of the case study analysis, this paper examines in detail the decision-making processes involved in providing the drinking-water facilities and services to slum areas in the Hyderabad region. It is revealing that the current decision-making support is only accounting for the technical and financial parameters—making it more a techno-bureaucratic approach, but not considering social, institutional, and economic aspects. In order to address these concerns, it is vital for the urban administrators to consider and analyze social, institutional, technical, and economic aspects pertaining to infrastructure facilities and services in an integrated fashion for their search of solutions during decision-making processes. This research paper provides insights on the proposed theoretical framework "S.I.T.E" valuation, a systematic and integrated valuation framework for the decision-making process to provide safe and adequate

A. K. Palakodeti (✉)
Erasmus University Rotterdam, Burgemeester Oudlaan 50, 3062 PA Rotterdam, The Netherlands
e-mail: anilkumar8030@gmail.com; palakodeti@ihs.nl

R. B. Singh et al. (eds.), *Urban Health Risk and Resilience in Asian Cities*,
Advances in Geographical and Environmental Sciences,
https://doi.org/10.1007/978-981-15-1205-6_2

infrastructure facilities and services to urban poor population in metropolitan cities in order to meet the set objectives and targets.

Keywords Slum areas · Urban infrastructure · Water infrastructure · Systematic · Integrated · Decision-making support · Valuation framework · Hyderabad · Telangana

2.1 Introduction

The urban environment of cities which is a multifaceted living unit comprises the components such as structure and infrastructure built to a definite extent, the existing natural assets and conditions that shape and enable the city to exist, and the inhabitants who reside and work in it. With rapidly persistent transformations and emergent urbanism in Indian cities, these aforesaid components are affected by the urban growth patterns and economic activities taking place within and around the cities (Harte 2007; Hogan et al. 2012; Jodha 1990; Lakshmana 2008, 2013; Ramachandraiah 2004; Shaw 1989). It is evident that the urban environmental challenges are increasing at an alarming rate in many cities due to population growth, economic development, and other related factors (Lakshmana 2008, 2013; Nickum 1997). In turn, these urban environmental challenges are increasingly affecting productivity, human health, and safety, amenity, and ecology (Bartone 1990). It is also apparent that different environmental challenges vary from region to region (Lakshmana 2013) and diverse income groups in different ways. In other words, the environmental challenges affecting the individuals/households/communities depend on their income levels. The poor are affected most sternly by deteriorated environmental conditions. Studies of urban health conditions in developing countries also disclose significant differences in terms of environmental effects between the rich and the poor; urban poor populace (especially women and children) in many cases are susceptible to adverse environmental conditions (Hardoy and Satterthwaite 1989).

The concentrated growth in certain regions and unfettered development in the urban areas, particularly over the last two decades, with deteriorated environmental conditions and without adequate infrastructure facilities and services, dwindled the habitable conditions in most of the metropolitan cities in India. The consequences of these development patterns resulted in the growth of slum areas. In India, it is estimated that about 46% of the total population, living in absolute poverty, are prone to these adverse living conditions (Ramachandraiah 2004). It was also argued that the inability of many state and local governments in developing countries to make available adequate infrastructure facilities and services has also led to the degradation of the living and natural environment in and around the cities (World Bank 1990a). The slum population in the cities was neglected by the governing agencies to a greater extent for various reasons that led to deteriorated conditions. However, considering the consequences of neglected slum population and realizing the importance of slum dwellers in contributions to GDP growth, policies and strategies were formulated during the 11th five year plan in India in order to make cities "inclusive" in nature, i.e., including the slum population in the process of growth and development and in improving the living conditions (viz., delivering basic facilities and services, access to affordable shelter, and employment opportunities) of the residents of the slum areas. Despite these policies and strategies, in Indian cities with the increasing population at geometric progression caused due to rapid urbanization, industrialization and agricultural advancements are creating stress on existing limited infrastructure and in particular on water access and consumption patterns (Central Pollution Control Board (CPCB) 2013). Millions of people, particularly the poor population, are still living in inhabitable areas, with a lack of basic needs—especially drinking water.

2.2 Objectives

The prime objective for any possible solution is to satisfy the thriving necessities and diminish the glitches to support the human activities, support sustainability of facilities and services, taking into account the environmental and social demands, and support the sustainability of institutions. Since the provision of safe and adequate infrastructure facilities and services for slum population is becoming more intricate due to several factors, there is a need for enhancements in the decision-making approach in providing infrastructure facilities and services to the slum population to prevent the negative impacts on the environment, society, and institutions. In order to accomplish these objectives, the decisions must take place in a systematic and integrated process and this should be made as a practice (Department for International Development (DFID) 1998). Considering the negative bearings on the environment and society caused due to the management practices in the past few decades, the need for a paradigm shift toward necessity of systematic, interdisciplinary, and integrated approach of natural and social sciences in planning and decision-making process of development policies and projects became inevitable (Soentoro et al. 2011). Based on the abovementioned, the current research paper emphasizes on the following overarching research question: Which are the values to be integrated into a systematic decision-making process to provide safe and adequate infrastructure facilities and services in cost-effective ways to slum areas in select metropolitan region in India. It is vital for urban administrators to consider and analyze factors in an integration fashion for their search of solutions and these should include physical, social, economic, and institutional aspects pertaining to infrastructure facilities and services (Meena et al. 2008).

2.3 Supposition

The absence of systematic and integrated decision-making support in government agencies are creating hurdles in providing the safe and adequate infrastructure facilities and services to urban poor population in a cost-effective manner. The current research paper positions itself to propose the involvement of stakeholders' (societal-institutional) values in policy-making along with economical-technical values of the issues that must be included in the decision-making processes to reduce the environmental spillovers. It should be noted that the current research paper does not aim to develop a new method or overlay the existing valuation methods. In addition, this does not aim to provide an improved measurement or identification of the right, or proper value of infrastructure facilities and services. It will only develop a new approach to the valuation process; perhaps a new structuring by the integration of existing and proven techniques and methodologies. It is just an addition to the existing methodologies of (e)valuation.

2.4 Methodology

The current research paper initially emphasizes on the ontological position "constructivism" for "understanding the values in the existing infrastructure conditions and decision-making processes in providing the same" through the epistemological notion "interpretivism". This process which is purely subjective in nature will lead to the construction of facts in the case study area, i.e., existing conditions of drinking-water facilities and services and decision-making process in providing the same to slum areas in the Hyderabad region. It further concentrates on the ontological position "constructivism" for "understanding the research gap and identifying the key theories/concepts" in the literature, through the epistemological notion "interpretivism" through literature review. This process, which is purely subjective or interpretive in nature, will lead to the construction of a theoretical framework based on the gaps identified in the existing decision-making process.

2.5 Study Area Analysis

2.5.1 Hyderabad City Profiling

Hyderabad is one of such Indian cities suffering from the mentioned challenges (refer Introduction section). It is one of the major metropolitan cities, which have been contributing to India's population (Dahiya 2012). It is the city with a steadily increasing population over the last 50 years (Mulligan and Crampton 2005) and rapidly growing metropolitan region in India in the last few decades (Dahiya 2012).

The Hyderabad region which was initially 55 km^2 with a population of 0.35 million in the mid-eighteenth century is now spread across 7228 km^2 with a population of 7.7 million (CoI 2011). Analyzing the trends of the growth rates, the city is the sixth largest metropolitan city[1] in India and the region is expected to be 11 million by the year 2020 and 19 million by the year 2041 (GoAP 2013). The growth patterns of the metropolitan region portray that the population is growing at significant rates in the neighboring municipal areas and the corresponding decline of population in the core area (erstwhile Municipal Corporation of Hyderabad (MCH) area) (CDS 2004). The city has a historical significance along with the process of industrial development and economic liberalization that influenced the changes in state economy,[2] population growth, socioeconomic composition, city growth patterns, urban agglomeration, and spatial-governance structuring patterns (Diganta. 2015; Kennedy 2007; Kirk 2005; Krueger 2002; Naidu and Ninan 2000; Rao 2007; Krishna. 2002; Rudolph and Rudolph 2001).

With the advent of increasing regional growth of the city, there was also the necessity of structuring and restructuring of municipal bodies responsible for providing better amenities and development bodies responsible for planning and development of the complete region. The Municipal Corporation of Hyderabad (MCH) which was initially the local body responsible for provision of civic and infrastructure facilities to the inhabitants of the city covering the area of only 172 km^2 was transformed to Greater Hyderabad Municipal Corporation (GHMC) functioning similarly to erstwhile MCH, but responsible for a larger area, i.e., of 650 km^2 (including core area (MCH region) and surrounding 12 municipalities). In addition to the transformation of civic bodies, planning body Hyderabad Urban Development Authority (HUDA) was also formed to function as preparation of land-use planning, zoning regulation, and infrastructure creation for the jurisdiction of GHMC. However, with the growth in the administrative boundaries of the civic body, the jurisdictions of HUDA also increased from 1861 km^2 to 7228 km^2 and replaced it with Hyderabad Metropolitan Development Authority (HMDA) to cater to the planning and development needs of an urban agglomeration of municipal area and its surrounding 55 peri-urban areas (Diganta. 2015; GoAP 2013). In addition to the formation of civic bodies, there were also special development authorities, viz., Cyberabad Development Authority (CDA), Hyderabad Airport Development Authority (HADA), and Buddha Purnima Project Authority (BPPA) for planning and development of special infrastructure in delineated zones of Hyderabad metropolitan region.

It is evident that the Hyderabad region became the focal point for rest of the state for new developments (especially in territory sector[3]), settlements, infrastructure,

[1]Next to Chennai, Delhi, Kolkata, Chennai, and Bangalore.

[2]During 1980 to 1995, the state's economy (which was predominately into agriculture) declined and forced the state government to procure loans, drastically cut down the subsidies on welfare programmes, introduction of new economic reforms to strengthen the structure of the economy to recover from the State GDP crisis and to regulate the fiscal deficit by giving priority to obtain funds for the development of tertiary sector.

[3]Include IT-related services, biotechnology, tourism, logistics, health care, and educational services, etc.

businesses, trading, and political systems (Alam and Khan 1972; Bunnell and Das 2010; Luther 2006; Moser 2010; Sen and Frankel 2005; Srinivasulu 2002). It is currently the capital city of Telangana State[4] and shared administrative capital of residual Andhra Pradesh for a period of the next 10 years (Diganta. 2015). This capital city which projects itself as smart, high tech destination in India (PTI 2014b) and world-class knowledge hub of information technology (Sen and Frankel 2005) functions as the center for administrative activities, scientific research, industrial, commercial, and IT hub of the state—facilitating employment prospects and acting as economic growth center of the state. All these promising status quo of the Hyderabad city, viz., industrialization pooled with new economic policies, resulted in rapid urbanization and urban sprawl in a haphazard way (Dupont 2007). The structuring and restructuring of Hyderabad served only political tenacities rather than administrative efficacy and balanced growth of the entire region (Naidu 1990; Ramachandraiah and Prasad 2008). These created adverse impacts on the environment and increasing burdens on existing infrastructure, especially on water supply and security (Rao 2007); and consequently the challenges in providing the water infrastructure facilities and services by the civic and planning bodies to the proliferating population and to the expanded city region (Diganta 2015), especially to the slum population and slum areas.

2.5.2 *Drinking-Water Facilities and Services in the Hyderabad Region*

The Hyderabad region, once popularly known as the "city of lakes" was blessed with thousands of water bodies serving the growing population for a long time. Since Hyderabad does not have any perennial rivers and is having low average annual monsoon rainfall,[5] the drinking-water requirements of the city residents are predominantly dependent on available resources like lakes, tanks, and rivers flowing at a distance away from the city (Lundqvist et al. 2003; Mukherjee et al. 2010). However, Narain (2006) argues that Hyderabad receives enormous rainwater and if strategically utilized, this can potentially cater to 35% of the domestic water demand (George et al. 2009). Nevertheless, in the past two decades, with the rapidly increasing urban population and the size of the city, the water deficit for the city population became inevitable. In addition to this, there are localities in the city wherein the existing infrastructure is below the required standards and drinking-water flow is contaminated (Ramachandraiah 2004). The discharge of untreated sewage wastes,

[4]Telangana officially became the 29th state of India bifurcated from Andhra Pradesh state with effect from June 2, 2014. The erstwhile Andhra Pradesh State was formed in 1956 with three distinct geographical regions (based on dialects) namely Telangana, Coastal Andhra, and Rayalaseema. Hyderabad was the capital of undivided Andhra Pradesh for nearly six decades. However, due to increasing economic backwardness of Telangana and cultural domination of people of coastal Andhra, people of Telangana region demanded separation from Andhra Pradesh.

[5]The average annual rainfall for Hyderabad is 78 cm, Mumbai is 225 cm, and Kolkata is 163 cm.

toxic wastes, and illegal encroachments of the water bodies drastically reduced the quality standards and caused insecurity of available water resources. It was also indicated that the drinking-water that is been supplied through taps is restricted to only two hours every alternate day (George et al. 2009; McKenzie and Ray 2009). Consequently, this inadequate amount of water and quality supplied by the service provider resulted in dependency on private and groundwater sources to cater to their day-to-day requirements; and the compulsions of the government agencies to look for new water sources (Sahu 2012). All the water policies and reforms paid less attention (for various reasons) (Chakrabarti 2001; Kumar 2004; Narain 2006) to innovative technological options in relation to water allocation and distribution in the city, like relying on rainwater harvesting approaches. The most prevalent idea of water management, which is, only associated with transferring long-distance water to the city and distributing it by means of pipes and tankers shall drastically increase not only the cost of production per unit, but also huge transmission losses and eventually increase burdens on city residents (George et al. 2009; Van Rooijen et al. 2005). It can be noted that it is environmentally and technologically unsustainable (Department for International Development (DFID) 1998). In addition to this, it can further be claimed that the proliferating demand for water in the region is aggravated with the rise in the number of slums and their population; and challenges in catering to the basic needs for this set of the population too.

2.5.3 Drinking-Water Facilities and Services in Slum Areas of the Hyderabad Region

The unfavorable conditions for agriculture in the surrounding areas of Hyderabad combined with the attraction of new employment-generating economy in different sectors of the city influenced the population of the adjoining districts to migrate to Hyderabad for their livelihoods (Ramachandraiah and Bawa 2000). However, the income levels[6] of the migrated population could not afford the available housing within the city that led to the formation of informal settlements having fewer or no basic infrastructure facilities and services (Afshar and Alikhan 2002; Goli et al. 2011; Rao 2007). It was argued that in Hyderabad there are areas with thorough infrastructure facilities as well (Nastar 2014). The aspiration to make Hyderabad the world-class information society led to the fragmented metropolitan region with a wide range of disparities in terms of infrastructure distribution, quantity, quality, and frequency of supply (Baud and Dhanalakshmi 2006; Diganta 2015). It was argued that the quality and quantity of water supplied is predominantly dependent upon the areas of economic and political importance (Sahu 2012). It can be noted that these raise the concerns for the effectiveness of the service provider in terms of achieving universal service (Department for International Development (DFID) 1998).

[6]It has been estimated average monthly income for the population living in the slum areas is INR 2000.

The disparities are just not in terms of provision of basic facilities, but also in pricing for availing the limited services. It was argued that the connection charges and tariff structure defined by the government agencies favored only the affluent population (Raghavendra 2006; Whittington 2003). This resulted in the supply of water at cheaper rates to enormous consumers than the common citizens (Nastar 2014; Sahu 2012). Several studies indicate that the slum population in Hyderabad is paying more per drop of water consumption though water is provided at subsidized rates. This shall be a high proportion of their income when compared with the proportion of income spent on water by middle- and high-income groups (Davis et al. 2008; Foster et al. 2002; Pattanayak and Stalker Prokopy 2003a, b; Kaminsky and Long 2011; Raghavendra 2006; Whittington 2003). The decreasing water resources, inadequate water supply infrastructure, and its quality forced the residents of the region to buy drinking water. The affluent population, however, is able to pay for their needs, but the poor population who are only dependent on water supplied by the water board is still affected by unsafe and insufficient drinking water (EBGH 2009). All these resulting in a record of productivity losses, several deaths, and diseased cases in various public and private hospitals due to unsafe water (McKenzie and Ray 2009; Nastar 2014; Ramachandraiah 2004). In addition to the expenditure toward access to water, there is a certain proportion of the income spent on externalities caused due to unsafe and insufficient water. It was further argued that though there were efforts made by the government agencies in terms of land tenure regularization,[7] improvement and development projects for the slum areas, lack of clarity in policy guidelines, cumbersome procedures, and political influences became the bottlenecks, which is limiting the provision of basic services like drinking water by the service provider (government agencies) to these slum dwellers (Baindur and Kamath 2009; Bakker et al. 2008; Davis et al. 2008; Greig et al. 2007; Raghavendra 2006; Robbins 2003) and consequently building pressure on the decision-makers in the concerned agencies. The conditions of slum areas in terms of availability of these services are still rudimentary and below the required standards. The gap between the formulation of strategies and implementation outcomes leaves us with an ambiguity on how the decisions are taken and what are the values considered during the decision-making process.

2.5.4 Decision-Making Process to Provide Drinking-Water Facilities and Services in Slum Areas of the Hyderabad Region

The Hyderabad Metropolitan Water Supply & Sewerage Board (HMWSSB) is a key stakeholder (or institution) responsible for planning, designing, constructing,

[7]According to this policy, the tenure of all the slum dwellers residing in the city for more than 5 years will be regularized. This shall allow the government to take up slum improvement and development projects/programmes.

organizing, executing, and managing water supply systems. It is run with a goal to provide safe and adequate drinking water to 100% of the population of Hyderabad metropolitan region. Though it is ambitious to have such a goal, it is predominately based on National Water Policy and National Urban Policy initiated by Ministry of Water Resources and Ministry of Urban Development, respectively (Ministry of Urban Development (MoUD) 2012; Ministry of Water Resources (MoWR) 2012; Nastar 2014) at the national level. The objectives of the water board to provide drinking-water services are also steered by City Development Plans prepared by Greater Hyderabad Municipal Corporation (Greater Hyderabad Municipal Corporation (GHMC) 2006). However, it was argued that there is less participation of the public and integration of various plans developed by different parastatal agencies in Hyderabad (GIZ 2013). It was also argued that the very little role of Greater Hyderabad Municipal Corporation (GHMC) and other government agencies has made water sector services in Hyderabad a techno-bureaucratic activity (Sahu 2012). However, the overall supervision of the developmental projects taken up by the water board in the Hyderabad metropolitan region and coordination of activities among different local bodies is taken care by Hyderabad Metropolitan Development Authority (HMDA) (Greater Hyderabad Municipal Corporation (GHMC) 2006; Hyderabad Metropolitan Development Authority (HMDA) 2010; Nastar 2014).

HMWSSB is one of the youngest institutions, which was formed under Act 15 of 1989 with administrative and financial independence from the Municipal Administrative Department[8] and state government[9] (Hyderabad Metropolitan Water Supply and Sewerage Board (HMWSSB) 2008). However, considering the composition of the board, it can be stated that there is a compromise on administrative independence in the decision-making powers of the water board. The state government retained the ownership of the board's assets; and the powers to formulate policies and strategies vested on the managerial committee, which is acting as a central administrative body consisting of bureaucrats.[10] It was further claimed that considering the challenges faced during the period of erstwhile water board[11] by the public representatives during the discussions and board's functioning, the notion of public representation and participation in the decision-making process has been removed from the Act 15 of 1989 (Sahu 2012). In addition to this, there is no direct involvement of the general/slum population in the decision-making process. It can be noted that the non-involvement of the public in the decision-making process leads to non-accountability and non-transparency, i.e., the notion of good governance is put at risk (Department

[8]Before HMWSSB, Water Works Department was responsible of supplying drinking water headed by Municipal Administrative Department.

[9]State government financially supported Water Works Department.

[10]Chief Minister of the State, Ministers, Secretaries to Government (MAUD, Finance, Irrigation), Commissioner (GMCH), Chairman (Pollution Control Board), Director (Health), Chief Engineer, Accountant General, IAS (Indian Administrative Service) officer appointed by Chief Minister, and other specialist officer.

[11]Formed in 1982 under Andhra Pradesh Act No. 6 of 1982 to ensure autonomy in planning effectively; this is headed by administrator/ engineer/ technical person and representations from the community.

for International Development (DFID) 1998). However, several studies indicate that the e-governance initiatives of HMWSSB like Metro Customer Care, Single Window Cell, Citizen Charter Center, and Lok Adalat are the only means of channel to communicate with the water board for processing new applications, user fee payments, and complaint registration (Caseley 2006; McKenzie and Ray 2009). Nevertheless, all these pertain to operational and maintenance issues, but not in the provision of required facilities and services. It was also argued that all these initiatives have had a constructive impact in delivering services to the affluent population only (Caseley 2006; Davis et al. 2008; Robinson 2007) and, however, were non-friendly to the slum population who have no access to computers and Internet facilities (Sahu 2012).

The issues pertaining to the provision of drinking-water services and improvements to the existing facilities and services at individual/community level are generally put forth by the public/resident welfare associations to the corporators[12]; and subsequently,[13] the proposals are prepared and submitted by the corporators to HMWSSB and/or state government. Though there is no distribution of duties dictated by official rulebooks of HMWSSB or any other parastatal agency per se, it was argued that in this regard, it rather takes the form of informal arrangements (Sahu 2012). These sorts of arrangements are largely dependent on personal relationships, party affiliations, political tensions surrounding the issue of water and expectations of the population. Therefore, the representation of his/her role differs from one stakeholder to another and from one community to another. On the other hand, there are Non-Government Organizations (NGOs), Civil Society Organizations (CSOs), and Voluntary Organizations (VOs) involved in the decision-making process. It was observed that these actors play the role of frontrunners in transition areas to confront the policy-makers seeking the creation of rights to basic needs and influence the policies through petitions, public hearings, campaigns, press meets, critical debates, resistances, disobedience movements, and political conflicts (Frantzeskaki et al. 2012; Rotmans and Loorbach 2009). These indirectly build pressure on the HMWSSB as well as on the government to provide better services.

Despite clear mandates, the board enjoys neither administrative powers nor budgetary autonomy. Given the board's difficulty in the decision-making process, it also affected the financial status (Sahu 2012). Though the notion behind the introduction of water tariff is to make the public aware of the economic value of water, the water in Hyderabad is treated as social good, which is restricting the Board's revenues and, in turn, providing adequate infrastructure. It was observed that any proposals by the water board for the revisions in the water tariffs are strongly opposed by the political actors. Considering the vested interests of these political actors, they also propose to treat water as "common good" and to provide it for no charge. Considering these factors, it is not surprising that water board's service levels have been declining over the past several years, especially in slum areas of the metropolitan region. It was argued that the HMWSSB, whose primary and major income is through user charges and partly through grants-in-aid, is not financially self-sustainable and is limiting the

---

[12]Elected members incharge of the development of their electoral constituencies.

[13]General representation (in the form of a request letter) stating the challenges and requirements.

improvements in the services to the poor (Hyderabad Metropolitan Water Supply and Sewerage Board (HMWSSB) 2008). Historically, it is evident that the consumption of water for domestic purposes is always more than 60% and the rest by commercial and industrial purposes. However, the revenues generated from domestic consumption are only 30% of the board's revenue (Hyderabad Metropolitan Water Supply and Sewerage Board (HMWSSB) 2011). The cross-subsidy pricing for water services combined with a lack of adequate investment funds is creating tremendous pressure on HMWSSB. In addition to this, there are also physical losses due to improper infrastructure, administrative losses due to illegal withdrawal of water and defaulters of user charges in slum areas. This is resulting in a substantial increase in levels of non-revenue water (Greater Hyderabad Municipal Corporation (GHMC) 2006; McKenzie and Ray 2009; Robbins 2003; Swyngedouw 2006). But, due to political influence, any preventive measures against these activities are hindered (Sahu 2012). It can be noted that the objective of equity and efficiency of services are jeopardized, i.e., treating all user groups equally and fairly and service delivery assessment by indicators such as non-revenue water and bill collection efficiency ratio (Department for International Development (DFID) 1998). With the current bulk source of water drying up and non-ability to meet the daily water requirements of city residents, the HMWSSB is now steering to develop new water projects that fetch water resources from rivers Krishna and Godavari which are approximately 110 and 200 km away from the city premises (Hyderabad Metropolitan Water Supply and Sewerage Board (HMWSSB) 2015). This shall not only increase the transmission losses but also production cost per unit, resulting in financial burdens on residents of the city (especially on slum population) in the form of user charges and eventually on HMWSSB in the form of non-revenue water (administrative losses).

In addition to the existing administrative setup of HMWSSB and role of informal actors, the operational and maintenance staff of the water board who are geographically spread across the metropolitan region is responsible for meter reading, bill distribution, identifying illegal connections, and carrying out feasibility studies (Hyderabad Metropolitan Water Supply and Sewerage Board (HMWSSB) 2008) for improving the water services in the Hyderabad region. These feasibility/project studies play a significant role in the decision-making process and further to procure funding assistance from state/central/international agencies (Hyderabad Metropolitan Water Supply and Sewerage Board (HMWSSB) 2008; Sahu 2012). It can be stated that HMWSSB in coordination with professional firms/companies[14] prepares these reports pertaining to the water sector services predominantly covering the technical and financial aspects. The preparation of project reports are primarily based on the standardized reference framework of Jawaharlal Nehru Urban Renewal Mission (JNNURM) for Detailed Project Report (DPR) Preparation (MoUD n.d.) which include details covering (a) sector background, context, and project rationale, (b) project definition, concept and scope, (c) project costing, (d) project institution framework, (e) project financial structuring, (f) project phasing, (g) project Operation & Maintenance (O&M) framework and planning, (h) project financial

[14]Generally appointed through tendering process.

viability/sustainability, (i) project benefits assessments, (j) engineering drawings, (k) specifications, etc. However, the source for these projects identification can be either/combination of (a) development/action plans, (b) priorities of local elected representatives, (c) demand from interest groups or beneficiaries, and (d) dialogues between various agencies (Sahu 2012). Despite having urban planning instruments steering ambitious plans and policy initiatives[15] (and vice versa) to improve the municipal water sector services in the city and living conditions of the slum settlements (Diganta. 2015; Nastar 2014; Sharma 2011), it was observed that lack of institutional (service provider) strengths, lack of inter-departmental coordination, lack of participation of slum population, and estimations of their requirements and acceptance combined with treating water as "social good" in decision-making processes led to these disturbing conditions of drinking-water infrastructure facilities and services in slum areas in the Hyderabad region.

2.6 Findings

The current research paper initially emphasizes on the ontological position "constructivism" for "understanding the values in the existing infrastructure conditions and decision-making processes in providing the same" through the epistemological notion "interpretivism". This process which is purely subjective in nature will lead to the construction of facts in the case study area, i.e., existing conditions of drinking-water facilities and services and decision-making process in providing the same to slum areas in the Hyderabad region. It further concentrates on the ontological position "constructivism" for "understanding the research gap and identifying the key theories/concepts" in the literature, through the epistemological notion "interpretivism" through literature review. This process, which is purely subjective or interpretive in nature, will lead to the construction of a theoretical framework based on the gaps identified in the existing decision-making process.

To date, there have already been various household and community-level projects to improve infrastructure facilities and services (especially, water supply[16]) in India and many are still ongoing to meet the Millennium Development Goals (MDGs) and proliferating demands. Though there was huge attention given to these issues both at the global and local level, the water-related challenges still persist, particularly in slum areas. The lack of basic infrastructure facilities to provide safe and adequate drinking water led to adverse consequences in terms of poor health, loss of productivity—making a living of the poor population even more miserable. All these consequences are against constitutional mandates and official proclamations

[15]National level—National Water and Urban Policy; State level—Andhra Pradesh Urban Services for the Poor Programme (APUSP), Jawaharlal Nehru Urban Renewal Mission (JNNURM), Slum Free City Programme, Rajiv Awas Yojana (RAY), and Institutional reforms.

[16]The improved water supply is defined as "piped water, public tap, borehole or pump, protected well, protected spring, or rain water. Improved water sources do not include vendor-provided waters, bottled water, tanker trucks, or unprotected wells and springs".

developed across the world (also in India) which have recognized the importance of "safe,[17] adequate,[18] affordable,[19] and accessible[20] drinking water" to all as "right to life" and necessity for the existence and well-being of any society (APPEN 1998; GoI 1986; ILR 2001; Petrella 2001; Ramachandraiah 2004; Sengupta 2001; Satyabrata 2001; Smets 1999; United Nations 2002; Venugopal 2003). Further, with the advent of structural adjustments and readjustments, there was also a paradigm shift toward private players' involvement in providing basic services (ISODEC 2002). In addition to this, a strategic instrument "Public Interest Litigations (PIL)" was also formulated that strengthened the power of citizens in enjoying their rights (ILR 2001). According to human rights perspective, the decreasing access to these services signifies the violation of human rights (Gleick 1999; ISODEC 2002; United Nations 1995) and it is the state's responsibility to enable laws and policies in a way that every citizen without any discrimination shall realize right to water (Gleick 1999; Lisbon Principles 2002). The right to water refrains from engaging any practice, activity, or third party that denies or limits equal access to adequate drinking-water, pollution and inequitably extracting water from resources (Ramachandraiah 2004). This indeed created the alertness and obligations for the state governments to take necessary steps in providing safe and adequate drinking water to all individuals at affordable levels (Ramachandraiah 2004). It was argued that having access to safe and adequate drinking water shall reduce the externalities caused due to water-borne diseases (Smets 1999). Accordingly, appropriate approaches, policies, institutional, technological, and economic interventions and instruments were adopted to address the challenges (Central Pollution Control Board (CPCB) 2013). Various forms of decision-making frameworks have been proposed for applications around the world (e.g., Analytical Hierarchy Processes, Water Safety Plans). However, it is desirable to create an easy-to-use model that incorporates factors that are specific to urban poor communities in order to provide facilities and services to these communities. Efforts to address water sector challenges have shown that no single technological solution, economic tool, or institutional setup in an isolated way be applied to all populations. Indeed, there is no shortage of potential technical solutions for improving the water supply conditions in urban poor communities. However, the criteria for selecting the appropriate solutions for a given community are not always clear. There should be a comprehensive analysis of the challenges pertaining to communities that may vary from place to place.

By drawing from the debates on the decision-making process in providing the drinking-water infrastructure facilities and services, the study area analysis has explored technical, institutional (formal and informal), and financial aspects of the

[17]Free from microorganism, with acceptable color, odor, and flavor.

[18]Be approximately 50 L or the minimum essential level of 20 L per day per individual.

[19]To be without direct and indirect costs.

[20]To be regular supply without any waiting time and must be within the premises or immediate vicinity.

urban water sector in Hyderabad, and their impacts on slum dwellers' access to drinking water. The research on the existing decision-making process to provide drinking-water facilities and services which was analyzed revealed that it failed to adequately serve water sector practitioners and achieve the objectives of sustainability, equity, effectiveness, efficiency, and good governance in the decisions made. It is indicative that the current decision-making support is accounting for regional variations, i.e., only technical parameters are taken into consideration. This analysis includes the aspects such as soil conditions, social structure, existing service levels (including infrastructure, water quality, and quantity aspects), requirements, and possible strategies to address the prevailing challenges and broad overview of institutional landscape responsible for implementation, operating, and monitoring. In addition to technical aspects, financial aspects such as estimates of costs to implement, operate, and maintain the project are also included. However, with these aspects included in the decision-making processes, the challenges are still glaringly rampant, especially in slum areas. It is indicative that the current urban water regime decision-making practices are social, institutionally and economically unsustainable, especially in providing drinking-water facilities and services for the urban poor population. It can be argued that most commonly missing aspects which play a crucial role in the decision-making process are consideration of social (impact), institutional, and economic aspects (Meena et al. 2008), along with technical (include financial) aspects (which are already taken into consideration in the current case study analysis).

2.7 Conclusions

The integration of the aforesaid aspects of the decision-making process is negligent (Meena et al. 2008), especially in the identified case study area pertaining to the provision of safe and adequate drinking-water facilities and services to the slum areas. Having identified the gaps in decision-making support systems in improving the drinking-water conditions, the new integrated approach shall help guide those organizations and individuals to make sustainable improvements to the community water systems (Meena et al. 2008). In order to resolve the burgeoning challenges effectively and efficiently in a comprehensive manner, the water sector practitioners in developing countries should design an unique approach by systematically consider "social appropriateness, economically feasibility, technically acceptable, institutionally capable…" (Meena et al. 2008) to cater to the needs of the communities. This shall address the needs and guide policy-makers to frame appropriate water policies and strategies, which are sustainable, equitable, effective, efficient, and good governance (Department for International Development (DFID) 1998).

The consideration of "Systematic" and "Integration" aspects in the decision-making process is critical to define clear objectives for any decisions taken in the drinking-water sector. The decision-making process in the services provision sector is said to be systematic and integrated unless and until the decisions taken are resulting in providing services to all the consumers which are sustainable in terms of financial status of the service provider, environmental aspects of the community, technological aspects of the services; equity considerations in terms of treating all users groups; effectiveness in terms of universal access to the services; efficient in terms of service delivery, and good governance in terms of accountability to users, transparency, and customer involvement in decision-making process (Department for International Development (DFID) 1998). Therefore, all the decisions taken should lead to the delivery of drinking water satisfying these following conditions with respect to the delivery of drinking-water services. Table 2.1 illustrates the factors influencing the elements of the "Systematic and Integrated decision-making process".

Table 2.1 also pronounces the correlation between the factors determining "Systematic and Integrated decision-making process" and "SITE valuation". Considering the above-discussed theoretical notion, in the current research paper, the decision-making process "SITE Valuation" in providing safe and adequate drinking water to slum areas is grounded on understanding and analyzing four vital aspects: the social, institutional, technical, and economic to achieve the status of the systematic and integrated decision-making process. The factors affecting the decision-making processes are considered from these four distinct but interrelated areas, as illustrated in Fig. 2.1. The first level of influence is the institutional aspects, which is the part of decision-making processes that reflects the interplay of various stakeholders, functions, and influencing factors in an organization responsible for providing safe and adequate drinking water. The technical aspects of providing safe and adequate drinking water are the second level of influence in the decision-making process. It refers to the immediate inside element of institutions and directly related to institutional aspects. This level of influence reflects the quality, quantity, and infrastructure conditions related to drinking-water facilities and services. The last and most significant level of influence is related to the "consumer end", i.e., the Social and Economic aspects, which refers to the immediate element and directly under the influence of technical aspects of drinking water. These aspects play a significant role in the decision-making process. This level, in itself, is the interplay between the social health conditions and economic (monetary) values caused due to drinking-water facilities and services. This level, which considers the socioeconomic health impacts caused due to drinking-water challenges, is under the influence of technical aspects. These social-economic-health conditions influence the economic values (monetary costs) caused due to drinking-water challenges, which directly influence the institutional aspects and, in turn, the decision-making processes.

The context of this research paper is based in slum areas in metropolitan cities in India where the challenges faced by the municipal service providers and policy-makers is the "poor quality and quantity of drinking water" and the opportunity realized is the "systematic and integrated approach to (e)value the challenge and collective actions in decision-making process for improving the deteriorating conditions in cost-effective way". The decision-making process in providing safe and adequate drinking water in India will be built on the abovementioned (refer Fig. 2.1) four aspects. They form the research circle (refer Fig. 2.2) with broad queries that need to be integrated in the decision-making process such as which are the technical values of drinking water, followed by which are the social health impacts caused,

Table 2.1 Factors influencing elements of the systematic and integrated decision-making process

Elements	Description	Factors influencing	Category
Financial sustainability	Financial sustainability of the service provider	Managerial, financial, governance, and external constraints	Institutional
Environmental sustainability	Environmental sustainability of the community	Physical infrastructure pertaining to drinking water suiting the conditions of the communities; spread of water pollution conditions in the communities; depleting water resources; social health conditions in the communities	Social and technical
Technical sustainability	Technical sustainability of the physical infrastructure in the community	Physical infrastructure pertaining to drinking water suiting the conditions and acceptance of the communities	Technical
Equitable	All the users are treated equally	Quality of drinking water, quantity of drinking water, coverage of physical infrastructure pertaining to drinking water without any discrimination in terms of the community's socioeconomic conditions	Technical and social
Effective and efficient	Universal services in a cost-effective and efficient way	Innovative technologies; damage costs and preventive costs and estimate caused due to unsafe and inadequate drinking water; communities' willingness to pay/accept	Technical, economic, and social
Good governance	Accountability to users, transparency, and customer involvement	Managerial governance, external constraints, and public involvement in the decision-making process	Social and institutional

Source Author 2018

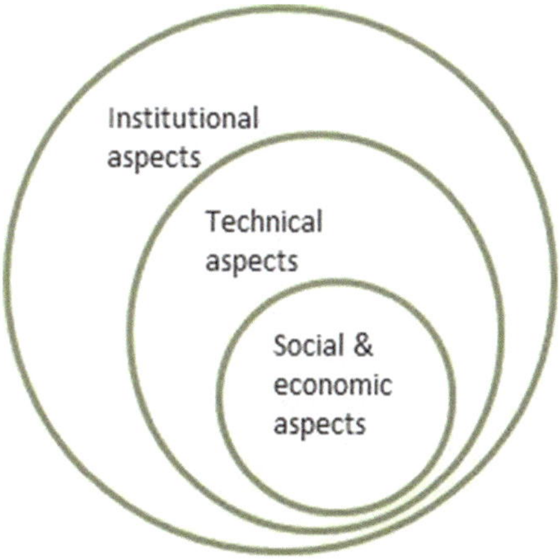

Fig. 2.1 Factors influencing the decision-making process (*Source* Author 2018)

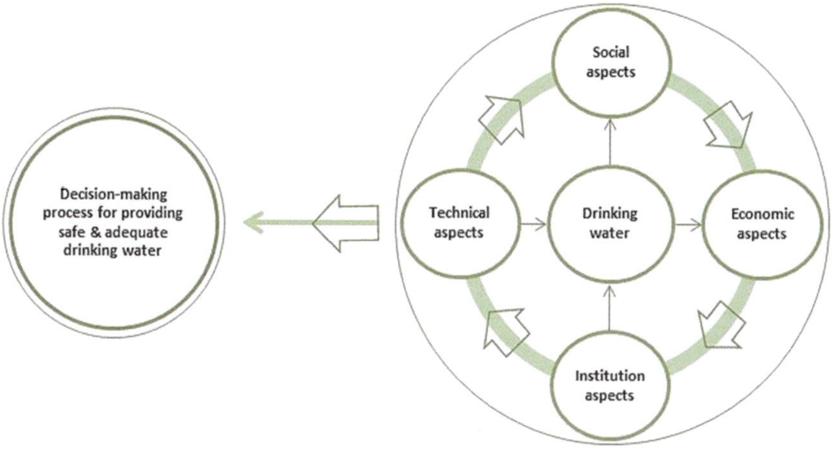

Fig. 2.2 Research circle (*Source* Author 2018)

further leading to identifying the economic (monetary) values of the impacts, and which are the values of institutions (service provider) responsible for better facilities and service provision. The relationship between these four levels of influence in the decision-making process is illustrated in Fig. 2.2. It shows that the technical aspects of drinking water affect the social health conditions of society. These, in turn, affect the economic (monetary) values. The economic (monetary) values of the impacts have an effect on institutions, which is, in turn, affect the technical aspects of drinking water. From Fig. 2.2, understanding and estimating the values of these aspects of drinking water will benefit in providing a comprehensive understanding of the prevailing challenges for taking decisions for providing safe and adequate drinking water.

It is further important to have a clear goal or objective for decision-making processes in the drinking-water sector. The "decision-making process" is considered to be "systematic and integrated" if the "decision taken" for providing safe and adequate drinking water is "sustainable", i.e., financially, environmentally, and technically; "equitable", i.e., all the users are treated equally with safe and sufficient drinking water; "effective and efficiency", i.e., universal services in a cost-effective way without any social health impacts and damage/preventive costs; and "good governance", i.e., accountability to users, transparency, customer involvement in the decision-making processes. Therefore, a good "decision" should have these states of affairs with respect to the provision of safe and adequate drinking-water services. The factors influencing the decision-making process in the drinking-water sector in Indian metropolitan cities will form the basis for the conceptual framework used in this study. The "technical aspects of drinking water, social health conditions, economic (monetary) values, and institutional values" of unsafe and inadequate drinking water are the functions of a "systematic and integrated approach to the decision-making process for providing safe and adequate drinking water" as shown in Fig. 2.3.

As shown in Fig. 2.3, the key technical values influencing the decision-making process are quality, quantity, and infrastructure parameters of drinking water. In addition, the key economic values of the slum population in terms of damage costs, preventive costs, and willingness to pay/accept for new or better drinking-water services play a critical role in decision-making processes. The social health impacts act as a bridge between technical and economic (monetary) values. The key social health values are to identify the socioeconomic statutes, and health impacts caused due to unsafe and inadequate drinking water. This provides vital information to understand the effects of poor technical values of drinking-water services and estimate the economic (monetary) values. In addition, any decision-making processes are under the influence of policies, legal and regulatory frameworks, political interferences, roles,

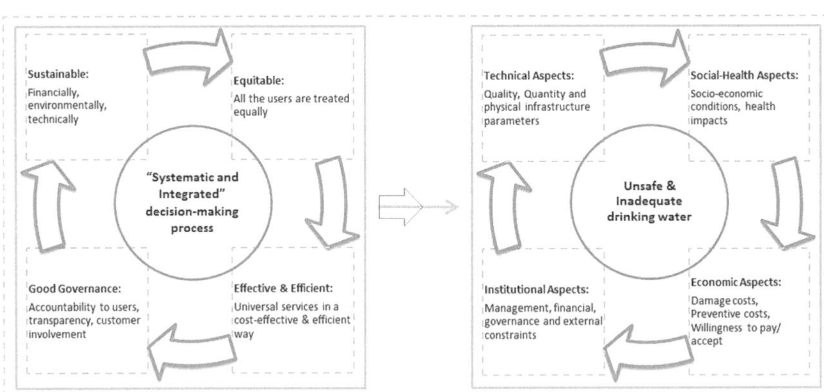

Fig. 2.3 Theoretical framework (*Source* Author 2018)

and responsibilities of all stakeholders, interactions between different agencies, community participation, availability of resources and feasibility of providing services. Broadly, the key factors relevant to institutional values are to understand the management, finance, governance, and external constraints, which should provide clear and appropriate policy and institutional framework, i.e., rules of the game for a comprehensive, systematic, and integrated approach to decision-making. This research paper is aiming at the same, for which institutional analysis plays an important role, in addition to the analysis of technical, social, economic (monetary) aspects.

The framework of "SITE valuation" will only comprehend the approach by providing the social, institutional, technical, and economic values in the decision-making process, regardless of what these values may be. It can be stated that it is important to assess the social (impact) to reduce the inequalities in providing drinking-water services, risks of building infrastructure facilities that people do not want and for which they are not willing to pay. The analysis shall consider local control over resources, end-user ability to pay, social cohesion, resource conflict, and impact on different age groups and genders. Further, it is also important to consider the economic aspects, which shall estimate the long-term sustainability of the strategies as well as the communities. It is important to note that if communities cannot afford, or are not willing to pay for the solutions, the water service infrastructure is more likely to fall into a state of disrepair (Meena et al. 2008). However, it is important not only to estimate the communities' willingness to pay for the solutions but also to estimate the burdens on communities in terms of costs that can be reduced. It was indicated that an ideal decision-making support system should include all these necessary elements to implement a water project to reduce the burgeoning challenges (Meena et al. 2008). It is also important to note that it is a significant challenge to include these aspects in the decision-making support system, which shall require the concrete efforts of institutions to implement. For the same to be successful, it is vital to understand, analyze, and transform the existing institutional setup, which is responsible for the service delivery. The analysis shall consider the existing formal and informal stakeholders, financial aspects, existing policies and projects, and their role in the decision-making process. Focusing on bottom-up approaches, this research paper draws attention to the possibility of building continuous process in decision making through integration of social, institutional, technological, and economic analysis and improve the living conditions of the entire metropolitan region in a balanced way through developing integrated solutions (Ballas 2013; Baud and Dhanalakshmi 2006; Pethe et al. 2014).

The wide range of perspectives empower the study of conceivable links, or embedding this "new" approach into the existing decision-making process in a systematic and integrated way—through "social perspective" for analyzing the impacts of existing/future policies and peoples' participatory role in decision-making process; through "institutional perspective" for analyzing the existing institutional processes and structures; through "technical perspective" for analyzing the quality and quantity aspects of the any infrastructure facilities and services; and through "economic perspective" for analyzing the monetary values of infrastructure-related challenges. The Author took further steps to test this new valuation framework proposed in this research paper in valuing the challenges in select population of select metropolitan region in India—to identify and analyze the stakeholder involvement in decision-making processes, to identify and analyze the social (impact) assessment in decision-making and of the decisions made, to analyze the technical values of quality of infrastructure facilities and services, and to analyze the economic values of the problems caused due to infrastructure facilities and services. This will result in support of the policies be framed in a way to reduce the environmental costs of unsafe and insufficient infrastructure facilities and services. The use of the above described new valuation framework in a scenario of process-to-be-developed and integrate the same into policy-making and infrastructure management for governing authorities in urban spaces accomplish the environmental objective, i.e., "good infrastructure status for all infrastructure, in the cost-effective manner".

References

Afshar H, Alikhan F (2002) Age and empowerment among slum dwelling women in Hyderabad. J Int Dev 14(8):1153–1161

Alam MS, Khan W (1972) Metropolitan Hyderabad and its region: a strategy for development. Asia Publishing House, Bombay

APPEN (1998) Water watch – a community action guide. Asia-Pacific People's Environmental Network, Penang, Malaysia

Baindur V, Kamath L (2009) Reengineering urban infrastructure: how the World Bank and Asian Development Bank shape urban infrastructure finance and governance in India. Bank Information Centre, South Asia, New Delhi

Bakker K, Kooy M, Shofiani NE, Martijn EJ (2008) Governance failure: rethinking the institutional dimensions of urban water supply to poor households. World Development 36:1891–1915

Ballas D (2013) What makes a 'happy city'. Cities 32(1):39–50

Bartone C (1990) Analytical framework for urban environmental problems. World Bank, Washington, DC

Baud I, Dhanalakshmi R (2006) Governance in urban environmental management: comparing accountability and performance in multistake holder arrangements in South India. Cities 24(2):133–147

Bunnell T, Das D (2010) A geography of serial seduction: urban policy transfer from Kuala Lumpur to Hyderabad. Urban Geogr 31(3):1–7

Caseley J (2006) Multiple accountability relationships and improved service delivery performance in Hyderabad City, Southern India. Int Rev Admin Sci 72:531–546

CDS (2004) City development strategy. Municipal Corporation of Hyderabad. Government of Andhra Pradesh, Hyderabad

Central Pollution Control Board (CPCB) (2013) Status of water quality in India 2012. Monitoring of Indian National Aquatic Resources. Series MINARS/36/2013–14. Ministry of Environment & Forests, Delhi

Chakrabarti PGD (2001) Urban crisis in India: new initiatives for sustainable cities. Dev Pract 11:260–272

CoI (2011) Census of India. The Registrar General & Census Commissioner. Ministry of Home Affairs. Government of India, New Delhi, India

Dahiya B (2012) Cities in Asia 2012: demographics, economics, poverty, environment and governance. Cities 29:45–61

Das D (2015) City profile Hyderabad: visioning, restructuring, and making of a high-tech city. Cities 43:48–58. Elsevier Ltd

Davis J, White G, Damodaron S, Thorsten, R (2008) Improving access to water supply and sanitation in urban India: microfinance for water and sanitation infrastructure development. Water Sci Technol– WST 58:887–891. IWA Publishing

Department for International Development (DFID) (1998) Guidance manual on water supply and sanitation programmes. Water Environmental Health at London and Loughborough (WELL). Department for International Development. 94 Victoria Street. London SW1E5JL. United Kingdom. ISBN 0906055 58 X

Dupont V (2007) Conflicting stakes and governance in the peripheries of large Indian metropolises – an introduction. Cities 24(2):89–94

EBGH (2009) Environmental building guidelines for Greater Hyderabad (EBGH). Ecology & Geology – E&GN3. Version 1.2

Foster V, Pattanayak S, Prokopy LS (2002) Do current water subsidies reach the poor?. World Bank, Washington, D. C., USA

Foster V, Pattanayak S, Stalker Prokopy L (2003a) Do current water subsidies reach the poor? Water tariffs and subsidies in South Asia Paper 4. Water and Sanitation Program–South Asia. South Asia

Foster V, Pattanayak S, Stalker Prokopy, L (2003b). Can subsidies be better targeted? Water tariffs and subsidies in South Asia Paper 5. Water and Sanitation Program–South Asia, South Asia

Frantzeskaki N, Loorbach D, Meadowcroft J (2012) Governing societal transitions to sustainability. Int J Sustain Dev 15:19–36

George BA, Malano HM, Khan AR, Gaur A, Davidson B (2009) Urban water supply strategies for Hyderabad, India - future scenarios. Environ Model Assess 14:691–704

GIZ (2013) GIZ sustainability report 2013. Living Sustainability

Gleick P (1999) The human right to water. Water Policy I(5):487–503

GoAP (2013) Approval of Metropolitan Development Plan-2031 for Hyderabad Metropolitan Region. Department of Municipal Administration and Urban Development. Government of Andhra Pradesh, Hyderabad

GoI (1986) Constitution of India. Government of India, New Delhi

Goli S, Arokiasamy P, Chattopadhayay A (2011) Living and health conditions of selected cities in India: setting up priorities for the National Urban Health Mission. Cities 28(5):461–469

Greater Hyderabad Municipal Corporation (GHMC) (2006) City development plan. GHMC, Hyderabad, Andhra Pradesh, India

Greig A, Hulme D, Turner M (2007) Challenging global inequality: development theory and practice in the 21st century. Palgrave Macmillan, New York, USA

Hardoy J, Satterthwaite D (1989) Squatter citizen. Earthscan, London

Harte J (2007) Human population as a dynamic factor in environmental degradation. Popul Environ 28:223–236

Hogan T, Bunnell T, Pow CP, Permanasari E, Morshidi S (2012) Asian urbanisms and the privatization of cities. Cities 29:59–63

Hyderabad Metropolitan Development Authority (HMDA) (2010) Hyderabad Metropolitan Development Authority

Hyderabad Metropolitan Water Supply and Sewerage Board (HMWSSB) (2008) Hyderabad Metropolitan Water Supply and Sewerage Board

Hyderabad Metropolitan Water Supply and Sewerage Board (HMWSSB) (2011) Hyderabad Metropolitan Water Supply and Sewerage Board

Hyderabad Metropolitan Water Supply and Sewerage Board (HMWSSB) (2015) Hyderabad Metropolitan Water Supply and Sewerage Board

ILR (2001) The Indian Law Reports, Volume II. Part I. Andhra Pradesh Series, Hyderabad

ISODEC (2002) Report of the international fact finding mission on water sector reform in Ghana. Integrated Social Development Center, Ghana

Jodha NS (1990) Depletion of common property resources in India: micro level evidence. Popul Dev Rev 15:261–284
Kaminsky AP, Long RD (2011) India today: an encyclopedia of life in the Republic. ABC-CLIO, California, USA
Kennedy L (2007) Regional industrial policies driving peri-urban dynamics in Hyderabad, India. Cities 24(2):95–109
Kirk JA (2005) Banking on India's states: the politics of world bank reform programs in Andhra Pradesh. India Rev 4(3–4):287–325
Krueger AO (ed) (2002) Economic policy reforms and the Indian economy. University of Chicago Press, Chicago
Kumar MD (2004) Roof water harvesting for domestic water security: who gains and who loses? Water Int 29:43–53
Lakshmana CM (2008) Effects of population growth on environmental degradation: with reference to India. Demogr India 37(1):63–79
Lakshmana CM (2013) Population, development, and environment in India. Chin J Popul Resour Environ 11(4):367–374
Lisbon Principles (2002) The local government water code. TUGI's Urban Links, 32, January
Lundqvist J, Appasamy P, Nelliyat P (2003) Dimensions and approaches for third world city water security. Philos Trans R Soc London Ser B: Biol Sci 358:1985–1996
Luther N (2006) Hyderabad - a biography. Oxford University Press, New Delhi
McKenzie D, Ray I (2009) Urban water supply in India: status, reform options and possible lessons. Water Policy 1:442–460
Meena P, Micah L, Gleick PH (2008) A review of decision-making support tools in the water, sanitation, and hygiene sector. Environmental change & security program. Woodrow Wilson International Center for Scholars, Pacific Institute
Ministry of Urban Development (MoUD). 2012. Urban reforms
Ministry of Water Resources (MoWR) (2012) Domestic water supply. MoWR, Government of India, New Delhi, India
Moser S (2010) Putrajaya: Malaysia's new federal administrative capital. Cities 27:285–297
MoUD (nd) Detailed project report: preparation toolkit. Ministry of Urban Development. Government of India
Mukherjee S, Shah Z, Kumar M (2010) Sustaining urban water supplies in India: increasing role of large reservoirs. Water Resour Manag 24:2035–2055
Mulligan GF, Crampton JP (2005) Population growth in the world's largest Cities. Cities 22(5):365–380
Naidu NC, Ninan S (2000) Plain speaking. Viking, New Delhi
Naidu R (1990) Old cities, new predicaments: a study of Hyderabad. Sage, New Delhi
Narain S (2006) Community-led alternatives to water management: India case study. Human Development Report Office, United Nations Development Programme (UNDP)
Nastar M (2014) The quest to become a world city: implications for access to water. Cities 41(1):1–9
Pethe A, Ramakrishna N, Gandhi S, Tandel V (2014) Re-thinking urban planning in India: learning from the wedge between the de jure and de facto development in Mumbai. Cities 39:120–132
Petrella R (2001) The water manifesto – arguments for a world water contract. Books for Change, Bangalore
PTI (2014) Telangana government working towards making Hyderabad global smart city: minister. The Economic Times, 8 Oct 2014
Raghavendra S (2006) Re-examining the 'low water tariff' hypothesis: Lessons from Hyderabad, India. Urban Water J 3(4):235–247. https://doi.org/10.1080/15730620601059199
Ramachandraiah C (2004) Right to drinking water in India. Working Paper No. 56. Center for Economic and Social Studies, Hyderabad, India
Ramachandraiah C, Bawa VK (2000) Hyderabad in the changing political economy. J Contemp Asia 30(4):562–574

Ramachandraiah C, Prasad S (2008) The makeover of Hyderabad: is it the 'model' IT city? In: Ramachandraiah C, Westen GV, Prasad S (eds) High-tech urban spaces: Asian and European perspectives. Manohar Publishers, New Delhi
Rao N (2007) Cities in transition: growth, change and governance in six metropolitan areas. Routledge, London
Rao YVK (2002) Development experience of India and Andhra Pradesh in the context of global trends. In: Rao YVK, Subramanyam S (eds) Development of Andhra Pradesh: 1956–2001 a study of regional disparities. N.R.R. Research Centre, Hyderabad
Robbins PT (2003) Transnational corporations and the discourse of water privatization. J Int Dev 15:1073–1082
Robinson M (2007) Does decentralization improve equity and efficiency in public service delivery provision? IDS Bull 38(1):7–17
Rotmans J, Loorbach D (2009). Complexity and transition management. J Ind Ecol 13(2):184–196
Rudolph LI, Rudolph SH (2001) Iconisation of Chandrababu: sharing sovereignty in India's federal market economy. Econ Polit Wkly 1541–1550
Sahu S (2012) Politics of access to drinking water in urban areas in India: state and market interventions – a case study of Hyderabad
Sen S, Frankel F (2005) Andhra Pradesh's long march towards 2020: electoral detours in a developmentalist state. In: Doing business in India: political social and cultural overview. Centre for the Advanced Study of India. University of Pennsylvania, Philadelphia
Sengupta A (2001) Right to development as a human right. Econ Polit Wkly XXXVI(27):2527–2536
Sharma S (2011) Rajiv Awas Yojana: mixed-scanning model for slum-free cities. The Economic Times, 8 Sept 2011
Shaw RP (1989) Rapid population growth and environmental degradation: ultimate versus proximate factors. Environ Conserv 16:199–208
Shin E, Hufschmidt M, Lee, Y-S, Nickum EJ, Umetsu C, Gregory R (1997) Valuing the economic impacts of urban environmental problems: Asian cities. Urban management and poverty reduction. Working Paper No. 13. UNDP/ UNCHS/ The World Bank – UMP, Washington DC
Sinha S (2001) Environmental protection: role of constitutional courts. The Andhra Pradesh High Court in the Development of Constitutional Law. High Court of Andhra Pradesh, Hyderabad, India
Smets H (1999) Implementing the right to drinking water in OECD countries. OECD seminar on social and environmental interface, 22–24 Sept 1999. OECD, Paris
Soentoro EA (2011) Integrated decision-making for urban raw water supply in developing countries. Department of Civil Engineering. Colorado State University. Fort Collins, Colorado
Srinivasulu K (2002) Caste, class and social articulation in Andhra Pradesh: mapping differential regional trajectories. Overseas Development Institute, London
Swyngedouw E (2006) Power, water and money: exploring the nexus. United Nations Development Programme, NewYork, USA
United Nations (1995) The United Nations and human rights 1945–95. The United Nations Blue Book series, vol VII. United Nations Publications, NewYork
United Nations (2002) Substantive issues arising in the implementation of the international covenant on economic, social and cultural rights. 29th Session. 11–29 Nov 2002. General Comment No. 15. Economic and Social Council. United Nations, Geneva
Van Rooijen D, Turral H, Biggs TW (2005) Sponge city: water balance of mega-city water use and wastewater use in Hyderabad, India. Irrig Drain 54:81–91
Venugopal KK (2003) Supreme court and the common man. The Hindu, 19 Dec 2003
Whittington D (2003) Municipal water pricing and tariff design: a reform agenda for South Asia. Water Policy 5:61–76
World Bank (1990a) Environmental assessment sourcebook, volume II: sectoral guidelines. World Bank Technical Paper No. 140. Washington DC

Chapter 3
Inequalities in Access to Water and Sanitation: A Case of Slums in Selected States of India

Anuradha Banerjee and Bidisha Chattopadhyay

Abstract India's urbanisation in recent decades has ushered in a proliferation of slums and the outcome of this has been a large number of poor being exposed to appalling living environment. The present study seeks to understand the inequalities in access to water and sanitation between the slum and non-slum households. The analysis has been carried out for some selected states of India. The main objective of the study is to delineate the inter-state differences between slum and non-slum households with respect to access to water and sanitation. Database for the study pertains to Census of India, 2011 and the analysis has been done for only statutory towns to maintain parity between slum and non-slum households. The major findings of the study reveal that the development level of the states has a bearing on the access to water and sanitation with respect to the refined indicators exemplifying municipal efficiency such as access to tap water within premises and not so much on the broader indicators like access to safe drinking water. The slums and the non-slums were found to be in comparable positions with respect to the broad indicators such as overall access to safe drinking water, but the disparity was found to increase when distance between the household and water point was taken as an indicator. In the case of access to latrines, level of development has a bearing on both the broad and the refined indicators. Rajasthan and Punjab emerged as two states with near parity between non-slum and slum households with respect to household access to treated tap water within premises and flush latrine connected to piped sewer network, respectively.

Keywords Inequality · Access · Water · Sanitation

A. Banerjee (✉) · B. Chattopadhyay
Jawaharlal Nehru University, New Delhi, India
e-mail: banerjee.anb@gmail.com

B. Chattopadhyay
e-mail: bichat@rediffmail.com

R. B. Singh et al. (eds.), *Urban Health Risk and Resilience in Asian Cities*,
Advances in Geographical and Environmental Sciences,
https://doi.org/10.1007/978-981-15-1205-6_3

3.1 Introduction

The importance of safe drinking water and sanitation for the benefit of human health is well established. In the past few decades, global institutions such as the United Nations and World Health Organisation have driven the agenda of increasing household access to safe drinking water and sanitation. While the coverage for household access to safe drinking water has increased from 76% in 1990 to 88% in 2015, the progress has been slow on the sanitation front. Globally, 2.5 billion people do not have access to improved sanitation facilities (WHO and UNICEF 2014). Despite the overall increase in coverage, inequalities exist at various levels: national, subnational, urban–rural, on the basis of wealth differences and communities. Inequalities in access to safe drinking water and sanitation based on wealth differences get manifested in the disparities between slum and non-slum households.

Nearly one-third of the population of developing countries is said to be residing in slums. In India, 17.4% of the urban households are found in slums. Even slums are not a homogenous group. There are differences in the provisioning of amenities depending on the status of the slums. Notified slums, housing 4.96 million households or 36.09% of the total slum population are the best placed legally to receive water supply and sanitation facilities from the Government (Satapathy 2014). While the percentage of slum population to urban population has declined from 18.3 to 17.4% (2001–2011), the absolute figures have increased from 5.23 to 6.54 crore (Census of India 2001a, b, 2011a, b). The decadal growth rate of slum population (25.1%) has also been similar to the urban population growth rate (31.8%) (Census of India 2001a, b, 2011a, b). Given the deplorable condition of the amenities in the existing slums, any increase in the slum population would only add to the pressure on the existing infrastructure and further lead to deterioration of the conditions prevailing in the slums.

Real development is an outcome of an inclusive society and infrastructure is one of the most important stepping stones towards development. Better access to amenities empowers the economically downtrodden by facilitating a better quality of life and thus it is imperative to understand the disparities that exist in amenities. Inter-state variation is also seen in the availability of water and sanitation amenities. Some states perform better than others with respect to provisioning of these amenities in both their slum and non-slum households. The present paper focuses on the prevailing inter-state differences with respect to certain selected indicators of household access to amenities for both slum and non-slum households.

3.2 Objectives

The major objectives of the paper are to understand (a) level of household access to some of the important water and sanitation indicators across some of the selected states of India for both slum and non-slum households; and (b) disparity in the level of access of slum and non-slum households regarding these indicators.

3.3 Database and Methodology

Household access to water and sanitation for both slum and non-slum for different states has been analysed through selected indicators for the major states of India. Some of the chosen indicators are broad indicators which further encompass other indicators within them and are more easily accessible to households such as safe drinking water and improved latrine. On the other hand, there are other refined indicators which are very specific and in this case, these can capture the best possible scenario with respect to water and sanitation. Household access to treated tap water and flush/pour-flush latrines connected to piped sewer network are two such indicators. The data for the present study has been taken from the Census of India, 2011. Census of India provides data only for the physical access to water and sanitation and thus only that aspect has been covered. Slums have been taken as defined and enumerated by the Census of India, 2011. Slums have been enumerated by Census of India only in the statutory towns. To maintain parity, only the non-slum households of statutory towns have been taken for each of the selected states. This would help in highlighting the differences between the slum and the non-slum households better as the average figure for the non-slum households in each state will not be pulled down by the census towns. Disparity in the percentage of household access to the selected amenities of non-slums and slums has been measured.

3.4 Discussions and Results

3.4.1 Share of Slum Population in Urban Population

Increase in share of urban population and slumification of cities has affected all major cities of the world, particularly in the countries of the Global South. The African countries are the worst affected with countries such as Central African Republic and Chad having more than 85% of their urban population living in slums (https://data.worldbank.org/indicator/EN.POP.SLUM.UR.ZS). Usually slums are located in the big, thriving cities as they are the ones with economic opportunities. Addis ababa (Ethiopia) and Manila (Phillipines) have more than 85 and 40% of population residing in illegal settlements. Among the Indian cities, Greater Mumbai (41.3%) has the highest proportion of slum population followed by Kolkata (29.6%) and Chennai (28.5%) (Census of India 2011b). The slum population and its share of the urban population also vary among the states. The slum population of the major states has been presented in Table 3.1.

Table 3.1 Slum population by selected states of India: 2011

S. No.	States	Slum population (No.)	Percentage of slum population to urban population
1	Andhra Pradesh	101,86,934	35.93
2	Bihar	12,37,682	10.55
3	Chhattisgarh	18,98,931	31.99
4	Gujarat	16,80,095	6.53
5	Haryana	16,62,305	18.84
6	Jharkhand	3,72,999	4.70
7	Karnataka	32,91,434	13.96
8	Kerala	2,02,048	1.27
9	Madhya Pradesh	56,88,993	28.36
10	Maharashtra	118,48,423	23.31
11	Odisha	15,60,303	22.30
12	Punjab	14,60,518	14.06
13	Rajasthan	20,68,000	12.11
14	Tamil Nadu	57,98,459	16.59
15	Uttar Pradesh	62,39,965	14.03
16	West Bengal	64,18,594	22.03
17	INDIA	654,94,604	17.37

Source Census of India (2011b)

India has a slum population of 6.54 crore comprising 17.37% of India's urban population. Some of the states such as Andhra Pradesh (35.93%), Chhattisgarh (31.99%), Madhya Pradesh (28.36%), Maharashtra (23.31%), Odisha (22.30%) and West Bengal (22.03%) have a higher slum population than the national average (17.37%). In these states, the big cities have a large share of the state slum population but considerable variation is seen in the distribution of slum population among the cities. For instance, six cities in Andhra Pradesh have 28% of the state slum population while six cities in Chhattisgarh have 80% of the state slum population. At the same time, 16 cities in Maharashtra have 83% of the state slum population. Primacy of urban centres seems to play a role here with slum population getting concentrated in few important centres of the state. On the other hand, Kerala has the lowest percentage share of only 1.27% followed by Jharkhand (4.70%) and Gujarat (6.53%). Kerala also has the lowest slum population in terms of absolute figures (2,02,048 No.).

3.4.2 *Household Access to Water: A Comparison of Slum and Non-slum Households Across Selected States*

Household access to water has been assessed through indicators such as access to safe drinking water and access to treated tap water. The characteristics of both the indicators are very different and gives an insight into the performance of the states and their slums.

Household Access to Safe Drinking Water: Safe drinking water as specified by World Health Organisation refers to water from improved sources such as taps, tube wells and borewells, covered wells, etc. Although water from these sources are assumed to be safe, they may not be so. Water from tube wells and borewells are known to be affected by arsenic, nitrate, fluoride which play havoc with human health. It is the same with covered wells, though covered wells reduce the chances of bacteriological contamination. Household access to safe drinking water largely depends on the locational advantage of the household with respect to water resources and initiatives taken by the household itself. This section gives an insight into the situation of household access to safe drinking water for both slum and non-slum households in the selected states of India.

The Census of India, 2011 results as seen Table 3.2 have challenged the existing notion of slum households faring worse than the non-slum households in the safe drinking water category. Among the major states, the statutory towns of Punjab (99.26%) and Uttar Pradesh (98.01%) have the highest percentage of non-slum households with access to safe drinking water. The slum households of these states also fare better than the other states with 98.60% of slum households in Punjab and 98.01% of slum households in Uttar Pradesh having access to safe drinking water. Although both the states have high household access to safe drinking water, the non-slum households of statutory towns of Punjab largely rely on tap water (80.33%), while Uttar Pradesh's statutory town households depend both on tap water (53.81%), tube well and borewell water (44.24%). On the other hand, Kerala had the lowest household access to safe drinking for both slums (74.39%) and non-slums (73.76%). This is largely because of a very high percentage of households (24.22%) using water from uncovered wells. There is a high positive correlation between household access to safe drinking water of slum and non-slum households with Pearson's correlation coefficient as 0.981; statistically significant at 0.05 levels implying that the general condition of access to safe drinking water also affects the slum dwellers and that slum households fare better in states with overall better household access to safe drinking water.

Households with access to safe drinking water away from the premises have been taken as a share of the total households. While not too much of inter-state variation is seen when complete household access to safe drinking water is taken, the scenario changes when the distance to the source of water is also taken into account as seen in Table 3.2. When the indicator of household access to safe drinking water away from premises or at a distance of more than 100 m was taken, among the non-slum households, Jharkhand (12.03%) and Odisha (11.05%) fared the worst in

Table 3.2 Household access to selected water indicators in slum and non-slum households across selected states in India, 2011

State	HH access to safe drinking water (percent)		HH access to safe drinking water at more than 100 m distance (percent)		HH access to treated tap water within premises (percent)	
	Slum HH	Non-Slum HH	Slum HH	Non-Slum HH	Slum HH	Non-Slum HH
Andhra Pradesh	95.25	95.78	8.92	5.29	52.7	66.35
Bihar	94.99	95.80	9.05	4.71	9.0	14.07
Chhattisgarh	94.03	95.18	14.21	10.30	18.4	32.18
Gujarat	95.32	97.78	7.46	2.85	53.7	65.44
Haryana	96.03	97.22	6.18	2.68	54.7	68.63
Jharkhand	79.94	83.34	19.35	12.03	13.2	26.86
Karnataka	94.74	93.88	12.29	5.52	37.9	59.42
Kerala	74.39	69.68	3.87	2.90	37.8	40.02
Madhya Pradesh	92.03	94.28	16.34	9.15	26.4	40.79
Maharashtra	97.06	97.15	5.87	3.20	61.1	78.83
Odisha	84.26	86.92	23.73	11.05	16.3	36.87
Punjab	98.60	99.23	2.27	1.15	56.0	65.68
Rajasthan	94.30	95.46	7.56	4.22	64.4	70.79
Tamil Nadu	94.78	94.65	5.49	3.65	29.9	45.96
Uttar Pradesh	98.01	98.44	6.40	4.20	33.2	42.54
West Bengal	96.29	96.16	14.86	10.26	29.6	44.97
INDIA	95.07	94.77	9.36	4.90	42.5	53.61

Source Computed from Census of India (2011c, d)

terms of a higher percentage of households having access to safe drinking water source at a distance of more than 100 m. On the other hand, Haryana (2.68%) and Gujarat (2.85%) had the lowest percentage of non-slum households with access to safe drinking water away from premises. With respect to slum households, Odisha (23.73%) and Madhya Pradesh (16.34%) had the highest share of households with access to safe drinking water away from the premises. The slum households of Punjab (2.27%) and Kerala (3.87%) fared the best with respect to this parameter. However, Punjab had the highest share of slum household access to safe drinking water within the premises (89.19%) which can be largely attributed to the heavy dependence on borewell/tube well (25.46%) as a source of water within the premises.

Household Access to Treated Tap Water: The refined indicators such as household access to treated tap water are more specific and are largely a function of municipal efficiency. They also behave very differently from the broad indicators such as the ones discussed in the previous section. Thus, with respect to water, treated tap water is an indicator that largely exemplifies the household coverage as provided by the water utility.

With respect to non-slum household access to treated tap water within premises (Table 3.2), Maharashtra (78.83%) and Rajasthan (70.79%) have the highest percentage share of non-slum households with access to treated tap water within premises. Maharashtra (61.08%) and Rajasthan (64.44%) also have the highest percentage share of slum households with access to treated tap water, while Bihar fares the worst for both slum (9.01%) and non-slum (14.07%) households. Similar to household access to safe drinking water, a statistically significant correlation is seen between the household access to treated tap water of slums and non-slums, although weaker than the former. (Pearson's correlation coefficient is 0.651, statistically significant 0.01 level). As seen earlier, household access to safe drinking water was not influenced so much by the level of development of the state but rather by the availability of water resources. In the case of household access to treated tap water, the level of development seems to be a clear defining factor with states such as Maharashtra, Punjab, Haryana faring better than Bihar, Jharkhand and Chhattisgarh. Rajasthan seems to be an abberration, performing even better than the developed states.

3.4.3 Household Access to Sanitation: A Comparison of Slum and Non-slum Households Across Selected States

Household access to sanitation has been assessed on the basis of households without access to any latrine facility captured through households practising open defecation, household access to improved latrines within premises and households connected to piped sewer network.

Household Access to Latrines: An analysis of the household access to latrines in the statutory towns of the major states of India was done for both slum and non-slum households as seen in Table 3.3. Even in statutory towns with urban local bodies, open defecation is an issue not only among the slum households but also among the non-slum households.

While the percentage share of slum households practising open defecation is high among most of the states, Odisha (48.34%) fared the worst followed closely by Bihar (42.49%) and Chhattisgarh (41.65%). On the contrary, 3.34% and 10% of the slum households of Kerala and Punjab, respectively, were reported to be defecating in the open, much below the national average of 18.9%. Among the non-slum households, Bihar and Chhattisgarh have the poorest figures with 26.76% and 29.05% of households followed closely by Odisha (24.28%) going for open defecation. The figures are much higher than the other states with Kerala (1.25%) and Punjab (4.05%) having the lowest percentage of non-slum households defecating in the open.

Table 3.3 Sanitation indicators in slum and non-slum households across selected states of India, 2011

State	HH practising open defecation (percent)		HH access to improved latrine (percent)		HH access to piped sewer network (percent)	
	Slum households	Non-Slum households	Slum households	Non-Slum households	Slum households	Non-Slum households
Andhra Pradesh	14.78	6.78	77.59	88.08	29.69	42.41
Bihar	42.49	26.76	47.50	65.63	4.36	7.68
Chhattisgarh	41.65	30.73	46.84	64.42	3.59	11.90
Gujarat	21.26	6.68	62.23	89.64	40.28	64.73
Haryana	17.31	5.73	74.05	88.96	45.89	62.52
Jharkhand	41.88	23.90	50.06	71.37	5.93	14.34
Karnataka	24.97	8.17	59.35	85.75	33.33	58.73
Kerala	3.34	1.33	87.06	93.54	15.58	19.79
Madhya Pradesh	31.65	17.59	59.29	77.49	13.57	23.58
Maharashtra	9.75	6.06	38.54	77.93	22.98	45.35
Odisha	48.34	24.28	38.38	70.38	5.43	14.05
Punjab	10.49	4.23	83.77	91.41	58.28	67.79
Rajasthan	26.26	13.33	62.80	79.29	22.01	27.89
Tamil Nadu	23.08	12.84	57.98	76.08	30.67	29.54
Uttar Pradesh	18.76	13.38	70.36	79.37	20.53	29.47
West Bengal	11.10	4.52	77.77	89.58	14.57	18.07
INDIA	18.90	10.26	61.40	81.72	24.51	37.33

Source Computed from Census of India (2011c, d)

Household Access to Improved Latrines: Improved latrines and not only latrines are important as they segregate the human waste from human contact hygienically. Improved sanitation comprises access to flush/pour-flush latrines connected to piped sewer system or septic tank and pit latrine with ventilated improved pit within premises.

Kerala has the highest percentage of slum households (87.06%) with access to improved latrines followed by Punjab (83.77%) and West Bengal (77.77%). Although, all these states have a higher percentage of slum households than the national average, the components within the improved latrine category differ significantly.

While Kerala and West Bengal show a higher percentage of households with ventilated pit latrine and pour-flush latrines connected to septic tanks, Punjab has a higher percentage of households connected to the piped sewer network. In Kerala and West Bengal, households with pour-flush latrines connected to septic tanks and ventilated pit latrines comprised 82 and 81.26% of the households connected to an improved latrine. In Punjab, households connected to piped sewer network and septic tanks comprised 91.63% of the households with access to improved latrine. Odisha (51.66%) and Bihar (57.51%) have the lowest percentage of slum households with access to improved latrines. With respect to non-slum households, again Kerala (93.54%) has the best figures followed by Punjab (91.41%) while Chhattisgarh (64.42%) and Bihar (65.63%) fare the worst.

There is high correlation between percentage of slum households and non-slum households with access to improved latrines (Pearson's coefficient of correlation is 0.834, statistically significant at 0.01 level) implying that development in the slum and non-slum areas often go hand in hand.

Household Access to Piped Sewer Network: Household access to piped sewer network is one of the most refined indicators for sanitation and it also represents the municipal coverage as households can be connected to a piped sewer network if the urban local body provides the service.

With respect to the slum households (Table 3.3), the developed states such as Punjab (58.28%), Haryana (45.89%) and Gujarat (40.28%) have a higher percentage of households with access to piped sewer network compared to the Empowered Action Group states such as Odisha (5.43%), Jharkhand (5.93%) and Bihar (4.36%). A similar situation is seen for the non-slum households with Punjab (67.79%), Gujarat (64.73%) and Haryana (62.52%) being ahead of the other states. The difference between the developed and the non-developed states is starker than in the case of percentage household access to improved latrines.

3.4.4 *Level of Disparity in Access to Water and Sanitation Across Selected States of India: A Case of Slums and Non-slums*

The level of disparity between the non-slum and the slum households varies among the indicators and the states. For the present analysis, only refined indicators have been selected that include percentage household access to (a) treated tap water (b) improved latrines and (c) piped sewer network. The disparity between the two has been taken as the difference between the percentage of non-slum household access to a particular indicator and percentage of slum household access to that indicator for a state. The disparity between the non-slum and slum households for each of the selected indicators have been presented for the selected states in Table 3.4 and have been further discussed in detail in the ensuing paragraphs.

It is seen that the disparity is much more in case of household access to treated tap water within premises. Karnataka (21.55%) followed by Odisha (20.57%) has the highest level of disparity while Bihar (5.06%) and Kerala (2.20%) have the least. It is noteworthy to mention Rajasthan as it is the only state which has a high percentage of households with access to treated tap water within premises for both slums and non-slums.

Table 3.4 Level of disparity between non-slum and slum households indicators across selected states of India, 2011

State	HH access to treated tap water within premises (percent)	HH access to improved latrine (percent)	HH access to piped sewer network within premises (percent)
Andhra Pradesh	13.69	10.49	12.72
Bihar	5.06	18.13	3.32
Chhattisgarh	13.77	17.58	8.31
Gujarat	11.72	27.41	24.45
Haryana	13.94	14.91	16.63
Jharkhand	13.70	21.31	8.41
Karnataka	21.55	26.40	25.40
Kerala	2.20	6.48	4.21
Madhya Pradesh	14.41	18.20	10.01
Maharashtra	17.75	39.39	22.37
Odisha	20.57	32.00	8.62
Punjab	9.68	7.64	9.51
Rajasthan	6.35	16.49	5.88
Tamil Nadu	16.10	18.10	– 1.13
Uttar Pradesh	9.34	9.01	8.94
West Bengal	15.38	11.81	3.50
INDIA	11.15	20.32	12.82

Source Computed from Census of India (2011c, d)

Household access to improved latrine is higher among the non-slum households in all the states. The disparity between the non-slum and slum households is the highest in Maharashtra (39.39%) and Odisha (32%) and the lowest in Kerala (6.48%) and Punjab (7.64%). A special mention is needed for Kerala and Punjab as they also have one of the highest percentage of household access to improved latrine. The figures do not seem to be influenced by the levels of development of the state with several of the developed states showing higher or similar levels of disparity as the less developed ones.

With respect to household access to piped sewer network within premises as seen in Table 3.4, slum households of Karnataka, Gujarat and Maharashtra fare the worst compared to their non-slum counterpart with the non-slum and slum household difference in access being 25.40%, 24.45% and 22.37%, respectively. On the other hand, the slum households of Bihar (3.32%) and Kerala (4.21%) fare the best vis-a-vis the non-slum households. An association between level of development of the state and the disparity level does not seem to be present, with some of the developed states such as Gujarat, Maharashtra having higher level of disparity while others such as Punjab (9.51%), Kerala (4.21%) and Tamil Nadu (–1.13%) having lower levels.

Punjab again stands out in this category for maintaining a near parity between slum and non-slum households, unlike the other developed states.

In this analysis, Rajasthan and Punjab stood out for having low levels of disparity between non-slum and slum households with respect to household access to treated tap water within premises and piped sewer network. The case of Rajasthan has been further delved into to understand the probable reasons for good performance despite being one of the less developed states of India.

3.4.5 *Case Study: Rajasthan*

Rajasthan despite being an arid, semi-arid and less developed state exhibits the second highest percentage of household access to treated tap water within premises among non-slum and highest among slum households. This appears as an anomaly and warrants further analysis to understand the reasons for high levels of access to treated tap water within premises among the slum households.

The present state of Rajasthan was formed in 1956. The Aravali range divides the state such that three-fifth lies to its north-west and two-fifth lies to its south-east. The north-west region is largely arid and semi-arid while the south-east is predominantly forested and hilly. The Thar desert covers nearly 60% of the state. Rajasthan, since its formation, was a part of the group of least developed states. The group was termed as BIMARU (Bihar, Madhya Pradesh, Rajasthan and Uttar Pradesh) by demographer Ashish Bose in early 1980s. Later, in the early 2000s, Rajasthan became a part of the Empowered Action Group comprising seven other states (Bihar, Madhya Pradesh, Uttar Pradesh, Jharkhand, Chhattisgarh, Odisha and Uttarakhand). Some of the selected socio-economic and infrastructure indicators of Rajasthan and India are given in Table 3.5.

Table 3.5 Select socio-economic and infrastructure indicators: India and Rajasthan

Indicators	Rajasthan	India
Population (2011)	6.85 Crore	121 Crore
Sex ratio (2011)	926	940
Percentage urban population (2011)	24.87%	31.15%
Literacy rate (2011)	66.10%	73.0%
Poverty rate (%) (2012)	15	22
IMR (2011–12)	52	44
Electrified villages (2013–14)	90.2	95.7
Road density per 100 km^2 (2012)	55.52	120.6
Per capita income (2011–12) 2004–05 constant prices (Rs.)	28851	38037

Source Directorate of Economics and Statistics 2013, World Bank 2016

In Rajasthan, 12.11% of the urban population resides in slums. A little less than half of the total slum households (43%) is concentrated in the three cities of Kota, Jaipur and Jodhpur. Among the slum households in Rajasthan, the dependence on groundwater through borewells, tube wells, wells and handpump is limited to only 11.9% compared to 13.41% for non-slum households. This is much lower than the all India figures (23.32%) of all statutory town households depending on groundwater for domestic purpose. This could be due to the poor quality of groundwater in most parts of Rajasthan and the deep groundwater levels which results in high costs of water extraction. Groundwater depth was found to be more than 40 m in many parts of western Rajasthan and varying from 5 to 20 m with pockets of depth 20 to 40 m in eastern parts of the state in May 2014. The presence of fluoride and nitrate in groundwater has been identified as areas of concern (Central Ground Water Board 2014). Slum household access to treated tap water within premises according to Census of India, 2011 has been further disaggregated to understand the pattern in percentage household access among the urban centres. The percentage access figures range from zero (Nagar, Gangapur city, Tijara, Bilara, Khairthal, Deogarh and Rajgarh) to 88.54% (Bhadra). Among the 108 urban centres which were found to have slums by Census of India, 2011, only 22% of the urban centres had percentage household with access to treated tap water more than the state average of 64.4% which also shows that the better off slums are concentrated in a few urban centres.

Statutory towns have been categorised as per the slum household access to treated tap water within premises in Table 3.6. An association between the number of slum households in the urban centre with percentage slum households has also been studied to understand whether the slum population size has any bearing on the water services rendered. Pearson's coefficient of correlation was found to be positive and low at 0.313 statistically significant at 0.01 levels. Urban centres with higher number of slum households also had higher percentage slum household access to treated tap water within premises.

Percentage slum household access to treated tap water within premises was found to have a negative correlation with groundwater use expressed as percentage slum household access to wells, tube well/borewell and handpump. Pearson's coefficient of correlation was found to be negative and medium at 0.702 statistically significant at 0.01 levels. This could either mean that areas which had poorer quality or inaccessible groundwater had been provided taps by Public Health and Engineering Department (PHED) to meet their daily needs or vice versa. This needs to be explored further to understand the dynamics better. The importance of the Indira Gandhi canal is evident from the fact that nearly 60% of the 22 urban centres which were found to have a higher figure than the state average are located in the command area of the canal system.

Rajasthan has had the highest capital expenditure for water and sanitation as percentage share of total development outlay from 2001–02 to 2015–16. In 2015–16, it was 23.18% for Rajasthan compared to 6.20% of national average (https://rbidocs.rbi.org.in). Besides the above-mentioned factors, programmes have also been introduced from time to time by the Government of India. Some of the programmes introduced in the decade prior to 2011 have been briefly discussed.

Table 3.6 Percentage household access to treated tap water within premises in slums in statutory towns in Rajasthan 2011

Percentage slum household access to treated tap water within premises	Urban centres
80–100	Bhadra, Sardarshahr, Nohar,Jalore, Sadulshahr, Karanpur, Bikaner, Hanumangarh, Ajmer
60–80	Raisinghnagar, Taranagar, Kota, Kekri, Pushkar, Dungargarh, Pilibanga, Sanchore, Jodhpur, Ganganagar, Nokha, Gajsinghpur, Udaipur, Kesrisinghpur, Bharatpur, Salumbar, Sangod, Jaipur, Kishangarh, Barmer, Kuchaman, Pali
40–60	Gulabpura, Churu, Sirohi, Kaithoon, Pindwara, Nainwa, Suratgarh, Jhalawar, Amet, Vijainagar, Sumerpur, SawaiMadhopur, Pipar, Jaitaran, Baran, Takhatgarh, Sarwar, Sheoganj, Phalodi, Bhinmal, Deshnoke, Sujangarh, Alwar, Rawatbhata, Beawar, Lalsot, Merta, Kapasan, Anupgarh, Nagaur, Keshoraipatan
20–40	Jahazpur, Todaraisingh, Bhilwara, Pratapgarh, Chhabra, Sagwara, Pokaran, Banswara, Kushalgarh, Rajakhera, Balotra, Pirawa, Jhalrapatan, Chittaurgarh, Falna, Ramganj, Weir, Bari, Jaipur, Sojat, Shahpura, Begun, Asind
0–20	Aklera, Mandalgarh, Fatehnagar, Rani, Antah, Jaisalmer, Chhoti Sadri, Dhaulpur, Abu, Nimbahera, Mangrol, Chaksu, Nadbai, Padampur, Makrana, Sikar, Tijara, Khairthal, Rajgarh, Nagar, Gangapur, Bilara, Deogarh

Source Census of India (2011d)

The thrust of Basic Services for Urban Poor (BSUP), sub-mission of Jawaharlal Nehru Urban Renewal Mission launched in 2005 was on integrated development of slums including provisioning of basic services. BSUP was limited to only two cities in Rajasthan, Jaipur and Ajmer–Pushkar. The funds were utilised for building in situ dwelling units (http://www.pib.nic.in/newsite/erelcontent.aspx?relid=37060). In a Press information released by Government of India in 2012, Rajasthan had one of the lowest (6.86%) completion rates among the Indian states with the national average being 40.76%. (http://pib.nic.in/newsite/PrintRelease.aspx?relid=81775). Urban centres other than the ones covered under BSUP are under another programme named Integrated Housing and Slum Development Programme (IHSDP). Under this programme, Rs.1035.48 crores have been sanctioned for 67 projects involving building of dwelling units, upgradation of water and sewerage infrastructure within slums (http://lsg.urban.rajasthan.gov.in/content/raj/udh/lsg-jaipur/en/schemes/intergrated-housing—slum-development-programme–ihsdp-/ihsdp-.html). As of 2015, Rajasthan (68.38%) fared below the national average (72.97%) with respect to the completion of construction of dwelling units (http://jnnurmmis.nic.in/jnnurm_hupa/jnnurm/Jnnurm_Ray_AHP_Progress_Report/State_Wise_IHSDP.pdf). Rajiv Awas Yojana was launched in 2009 and continued till 2014 after which it was subsumed in Pradhan Mantri Awas Yojna. This was also for building dwelling units and slum upgradation and improvement. Nearly 31% of the sanctioned amount has been utilised so far (http://urban.rajasthan.gov.in/content/raj/udh/rudsico/en/sectors-programmes/ray.html). Rajasthan Urban Infrastructure Development Programme (RUIDP) is another of such programmes focussing on urban infrastructure including water and sanitation. The first phase included the cities of Ajmer, Bikaner, Jaipur, Jodhpur, Kota and Udaipur. It began around the year 2000 with a $250 million loan from ADB; nearly 69% of the total project cost (WaterAid India 2006). Around 3.18% of the project cost was dedicated to slum improvement. The main water infrastructure facilities provisioned within the slums involved rehabilitation of main pipelines and installation of community taps. The project cost was increased to Rs.18,480 million for the Bisalpur Jaipur water supply scheme. In the next phase, Alwar, Baran, Barmer, Bharatpur, Bundi, Chittorgarh, Churu, Dholpur, Jaisalmer, Jhalawar, Karauli, Nagaur, Raj Samand, Sawai Madhopur and Sikar were selected (http://ruidp.rajasthan.gov.in/Introduction.aspx). Slum Improvement schemes seem to have met with moderate success with the completion rates being rather low. This is reiterated by National Sample Survey, 2012 data (National Sample Survey Organisation 2014), according to which 23.1% of the notified slums in Rajasthan seem to have benefited from any such scheme compared to the national average of 32.3%. The case of non-notified slums is worse with only 3.2% benefiting compared to 18% for all India. The stark difference in the reach of slum improvement programmes between the notified and non-notified slums is another major area of concern.

It is important to mention that although Rajasthan has one of the higher percentage access to treated tap water within premises, the other water-related parameters such as duration and frequency of water supply are very poor. The case of Rajasthan needs to be taken up for further research to understand the true picture beneath the high household access figures.

3.5 Summary and Conclusion

With respect to access to broad indicator for household access to water such as safe drinking water, there is less variation among the states and less disparity between the non-slum and slum households as compared to the refined indicators such as household access to treated tap water due to inclusion of borewell/tube well water and covered wells in the category of safe drinking water. Nearly one-fifth of households, both in the slum and non-slum category were found to be dependent on groundwater and states with readily available groundwater also fared better. The slum households also fared marginally better than non-slum households in this category. With respect to household access to treated tap water within premises, the level of development of the states along with the emphasis on water and sanitation programmes played an important role in determining the household access. Slums fared better in states with high non-slum household access to treated tap water within premises. Analysis of the broad indicator of households practising open defecation revealed that a higher percentage of households without access to latrines were more common in the backward states. A similar status is seen for household access to improved latrines, except in the case of Maharashtra where the household access to improved latrines is below the national average. With respect to the refined indicator of household access to piped sewer network, the developed states considerably outperform their less developed counterparts. Rajasthan, in the case of household access to water and Punjab for household access to improved latrines and piped sewer network are the only two states in which there is near parity between the non-slums and the slums coupled with overall high levels of percentage access.

Rajasthan being a water parched state places a lot of importance on water procurement and distribution. Urban centres located in areas with poor quality of groundwater and low access to surface water might have been the first to benefit from water network expansion projects. It seems that the presence of Indira Gandhi canal has been beneficial for the slum residents, besides benefiting the non-slum residents. The Government-led slum development programmes seem to have met with moderate success and their contribution to the low levels of slum and non-slum disparity is not very clear. There is a need to further understand the peculiar situation of the state where taps might be there in most households but without water.

References

Census of India (2001a) A series (Primary Census Abstract), office of the Registrar General & Census Commissioner (India)

Census of India (2001b) A series (Primary Census Abstract: Slums), office of the Registrar General & Census Commissioner (India)
Census of India (2011a) A series (Primary Census Abstract), office of the Registrar General & Census Commissioner (India)
Census of India (2011b) A series (Primary Census Abstract: Slums), office of the Registrar General & Census Commissioner (India)
Census of India (2011c) H series (Tables on Houses, Household Amenities and Assets), office of the Registrar General & Census Commissioner (India)
Census of India (2011d) H series (Tables on Houses, Household Amenities and Assets: Slums), office of the Registrar General & Census Commissioner (India)
Central Ground Water Board (2014) Ground water year book 2012–2013, Rajasthan state, ministry of water resources, Government of India
Directorate of Economics and Statistics (2013) Some facts about Rajasthan, Government of Rajasthan
Local Self-Government, Government of Rajasthan http://lsg.urban.rajasthan.gov.in/content/raj/udh/lsg-jaipur/en/schemes/intergrated-housing—slum-development-programme–ihsdp-/ihsdp-.html
Ministry of Urban Development, Government of India, http://jnnurmmis.nic.in/jnnurm_hupa/jnnurm/Jnnurm_Ray_AHP_Progress_Report/State_Wise_IHSDP.pdf
National Sample Survey Organisation (2014) Urban slums in India, NSS 69th round, July 2012–December 2012
Public Information Bureau. http://www.pib.nic.in/newsite/erelcontent.aspx?relid=37060
Public Information Bureau. http://pib.nic.in/newsite/PrintRelease.aspx?relid=81775
Rajasthan Urban Drinking Water Sewerage And Infrastructure Corporation. http://urban.rajasthan.gov.in/content/raj/udh/rudsico/en/sectors-programmes/ray.html
Rajasthan Urban Infrastructure Development Programme, Government of Rajasthan. http://ruidp.rajasthan.gov.in/Introduction.aspx
Reserve Bank of India, State finances: a study of budgets. https://rbidocs.rbi.org.in
Satapathy BK (2014) Safe drinking water in slums: from water coverage to water quality, Econ Polit Wkly 49(24)
WaterAid India (2006) Water for all? Implementation of ADB's water policy in India:A review
WHO, UNICEF (2014) Progress on drinking water and sanitation, Geneva, Switzerland
World Bank (2016). http://www.worldbank.org/en/country/india/brief/india-states-briefs-rajasthan
World Bank. https://data.worldbank.org/indicator/EN.POP.SLUM.UR.ZS

Chapter 4
Chennai Floods 2005, 2015: Vulnerability, Risk and Climate Change

T. Vasantha Kumaran, O. M. Murali and S. Rani Senthamarai

Abstract The purpose of the paper is twofold: the vulnerability and risk in climate change in Chennai, India; and the difficulty in communicating climate change in the city of our concern. The paper fulfils this purpose through cases in point: the Adyar basin and the megacity of Chennai and experiences over time and space of rainfall events, particularly Chennai Rains/Floods 2015. For the space–time experiences, the study takes on the flood experiences of Adyar sub-watershed, including Chembarambakkam lake, which triggered the Chennai Floods 2015, and provides for rainfall analysis for the period 1984–2013 and the perspectives we gain from it. Chennai Rains/Floods 2015 are taken to illustrate the nature of vulnerabilities and risks in Chennai of a climate change event and how such vulnerabilities and risks can be communicated to the people and the communities for safeguarding them from impacts. The most important climate change risk is increased riverine (Adyar river) and inland (Chennai city) flooding and tens of millions of people are affected by floods during the monsoon months (November–December 2005, 2015). Excessive and exceptional rainfalls and hence higher peak monsoon river flows could exacerbate the risk for hundreds of communities in both rural (the Adyar basin) and urban areas (Chennai city). Reasons for the impacts are in the high population densities, in association with very high vulnerability because of poorly designed and executed flood management systems, or their poor maintenance, complex land and water regimes and high levels of poverty, which together have severely degraded the coping capacity of residents of the region. Above all, the one most important reason is that the people are generally not aware of climate change nor are the communities committed to mitigating the risk of climate change unless a catastrophe occurs to kick-start community engagement.

Keywords Vulnerability · Risk · Independent researchers · Climate change · Flood management systems · Chennai

T. Vasantha Kumaran (✉) · O. M. Murali · S. Rani Senthamarai
Formerly Department of Geography, University of Madras, Chennai, India
e-mail: thangavelukumaran@gmail.com

O. M. Murali · S. Rani Senthamarai
Independent Researchers, Chennai, India

R. B. Singh et al. (eds.), *Urban Health Risk and Resilience in Asian Cities*, Advances in Geographical and Environmental Sciences,
https://doi.org/10.1007/978-981-15-1205-6_4

4.1 Introduction

We—the authors of this short paper—are '*insiders*' to an experience of rains and floods in Chennai in both 2005 and 2015. We have experienced the impact of widespread rains of 2005 and the devastating rains and floods of Chennai 2015. On two occasions last year, about the festival of lights known as Deepavali and in the early days of December, floods entered the houses of two of us and we could perceive/relate therefore to the horrid experiences of Chennai Rains/Floods 2015. Climate change is an important risk in the Indian subcontinent over time and space but there is still considerable uncertainty as to the precise mechanisms and impacts, particularly those related to rainfall and sea-level rise (Aggarwal and Lal 2001).

The purpose of the paper, as the title indicates, is twofold: It speaks of the vulnerability and risk in climate change in Chennai, India; and it talks about vulnerability and risk in communicating climate change in the city of our concern. The paper strives to fulfil this purpose through a case in point, that is, the megacity Chennai and the space–time experiences of the people and communities of the city, particularly with reference to Chennai Rains/Floods 2015. For the space–time experiences, the study takes the flood experiences of Adyar sub-watershed, which includes Chembarambakkam lake, which triggered in a way the Chennai Floods 2015, and this provides for rainfall analysis for a period of several years from 1984 to 2013 and the perspectives we may gain from it. Chennai Rains/Floods 2015 are taken to illustrate the nature of vulnerabilities and risks in Chennai out of a climate change event and how the vulnerabilities and risks may be communicated to its people and the communities for safeguarding them from climate change impacts.

The most important climate change risk is increased riverine (Adyar river) and inland (Chennai city) flooding and tens of millions of people are affected by floods during the monsoon months (November–December 2005, 2015). Excessive and exceptional rainfalls and hence higher peak monsoon river flows could exacerbate the risk situations for hundreds of thousands of communities in both rural (the Adyar basin) and urban areas (Chennai city). Reasons for this could be found in the high population densities across this region, in association with very high vulnerability because of a mix of poorly designed and executed flood management systems, complex land and water tenure regimes and high levels of poverty, which together have severely degraded the coping capacity of millions of residents of the region, rural as well as urban.

As risk of climate change in Indian cities is associated greatly with 'vulnerability' than with 'exposure to hazard', it is important to understand processes that are rapidly changing India's urban landscape, altering opportunities for livelihoods and affecting the vulnerability of many communities and stakeholders and their capacity to adapt to long-term risks. It is our understanding that to help the people of the urban communities in specific contexts of climate change, the very climate change has to be communicated in the perspective of a three-part transition, namely, (a) a demographic transition that would stabilize India at 1.6 billion in 2060; (b) a rural-to-urban transition, which would add 500 million people to the urban places over the same period;

and (c) a simultaneous environmental transition that comprises 'brown' (water, sanitation and environmental health), 'grey' (air and water pollution) and 'green' (climate change). India would need multiple, sub-regionally nuanced strategies to respond to the climate crisis, based on local experiences of coping with uncertainty and with systems far from equilibrium (Revi 2008; De Vries et al. 2007).

4.2 Global Warming, ENSO Events and Indian Monsoon

A warmer Earth may lead to changes in rainfall patterns. This means that global warming could trigger rainfall changes in traditionally drier parts of the country. Indian Monsoon is part of a large-scale tropical monsoon system, and it is confined not just to South Asia, North Australia and Tropical Africa but also to the America as well (Asnani 1999). The monsoon is influenced by large-scale atmospheric and oceanic circulations and phenomena such as the ENSO event in the equatorial eastern and central Pacific, Eurasian winter snow cover and the phase of the Quasi-Biennial Oscillation (QBO) before and during the monsoon season (Rao 2004: 55). The correlation between an El Nino's intensity and the monsoon seems weaker if its impact is also gauged by the location of warm waters in the Pacific. The inter-annual variability of Indian summer monsoon rainfall (ISMR) has been linked to variations of Sea Surface Temperatures (SST) over the equatorial Pacific and Indian Oceans, and Eurasian snow cover. The 126-year record of rainfall of the Indian monsoon (1880–2005) suggests that only less than half of the El Nino events are associated with deficient rainfall over India. In other El Nino years, ISMR was normal or excess.

Familiarity with the El Nino and Southern Oscillation is more than a century old. In recent times, the SO has become well known because it causes dramatic climatic anomalies. The two opposite phases of the SO, when the largest climatic anomalies occur, are known as the El Nino and La Nina. El Nino is an invasion from time to time of warm surface water from the western equatorial parts of the Pacific basin to the eastern equatorial region and along the coasts of Peru, Ecuador and Northern Chile. La Nina is the enhancement of the normal pattern along the equator within which easterlies cause upwelling of deep cold water rich in inorganic plant nutrients along the coasts of South America and Equator, which results in high rate of primary production.

It was Sir Gilbert Walker, the then Director General of Observatories in India, who was the first to document the influence of the ENSO in 1932 as '*when pressure is high in the Pacific Ocean, it tends to be low in the Indian Ocean, from Africa to Australia, with the rainfall varying in the opposite direction to pressure*'. During El Nino events (warm events), convergence over tropics is reduced leading to droughts in Australia, Indonesia, India and parts of Africa and heavy rain and floods in the west coast of South America. During the La Nina events of the ENSO (cold events), convergence is enhanced over Indonesia, New Guinea and also over Africa (Zaire) and Amazon basin. Many Indian researchers (for example, Sikka 1980; Bhalme et al. 1990) have related monsoon droughts over India with the occurrence of El Nino events. Jayanthi and Govindacharya (1999) have established that the northeast monsoon rainfall of Tamil Nadu is positively correlated with the El Nino phase of the ENSO.

4.3 Vulnerability, Risk and Climate Change

Vulnerability (Fig. 4.1) is a multi-layered and multidimensional social space defined by the determinate political, economic and institutional capabilities of people in specific places and specific times (Bohle et al. 1994: 39). In this sense, a theory of vulnerability should be capable of mapping the historically and socially specific realms of choice and constraint, which determine risk exposure, coping capacity and recovery potential.

In the light of our Chennai Floods 2015 experience, and the widespread nature of devastation, deaths and suffering in the whole of the city, climate change is perceived as an uncertain but potentially serious threat to vulnerable populations of the city. It is very much so also in the context of the rural areas as evidenced during the floods of 2005 in Tamil Nadu, with only 3 districts of the total 32 districts spared.

In the words of Bohle et al. (1994: 37–38), *vulnerability is best defined as an aggregate measure of human welfare that integrates environmental, social, economic and political exposure to a range of potential harmful perturbations. Vulnerability can*

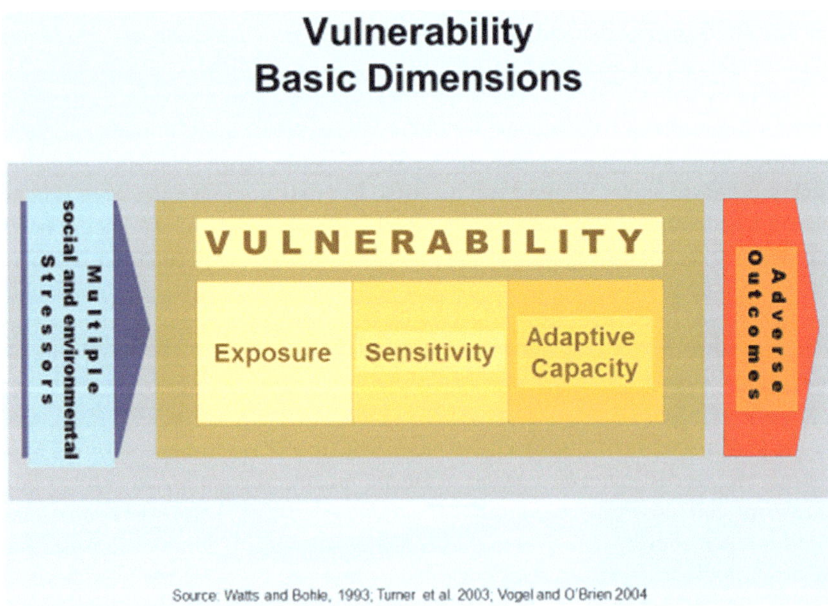

Fig. 4.1 Vulnerability basic dimensions

be, in other words, defined in terms of exposure, capacity and potentiality. Accordingly, the prescriptive and normative response to vulnerability is to reduce exposure, enhance coping capacity, strengthen recovery potentiality and bolster damage control (that is, minimize destructive consequences) via *private and public means.* Quoting Chambers (1989), Bohle et al. (1994) elaborate three basic coordinates of vulnerability, namely,

- The risk of exposure to crises, stress and shocks (extreme rainfall events, impacts such as loss of property, deaths and suffering);
- The risk of inadequate capacities to cope with stress, crises and shocks (which implicitly subsumes timely and effective external interventions); and
- The risk of severe consequences of, and the attendant risk of slow or limited recovery from, crises, risk and shocks (people in several parts of the city and in Tamil Nadu are yet to recover from the impacts of the floods of 2015, even after a year).

For our study, however, we recognize two types of vulnerability, namely, the *physical and social vulnerability*. The methodology used in determining the physical and social vulnerabilities is explained in a larger, doctoral study of O. M. Murali (completed 2019 currently ongoing) and, considering space restraints, it is not elaborated here expect for providing some glimpses of the processes.

To understand the vulnerability of an area to floods, it is essential to know the physical factors which play a major and potential role in increasing flooding. It is required therefore to prepare the thematic layers of annual rainfall, size of the watershed, slope and gradient profile of the mainstream, drainage density, soil types, land uses and infrastructures like the road network (Fig. 4.2).

A framework for methodology for Community-Based Flood Preparedness, as shown below, has been developed for a larger study for a case study of the Adyar basin to describe the relationships between various elements of the methodology and the results of statistical and GIS analysis. The framework for analysis shows the entire methodology adopted for the study and it is good enough to understand the aspects dealt with below: physical and social vulnerabilities for floods in Chennai.

By overlaying the thematic layers, flood hazard areas with different probabilities can be generated. Each factor is, however, divided into classes, and each class is given weight based on the significance of the factor for causing floods. The total weight used for considering the rate of flooding is given by the equation:

$$\text{Total weight of each factor} = \text{factor weight} * \text{weight of factor class}$$

Data inputs used in the study are given in Table 4.1. Figure 4.3 shows the distribution of physical vulnerabilities in the Adyar basin. Physical vulnerability to flooding ranges from very low (weightage 2), low (weightage 3), high (weightage 4) to very high (weightage 5) and has been mapped. Such analysis helps the authority to identify the flood-prone areas to take appropriate measures to minimize the damage incurred by floods. From this study, it is clear that the areas falling within the Chennai Corporation are exposed to high flood frequency due to factors such as high population

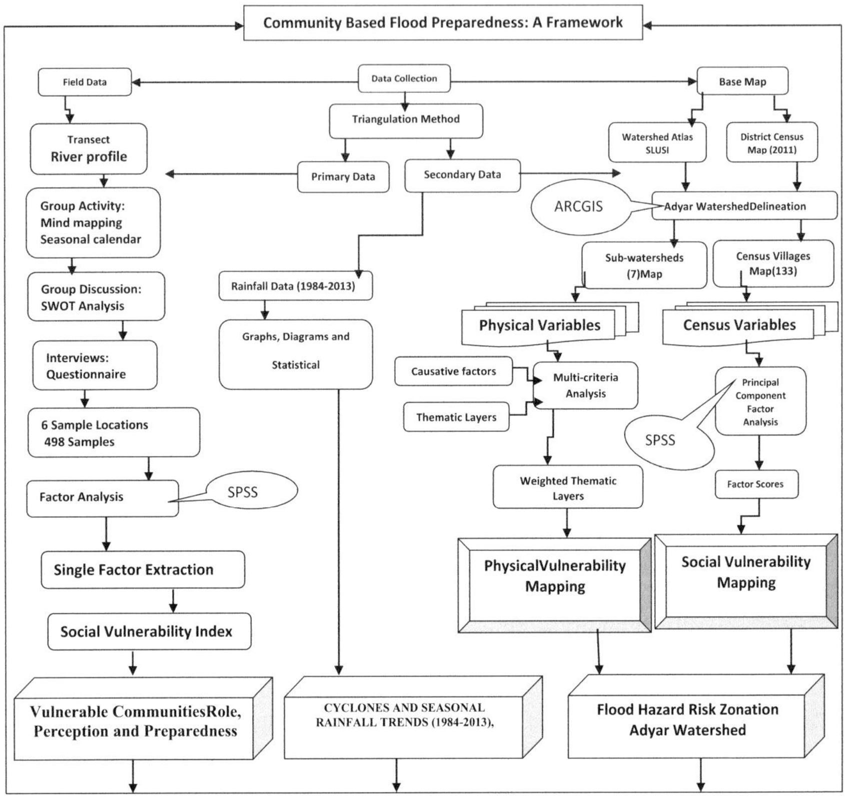

Fig. 4.2 A framework for community-based flood preparedness (physical and social vulnerabilities)

Table 4.1 Physical vulnerability data inputs

No#	Layers	Department/Source
1	Rainfall (1980–2009)	Statistical Department, Tamil Nadu
2	Micro-watershed and Soils	IRS, Anna University, Chennai and Suriya (2012)
3	Slope	Shuttle Radar Topographic Mapping (SRTM), NASA
4	Land use	AWIF Sand LISS III merged, 2009, NRSC
5	Road network	Open Street Map (OSM), 2014

Source Recreated with permission from the Departments/Sources (2014)

density, flat terrain and large number of populations living along the river banks. They are thus flood-prone and are vulnerable for floods.

Social vulnerability is defined as the characteristic of a person or group and their situation that influences their capacity to anticipate, cope with, resist and recover from the impact of a natural hazard. It involves a combination of factors that determine the degree to which someone's life, livelihood, property and other assets are put

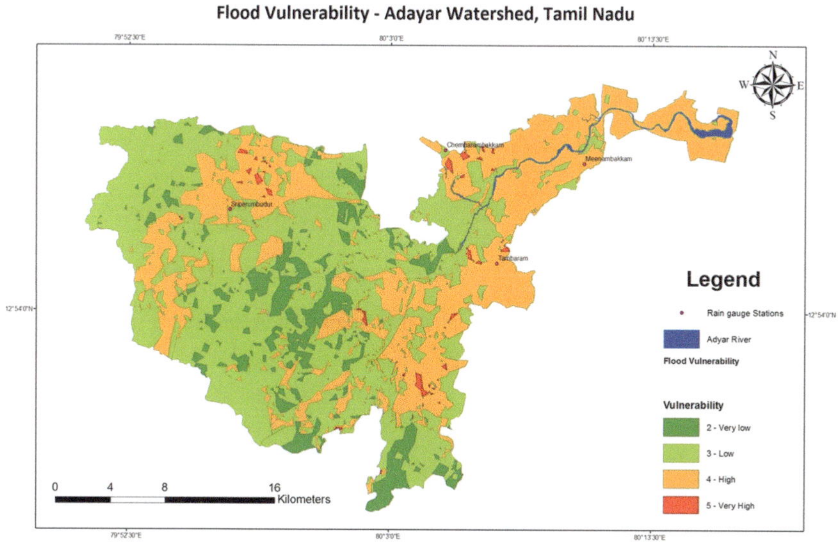

Fig. 4.3 Flood vulnerability—Adyar Watershed, Tamil Nadu

at risk by a discrete and identifiable event in nature and in society. Thus, social vulnerability refers to the communities' inability to cope with, or be resilient to, the natural hazards, given the socio-economic-demographic factors not suitable for safe living. So, addressing the social vulnerability helps to reduce the human sufferings and drastically reduces the risk of potential loss to human lives and property damages (Fig. 4.2 for factors of social vulnerability as well).

As for social vulnerability indices we could gather out of the 85 census variables available at the city ward level, only 11 variables are considered relevant from the social vulnerability point of view. The data inputs used in the social vulnerability index are given in Table 4.2. Figure 4.4 shows the result of social vulnerability zonation performed using factor analysis of the data inputs at the village and city ward levels for the Adyar basin.

Set of locations categorized as 'high' under social vulnerability includes Kattankolathur (1.19), Maraimalainagar (1.19), Tambaram (1.58), Pallavaram (1.85) and Velachery (2.052). All the locations are in the southern parts of the Adyar basin and contribute significantly to the source of water feeding the southern arm of the Adyar river. Similarly, Sathyamurthy Nagar (0.55) located within the Chennai Corporation, Porur (0.72) and Maduravoyal (0.74) are categorized under 'low' vulnerability. Lastly, 21 locations are categorized under the 'very low' vulnerability category. A few locations such as Anakaputhur (0.15), Meenambakkam (0.04), Saidapet (0.17) and Pammal (0.49) have a very low social vulnerability but physical vulnerability from the flood impact could be more as they are located very close to the Adyar river.

Table 4.2 Social vulnerability data inputs

No#	Census variables
1	Male 0–6 (age group)
2	Female 0–6 (age group)
3	Male illiterate (uneducated)
4	Female illiterate (uneducated)
5	Main cultivators (Male)
6	Main cultivators (Female)
7	Main agricultural labourers (Male)
8	Main agricultural labourers (Female)
9	Non-workers (Male)
10	Non-workers (Female)
11	Population density

Source Census of India 2011

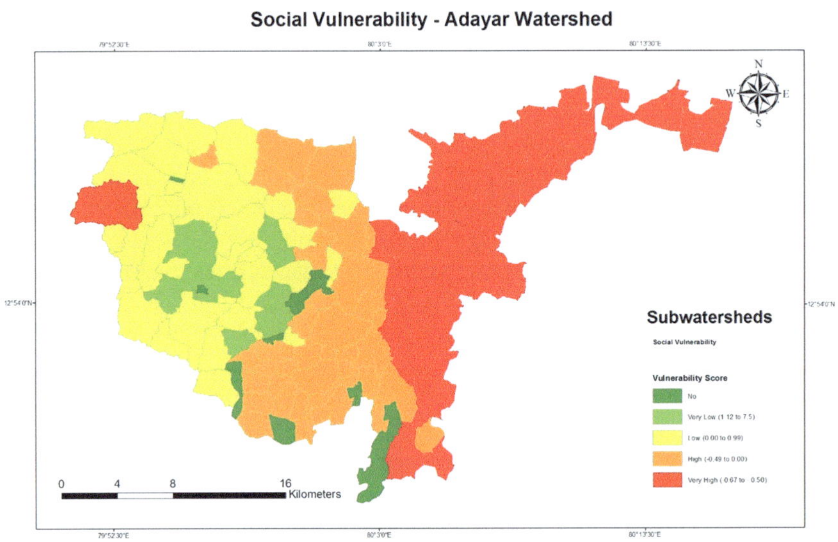

Fig. 4.4 Social vulnerability—Adyar Watershed

4.4 Chennai, the Megacity

The Greater Chennai is the study area, even as the Adyar basin has been chosen to demonstrate the rainfall and flood events of a period of 1984–2013. However, the floods of 2005 and 2015 are a focus of the study because of their widespread and devastating impacts on the state of Tamil Nadu, the capital city of Chennai and the rural areas of the state with the exception of three districts in 2005. Hence, the megacity of Chennai and the Adyar basin with its sub-watersheds have been taken as the case studies to demonstrate the vulnerability and risk of climate change and the difficulties in communicating climate change and its impacts to the people, even the educated.

Chennai, the State capital of Tamil Nadu, lies on the eastern coast of South India. Three watercourses meander through the city, namely, the Cooum River, the Adyar River and the Buckingham Canal. Chennai is the fourth largest metropolis of India with a total population of nearly 7.1 million spread over an area of 426 km^2, with a growth rate of 1.3% per annum and a density of 26,903/km^2. The city alone has a population of about 4.7 million (Census 2011). Within a century, however, Chennai has grown eightfold in population. The megacity of Chennai stretches along the Coromandel coast, with a superb sandy beach, which is the second largest in the world. The current population of the city is estimated at 8.2 million (2016). Figure 4.5 shows the city Corporation of Chennai, what is also known as the Greater Chennai. The Corporation of Chennai has 15 administrative divisions divided into 200 wards.

Chennai is a plain terrain, bounded by the Bay of Bengal on the east with an average elevation of 6.7 m from the mean sea level. Chennai experiences most of its rainfall during October–December, associated with the depressions and frequent cyclones, some very severe, during this period. Average annual rainfall is about 1,200–1,300 mm.

The geology of Chennai comprises mostly clay, shale and sandstone. The city is classified into three regions based on geology: (a) sandy areas, (b) clayey areas and (c) hard rock areas. Sandy areas are along the river and canal banks and coasts. Clayey regions cover most of the city. Hard rock areas cover Guindy, Velachery, Adambakkam and a part of Saidapet. In sandy areas such as Thiruvanmiyur, Adyar, Kottivakkam, Santhome, George Town, Tondiarpet and the rest of coastal Chennai, rainwater run-off percolates very quickly. In clayey and hard rock areas, on the other hand, rainwater percolates slowly, but it is held by the soil for a longer time. The city's clayey areas include T. Nagar, West Mambalam, Anna Nagar, Perambalur and Virugambakkam (Saravanan and Chander 2015: 1575). Figure 4.6 shows the river and drainage systems of Chennai.

4.5 The Adyar Sub-basin

The Chennai Basin is one of the 17 river basins of Tamil Nadu state and situated at the north-eastern corner of Tamil Nadu. Geographically, it is located between 12° 40′N and 13° 40′N latitudes and 79°10′E and 80° 25′E longitudes. The Chennai basin consists of eight sub-basins, namely, Gummidipundi, Araniar, Kosasthalaiyar, Cooum, Adyar, Kovalam, Nagari Ar and Nandhi Ar. The total area of the basin is about 7,282 km^2; out of this, 5,542 km^2 lies in the state of Tamil Nadu and the rest is in the state of Andhra Pradesh.

In Chennai basin, Adyar river covers approximately 704 km^2 and the rest is covered by other rivers. The Chennai basin of Tamil Nadu is an interdependent river basin, consisting of Araniar, Kosasthalaiyar, Cooum and Adyar sub-basins.

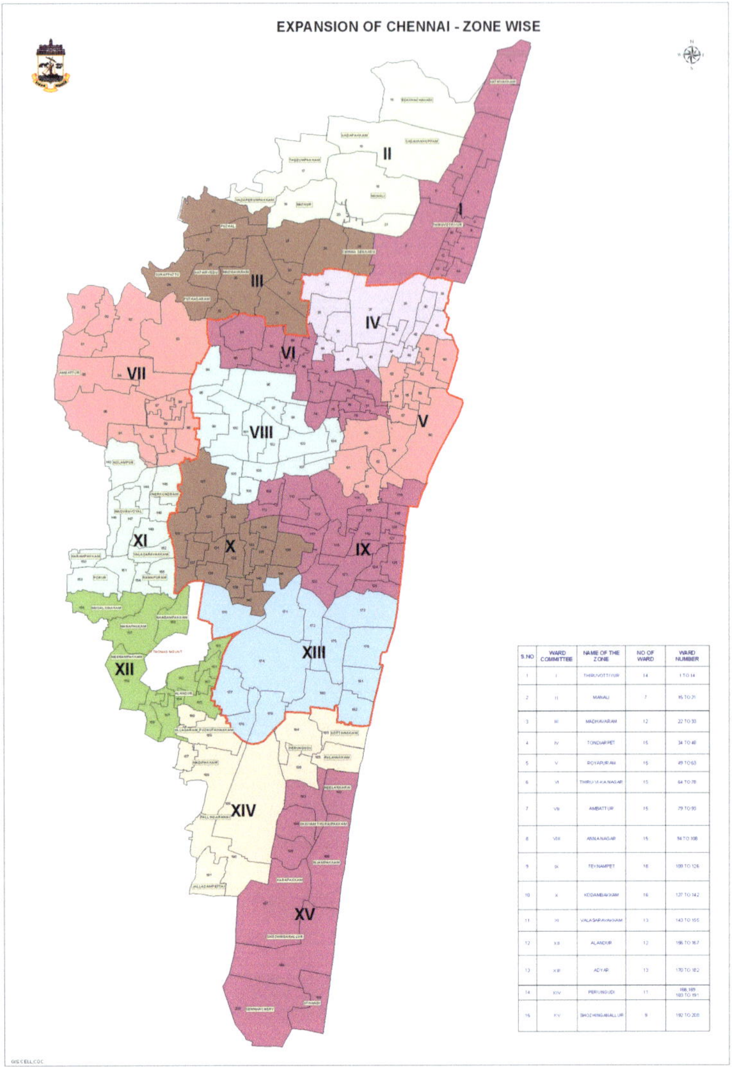

Fig. 4.5 The Greater Chennai

Geographically, the Adyar sub-basin extends between 12° 47′ 6″ N and 13° 3′ 22″ N latitudes and 79° 52′ 36″ E and 80° 17′ 1″ E longitudes. The Adyar sub-basin (Fig. 4.7) includes three major tanks, namely, Chembarambakkam, Sriperumbudur and Pillaipakkam. Additionally, this sub-basin has over 100 minor tanks (Institute of Water Studies, PWD, 2005). Approximately, 40% of the sub-basin is controlled by Chembarambakkam tank and hence important in generating floods. The upper catchments comprise about 75% of the basin and hence important in generating floods.

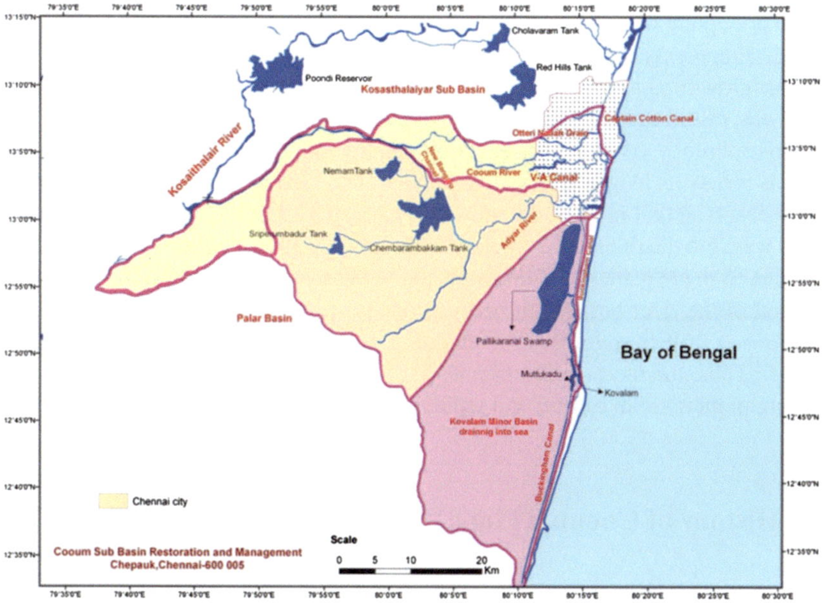

Fig. 4.6 River and drainage systems of Chennai. *Source* http://www.cmdachennai.gov.in

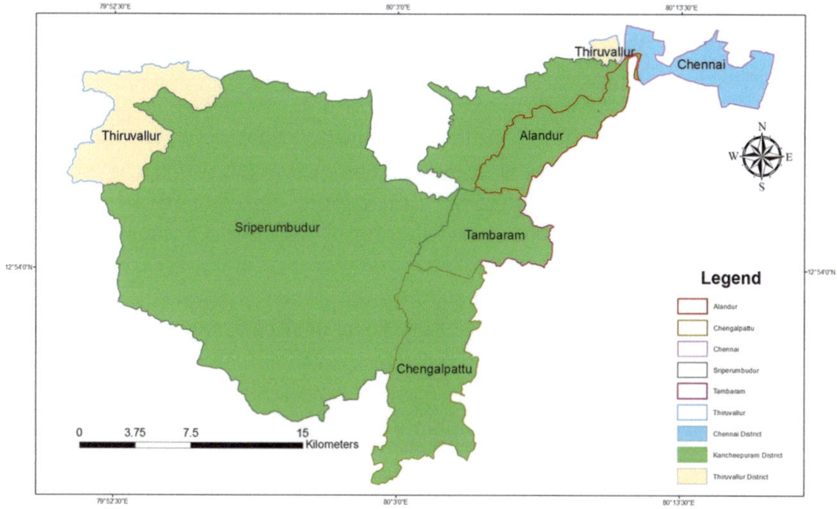

Fig. 4.7 The Adyar sub-basin

The floodplain of the Adyar River is stressed due to rapid, unplanned urban and slum developments and encroachments. The change in the land use pattern due to urban growth and migration adversely affects the hydrological processes of the catchment, resulting in the deterioration of the river course. On account of change in urban morphology, the Adyar river is polluted by both treated and untreated sewage effluents across its length, especially in and around the Chennai Metropolitan Area.

It is observed that most of the sewage generated by municipalities and industries find its way into the river either directly or through stormwater drains in the suburban areas due to inadequate storm drainage and sewer systems. This results not only in pollution of the river but also the deposition and consolidation of sludge along the river. This is further aggravated by dumping of solid wastes and building debris in the river. Thus, the increase in impervious areas disrupts the natural water balance, resulting in increase in run-off and reduction in infiltration.

4.6 History of Chennai Floods and Floods of 2005 and 2015

A river flood is one of the most common forms of natural disaster. It occurs when a river fills with water beyond its capacity. The surplus water overflows the banks and runs into adjoining low-lying lands. Several factors can cause a river flood. The most common reason for river flooding is heavy rainfall. The floods of Adyar in 2005 and 2015, for example, were caused by extreme rainfall events in the northeast monsoon of those years and also as a result of the tremendous flow of water in the Adyar river, primarily from the excesses released from the Chembarambakkam lake.

Chennai frequently experiences flooding due to heavy rains associated with depressions and cyclones of the monsoon period. A few catastrophic floods in the recent years of 1976, 1985, 1996, 1998, 2005, 2008, 2010 and 2015 have caused heavy damages, devastation and human sufferings. November is normally the month of heavy to very heavy rains associated often with the severe and very severe cyclones. Chennai is also a city of severe water shortage in summer months, although the city does not starve of rains; and it is mainly because of the mismanagement of water storages. In fact, the waterbodies and the swamps/wetlands of the city have been built over that; there are no proper storages for water. Chennai has Pallikaranai swamps, Madhavaram and Manali jhils, Adyar and Cooum estuaries as wetland sources apart from natural and man-made waterbodies.

In 2010, for example, Chennai received 760 mm of rainfall only during October–December. The highest rainfall in a day was 423 mm on October 27, 2005. The year 2015 has been a year of some very exceptional rains in November and December, around the festival times (Deepavali) and in the early days of December. The experiences of these rainy years have shown that the people of Chennai, and the Governments of day, were/are not prepared for flash floods. With the season of depressions and cyclones at present are in constant fear of floods. The Chennai Rains/Floods of 2015 have found mention in the climate change conferences and meetings, including the climate change meet of Paris, France last year (2015).

Extreme rainfall events last only for a few hours at the most but can generate terrifying and destructive floods. Their impact can be affected by a wide range of factors (or processes) such as the location and intensity of rainfall, the shape and steepness of the catchment it falls on, how much sediment is moved by the water and the vulnerability of the communities in the flood's path. In recent years, heavy precipitation events have resulted in several damaging floods in India. The consecutive flash floods over three major metro cities in the same year, that is, Mumbai in July 2005, Chennai in October 2005 and again in December 2005, and Bangalore in October 2005, caused heavy damages to the economy and loss of life. India is one of the highly flood-prone countries of the world. Around 40 million hectares of land in India is prone to floods, according to the latest National Flood Commission Report.

4.7 Causes of Chennai Floods

Table 4.3 shows the causes, factors and elements of floods in Chennai. Chennai accounts for frequent floods due primarily to consistent increase in rainfall. The floods of 2005, 2008 and 2010 are of considerable impacts on lives, properties and infrastructures. Changes in climate and micro-regional environmental factors shown in Table 4.1 are the major factors causing floods in the megacity.

Much of the flood damages are the consequence of Chennai's geography. The city is spread across a low-lying area and is flat like a pancake. The average elevation of the city is only about 6.7 m above mean sea level, with many neighbourhoods actually at sea level, which makes drainage a challenge even under normal circumstances (Saravanan and Chander 2015).

Encroachment of nearly all waterbodies for construction of residential, institutional and industrial buildings, development of slums and also of urban infrastructure has reduced the rainwater carrying capacity of the existing waterways. Development of transportation facilities such as the MRTS and subsequent developments along the train tracks has aggravated the flood situation in the city.

Most of the vegetal covers have already been reduced to non-vegetative or a concrete jungle. Surface run-off has also become too high while the infiltration capacity of land has declined, drastically. Because of the increase in impervious areas, therefore, Chennai has experienced severity of floods during every heavy rainfall.

More than half the wetlands have been converted to other uses, particularly residential, industrial and institutional. Chennai had about 6,501 small and large waterbodies in and around the city, but the number has dwindled to mere 30 or less. Ownership of waterbodies is scattered among the various government departments and has been the root cause for lack of proper water storage and management. The Protection of Tanks and Eviction of Encroachment Act (October) 2007 has never been implemented strictly at all.

Table 4.3 Causes of Chennai Floods 2005

Causes	Types of factors	Elements
Direct Factors	Increase in rainfall	Due to global climatic change
Urbanization	Encroachment of all waterbodies, wetlands and other land areas	
Construction of transportation networks all along major watercourses		
Increase in concrete spaces stopping or hindering percolation of water into ground		
Decrease in open areas/green spaces		
Lack of transportation facilities (especially in/for slums and poorer localities)		
Topography	Plain terrain lacking natural gradient for free run-off	
Indirect factors	Inadequate and poor drainage systems	Sewage systems were planned originally before 4 decades and only few minor modifications made which is far below the required capacity
Heavy siltation all along the drainage channels		
Lack of coordination between the agencies		
Disposal of solid waste and other debris	Attitude of people	
Lack of management measures by the agencies		
Vehicle parking on roads	Increase in concrete spaces	
Discrepancies between public and local authorities		

Source Lavanya (2012)

Inadequate and very poor drainage systems have been another cause for floods. The plain terrain lacks the natural gradient necessary for free run-off. Sewage systems, originally designed for 0.65 million people at 114 lpcd of water supply, was later modified during 1989–1991 but this has not achieved required capacity. Dumping of garbage and massive reclamation of marshlands in Pallikaranai has been reduced to one-tenth of its earlier size. The government has also built slum resettlement colonies along the river channel and multi-storeyed housing estates have come up blocking the natural drainage of the river basins into the sea. The original silting pattern has also shifted due to such developments. Sewage and industrial effluents dumped into the canals and the consequent silting have left the waterways stagnant. Planning of individual division-oriented projects without involving concerned authorities has also been a major drawback which has been reflected as a major gap from the institutional side (Lavanya 2012).

4.8 Historic Depressions, Cyclones and Rainfall of 2005 and 2015

Between 1984 and 2013, there were 273 cyclones, depressions and severe cyclones which developed over the Bay of Bengal. The southwest monsoon recorded 38% of the total followed by the northeast monsoon at 46%. Summer season recorded 13% of the disturbances with a meagre 2% for the winter season. In this period, 7 years recorded over 10 disturbances with 1985 (flood year in Chennai) recording 13 cyclonic disturbances with 6 each for the northeast and southwest monsoons, respectively. Likewise, 2005 (flood year) recorded 5 disturbances in the northeast monsoon alone. On the other hand, 1996 (flood year) recorded 4 disturbances during the northeast monsoon. The year 1982 recorded the maximum of 15 cyclonic disturbances with a record 9 disturbances forming during the southwest monsoon alone (Figs. 4.8 and 4.9).

A special focus is drawn here to the rainfall of 2005 which was very phenomenal in the distribution and extent of damage by flood inundation. The year 2005 was a year of **near-normal cyclonic activity** over the north Indian Ocean. This basin witnessed the formation of **12 disturbances,** against a normal of 15. Out of the 12 disturbances, **four** (against a normal of five to six) intensified into cyclonic storms and three concentrated only up to deep depressions. There was one land depression during this year (Fig. 4.10).

The Arabian Sea was less active and was devoid of any cyclonic storm. Only two depressions formed over that basin. The Bay of Bengal was more active with formation of four cyclonic storms, three deep depressions and two depressions during the year. The first system, Cyclone **HIBARU** developed from a low-pressure area which concentrated into a depression over the southwest Bay of Bengal and adjoining

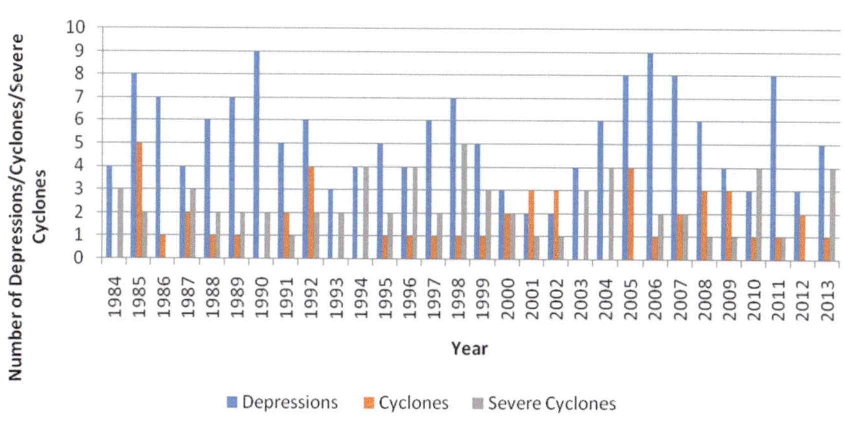

Fig. 4.8 Depressions, cyclones and severe cyclones—Bay of Bengal (1984–2013)

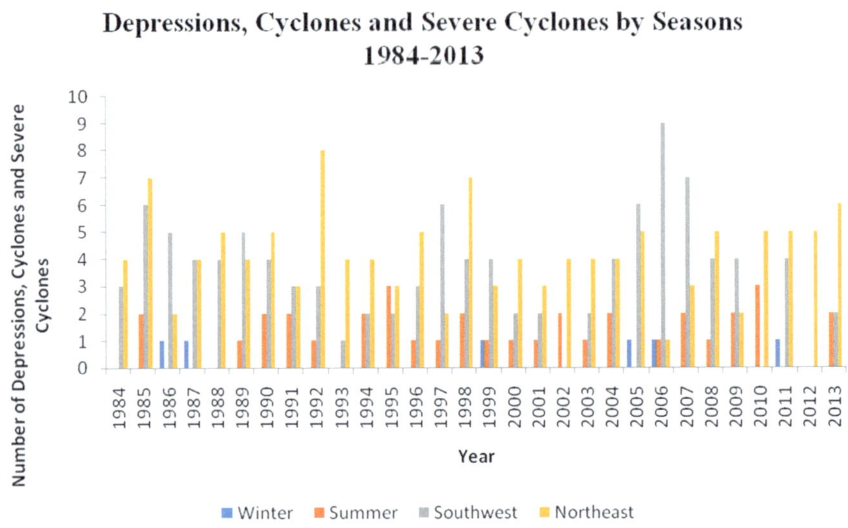

Fig. 4.9 Depressions, cyclones and severe cyclones by seasons (1984–2013)

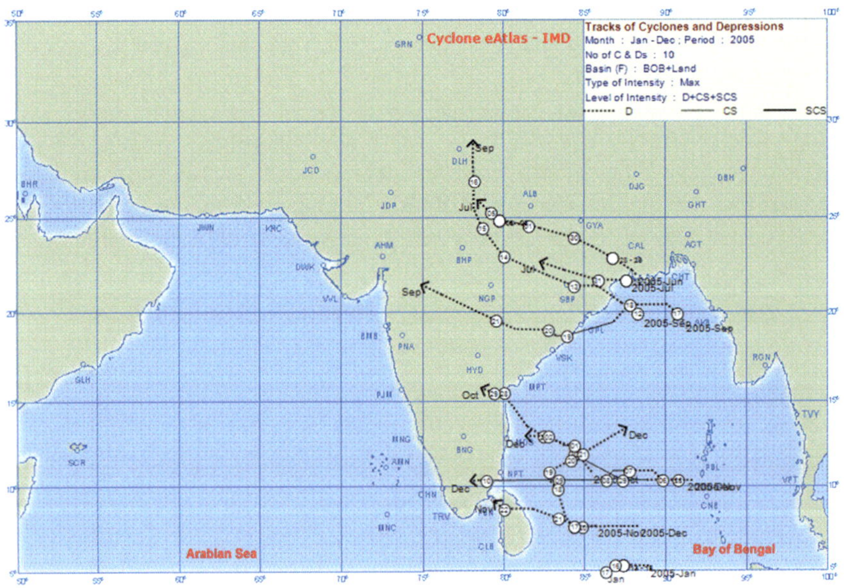

Fig. 4.10 Cyclonic disturbances 2005

Indian Ocean on the evening of 13 January 2005. It remained practically stationary and intensified into a deep depression on the morning of the next day and into a cyclonic storm on the morning of the third day. Moving slowly in a westerly and then in a west-south-westerly direction, it maintained its intensity till the morning of 17 January 2005 when it weakened into a deep depression and then into a depression on the same afternoon. On the same night, the system further weakened into a low-pressure area over southwest Bay of Bengal and adjoining Indian Ocean.

The cyclone **PYARR** developed from a low-pressure area over east-central Bay of Bengal on 16 September 2005. Subsequently, it intensified into a depression on the morning of 17 September 2005. The system initially moved in a west-north-westerly direction and then in a westerly direction and intensified into a deep depression in the morning of 18 September 2005. Further, the system moved in a south-westerly direction and intensified into a cyclonic storm in the same evening. Continuing to move in a south-westerly direction, the cyclone **PYARR** crossed north Andhra Pradesh coast close to Kalingapatnam in the morning of 19 September 2005. It remained as a cyclonic storm over north coastal Andhra Pradesh close to Kalingapatnam till the evening of the same day. It then moved west-north-westwards and weakened gradually into deep depression and then a depression. On the morning of 22 September 2005, it was seen as a well-marked low-pressure area over western parts of central India.

The third system formed from a well-marked low-pressure area which was seen over south Andaman Sea and adjoining southeast Bay of Bengal on the morning of 27 November 2005. It concentrated into a depression on the morning of the next day. Moving in a westerly direction, it intensified into a cyclonic storm **BAAZ** around midnight of the same day. Thereafter, it moved swiftly in a north-westerly direction till the evening of 29 November 2005. Then **BAAZ** became sluggish in its movement and hovered around the area till the morning of 1 December 2005. Thereafter, the system, moving in a north-westerly direction, gradually weakened and dissipated over the sea on the morning of 2 December 2005.

The cyclonic storm **FANOOS** developed from a low-pressure area over south Andaman Sea. It intensified into a depression and lay over the southeast Bay of Bengal in the morning of 6 December 2005. Moving in a north-westerly direction, it further intensified into a deep depression in the same afternoon. Thereafter, it took a steady westerly direction and intensified into a cyclonic storm in the morning of 7 December 2005. It had south-westward movement till the morning of the next day. Thereafter, it moved westwards till the morning of 10 December 2005. On the same morning, due to proximity to land, it weakened into a deep depression and crossed north Tamil Nadu coast south of Nagapattinam (close to Vedaranyam) around 0530 UTC. After landfall, it rapidly weakened into a depression at 0600 UTC of the same day. Moving in a westerly direction, it weakened further into a low-pressure area in the morning of 11 December 2005. The sequence of the cyclonic disturbances and the depressions are shown below in a chronological order.

1. Cyclonic Storm HIBARU over the Bay of Bengal, 13–17 January;
2. Depression over the Arabian Sea, 21–22 June;

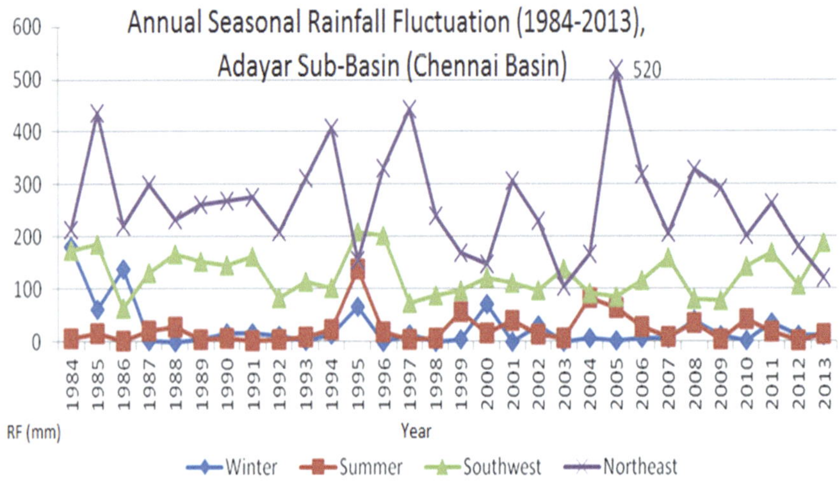

Fig. 4.11 Annual seasons rainfall fluctuation (1984–2013), Adyar sub-basin (Chennai Basin)

3. Land depression 27 June–5 July;
4. Deep depression over the Bay of Bengal, 29–31 July;
5. Depression over the Bay of Bengal, 12–16 September;
6. Depression over the Arabian Sea, 14–16 September;
7. Cyclonic storm PYARR over the Bay of Bengal, 17–21 September;
8. Deep depression over the Bay of Bengal, 26–29 October;
9. Depression over the Bay of Bengal, 20–22 November;
10. Cyclonic storm BAAZ over the Bay of Bengal, 28 November–2 December;
11. Cyclonic storm FANOOS over the Bay of Bengal, 06–10 December;
12. Deep depression over the Bay Bengal, 15–22 December.

It is evident from Fig. 4.11 that the annual seasonal rainfall for the 30-year period in the Adyar sub-basin could best be described as oscillatory. The trend is not regular but irregular. Even though the mean annual rainfall for the 30-year period (1,418 mm) is moderate, extreme cases of very high rainfall during the northeast season of about 520 mm was recorded in the year 2005 in the Adyar sub-basin. However, comparing the four seasons, it can be observed that the northeast monsoon winds brings heavy rainfall coupled with the cyclones in the sub-basin. Higher precipitation has been recorded during the season that generally could be attributed to the depressions and cyclones of the Bay of Bengal on the east coast of India which supplies the bulk of the atmospheric moisture for precipitation. Low rainfall has been recorded during the winter (Post-monsoon) and the summer (Pre-monsoon) in the Adyar sub-basin.

Table 4.4 gives descriptive statistics for annual rainfall of the Adyar sub-basin, particularly for the four rain gauge stations. Note the maximum rainfall recorded for the stations: 2,886 mm in Chembarambakkam, 2,411 mm in Meenambakkam, 1,805 mm in Sriperumbudur and 2,013 mm in Tambaram. Standard deviation, for example, is a measure that summarizes the amount by which every value within a

Table 4.4 Descriptive statistics for annual rainfall: Adyar sub-basin 1984–2013

Rain Gauge Stations	Mean (mm)	Variance (mm)	Standard deviation (mm)	CV	Skewness	Kurtosis	Maximum (mm)
Chembarambakkam	1,672.41	296,315.99	544.35	0.33	0.22	−0.64	2,886.4
Meenambakkam	1,439.11	140,237.34	374.48	0.26	1.13	1.01	2,410.5
Sriperumbudur	1,205.16	86,510.62	294.13	0.24	0.34	−0.58	1,805.0
Tambaram	1,357.69	59,987.48	244.92	0.18	0.75	0.08	2,013.0

Source The authors' computation (2015)

dataset varies from the mean and indicates effectively how tightly the values in the dataset are bunched around the mean value. When the values in a dataset are pretty tightly bunched together, the standard deviation is small. When the values are spread apart, the standard deviation is relatively large.

The highest standard deviation with the value of 544.35 mm per annum is observed at Chembarambakkam which indicates the large variation in the annual rainfall amount in the series, while Tambaram receives the lowest standard deviation with the value of 244.92 mm per annum which implies the small variation in the annual rainfall amount in the series. Furthermore, it is clear that the standard deviation for each station is lesser than the mean by referring to the fact that the data set is pretty tightly bunched together.

The coefficient of variation (denoted as COV) is a measure of spread that describes the variations in the data relative to the mean. The coefficient of variation is adjusted so that the values are on a unit-less scale. Because of this adjustment, the coefficient of variation is used instead of the standard deviation to compare the variation in data that have different units or that have very different means. If the data set have larger coefficient of variation, the greater is the spread in the data. The coefficient of variation represents the irregularity of the distribution of annual rainfall between stations. However, the results of the coefficient of variation among the rain gauge stations show that the values for each station range between 0.18 and 0.33. Chembarambakkam displays the highest value of coefficient of variation, which is 0.33 while Tambaram has the smallest value of coefficient of variation, which is 0.18.

The coefficient of skewness is used to verify the degree of asymmetric of a distribution around the mean. The values of skewness lie in between 0.22 and 1.13 where the positive results indicate that all skewness is positively skewed for all stations. Meenambakkam shows the highest value of skewness with the value 1.13 which gives a clear indication that this station is very obviously skewed and the asymmetric tail is extending to the right, while Chembarambakkam with the smallest skewness value of 0.22 indicates that the asymmetric tail is also extending to the right.

The values of kurtosis in the study are both positive and negative in the range of −0.058 to 1.01. Meenambakkam obtains the highest value of kurtosis with the value 1.01 which implies the possibility of a leptokurtic distribution where the data set tend to have a distinct peak near to the mean with a heavy tail since the peak of distribution is too high around the mean. Chembarambakkam and Sriperumbudur both register kurtosis value −0.64 and −0.58, respectively, which indicates that the data set tend to have a flat peak near to the mean.

During the northeast monsoon of 2015, Chennai and other parts of the rain-ravaged Tamil Nadu were left marooned and flooded with signs of devastation caused by the nature's fury. Over 400 lives were lost in the widespread flash floods. Normal life came to a grinding halt as schools, colleges, offices and IT firms for several days and businesses everywhere had to face a complete shutdown. The total rainfall in Chennai was 119.73 cm till November 30, midnight (2015) and broke the 1918 record of 108.8 cm which had stood as the highest record until then. Chennai received 539 mm of rain in December as against the monthly average of 191 mm. Chennai rain in December was almost 3 times more than the normal rainfall in December. It also broke the record of last 10 years when the capital city received 421 mm of rain in 2005 (Source: Skymet Meteorology, India Division).

Figure 4.11 shows a NASA map showing the intensity of rainfall in and around Chennai between 28 November and 4 December 2015 (PMM, NASA), while Fig. 4.12 shows the distribution of exceptional rainfall in Chennai metropolitan area on 2 December 2015, and Fig. 4.13 shows the exceptional rainfalls in Chennai basin and Adyar sub-basin.

Table 4.5 shows the distribution of rainfall by stations and by seasons in the Flood Year 2005 in the Adyar sub-basin. As can be observed, the maximum seasonal (northeast) rainfall for the 30-year period has been in 2005, which has recorded a total rainfall of 6,236.90 mm with a corresponding highest mean value of 519.74 mm. The record indicates that the standard deviation correlating the highest seasonal rainfall is 201.3 mm and the data is skewed left, meaning the rainfall distribution is flat. The

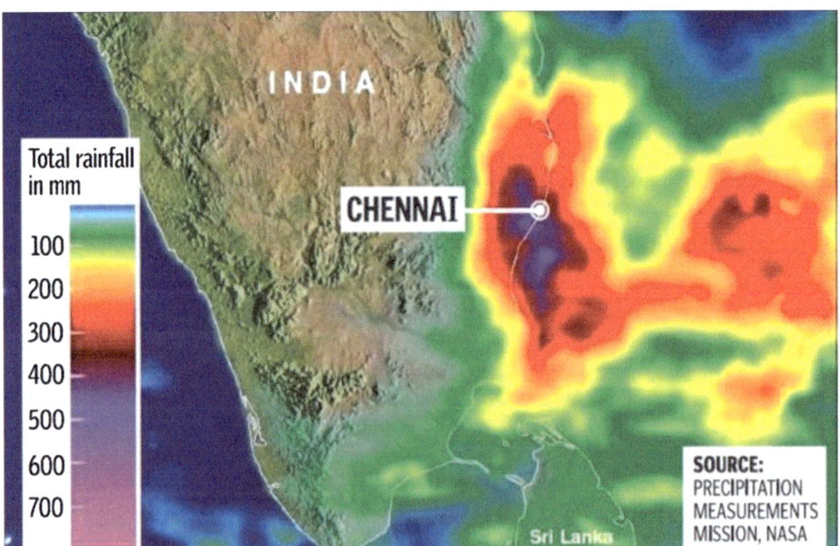

Fig. 4.12 Total rainfall

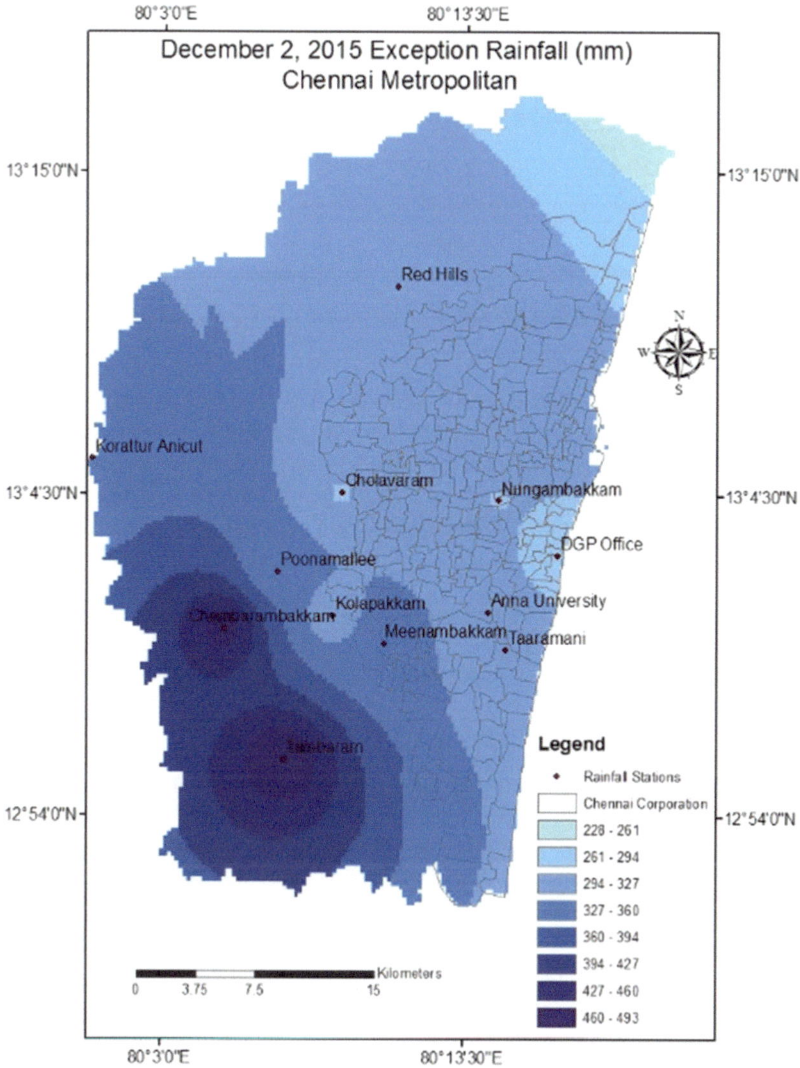

Fig. 4.13 Exceptional rainfall

highest variance and coefficient of variation figures recorded show the inconsistency in the rainfall pattern. The coefficient of variation values indicates the existence of strong variability in the rainfall of the study area. Nonetheless, the maximum monthly rainfall (120.60 mm) during the northeast monsoon has occurred in November 1985. In addition, "0" rainfall has been reported in the month of December 1992 and 1995 and a corresponding seasonal mean value of 210.13 mm and 154.43 mm, respectively. The high standard deviation value in the year 2005 can be easily correlated with the high rainfall range. The rainfall range signifies the difference between the maximum

Table 4.5 Rainfall in Adyar sub-basin by seasons in flood Year 2005

Seasons	Chembarambakkam	Meenambakkam	Sriperumbudur	Tambaram
	Rainfall (mm)	Rainfall (mm)	Rainfall (mm)	Rainfall (mm)
Winter	0	8.8	0	5
Summer	230	216.5	2	302.2
Southwest	483	318.8	314	236.9
Northeast	1723	**1866.4**	1489	1158.5
Annual	**2436**	2410.5	1805	1702.6

Source TN Statistical Department (2014)

and minimum annual rainfall. The year 2005 flood was one of the worst floods to have hit the city of Chennai and was considered as '100-year flood'. Exceptionally heavy rainfall during 2005 was associated with the formation of depressions or cyclones in the Bay of Bengal leading to flood (see Table 4.3) during northeast monsoon season. Among the four stations, Meenambakkam received the highest followed by Chembarambakkam, Sriperumbudur and Tambaram during the northeast monsoon

4.9 Flood Mitigation in Chennai

The Flood Alleviation Scheme of 1998, funded by the Central Government at a cost of Rs. 3,000 million, was launched focusing mainly on the structural measures with objectives such as the adequacy of flow in the arterial drainage systems, safeguards against tidal and fluvial flooding, removal of impediments, relocation and rehabilitation of encroachers. Cleaning of waterways and lakes was also undertaken under packages 2 and 3 of the scheme. The Chennai City River Conservation Project was launched in 2000, aiming to improve the waterways, with an estimated outlay of Rs. 17,000 million. The Master Plan 1992–1993 incorporated Madras Metro Flood Relief/Storm Water Drainage in the form of structural and non-structural measures. Funds under the Jawaharlal Nehru National Urban Renewal Mission (JNNURM) project were visualized for implementation of underground sewerage schemes, and detailed project reports were also developed. In 2010, the State Government launched a massive flood mitigation project for the city, involving construction of new micro- and macro-drainage systems in four basins and making improvements to existing drainage, at a cost of Rs. 14,470 million under the JNNURM. The works such as improvements to divert the surplus water, desilting and strengthening existing city drainage network were also carried out. But these have been of no particular use/mitigation of floods in Chennai as 2005 and 2015 flood events have shown.

This in fact brings us to the crux of the problem the present paper has set out to discuss: vulnerability and risk in (communicating) climate change in Chennai with a focus also on the Adyar basin which includes areas far beyond the boundaries of the metropolitan city or the Greater Chennai. The discussion now turns to a rainfall analysis for the Adyar basin for 1984–2013 and the floods of Chennai in 2005 and 2015.

4.10 Drivers of the Extreme Rainfall Events

The extreme rainfall event of 2005 may be considered as a forerunner of the extreme rainfall event of 2015 and the ones that may happen in the near and distant future.

Understanding the drivers of extreme rainfall events needs information of large-scale atmospheric and oceanic conditions *before, during and after* the extreme events. While the impacts of ***global climate change versus local issues*** like urbanization on the extreme events will remain an important scientific question. However, addressing such a question is not possible without the insight on the mechanism and processes associated with the events. The important large-scale characteristics of 2005 and 2015 events are the very strong El Nino phenomenon and very warm Bay of Bengal, and possibly Arabian Sea, which have probably caused the extreme events. Taking the case of extreme rainfall and flood event of 2015, we may say this: the ***Strong El Nino of 2015 had a possible connection*** to the extreme rainfall events known as Chennai Rains and the ensuing and widespread floods in the city and its environs.

The El Nino of 2015 has been one of the strongest reported, which has started developing in 2014. El Ninos are reported to have impacts on the northeast monsoon (Zubair and Ropelewski 2006) by modestly intensifying it. A possible hypothesis would be that a stronger El Nino led to a strong easterly during November–December 2015 (Fig. 4.14) which probably brought moisture to the east coast of India over Chennai with a much intensified rainfall. However, the scientific question remains if the easterlies (possibly) generated by El Nino are the only reason or some other

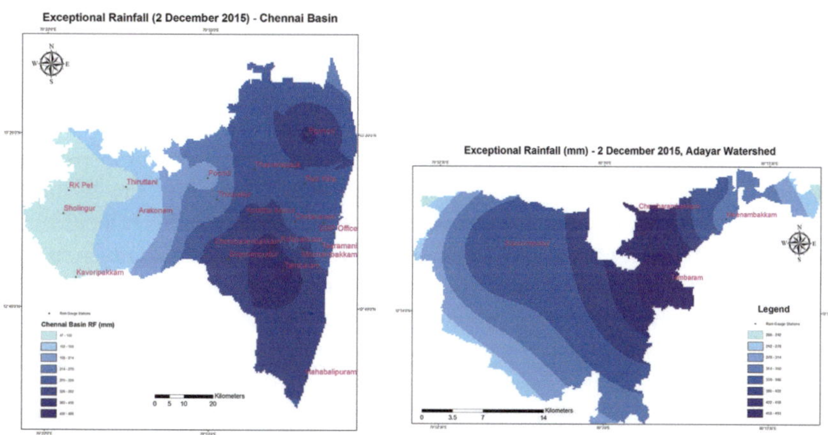

Fig. 4.14 Exceptional rainfall

factors/drivers have caused heavy precipitation over Chennai. This hypothesis needs, however, a climate model-based verification (Narasimhan et al. 2016).

The potential destructiveness of cyclones, defined as the total dissipation of power, integrated over the lifetime of the cyclone has increased, is increasing and will increase in a warming environment (Emanuel 2005), and this is primarily associated with the warming of SST (Knutson et al. 2010). It is true that the warming of Bay of Bengal is attributed to the global warming, which is hypothesized to be a potential cause of the extreme events over Chennai; however, to confirm this, it again needs an event-based attribution study.

Urbanization is yet another factor reportedly intensifying the extreme precipitation, either through generation of convection due to urban heat island (UHI) or through the uneven urban terrain resulting in wake diffusion and turbulence (Shastri et al. 2015). Extremes in Southern and Central Region of India are observed to be affected by urbanization during summer monsoon season (Shastri et al. 2015). But the specific impacts during winter monsoon and cyclones have not yet been fully explored. Chennai is reported to have a significant urban heat island (Swamy and Nagendra 2016), and there is a possibility of such UHI-extremes link.

4.11 The Spatial and Temporal Range of Issues that Needs Addressing

According to Narasimhan et al. (2016), the spatial and temporal issues that need addressing are visualized in Fig. 4.15. The schematic sets down the short-, medium- and long-term issues that are to be addressed towards a solution for floods and the consequent devastation in the city of Chennai.

There are literally hundreds of things to do to resolve problems arising out of floods in Chennai and in other areas, including the Adyar basin of our interest. But there are some very important things to do immediately:

- **Better understanding of weather**: There is need to understand the weather and weather forecasts better so that we could predict in advance the extreme weather events and also take precautionary measures to safeguard people, properties and public utilities from the risk of impacts. Understanding the weather better is important also for the common man who could then understand what his responsibility as a citizen of the rural and city communities. Individuals could do much in preventing the impacts of flood risks, if only they know and understand what to do in such circumstances as extreme rainfall and floods. The local governments have a responsibility to inform when they know in advance what is likely to happen, as well. As far as the weather forecasting systems are concerned, it is the responsibility of the Indian Meteorological Department to modernize and equip themselves with the latest tools of forecast (Fig. 4.16).

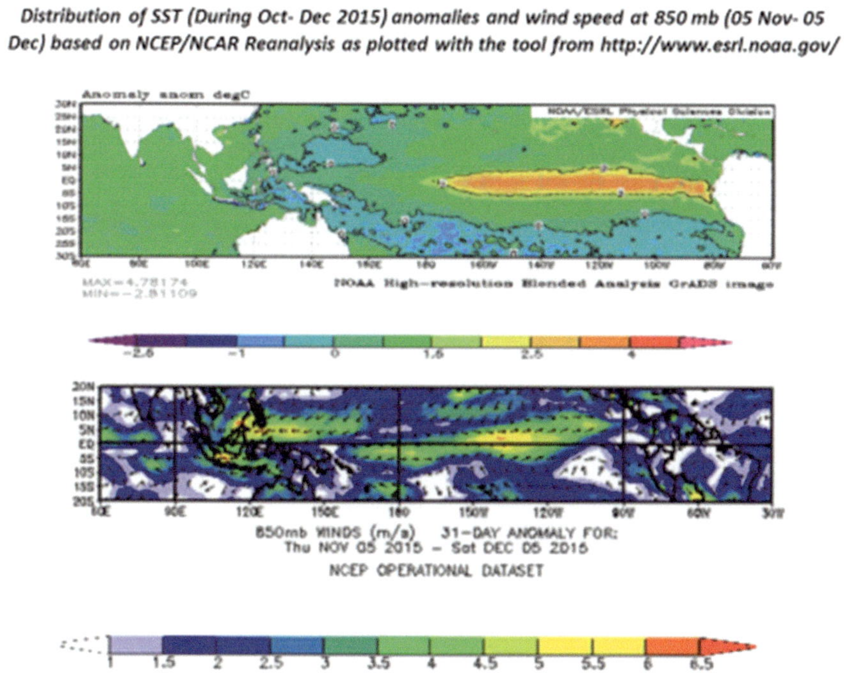

Fig. 4.15 Some very important things to do to adapt to floods in Chennai and other areas

- **Placement water gauging systems**: The one problem that caused the flooding of the megacity was the release of enormous amount of water from the Chembarambakkam lake system. It is well known that it is the release that always causes floods in the floodplains of the Adyar river. The worst hit is the city even as rural areas are also hard hit many a times. Gauging systems and flood sensors are the need of the hour. The concerned Department must on a war footing that acquires the gauging systems and sensors so that there is information available on the inflow and outflow at the lake end and also at various river and waterbodies which hold water and cause the floods, regularly. It would be worthwhile to acquire robust rural and urban flood models which could be helpful in assessing in advance the potential calamities such as the flood of 2005/2015 or worse.
- **Construction of storages**: It is common knowledge that there are currently no proper storages for excess of rainwaters because waterbodies have been insanely built over and even the marshlands which were once the flood holders for the city have been encroached upon/taken over and have become part of the real estate/institutional building enterprises. At least now, this madness of building over the flood holders everywhere and the waterbodies must stop. Otherwise, there will be no redemption from the vulnerability and risk of floods. Some means of constructing storages for the floodwaters must be found, here and now.

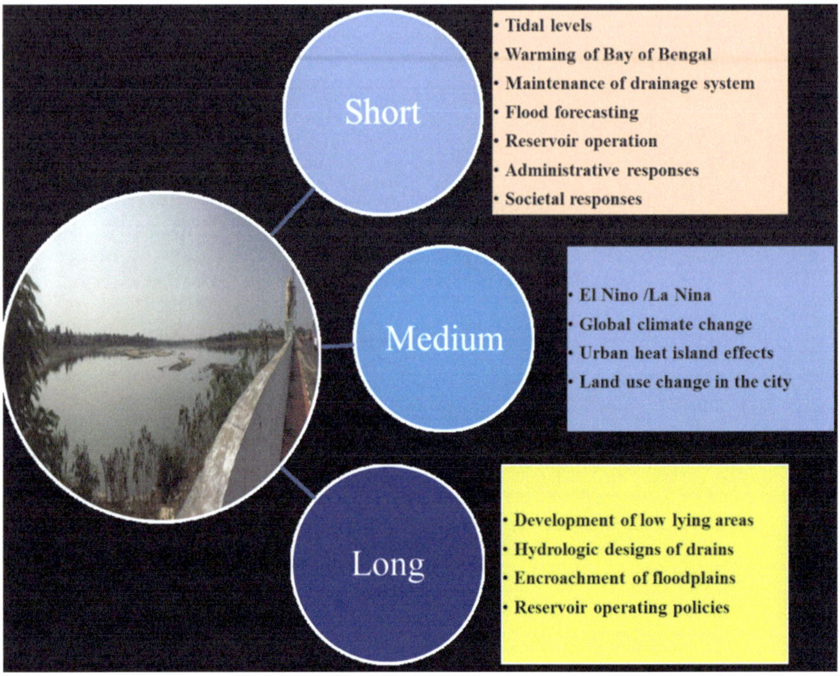

Fig. 4.16 Spatial and temporal range of issues that need addressing towards a solution. *Source* Narasimhan et al. 2016. Chennai floods 2015, a rapid assessment, p. 2

- **Public awareness**: People should be made aware of their vulnerability and risk involved in flooding of an area, about the flood preparedness, responses and mitigation measures. They should also be made aware of climate change and the triggers of climate change and how the humans, including themselves, are the causes for it. This knowledge should be made mandatory in schools of all levels, colleges and universities and in the public forums. Community-based/oriented/involved climate change assessments must be made to make common people aware of climate change prospects in the future. If farmers could be involved in such assessments, why not people of the small towns, cities and megacities?
- **Clearing off encroachments in and near the waterbodies**: All encroachments in and around the waterbodies and flood plains should be cleared and the people involved in such encroachments should be rehabilitated away from the flood-prone areas subsequently to reduce flood crisis.
- **Desilting and conservation of waterbodies**: All waterbodies, small and large—ponds, lakes and tanks; channels and canals—must be cleared off weeds and vegetation obstructing flow of waters and they may be desilted to improve the flow and storage of waters during rains. In off-seasons, rainwater harvesting may be practiced to save water.

- **Rapid assessment**: A rapid assessment of flood inundation through community mapping is also required before the monsoon starts every year so as to effectively alleviate the flood risks to the rural and urban communities.
- **Flood mitigation programmes**: Construction of raised platforms, floodwalls, town protection works should be carried out and monitored for effective mitigation of flood menace.
- **Watershed management**: Timely cleaning, desilting and deepening of natural water reservoirs and drainage channels (both urban and rural) have to be taken up and it has to be a regular, continuous process because silting is an everyday affair.
- **Increasing green cover**: It requires more of planning and regulatory control over the open space and ground cover in the villages, towns and cities on a war footing.
- **Scientific study of flooding patterns**: There is need to study rural and urban flooding patterns at least for a period of 30 years using hydrological and hydraulic modelling techniques which could guide us with strategic planning collectively.
- **Riverfront water development**: A riverfront plan for every river could help stakeholders in managing flood control effectively and also create awareness to the public about advantages in managing and maintaining water resources.

References

Aggarwal D, Lal M (2001) Vulnerability of Indian coastline to sea-level rise. Centre for Atmospheric Sciences, Indian Institute of Technology, New Delhi

Asnani GC (1999) Monsoon prevails over global tropics, *Vayu Mandal*, Special Issue on Asian Monsoon and Pollution over Monsoon. Environment 29(1–4):37–41

Bhalme HN, Sikder AB, Jadhav SK (1990) Relationships between planetary scale waves. Indian Monsoon Rainfall and ENSO. Mausam 41(2):279–284

Bohle H-G, Downing TE, Watts M (1994) Climate change and social vulnerability: toward a sociology and geography of food insecurity. Glob Environ Change 4(1):37–49

Census of India (2011)

Chambers R (1989) Vulnerability, coping andpolicy. IDS Bulletin 20(2):1–7

De Vries HJ, Revi A, Bhat GK, Hilderink H, Lucas P (2007) India 2050: scenarios for an uncertain future. Netherlands Environment Assessment Agency (MNP). Bilthoven

Emanuel K (2005) Increasing destructiveness of tropical cyclones over the past 30 years. Nature 436(7051):686–688

http://www.cmdachennai.gov.in

Jayanthi N, Govindacharya S (1999) El Nino and northeast monsoon rainfall over Tamil Nadu. Mausam 50(2):217–218

Knutson TR et al (2010) Tropical cyclones and climate change. Nat Geosci 3(3):157–163

Lavanya K (2012) Urban flood management: A case study of Chennai city. Architecture Research. 2(6):115–121

Narasimhan B, Bhallamudi SM, Mondal A, Ghosh S, Majumdar P (2016) Chennai floods 2015: a rapid assessment. Interdisciplinary Centre for Water Research, Indian Institute of Science, Bangalore

Rao PC (2004) Spatio-temporal variations of monsoon performance over Rayalaseema region of Andhra Pradesh. Doctoral Thesis (unpublished). Department of Geography, University of Madras, Chennai (August): pp. 197

Recreated with permission from the Departments/Sources (2014)

Revi A (2008) Climate change risk: an adaptation and mitigation agenda for Indian cities. Environ Urban 20(1):207–229

Saravana J, Chander KN (2015) Chennai floods (2015) and possible solutions from developed countries. Int J Sci Res 4(12):1575–1580

Shastri H et al (2015) Impacts of urbanization on Indian summer monsoon rainfall extremes. J Geophys Res Atmos 120(2):496–516

Sikka DR (1980) Some aspects of the large scale fluctuations of summer monsoon rainfall over India in relation to fluctuations in the planetary and regional scale circulation parameters. Proc Indian Acad Sci (Earth and Planet Science) 89:179–195

Swamy GSNVKSN, Nagendra SMS (2016) Study of Urban Heat Island (UHI) formation in Chennai city area. Poster presented at Indo-German Conference on Sustainability: Exploring planetary boundaries and their challenges and opportunities. February 27–28, 2016. Indo-German Centre for Sustainability, IIT Madras, Chennai, India

Turner BL et al (2003) A framework for vulnerability analysis in sustainability science. PNAS 100(14):8074–8079

Vogel C, O'Brien K (2004) Vulnerability and global environmental change: rhetoric and reality. AVISO (13)

Watts MJ, Bohle HG (1993) The space of vulnerability: the causal structure of hunger and famine. Prog Hum Geogr 17(1):43–67

Zubair L, Ropelewski CF (2006) The strengthening relationship between ENSO and northeast monsoon rainfall over Sri Lanka and southern India. J Clim 19(8):1567–1575

Chapter 5
An Environmental Study of Solid Waste Management System in Chandrapur City, Maharashtra, India

Y. Y. Dudhapachare

Abstract Municipal solid waste management system in the cities of India are controlled and regulated at present by the Municipal Solid Waste (Management and Handling) Rules, 2000; Solid Waste Management Rules, 2016; as well as Plastic Waste Management Rules, 2016. These are the comprehensive guidelines and ready reckoner documents for the municipalities and corporations for 4,041 statutory towns and 3,894 census towns in India. The municipalities are preparing the Environmental Management Plan for the cities for better environmental management within and around the city area while all the municipalities, corporation, cantonments and every urban authority is bind to follow the strict environmental rules for maintaining the ecological stability in the vicinity of towns. Chandrapur City had been considered as the third most polluted city in India by the CPCB in 2009, having population of 320379 (2011). At present it is the sixth most polluted city and industrial cluster in a country. There was a moratorium on industrial establishment in the city since 2009–2016 because of pollution level. The pollution can be classified into air pollution, soil pollution, water pollution and other categories too. Solid waste management is a solution to curbing the pollution level. Many of the cities are not following the norms given by ministry. This paper is an effort to study the practice and plan of Chandrapur Municipal Corporation.

Keywords Solid waste · Violation of rules · Environment · Solutions · Chandrapur

5.1 Introduction

Solid waste or semi-solid domestic waste, sanitary waste, commercial waste, institutional waste, catering and market waste and other non residential wastes, street sweepings, silt removed or collected from the surface drains, horticulture waste, agriculture and dairy waste, treated bio-medical waste excluding industrial waste, bio-medical waste and e-waste, battery waste, radio-active waste were generated in

---

Y. Y. Dudhapachare (✉)
Department of Geography, Janata Mahavidyalaya, Chandrapur, Maharashtra, India
e-mail: yogiraj7@gmail.com

R. B. Singh et al. (eds.), *Urban Health Risk and Resilience in Asian Cities*,
Advances in Geographical and Environmental Sciences,
https://doi.org/10.1007/978-981-15-1205-6_5

the area under the local authorities and other entities.[1] The municipal councils and corporations arc bound with certain rules to manage in scientific method the solid waste produced in the urban areas. The nature of solid waste varies from city to city, according to its geography. The nature of waste also varies in western world and India as the Europeans mostly use packed and junk food while in India household cooking is common in practice; consequently the packages and boxes are high in proportion in western world as compared to India in the solid waste.[2] The solid waste dumping yards are normally established outside of the city to insure that the bad odor will not come to the city and it will not influence the health of the citizens. In many cities of India there is a conflict between the city and surrounding villages' authorities.[3] There is a disagreement between Pune Municipal Corporation and surrounding village, i.e., Uruli Devachi and Fursungi on the dumping yard; the case moved to the National Green Tribunal and lastly the NGT has given relief to the villages.[4] This victory came after 3 decades of battle by the villagers. Similar conflict has also appeared in the case of NOIDA and its surrounding villages, i.e., Gadi Chaukhandi and Mangrauli are ready to fight with the NOIDA authority.[5] These fights between villagers and city dwellers are common today. The main cause behind this conflict is that the dump yards are not managed properly and consequently the health in the villages became a jeopardy. Improper usage, uneven distribution and absences of a proper implantation plan are some of the major problems leading many resources to become waste in large amounts. Garbage in public spaces is one of the biggest problems to aid an unhealthy society that introduce pathogens (bacteria and virus) into environment. Humans may be exposed to these pathogens that cause many diseases like diarrhea, cholera, skin diseases, respiratory allergies, malaria, tuberculosis, jaundice and cancer, etc.[6] So far, it is important to study the practices of municipal solid waste dumps. The present paper is an effort to bring in front the solid waste management system in Chandrapur city especially in the light of MSW rule 2000 and 2016.

5.2 Objectives of the Study

The main objective of this paper is to evaluate and explain the solid waste management system of Chandrapur Municipal Corporation. Other aims are as follows:

---

[1] Municipal Solid Waste Rules, 2016, p. 54.

[2] Satyamev Jayate, TV serial directed by Amir Khan, Season two Episode 3 available at https://www.youtube.com/watch?v=lS0_FCBzI_w.

[3] http://www.hindustantimes.com/noida/villagers-prevent-noida-authority-from-dumping-solid-waste-in-their-locality/story-lvzWy1iHkGiu4xn7X2GyUO.html.

[4] http://www.thehindu.com/news/national/other-states/ngt-prohibits-waste-dumping-in-pune-villages/article7278541.ece.

[5] Ibid.

[6] Maheshwari R., Gupta S, and Das K., Impact of Landfill Waste on Health: An Overview, IOSR Journal of Environmental Science, Toxicology and Food Technology (IOSR-JESTFT), ISSN: 2319-2402, ISSN: 2319-2399, Volume 1 Issue. 4, p. 17.

1. To critically analyse the present system of dumping the solid waste in dump yard.
2. To search the violations of guidelines given by MOEF through the solid waste rules 2000 and 2016.
3. To study the impact of present dump yard on environmental aspects and people.

5.3 Hypothesis

The speculations regarding the present dumping system are as follows:

1. The present dumping system of the solid waste in Chandrapur does not meet with the standard guidelines put by the Ministry of Environment, Forest and Climate Change.
2. There is an environmental impact on the various ecological aspects due to present dumping system.
3. The present solid dumping system may be creating conflict between residents and municipal authority.

5.4 Methodology

This study is based on personal observations and field visits to the dumping site of solid waste in Chandrapur along with the interviews and meeting with the residents in surrounding of the dump yard. The various official documents from Municipal Corporation and Maharashtra Pollution Control Board were extracted to study the present dumping system. This author is also continuously corresponding with the municipal authority regarding the anti-environmental dumping system; the references of these communication are also a base of this paper.

5.5 Review of Literature

Solid waste is an exciting topic among not only researchers but even environmentalists and environmental activists, city planners, judiciary, social workers, and multinational companies too. Many of the researchers have studied on different angles of solid waste like design of the solid waste dump yard, effects of solid waste, composting and fertilizers from solid waste, thermal power generation from waste, etc. It was a first call for the establishment of a municipal authority with waste removal powers were mooted as early as 1751 by Corbyn Morris in London.[7] Mr. Morris says that as the preservation of the health of the people is of great importance, it is proposed

[7]https://en.wikipedia.org/wiki/History_of_waste_management.

that the cleaning of this city, should be put under one uniform public management, and all the filth be... conveyed by the Thames to proper distance in the country.[8] But before this, in 1388 English Parliament had bars on waste dispersal in public waterways and ditches.[9] In 1885 the first garbage incinerator was built in USA (on Governor's Island in NY)[10] and there were about 300 incinerators in USA till 1914, and first federal solid waste management laws were enacted in 1965. The supreme court of India ordered in its judgment that, "It is commonly accepted that door to door collection of wastes and its segregation at the source is one of the viable factors of proper management of municipal solid waste. In this regard, public awareness programs should be encouraged."[11] It further ordered that, "We specifically direct that there shall be complete prohibition on open burning of waste on lands, including at landfill sites. For each such incident or default, violators including the project proponent, concessionaire, ULB, any person or body responsible for such burning, shall be liable to pay environmental compensation."[12]

5.6 Study Area and Its Geographical Setup

The city of Chandrapur is located on 19° 57′ north latitude and 79° 18′ east longitudes in the eastern Maharashtra,and it is situated at 189.90 m above the mean sea level[13] (Fig. 5.1). The area of the city is about 70 km^2; north south length of the city is about 10.6 km, while east west length is 7.6 km. Municipal Council of Chandrapur was established on May 17, 1867. Chandrapur city is located on Delhi–Chennai railway route having Municipal Corporation with a population of 320379, according to census 2011.[14]

With an average number of occupants per house as 5 persons,[15] the city is divided into 66 wards and three zones. 120 tons of solid waste is being generated in Chandrapur city per day.[16] There are 400 waste collection centers in the city. Due to rapid industrialization, population of Chandrapur city of Maharashtra State of India is increasing day by day, hence increasing the solid waste generation too. Around

[8]http://www.ciwm.co.uk/web/files/about_ciwm/100_yrs_london_and_se_centre.pdf.

[9]http://www.ewp.rpi.edu/hartford/~ernesto/S2014/SHWPCE/Papers/SHW-Introduction/CrowellHistoryofWaste.pdf.

[10]Ibid.

[11]Judgment of NGT dated 22/12/2016, in OA of 199 2016, Almitra Patel Vs Union of India.

[12]Ibid.

[13]Topographical Map no. 56 M/5.

[14]http://www.census2011.co.in/census/city/355-chandrapur.html.

[15]Samidha Siddam, Isha Khadikar, and Anil Chitade, Route Optimizations for Solid Waste Management Using Geo-Informatics, IOSR Journal of Mechanical and Civil Engineering (IOSRJMCE) ISSN: 2278-1684 Volume 2, Issue 1 (July-August 2012), pp. 78–83.

[16]http://cpcb.nic.in/divisionsofheadoffice/ess/Action%20plan%20CEPI-Chandrapur.pdf.

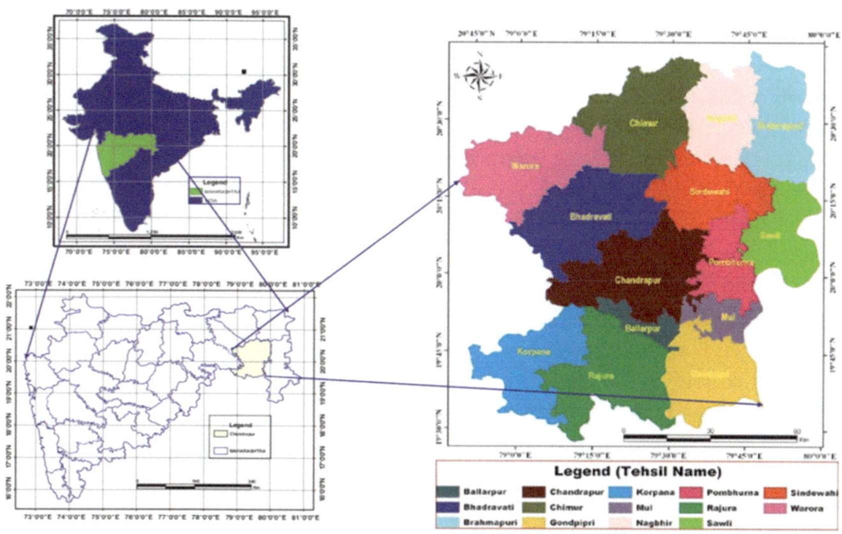

Fig. 5.1 Study area

0.25 kg/person/day solid waste is generated in the city of Chandrapur.[17] The city has its Bio-gas plant, but it is shut down due to technical problems.

5.7 Solid Waste in Chandrapur

There are a total of 400 MSW collection centers in the city. For transportation of MSW, the work is awarded to a private organization for a period of seven years. There are about 73000 families in the city of Chandrapur, and 84 self-help groups are collecting the solid waste. Each woman (per rickshaw vehicle) collects the waste from 200 houses. Chandrapur Municipal Corporation has estimated Rs 6.30 crores for collection of waste.[18] Total 22 vehicles are employed for collection and transportation of MSW. For transportation of MSW from three zones, Chandrapur municipal council is paying Rs 755 (US$ 11.61) per 1000 kg.[19] The solid waste activity is managed by a Chief Sanitary Inspector working under the Medical Officer of Health. A staff of 5 sanitary inspectors, 65 Mukadams and 545 sanitary workers support the Chief Sanitary Inspector apart from 678 sweepers and 19 drivers; all these workers are responsible for drain cleaning as well as solid waste management in the city of

[17] Ibid.

[18] Ibid.

[19] R K Kamble1, Namrata R Nimgade and Priyanka V Patil, Stakeholders behavior toward clean India mission's new municipal solid waste collection system in Chandrapur city, central India, international journal of environment, Volume-5, Issue-4, Sep-Nov 2016 ISSN 2091-2854, p. 34.

Chandrapur.[20] The organizational structure of solid waste management in Chandrapur is as under. In Chandrapur city mostly mixed type of waste is generated. The method followed for the collection of MSW is door to door collection pattern carried by mostly the women workers related with the various self-help groups. In an average, one woman goes to 200 houses for the MSW collection and collects the waste. Lastly the whole waste is collected and deposited at the disposal site, prepared by the Municipal Corporation of Chandrapur at Survey No. 369/11, 503/1, Mauza-Chanda Rayyatwari in 31 ha of land. The waste disposal in the dumping site is done in an unscientific manner in the depot. The waste is deposited on the ground without following any specific order.

The waste is also not leveled and remains in the form of heaps, scattered over the site. It is also being burnt many times and the uncompleted combustion of these emits obnoxious gases.[21] So far, there are many loopholes in the methods of handling the solid waste in dump yard. It is seen that many of the environmental norms have been violated in this dump yard and consequently the residents nearby the site are angrier with the solid waste management. At present the Medical Officer is managing the whole system of solid waste disposal and under his control Chief Sanitary Inspector, sanitary inspector and Mukdams are working but on the ground Sanitary Workers collect and dump the waste on the dumping yard (Fig. 5.2).

5.8 Violation of Environmental Norms Found in Present Dumping System in Chandrapur MSW Yard

The standard environmental norms established by the Central Pollution Control Board are enforced by Municipal Solid Waste Rule 2016 and are mandatory to follow by all Municipal Councils and Corporations within India. But the corporations are not complying properly the norms throughout the country. As Chandrapur is a sixth most polluted city in the country having CEPI 81.90 and its land CEPT is highest in the country having 75.50 score,[22] its municipality must concentrate on the land pollution issues and solid waste management, but this is not happening in Chandrapur case. This author had a correspondence with the municipal corporation regarding flouting the rules in solid waste yard, Chandrapur since last 3 years,[23] but

[20]Environment Status Report of City of Chandrapur, MC Chandrapur 2014, p. 6.2.

[21]Ibid, p. 6.5.

[22]Table 1: EPIs of various Industrial areas/clusters for Air Environment [Alphabetical order] (Interim assessment-2013) CPCB, New Delhi.

[23]Author's letters to MC Chandrapur dated December 22, 2014, January 17 2015, March 20, 2015, April 26, 2016 regarding MSW issue.

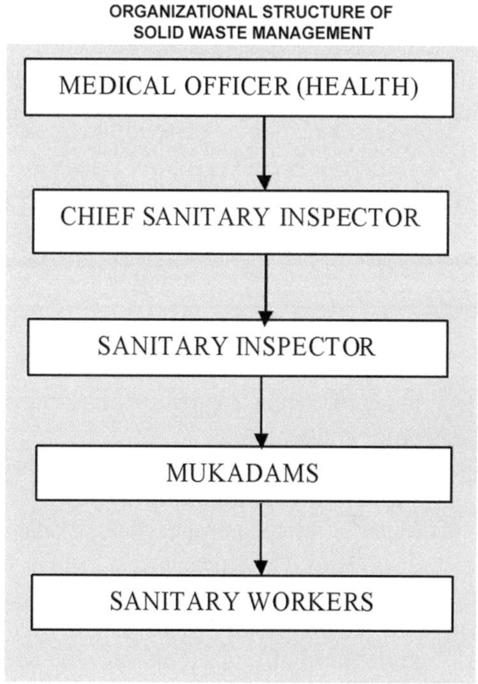

Fig. 5.2 Organizational Structure of solid waste management

the corporation replied that they are fulfilling the norms.[24] The Municipal Corporation of Chandrapur is breaching the following norms given by MSW rule, 2000 and 2016.

5.9 Findings

i. The compound wall of the solid waste dump yard is broken at many places and anybody can enter into the dump yard without prior permission.[25]
ii. There must be a security guard on the gate, who must enquire every person before allowing them, register the numbers of the vehicles which enter the yard, but this is not the case. There is no security guard and anybody can enter in the dump yard.
iii. Cattles from the city enters into the dump yard and eats the thrown material like food from city, catering wastage, etc. (Fig. 5.3). This creates danger to the cattles. Author has seen about 50 cattles on the dump yard at frequent observations.

---

[24]Letter of MC Chandrapur to author, no 555/2016 dated May 27, 2016, 503/2016 dated May 25, 2016.

[25]Authors personal observation and visit to the Solid Waste Dump Yard on many times.

Fig. 5.3 Cattle eating the material at solid waste dump yard

iv. Many of the cattle egrets and other birds are seen in the dump yard and are prone to diseases.
v. Many of the pigs are always digging the yard in the dump site and stray dogs from the city are following the pigs.
vi. Many of the scrape smuggler and other illegal workers are also seen on the dump yard. These people also include small boys of age 12–16 years which tends danger to their health.
vii. The solid waste is not bifurcated on the yard and whole waste is just depositing in the form of heaps. This situation does not help in the composting of the bio-degradable waste.
viii. The green vegetable, waste food and other recyclable and combustible waste is being dumped along with the non-degradable waste and hazardous waste. The plastic, glasses, electronic waste, medical waste, sanitary napkins, diapers, catering waste and many other things is being dump on the same site in the form of heaps.
ix. The waste is being burnt many times and smoke from the yard spreads toward all the sides of the residential areas around the dump yard. This burning creates bad odors and nuisance to the citizens.
x. The required buffer zone of the 500 m is not maintained and residential quarters as well as huts are around the core zone. Even the primary school is being run on just 60 m distance of the dump yard. The health of the small kids of the school is not well and they are victims of various diseases.
xi. MC is not covering the waste on daily basis by soil and the waste remains always open to the atmosphere and spreads bad odors and viral contamination.
xii. There are no built drainages so that the rain water will not meet the ground water and make contamination to the drinking water; the rainwater from the solid waste yard meets the other runoff.
xiii. The Leachate of the dump yard is not being treated properly and it meets the runoff; this is extremely dangerous to the health of people residing in the surrounding.

xiv. MC Chandrapur has not provided protection equipments like fluorescent jacket, hand gloves, raincoats, masks, etc. to handle the solid waste for the workers which makes their health endangered.

xv. There must be a doctor according to the rule on the dumping site, who keeps watch on the health of the workers but this is not the case in Chandrapur dumping site.

5.10 Conclusions

The Municipal Corporation of Chandrapur City (MCC) is not following the norms put by the Ministry of Environment, Forest and Climate Change in its regulations, i.e., MSW rule 2000 and MSW Rule 2016. The corporation is not complying to the order by NGT too given in the case; many times the solid waste is burnt in the waste yard. The present dumping yard is making the life of the people residing around dump vulnerable. Bad odor, burning of waste, stray dogs, pigs along with the percolation of the water from dumping site to the ground water makes their health a jeopardy. The Article 21 of the Constitution of India embraces the right to clean and well-maintained city, streets, highways and environs, but the constitutional provisions are too not being fulfilled.

5.11 Suggestions

It is indispensable to follow the environmental norms by the Municipal Corporation Chandrapur. MSW rule 2000 and 2016 which are the important guidelines to the local authorities in India and must be followed. It is suggested that citizens should be made aware through various campaigns so that they will segregate the waste at source and handover the waste to collectors. It is suggested further that the compound wall of the dump yard should be repaired and restrict the entry of animals as well as unauthorized person to the waste yard. The MPCB should continuously take the samples of air, water, soil for monitoring and check if there is any impact of the dump yard on these aspects. The Municipal Corporation should adopt the new techniques of decomposition of the solid waste to reduce the heaps of waste; it is further suggested to establish the research and development regarding the solid waste or make MOUs with the expert organizations to reduce the pollution.

Chapter 6
Explaining the links Between Purdah Practice, Women's Autonomy and Health Knowledge in India

Pintu Paul

Abstract The practice of purdah (veil) is an instrument of secluding women from their home as well as public spaces which may lead to restriction on women's physical mobility, control over their decisions and lower access over economic resources. This study has used recent round of the India Human Development Survey (IHDS) data, conducted during 2011–12, to depict the regional pattern of purdah practices across Indian states. This study reveals that the practice of purdah is highly prevalent in the north and north-western parts of the country, while the observance of purdah practice is limited/or forbidden in south and north-eastern (except Assam) region. This study examines the influence of purdah practice on women autonomy. Women autonomy comprises three indices such as women's physical mobility index, household decision-making index and access to control over economic resources index. Furthermore, this study assesses the links between purdah practice, women autonomy and women's health knowledge. Women's health knowledge index is composed of five sets of binary items: own health, male sterilization, child health and awareness of HIV/AIDS. Results from the binary logistic regression reveal that women's purdah practice is determined by their place of residence, caste, religion, age, educational level and current working status. Women's purdah practice is negatively associated with all three indices of women's autonomy. Furthermore, ordinal logistic regression discloses that purdah practice has a negative impact on women's health knowledge. Women's physical mobility and control over financial resources have a positive influence on women's health knowledge. However, women's household decision-making has no significant association with health knowledge of women. The findings of this study suggest that changing social norms, increasing education and employment opportunities could improve the health knowledge of women. Moreover, enhancing women's status in the society could make positive influence on women's health status.

P. Paul (✉)
Centre for the Study of Regional Development, School of Social Sciences, Jawaharlal Nehru University, New Delhi 110067, India
e-mail: pintupaul383@gmail.com

R. B. Singh et al. (eds.), *Urban Health Risk and Resilience in Asian Cities*, Advances in Geographical and Environmental Sciences,
https://doi.org/10.1007/978-981-15-1205-6_6

Keywords Purdah · Physical mobility · Household decision-making · Access to economic resources · Health knowledge

6.1 Introduction

Purdah is a Persian word meaning curtain, and it is the most common system of secluding women which limits women's mobility outside her home in South Asia, particularly in India and Pakistan (Jeffery 1979, Papanek 1971, 1973). The essential aspect of purdah system is that it restricts interaction between men and women (Papanek 1973). Purdah is characterized by the system of social subordination of women, rigidly structured division of labour, exclusion of women from visible social role. Moreover, it is historically and culturally specific forms of social control over women (Feldman and McCarthy 1983). However, the phenomenon of purdah practices among Muslims is different from Hindus. Muslim women practices complete veiling system, which functions after puberty in relation to men except very close kin, whereas Hindu women cover only head and some portion of face begins with marriage and in relation to husband's older male kin (Jeffery 1979, Papanek 1973).

Women's physical mobility, decision-making ability in the household and access to economic resources are three most important dimensions of women's autonomy. Previous studies have documented that women autonomy is influenced by a number of socio-economic and demographic characteristics of women (Dyson and Moore 1983, Jejeebhoy and Sathar 2001, Bloom et al. 2001). Bloom et al. (2001) has found that high economic status, education and women's currently working status have a positive influence on women autonomy. Jejeebhoy and Sathar (2001) in their study of women's autonomy in India and Pakistan asserted that region play a strong role in determining autonomy of women, whereas the influence of religion has found modest and inconsistent with women's autonomy. It is noteworthy that women's physical mobility is restricted by the practices of purdah, particularly in rural areas (Hussain and Smith 1999).

This chapter attempts to investigate the influence of purdah practice on women's physical mobility, household decision-making and access to economic resources. Furthermore, this study examines the impact of purdah practice and women's autonomy (physical mobility, household decision-making and access to control over economic resources) on women's health knowledge. Women's health knowledge have been derived from women's correct health beliefs about their own health, child health, male sterilization and HIV/AIDS awareness.

6.2 Objectives

(i) To depict the regional pattern of purdah practices across Indian states.

(ii) To examine the influence of purdah practices on women's physical mobility, household decision-making and access to control over economic resources. This study, further, assesses the impact of purdah practices and women autonomy on women's health knowledge.

6.3 Methods

6.3.1 Data Source

This study used data from the recent round of the India Human Development Survey (IHDS-II), conducted in 2011–12 by the University of Maryland and National Council of Applied Economic Research (NCAER), New Delhi. The IHDS-II is a nationally representative sample survey of 42,152 households spread across 33 states and union territories, 384 districts, 1420 villages and 1042 urban blocks located in 276 towns and cities. The information on purdah practices, different dimensions of women's autonomy and women's health knowledge have been collected from the sample of 39,523 ever-married women aged 15–49 in the households.

6.3.2 Variables

This empirical study explored the linkages between purdah practice, women autonomy and women's health knowledge using information from women's questionnaire. The detailed descriptions of independent and dependent variables are provided in the following:

6.3.3 Independent Variables

Women's purdah practice is considered as the key independent variable to examine the influence of purdah practice on women's autonomy and health knowledge. In IHDS-2, ever-married women were asked about their seclusion practices such as: do you practice ghungat/burkha/purdah/pallu? If yes, do you practice in front of relatives only or other people too? Therefore, practice of purdah has been used as a dichotomous variable in this study.

Socio-economic and demographic characteristics of women are included as control variables. These variables are: place of residence (rural and urban), caste (Brahmin, forward caste [except Brahmin], Other Backward Classes [OBC], Scheduled Caste [SC], Scheduled Tribe ST] and others), religion (Hindu, Muslim, Christian, Sikh, Buddhist, Jain and Tribal), age of women (15–19, 20–24, 25–29, 30–34 and 35 + years), women's education (no education, primary, upper primary and secondary, higher secondary and higher) and women's current working status (not working and working).

6.3.4 Dependent Variables

Women's autonomy and health knowledge have been considered as dependent variables in this study. Different dimensions of women autonomy have been calculated in the form of index. Women autonomy has comprised of three dimensions: (1) household decision-making, (2) women's physical mobility and (3) access to control over resources.[1]

The index of Household decision-making is constructed from 8 binary items: what to cook on a daily basis; whether to buy an expensive item such as TV or fridge; how many children you have; what to do if you fall sick; whether to buy land or property, how much money to spend on a social function such as marriage; what to do if a child falls sick; and to whom your children should marry. The respondents are scored on a scale from 0 to 8, where the score of '0' indicates no autonomy on household decision-making and the score of '8' implies full autonomy on household decision-making. Women's physical mobility index is composed of three items: whether they can go alone to the local health centre, friends/relatives homes, kirana shop and short distance by train or bus. The number of places where they can go alone without needing permission have been summed up to construct women's physical mobility index. The value of mobility index ranges from 0 to 4. Women with mobility index value of '0' are least mobile and mobility score of '4' indicates most mobile women.

Access to control over resources is measured from three factors: do they have any cash in hand to spend on household expenditure, name on any bank account and name on the ownership or rental papers of home. An index of access to control over resources has been constructed on the basis of these three items and the score of index ranges from 0 to 3 in which the value of '0' indicates least access over finances and the value of '3' indicates highest access to control over economic resources.

The most important outcome variable of this study is health knowledge of women. A health knowledge index (HKI) has been assessed from the questions on women's health beliefs about their own health, male's sterilization, child health and awareness of HIV/AIDS. The questions on women's health beliefs are as follows: Is it harmful

[1]The three indices of women's autonomy have also been used as explanatory variables, and categorized into low, medium and high to examine the influence of women autonomy on health knowledge.

to drink 1–2 glasses of milk every day during pregnancy? Do men become physically weak even months after sterilization? Is smoke from a wood/dung burning traditional chulha good/harmful for health or it really doesn't matter? Do you think that the first yellow milk that comes out after a baby is born is good for the baby, harmful for the baby or it really doesn't matter? When children have diarrhoea, do you think that they should be given less to drink than usual, more to drink than usual, about the same or it doesn't matter? Have you ever heard of an illness called HIV/AIDS? The correct response from the women's health beliefs was coded as '1' and the wrong beliefs was assigned as '0'. Then this binary information was added up to generate HKI and categorized into low or no knowledge, medium knowledge and high knowledge.

6.3.5 *Statistical Analyses*

The statistical analyses of this study has carried out in several stages. First, the proportion of purdah practices by socio-economic and demographic characteristics of women have been estimated. Secondly, the influence of women's socio-economic and demographic characteristics such as place of residence, caste, religion, age of the women, education and current working status on practices of purdah have been assessed using binary logistic regression Model. In the third stage, the impact of purdah observance on different dimensions of women's autonomy have been estimated by ordered logistic regression models. Lastly, this study assesses the links between purdah practice, women autonomy and health knowledge of women using ordered logistic regression models. The results of binary and ordered logistic regression are presented by odds ratios (ORs) with 95% confidence intervals (CIs). Sample weight has been used to estimate the results. All statistical analyses are carried out using STATA version 12.1 (StataCorp LP, College Station, TX, USA).

6.3.6 *Binary Logistic Regression Model*

Binary logistic regression is generally used when dependent variable is dichotomous. In this study, socio-demographic determinants of purdah practices have been examined by binary logistic regression. The mathematical form of binary logistic regression can be written as:

$$\text{logit } P = b_0 + b_1X_1 + b_2X_2 + \cdots + b_kX_k$$
$$\log = b_o + b_1X_1 + b_2X_2 + \cdots b_kX_k$$

or

$$\log \Omega = b_0 + b_1X_1 + b_2X_2 + \cdots + b_kX_k$$

where b_0 is intercept and $b_1, b_2 \ldots b_k$ are the coefficient of each explanatory variable and log Ω (uppercase omega) is called the log odds.

6.3.7 *Ordinal Logistic Regression Model*

The ordinal logistic regression model is commonly used when the dependent variable (outcome variable) is in ordered or likert scale and independent variables can be quantitative, categorical or the mixture of both. Likewise, in this study, the outcome variables such as indices of women autonomy (women's physical mobility [0–4], household decision-making [0–8] and access to economic resources [0–3]) incorporates ordered numerical value and women's health knowledge encompasses low or no knowledge, medium knowledge and high knowledge.

6.4 Results

6.4.1 *Purdah Practices by Socio-Economic and Demographic Characteristics of Women*

Table 6.1 shows that 66.6% of rural women are observed purdah, whereas 48.4% of women practices purdah in urban areas. A significantly higher proportion of Brahmin women are observed purdah compared with non-Brahmin castes. Women belonged to ST (50.5%) and other caste group (42.3%) are the lowest proportion in purdah observance.

The practice of purdah is significantly higher among Muslim women compared with women belonged to Hindu religion (88% vs. 59%). Moreover, purdah practices among the women affiliated to Christian (5.6%) and Sikh (19%) religion are limited. The observance of purdah among tribal women (38.5%) are also very less. Age cohort wise distribution has found that the practices of purdah decreases from younger age group to older age group. For instance, women in 15–19 years age group practices purdah in much greater degree compared with the women aged 35+ years (72% vs. 56%).

It is found that the practices of purdah lowering down with increasing level of education. Compared to women with no education, women with secondary or higher educational level significantly have lower prevalence of purdah practice. Women's current working status also have a strong negative influence on purdah practices. The observance of purdah is substantially lower among current working women compared with those who are not currently working (49% vs. 64%).

Table 6.1 Purdah practices by socio-economic and demographic characteristics of women, India, 2011–12

Characteristics	Purdah practices (%)	
	No	Yes
Place of residence		
Rural	33.39	66.61
Urban	51.58	48.42
Caste		
Brahmin	32.26	67.74
Forward/General (except Brahmin)	37.44	62.56
OBC	38.10	61.90
SC	39.47	60.53
ST	49.48	50.52
Others	57.67	42.33
Religion		
Hindu	41.12	58.88
Muslim	12.07	87.93
Christian	94.42	5.58
Sikh	81.10	18.90
Buddhist	43.00	57.00
Jain	56.52	43.48
Tribal	61.51	38.49
Others	63.04	36.96
Women's age in years		
15–19	28.19	71.81
20–24	30.11	69.89
25–29	35.28	64.72
30–34	38.93	61.07
35+	43.58	56.42
Women's education		
No education	30.86	69.14
Primary	36.40	63.60
Upper primary and secondary	43.52	56.48
Higher secondary	52.05	47.95
Higher education	60.08	39.92
Currently working for pay/wages		
No	36.31	63.69
Yes	51.32	48.68

Note Estimated from IHDS-II (2011–12) at all India level

6.5 Regional Patterns of Purdah Practice

Purdah is a crucial instrument of female seclusion, sex segregation and subordination of women. The practice of purdah is predominantly found in the most gender discriminated and less empowered areas of north and less practiced in the more gender-egalitarian southern India (Agarwal 1994; Desai and Andrist 2010).[2] Purdah is not only a rural phenomenon but also a considerable proportion of urban women practice purdah. Besides, wearing veil is almost absent among rural Muslim women and presumably greater among the urban Muslim women (Agarwal 1994, Faridi 1960). IHDS-II (2011–12) estimated that about 60% of the ever-married women are practicing purdah in India, in which 66.6% of purdah practices are found in rural India. IHDS-II also estimated that about 15% of women reported that they are practicing purdah in front of relatives only, about 10% women responded that they are practicing purdah in front of other people and about 75% of women are observed purdah in front of both relatives and others. The conceptual framework for examining the links between purdah practice, women's autonomy and health knowledge is depicted in Fig. 6.1.

Therefore, it is found that large proportions of women are secluded from both private and public domain which may affect their participation in economic activities, education and health outcomes. Figure 6.2 depicts regional patterns of purdah practice among ever-married women in India during 2011–12. It is found that the observance of purdah is greater in the north and north-western parts except *Punjab*[3], while southern and north-eastern states considerably have lower observance of purdah. Mizoram is the only state where purdah is not practiced at all (Fig. 6.2).

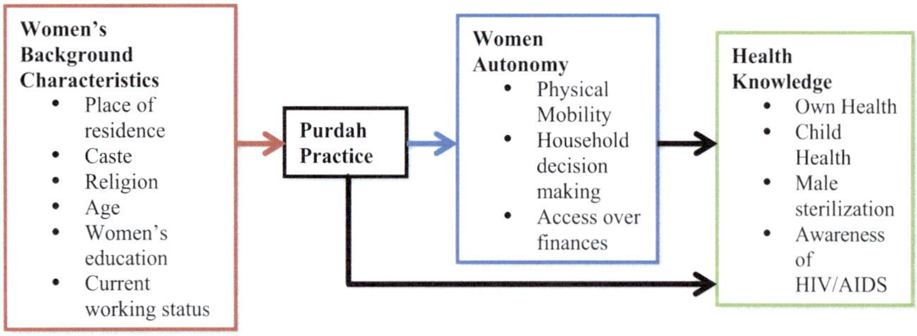

Fig. 6.1 Conceptual framework showing linkages between purdah practice, women autonomy and health knowledge of women

[2]See also Dyson and Moore (1983), Karve (1953), Kishor (1993), Raju and Bagchi (1993), Uberoi (1994), Raju (2011).

[3]Punjab is characterized by large proportion of Sikh population. Practicing Purdah among the Sikh women is much lower than the Hindu and Muslims. See also Table 6.1.

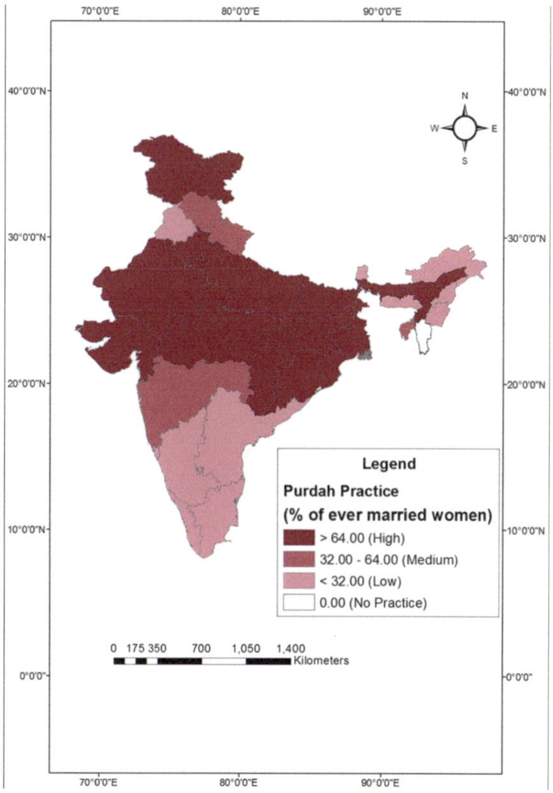

Fig. 6.2 Purdah practices among ever-married women aged 15–49 in India, 2011–12

6.6 Socio-Economic and Demographic Determinants of Purdah Practices

Table 6.2 shows the results of binary logistic regression for socio-economic and demographic determinants of purdah practices. In this model, place of residence, caste, religion, women's age, women's education and current working status are significantly associated with purdah observance. Urban women are less likely to practice purdah (OR = 0.66, 95% CI = 0.61, 0.71) compared with women from rural areas. Women belonged to forward/General caste (other than Brahmin), OBC, SC and ST are less likely to observe purdah than the Brahmin women. The practice of purdah is more common among Muslims compared with Hindu women (OR = 5.26, 95% CI = 4.46, 6.20). Moreover, the prevalnce of purdah practices is limited among tribal women. Older women are less likely to observe purdah (35 + age: OR = 0.53, 95% CI = 0.39, 0.73) than the younger age women. Women's education is significantly associated with lower likelihood of purdah observance. It has found that the prevalence of purdah is diminishing with increasing level of education (primary: OR = 0.77, 95% CI = 0.70, 0.85; higher education: OR = 0.38, 95% CI = 0.32, 0.45).

Table 6.2 Determinants of purdah practices (odds ratios from binary logistic regression model), India, 2011–12 (n = 39333)

Determinants	Odds ratio	95% CI
Place of residence		
Rural®	1.00	
Urban	0.66**	0.61, 0.71
Caste category		
Brahmin®	1.00	
Forward/General (except Brahmin)	0.59**	0.48, 0.73
OBC	0.60**	0.49, 0.73
SC	0.71**	0.57, 0.87
ST	0.54**	0.44, 0.68
Others	0.21**	0.14, 0.30
Religion		
Hindu®	1.00	
Muslim	5.26**	4.46, 6.20
Christian	0.06**	0.04, 0.10
Sikh	0.15**	0.11, 0.21
Buddhist	1.72**	1.18, 2.49
Jain	0.68	0.26, 1.7
Tribal	0.32**	0.20, 0.50
Women's age in years		
15–19®	1.00	
20–24	0.86	0.62, 1.19
25–29	0.73	0.53, 1.01
30–34	0.71*	0.51, 0.97
35+	0.53**	0.39, 0.73
Women's education		
No education®	1.00	
Primary	0.77**	0.70, 0.85
Upper primary and secondary	0.58**	0.53, 0.63
Higher secondary	0.49**	0.41, 0.58
Higher education	0.38**	0.32, 0.45
Currently working for pay/wages		
No®	1.00	
Yes	0.52**	0.48, 0.56
Constant	6.44	4.44, 9.35

Notes ®—reference category; *p <0.05; **p <0.01; *CI,* confidence interval

Currently working women are less likely to observe purdah compared with those who are not working (OR = 0.52, 95% CI = 0.48, 0.56).

6.7 Influence of Purdah Observance on Women's Autonomy

Table 6.3 presents the association between purdah prevalence and the three dimensions of women autonomy by fitting ordered logistic regression models. Crude analysis has found that purdah practice has a significant negative association with women's autonomy. The result reveals that the women who practices purdah have limited physical mobility (OR = 0.79, 95% CI = 0.76, 0.82), lower in household decision-maker (OR = 0.62, 95% CI = 0.59, 0.65) and have lower access to control over economic resources (OR = 0.62, 95% CI = 0.60, 0.65) than those women who are not observing purdah.

Adjusted models include women's social background such as place of residence, caste, and religion. The association between women's purdah practice and women's autonomy remained significant even after controlling for women's social background variables. Women from urban areas enjoy greater autonomy in all the three dimensions than the women residing in rural areas. Women from forward/General caste (other than Brahmin), OBC, SC and ST are less likely to have autonomy in physical mobility and access to economic resources and more likely to have autonomy in household decision-making compared with women belonged to Brahmin caste. Muslim woman significantly have less autonomy in all three indices than those of Hindu women. Women belonged to Sikh and Buddhist community are significantly more likely to have autonomy in physical mobility and less likely to have autonomy in decision-making and access to control over economic resources compared with Hindu women. Moreover, women from Christian, Jain and tribal community are more likely to have financial autonomy than the Hindu women.

6.8 Influence of Purdah Practice and Women Autonomy on Women's Health Knowledge

Table 6.4 presents the influence of purdah practice and indices of autonomy on women's health knowledge. In the first model, the influence of purdah practice on women's health knowledge has been examined. It is found that women's purdah practice has a negative impact on women's health knowledge (OR = 0.73, 95% CI = 0.70, 0.76). In the second model, association between the three indices of women autonomy and women's health knowledge have been examined and it is found that women's physical mobility and access to economic resources are positively

Table 6.3 The influence of purdah prevalence on dimensions of women autonomy (odds ratios from ordered logistic regression models), India, 2011–12

Dependent variable						
Determinants	Physical mobility (0–4)		Household decision-making (0–8)		Access to economic resources (0–3)	
	UOR (95% CI)	AOR (95% CI)	UOR (95% CI)	AOR (95% CI)	UOR (95% CI)	AOR (95% CI)
Purdah practice						
No®	1.00	1.00	1.00	1.00	1.00	1.00
Yes	0.79*** (0.76, 0.82)	0.84*** (0.81, 0.88)	0.62*** (0.59, 0.65)	0.62*** (0.59, 0.65)	0.62*** (0.60, 0.65)	0.67*** (0.63, 0.70)
Place of residence						
Rural®		1.00		1.00		1.00
Urban		1.36*** (1.30, 1.42)		1.18*** (1.12, 1.24)		1.42*** (1.35, 1.49)
Caste						
Brahmin®		1.00		1.00		1.00
Forward/General (other than Brahmin)		0.84*** (0.76, 0.93)		1.15*** (1.04, 1.28)		0.89** (0.80, 0.98)
OBC		0.60*** (0.55, 0.66)		1.24*** (1.12, 1.37)		0.76*** (0.69, 0.84)
SC		0.67*** (0.61, 0.74)		1.25*** (1.13, 1.40)		0.89** (0.80, 0.99)
ST		0.64*** (0.57, 0.71)		1.48*** (1.30, 1.68)		0.83*** (0.73, 0.94)
Others		0.49*** (0.41, 0.49)		1.23* (0.98, 1.53)		0.98 (0.78, 1.23)
Religion						
Hindu®		1.00		1.00		1.00
Muslim		0.82*** (0.77, 0.87)		0.87*** (0.81, 0.93)		0.82*** (0.76, 0.89)
Christian		0.96 (0.84, 1.09)		1.11 (0.94, 1.31)		1.27*** (1.09, 1.47)
Sikh		1.35*** (1.18, 1.54)		0.42*** (0.37, 0.47)		0.64** (0.56, 0.74)

(continued)

Table 6.3 (continued)

Dependent variable						
Determinants	Physical mobility (0–4)		Household decision-making (0–8)		Access to economic resources (0–3)	
	UOR (95% CI)	AOR (95% CI)	UOR (95% CI)	AOR (95% CI)	UOR (95% CI)	AOR (95% CI)
Buddhist		1.40*** (1.10, 1.79)		0.77** (0.60, 0.99)		0.69*** (0.52, 0.92)
Jain		0.97 (0.66, 1.43)		0.75 (0.49, 1.17)		1.57** (1.05, 2.36)
Tribal		0.79 (0.59, 1.06)		1.33 (0.86, 2.08)		1.71** (1.12, 2.60)

Notes ®—reference category; *$p < 0.10$; **$p < 0.05$; ***$p < 0.01$; *UOR*, unadjusted odds ratio; *AOR*, adjusted odds ratio; *CI*, confidence interval

associated with women's health knowledge. However, women's household decision-making has no significant association with their health knowledge. In the third or final model, place of residence, caste and religious affiliation of women have been included with purdah practice and indices of women autonomy. The association of women's purdah practice and physical mobility with health knowledge of women have found to be remained significant even after controlling for place of residence, caste and religion variable. However, women's access to economic resources has lost its significant relationship with their health knowledge. Women living in urban areas are more likely to have health knowledge (OR = 1.87, 95% CI = 1.76, 1.99) compared with its rural counterparts. Women belonged to OBC (OR = 0.55, 95% CI = 0.48, 0.62), SC (OR = 0.52, 95% CI = 0.45, 0.59), ST (OR = 0.41, 95% CI = 0.35, 0.49) and other caste category (OR = 0.61, 95% CI = 0.45, 0.83) are less likely to have health knowledge than the Brahmin women. Muslim women (OR = 0.70 95% CI = 0.63, 0.78) are significantly less likely to have health knowledge compared with Hindu women. However, Christian (OR = 1.72, 95% 1.41, 2.11), Sikh (OR = 1.25, 95% CI = 1.05, 1.50) and Buddhist (OR = 2.09 95% CI = 1.46, 3.01) are more likely have health knowledge compared with Hindu women.

6.9 Summary and Conclusion

This study has examined the association between women's purdah observance with different indices of women's autonomy and their health knowledge. The result exhibits that the practice of purdah is negatively associated with all the three indices of autonomy and women's health knowledge. Women's autonomy has a strong positive

Table 6.4 Influence of purdah practice and women autonomy on health knowledge of women (odds ratio from ordered logistic regression), India, 2011–12

Determinants	Health knowledge index (odds ratio)		
	Model 1	Model 2	Model 3
Purdah practice			
No®	1		1
Yes	0.73*** (0.70, 0.76)		0.90*** (0.85, 0.96)
Indices of women autonomy			
(1) Women's physical mobility			
Low®		1	1
Medium		0.95 (0.87, 1.04)	0.96 (0.88, 1.04)
High		1.36*** (1.26, 1.47)	1.22*** (1.23, 1.33)
(2) Household decision-making			
Low®		1	1
Medium		0.99 (0.88, 1.12)	0.92 (0.82, 1.04)
High		1.07 (0.97, 1.18)	1.04 (0.95, 1.15)
(3) Access to economic resources			
Low®		1	1
Medium		1.12*** (1.05, 1.19)	1.03 (0.97, 1.10)
High		1.05 (0.96, 1.15)	0.89** (0.82, 0.98)
Place of residence			
Rural®			1
Urban			1.87*** (1.76, 1.99)
Caste			
Brahmin®			1
Forward/General (except Brahmin)			1 (0.88, 1.15)
OBC			0.55*** (0.48, 0.62)
SC			0.52*** (0.45, 0.59)
ST			0.41*** (0.35, 0.49)

(continued)

Table 6.4 (continued)

Determinants	Health knowledge index (odds ratio)		
	Model 1	Model 2	Model 3
Others			0.61*** (0.45, 0.83)
Religion			
Hindu®			1
Muslim			0.70*** (0.63, 0.78)
Christian			1.72*** (1.41, 2.11)
Sikh			1.25** (1.05, 1.50)
Buddhist			2.09*** (1.46, 3.01)
Jain			0.8 (0.47, 1.36)
Tribal			1.08 (0.62, 1.89)

Notes ®—reference category;*$p < 0.10$; **$p < 0.05$; ***$p < 0.01$; 95% confidence interval in parentheses

relationship with their health knowledge. Although women's physical mobility and access to control over financial resources are positively associated with their health knowledge, household decision-making has no significant impact on health knowledge of women in adjusted model. Besides, this study also examined the influence of socio-economic and demographic characteristics of women on purdah practice. The findings indicate that women's educational attainment and current working status are negatively associated with purdah prevalence. Rural-urban residence, caste, religion and age also have significant influence on purdah practice. Likewise, socio-demographic characteristics of women significantly associated with their autonomy and health knowledge.

Although some of the scholars argued that women's purdah practice increases physical mobility and have a postive role in access to public goods and services, the present study suggests that women's purdah observance has a negative impact on their physical mobility and access to financial resources because most probably seclusion practices by itself is a characterization of segregation of female, subordination of women and division of labour.

A study conducted in India documented that women's health knowledge has a positive impact on reducing short term morbidities (fever, cough and diarrhoea) and improving health status of children (Patra et al. 2013). Moreover, the women who have adequate health knowledge are significantly more likely to report maternal health problems (pregnancy complications and postnatal complications), and receive antenatal care, delivery care and postnatal check-ups (Patra et al. 2016). Therefore, adequate health knowledge of women improves maternal and child health outcomes.

The present study suggests that improving female education and increasing employment opportunities of women could enhance the empowerment of women. Empowering women is the key to break the traditional norms of the society. This study also calls for community engagement to change the harmful norms for better women's position in the society. This study has found that women's increasing level of autonomy improves their health knowledge. Therefore, policymakers and stakeholders should focus on women empowerment by increasing opportunities for female education and that could eventually reduce maternal and child health vulnerabilities.

References

Agarwal B (1994) A field of one's own: gender and land rights in South Asia, vol 58. Cambridge University Press

Bloom SS, Wypij D, Gupta MD (2001) Dimensions of women's autonomy and the influence on maternal health care utilization in a north Indian city. Demography 38(1):67–78

Desai S, Andrist L (2010) Gender scripts and age at marriage in India. Demography 47(3):667–687

Dyson T, Moore M (1983) On kinship structure, female autonomy, and demographic behavior in India. Popul Dev Rev 35–60

Faridi T (1960) The changing role of women in Pakistan. Department of Advertising, Films and Publications, Government of Pakistan

Feldman S, McCarthy FE (1983) Purdah and changing patterns of social control among rural women in Bangladesh. J Marriage Fam 949–959

Hussain TM, Smith JF (1999) Women's physical mobility in rural Bangladesh: the role of socio-economic and community factors. Contemp South Asia 8(2):177–186

India Human Development Survey-II (IHDS-II), Desai, Sonalde; Vanneman, Reeve, 2011–12

Jeffery P (1979) Frogs in a well: Indian women in purdah. Manohar, New Delhi

Jejeebhoy SJ, Sathar ZA (2001) Women's autonomy in India and Pakistan: the influence of religion and region. Popul Dev Rev 27(4):687–712

Karve I (1953) Kinship organization in India. Asia Publishing House, Bombay

Kishor S (1993) May God give sons to all: gender and child mortality in India. American Sociological Review 58(2):247

Papanek H (1971) Purdah in Pakistan: seclusion and modern occupations for women. J Marriage Fam 517–530

Papanek H (1973) Purdah: separate worlds and symbolic shelter. Comp Stud Soc Hist 15(03):289–325

Patra S, Arokiasamy P, Goli S (2016) Relevance of health knowledge in reporting maternal health complications and use of maternal health care in India. Health Care Women Int 37(5):531–549

Patra S, Perianayagam A, Goli S (2013) Mother's health knowledge and practice and their linkage with childhood morbidity, medical care and medical care expenditure in India. In: 27th IUSSP international population conference, pp 26–31

Raju S (2011) Gendered geographies: space and place in the South Asia. Oxford University Press, New Delhi

Raju S, Bagchi D (1993) Women and work in Asia: regional patterns and perspectives. Routledge, London

Uberoi P (1994) Family, kinship, and marriage in India. Oxford University Press

Chapter 7
Development of Urban Heat Island and Its Relation to Heat Waves

Rajashree Borah

Abstract Urbanisation is a process common in the today's world where the shift from the agriculture-based economy to the tertiary sector-based economy has been in a rapid pace. The reason behind this is the development race. Rapid urbanisation is characterised growth of urban areas which change the environmental settings which results in some adversaries affecting human health. One such is the Urban Heat Island. This phenomenon results due to an increase in the number of concrete structures at the cost of depleting green cover. In this phenomenon, the temperature (both atmospheric and surface) of the urban areas remains higher than their surrounding rural areas. The effect of the Urban Heat Island is thermal discomfort felt by the urban people which even results in other medical conditions. The situation of Urban Heat Island is more intensified by another phenomenon, heat wave. It is a climatic condition of high to extreme temperature and occurs when temperature departs positively from the normal average maximum temperature. To be considered as a heat wave, the temperature has to depart from the daily maximum temperature by a definite threshold. The study area is Kota District of Rajasthan which is a heat wave prone region along with rapid urbanisation (one of four cities suggested by the Rajasthan Government for smart cities).

Keywords Urbanisation · Temperature · Urban Heat Island · Heat wave

7.1 Introduction

Urbanisation is a process common in today's world where the shift from the agriculture-based economy to the tertiary sector-based economy has been in a rapid pace. The reason behind this is the development race that every country and region running where one can score only by making development in the tertiary sector. This is resulting in rapid urbanisation, characterised by the growth of the urban areas. *Most probably 70% of the global population will be living in urban areas by 2050.*

R. Borah (✉)
Jawaharlal Nehru University, New Delhi, India
e-mail: rajashree23luna@gmail.com

R. B. Singh et al. (eds.), *Urban Health Risk and Resilience in Asian Cities*,
Advances in Geographical and Environmental Sciences,
https://doi.org/10.1007/978-981-15-1205-6_7

(*UN-HABITANT Report, 2010 and 2012*). Urban area growth is changing the environmental settings which results in some adversaries affecting human health. One of such is the Urban Heat Island. This phenomenon results due to an increase in the number of concrete structures at the cost of depleting green cover. Urban Heat Island is defined as a phenomenon in which the temperature (both atmospheric and surface) of the urban areas remains higher than their surrounding rural area. The concrete surfaces of the urban areas generally have lower solar reflectivity or albedo resulting in greater absorption capacity of solar radiation making them extremely hot and thereby raising the surrounding air temperature. The depletion of the natural vegetation due to the urbanisation process also results in the augmentation of the Urban Heat Island phenomenon because the surfaces with natural vegetation have cooler surfaces due to the evapotranspiration of plants. *The Urban Heat Island is a significant, man-made and inadvertent local climatic phenomenon* (Walawender et al. 2014). The effect of the UHI is the thermal discomfort felt by the urban people which even results in severe medical conditions and heat-related deaths. The Urban Heat Island effect is intensified by another phenomenon, heat wave. It is a climatic condition of high to extreme temperature and occurs when temperature departs positively from the normal average maximum temperature. To be considered as a heat wave, the temperature has to depart from the daily maximum temperature by a definite threshold. This threshold is different in different places because of the different climatic conditions of different places. A temperature departure which is appropriate to be considered as a heat wave for a place in particular climate regime will not be so for another place which is not under that particular climatic regime. Hence, heat wave is a relative term.

Surface Urban Heat Island can be identified by the difference in surface temperatures. Here, in this paper, the difference in surface temperatures is studied with the help of remote sensing. Land Surface Temperature method is utilised for this purpose. LST is a popular method for studying surface thermal conditions especially of the urban areas. Since the heat wave is a relative phenomenon, different regions have different criteria and thresholds for its identification. Now, as the study area is in India, the Indian Meteorological Department criteria for heat wave are used.

7.2 Objectives

The objectives of the study are

(i) To study the development of the Urban Heat Island over time.
(ii) To analyse the occurrence of heat waves along with the development of Urban Heat Island.

7.3 Study Area

The study area is Kota which is a district on the southern part of Rajasthan, a state on the western part of India. It is located at 25.18 °N and 75.83 °E. It is the third largest city of Rajasthan after Jaipur and Ajmer. The total area of the district is 5217 km^2 (25th largest district of Rajasthan) out of which 4906.95 km^2 is rural and 310.05 km^2 is urban. Its average elevation is 271 m. Kota is bounded by Sawai Madhopur, Tonk and Bundi districts on the north and north-west. Kota is situated along the banks of Chambal River on a high sloping tableland forming a part of the Malwa Plateau. The Mokandarra hills run from south-east to north-east axis of the town. The climate of Kota is semi-arid with high temperatures all throughout the year. Summers are long and dry extending from March to June. Monsoon season has lower temperatures with high humidity and torrential rains. Temperature again rises in October after the end of monsoon. Winters are mild and extend from late November to February. It receives an average rainfall of 660.6 mm.

Kota has fertile land with enough irrigation facilities through the canals. It is also an industrial hub in northern India with many industries such as fertilisers, chemical, engineering, power plants, etc. The city even acts as the trade centre for the area where millets, rice, pulses, coriander and oilseeds are grown. It is the 47th most populous district of India and has a density of population of 370 km/km^2. 60% of the population lives in urban areas. The city is famous for its coaching institutes for medicine and engineering and for this reason the city is also called the "Education City of India".

7.4 Methodology

This paper tries to study the Surface Heat Island development with the help of remote sensing. The study of the urban thermal conditions with the help of remotely sensed data is popular nowadays because of its advantages over conventional methods. Firstly, remotely sensed data helps visualise the temperature differences between the urban and the surrounding areas. It also helps to compare and visualise the development of Urban Heat Islands over time. Secondly, it is helpful in gathering data even for the inaccessible areas and is time and cost effective. Thirdly, it is also helpful in finding the effect in the development of Urban Heat Island due to the change in the factors which affect its development and visualise their relationship. For instance, we can see the thermal conditions of an area over time and also see the changes in vegetation cover in that area over time. With the help of both the resulting maps, we can easily visualise what effect the change in vegetation cover has on thermal conditions change. In the study, the Land Surface Temperature method of remote sensing is employed over the LANDSAT images to analyse the development of UHI in Kota, Rajasthan. Landsat TM (Thematic Mapper) and ETM + (Enhanced

Thematic Mapper) have been utilised to calculate the Land Surface Temperature over Kota.

The formulas that are followed for the calculation LST from TM and ETM + images of Landsat 5 and Landsat 7 images are as follows:

Step 1: Conversion to Radiance

$$L\lambda = \text{Grescale} * \text{QCAL} + \text{Brescale} \tag{7.1}$$

which is also expressed as

$$\begin{aligned} L\lambda =& ((\text{LMAX}\lambda - \text{LMIN}\lambda)/(\text{QCALMAX} - \text{QCALMIN})) \\ & * (\text{QCAL} - \text{QCALMIN}) + \text{LMIN}\lambda \end{aligned} \tag{7.2}$$

where

Lλ	Spectral Radiance at the sensor's aperture in watts/(meter squared * ster * μm)
Grescale	Rescaled gain (the data product "gain" contained in the Level 1 product header or ancillary data record) in watts/(meter squared * ster * μm)/DN
Brescale	Rescaled bias (the data product "offset" contained in the Level 1 product header or ancillary data record) in watts/(meter squared * ster * μm)
QCAL	The quantized calibrated pixel value in DN
LMINλ	The spectral radiance that is scaled to QCALMIN in watts/(meter squared * ster * μm)
LMAXλ	The spectral radiance that is scaled to QCALMAX in watts/(meter squared * ster * μm)
QCALMIN	The minimum quantized calibrated pixel value (corresponding to LMINλ) in DN
QCALMAX	The maximum quantized calibrated pixel value (corresponding to LMAXλ) in DN

Step 2: Conversion of Radiance to Temperature (in Kelvin)

$$T = K2/\ln(K1/L\lambda + 1) \tag{7.3}$$

where

T	Effective at satellite temperature in Kelvin
K2	Calibration constant 2 (from Table 7.1)
K1	Calibration constant 1 (from Table 7.1)

Table 7.1 Calibration constants

	Constant 1 (K1)	Constant 2 (K2)
Landsat 7	666.09	1282.71
Landsat 5	607.76	1260.56

L Spectral radiance in watts/(meter squared * ster * μm)

Step 3: Conversion of Temperature in Kelvin to Temperature in °C

$$C = T - 273.15 \tag{7.4}$$

where

C Effective at satellite temperature in degree Celsius
T Effective at satellite temperature in Kelvin.

In this paper, the occurrence of heat waves in the study area, that is, Kota will also be analysed. The general condition to be recognised as a heat wave is that the temperature has to depart from the daily maximum temperature by a definite threshold. But different regions of the world are under different climatic regimes. Due to the difference in climatic conditions, the thresholds for heat waves are also different. Because of this difference in thresholds, heat wave is considered as a relative term meaning that a temperature condition which will be considered as a heat wave in a particular region may not be considered as a heat wave in another region. Hence, the criteria for the identification of heat waves are different in regions. In India, the criteria for the identification of heat waves have been developed by the Indian Meteorological Department. This criterion of heat wave by I.M.D. has been used for studying the occurrences of heat waves in Kota. The I.M.D. criterion for heat wave is as follows:

(a) When normal maximum temperature of a station is less than or equal to 40 °C.
    Heat wave departure from normal is 5–6 °C.
    Severe heat wave departure from normal is 7 °C or more.
(b) When normal maximum temperature of a station is more than 40 °C.
    Heat wave departure from the normal is 4–5 °C.
    Severe heat wave departure from normal is 6 °C or more.
(c) When the actual maximum temperature remains 45 °C or more irrespective of normal maximum temperature, heat wave should be declared.

7.5 Database

For the calculation of Land Surface Temperature, the TM and ETM+ images are used of Landsat 5 and 7. The images are collected for 2000 and 2010 and the images belonged to the October month from the USGS Earth Explorer. The thermal bands of the images are used for LST calculation.

For the study of heat waves, the Daily Maximum Temperature and its Departures are collected from Indian Daily Weather Report (I.D.W.R.) and State Daily Weather Report (S.D.W.R.) of Indian Meteorological Department for the months of April, May, June and July for the time period 2000–2010. These months are selected because heat waves generally occur during these months.

The Bhuvan Portal Land Use and Land Cover map and data and Google Earth Satellite Image are used as a reference while doing analysis.

7.6 Results and Discussion

7.6.1 Study of Development of Urban Heat Island on the Basis of Land Surface Temperature

The Land Surface Temperature has been calculated for Kota, Rajasthan for 2000 and 2010 to see the development of Urban Heat Island over time. From the 2000 calculations of LST, it has been found that the highest surface temperature recorded over Kota is 41.5 °C and the lowest surface temperature recorded is 21.7 °C. High temperatures can be seen in the south and south-western part of the Kota district. Almost all the south and south-western regions are consistently showing a relatively high surface temperature. In this region lies the Kota district headquarters and the three major towns Ladpura, Sangod and Ramganj Mandi (out of four major towns of Kota). The high temperatures are found around the Kota district headquarters and around major towns of Ladpura, Sangod and Ramganj Mandi (Fig. 7.1). From the Land Use and Land Cover map (Bhuvan Portal), it has been found that most part of this region is under forest, scrub forest, deciduous forest and forest plantations. And the second highest type of land use found in this region is agriculture and crop lands. Though from the LULC map it has been seen that the majority of the area of this region is under scrub forest and cropland, from the satellite image (Google Earth Image) it has been seen that most parts of this region are in cleared areas. Heavy urban built up can be seen in and around the Kota district headquarters and little in Ramganj Mandi. The heavy built up in the Kota district headquarter area is the reason for the high surface temperature on the entire south-western part of the Kota district. Within this region relatively comparatively lower temperature is found in the south-eastern parts. This part of the region is still under agriculture and cropland, and heavy built up have not started in the area even if the area has sparse vegetation. Low temperature can be seen all along the track of Ahu River and also in pockets within the cropland. In the middle part of the district the surface temperature was seen low during 2000. The entire central part is agricultural land with some portion in the upper part as barren, uncultivable, gullied and scrub lands. Also the rivers Chambal and Kali Sindh are flowing across the central part. Though in the northern

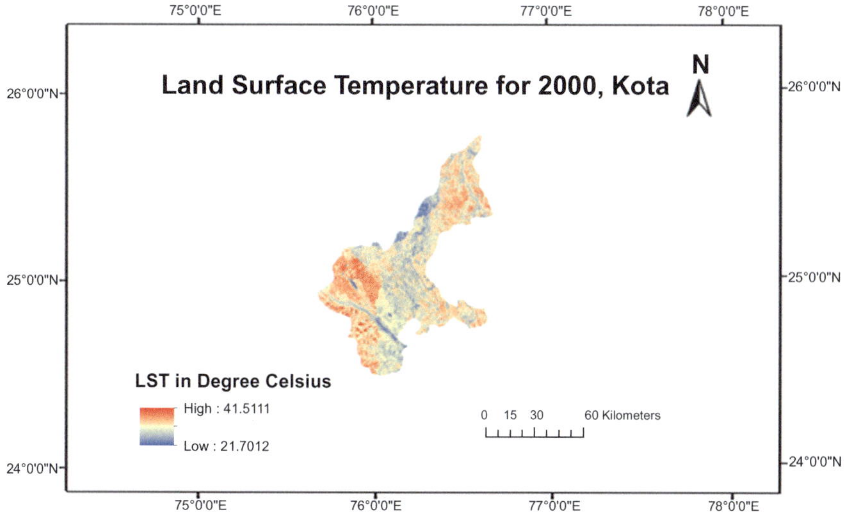

Fig. 7.1 Land Surface Temperature of Kota for 2000

part of the district there are pockets of low temperatures, but, most of the region is experiencing high to medium temperatures (36–27 °C). The surface temperature in the part is not so intense, even if high to medium temperatures are prevailing all over the entire part. The LULC map shows a majority of the area of this part as cropland with some portions as barren, uncultivable, gullied and scrub lands.

In the 2010 LST calculations of Kota district, it has been that the maximum LST is 41.1 °C and the minimum is 22.8 °C. Thus, both the maximum and the minimum surface temperatures of Kota have increased. The maximum temperature has increased by 0.4 °C and the minimum temperature has increased by 1 °C. The entire southern and the south-eastern region are having high temperatures with a few pockets of medium temperatures and negligible pockets of low temperature. The parts with high temperature are showing temperatures ranging 41–35 °C. The medium temperature parts are showing temperature ranging from 34 °C to 27 °C. The southern and south-eastern part of Kota covers the Kota district headquarter and the three major towns of Ladpura, Sangod and Ramganj Mandi (out of four major towns of Kota). Thus, it suggests that the surface temperature in this region is increasing because of the increase and expansion of the built ups in this regions. Even along the track of Ahu River which previously in 2000 was showing low surface temperature, there is an increase in temperature of up to 34 °C and almost no pocket of low temperature. The central part of the Kota is still having a predominance of low temperatures but they have not remained so intense. In fact, the entire western side of the central part of Kota has transformed to high-temperature region with temperatures ranging between 36 and 34 °C with many pockets of medium temperature (see Fig. 7.2). However, the northern part has transformed into a low-temperature region. The reason for this transformation may lie in the extension of agriculture to this

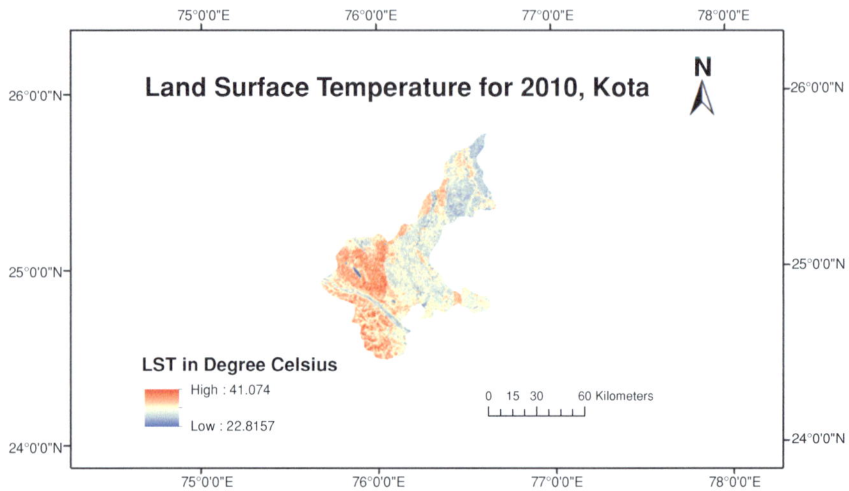

Fig. 7.2 Land Surface Temperature of Kota for 2010

region on an intensive scale. From the satellite image also it can be seen that this part has extensive agricultural fields, thus, lowering the temperature.

7.6.2 Frequency and Type of Heat Waves During the Study Period

For the analysis of the heat waves that occurred in Kota (station-Kota AP), the daily maximum temperature readings of 4 months, i.e. April, May, June and July, for 30 years from 1981 to 2010 are taken. The frequency of the heat waves will be analysed month-wise. However, an important consideration should be kept in mind during the whole analysis that even after data calculations went missing some dates still are missing from some months. Thus in the case of such months, the frequency of the heat waves does not represent the whole month.

7.6.2.1 April

In 2000, two heat waves occurred; one belonging to the category when the normal maximum temperature of a station is less than or equal to 40 °C and the departure from normal is 5–6 °C and the other belonging to the category that when the actual maximum temperature remains 45 °C or more—irrespective of normal maximum temperature, heat wave should be declared. Then in 2001 and 2002, two heat waves occurred in the category when the actual maximum temperature remains 45 °C or more—irrespective of normal maximum temperature, heat wave should be declared,

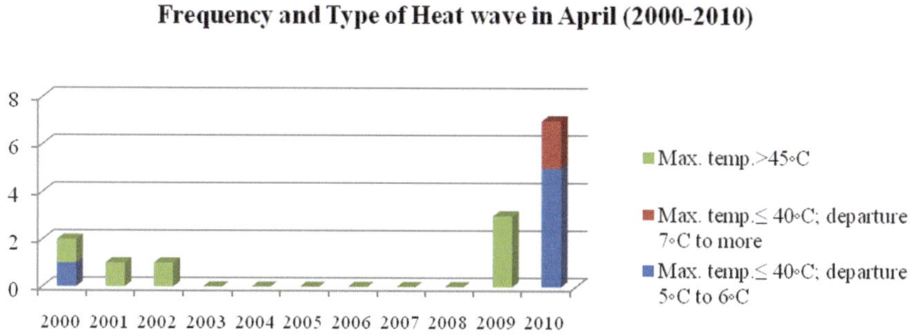

Fig. 7.3 Frequency and type of heat waves in April, Kota (2000–2010)

one in each year. After that from 2003 to 2008, no heat wave occurred in the April month. In 2009, three heat waves occurred and belonged to the category when the actual maximum temperature remains 45 °C or more irrespective of normal maximum temperature. In 2010 April, seven heat waves occurred. Five of them belonged to the category of heat wave when the normal maximum temperature of a station is less than or equal to 40 °C and the departure from normal is 5–6 °C. The other two were severe heat waves belonging to the category of when the normal maximum temperature of a station is less than or equal to 40 °C and the departure from normal is 7 °C or more (see Fig. 7.3).

7.6.2.2 May

During the years, 2000 and 2001, the number of heat waves was three and nine, respectively, belonging to the category of heat wave when the normal maximum temperature of a station is less than or equal to 40 °C and the departure from normal is 5–6 °C. During 2002, 2003, 2004 and 2005 the numbers of the heat waves were fourteen, five, three and three, respectively, belonging to the same category to which the previous waves belong. The number again increases to 11 in 2006. In the succeeding 2 years, i.e. in 2007 and 2008 there was no heat wave. In the last 2 years of the period, i.e. 2009 and 2010, the numbers of heat waves were 12 and 16, respectively (see Fig. 7.4).

7.6.2.3 June

In 2000–2001, there was no heat wave. After that in the years 2002, 2003, 2004 and 2005, the numbers of heat waves were four, seven, seven and six, respectively. A majority of them belonged to the category of when the actual maximum temperature remains 45 °C or more—irrespective of normal maximum temperature, heat wave should be declared. A few belonged to the category of when the normal maximum

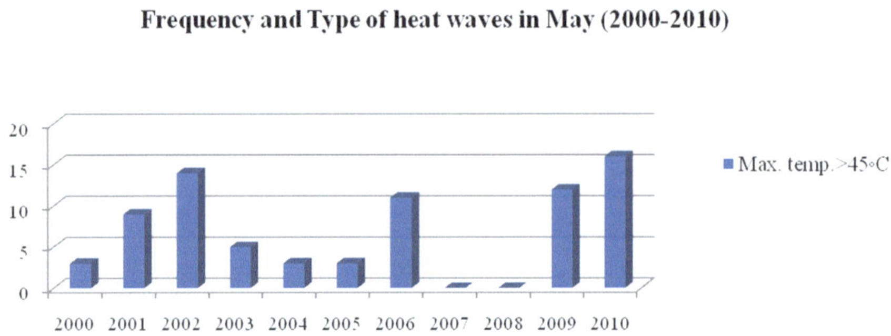

Fig. 7.4 Frequency and type of heat waves in May, Kota (2000–2010)

temperature of a station is more than 40 °C and the heat wave departure from the normal is 4 to 5 °C. In 2006 and 2007, one and three heat waves occurred, respectively, belonging to the category of when the actual maximum temperature remains 45 °C or more irrespective of normal maximum temperature. In 2008, there was no heat wave. In 2009 there were two heat waves, one belonged to the category of when the actual maximum temperature remains 45 °C or more irrespective of normal maximum temperature and other belonged to the category of when the normal maximum temperature of a station is less than or equal to 40 °C and the departure from normal is 5 to 6 °C. In 2009, 11 heat waves occurred in three different categories. Majority belonged to the category of when the actual maximum temperature remains 45 °C or more irrespective of normal maximum temperature followed by the heat waves of the category of when the normal maximum temperature of a station is less than or equal to 40 °C and the departure from normal is 5 to 6 °C. One belonged to the category of severe heat wave belonging to the category of when the normal maximum temperature of a station is less than or equal to 40 °C and the departure from normal is 7 °C or more (see Fig. 7.5).

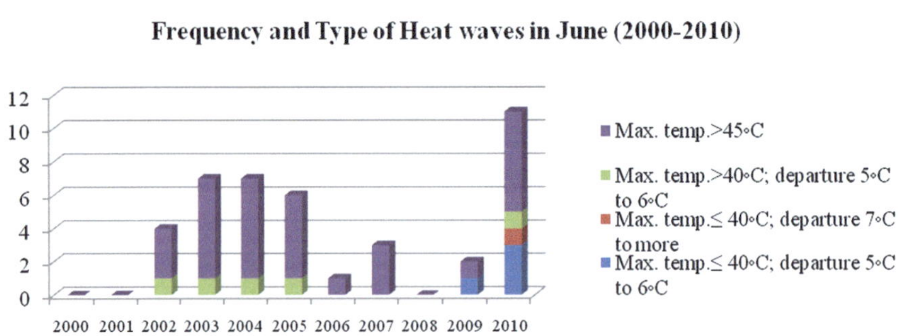

Fig. 7.5 Frequency and type of heat waves in June, Kota (2000–2010)

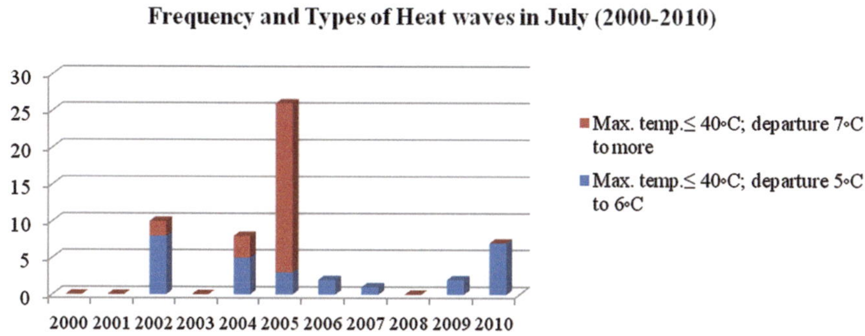

Fig. 7.6 Frequency and type of heat wave in July, Kota (2000–2010)

July

In 2000, 20001 and 2003 there were no heat waves. In 2002 and 2004, there were ten and eight heat waves, respectively. The waves belonged to two categories, one when the normal maximum temperature of a station is less than or equal to 40 °C and the departure from normal is 5 to 6 °C and other one is the severe heat wave when the normal maximum temperature of a station is less than or equal to 40 °C and the departure from normal is 7 °C or more. In 2005 the number of heat waves swung to 25 with the majority belonging to the category of severe heat waves when the normal maximum temperature of a station is less than or equal to 40 °C and the departure from normal is 7 °C or more. In 2006 and 2007, two and one heat waves were experienced belonging to the category of when the normal maximum temperature of a station is less than or equal to 40 °C and the departure from normal is 5 to 6 °C. It was zero in 2008 but the number increased to four and seven in 2009 and 2010, respectively, and belonging to the two categories that were found in the previous years (see Fig. 7.6).

7.7 Conclusion

From the calculations of Land Surface Temperatures for two points of time, that is, 2000 and 2010, it has been found that the Urban Heat Island is developing in Kota. The urban centres are experiencing temperatures as high as 41 °C while its surrounding rural areas are experiencing temperatures as low as 23 °C. Very high temperatures were recorded in the Kota district headquarter located in the south-western part of Kota. It is the largest area with high temperature and within the region, there are negligible pockets of low temperatures. Thus, it signifies that the Urban Heat Island effect is very intense over the region. It is also spreading very fast, converting the low-temperature regions to high-temperature regions. The three major towns of Kota—Ladpura, Sangod and Ramganj Mandi (out of four major towns of Kota) are also transforming into intense high-temperature regions and are also converting the surrounding areas to the same at a quick rate. It is very much evident from the comparison of two LST maps of 2000 and 2010. These three major towns and the Kota district headquarter are located in the southern part of the district. In the 2000 map, they were having high temperature but their surrounding areas were having low temperatures, except for Kota H. Q. area. But in the 2010 map, these *Urban* Heat Islands have spread to the extent that the entire southern part is showing high temperatures. The Urban Heat Island effect is high in the southern part of Kota with the highest intensity in the Kota H. Q. area and it seems to be spreading to the relatively low-temperature parts in the central and northern parts of Kota. Even along the rivers, the temperatures have risen. In the 2000 LST map, it has been seen that the areas along the rivers Chambal, Ahu and Kali Sindh were having low temperatures. But in 2010, areas along Ahu River are seen to be showing high temperatures. Even along Chambal River on the western edge of the district, high temperatures were recorded in 2010. Hence, it can be said that the Urban Heat Island effect is increasing over the Kota district and the effect is prominently high over the southern part of the district where the district H. Q. and three major towns of the district are located.

In case of heat waves, there are only four types of heat waves, that is, (a) when normal maximum temperature of a station is less than or equal to 40 °C and heat wave departure from normal is 5 to 6 °C and severe heat wave if departure from normal is 7 °C or more, (b) when normal maximum temperature of a station is more than 40 °C and heat wave departure from the normal is 4 to 5 °C and (c) when the actual maximum temperature remains 45 °C or more—irrespective of normal maximum temperature, heat wave should be declared and are seen in the study. However, the category of heat waves when the actual maximum temperature remains 45 °C or more irrespective of normal maximum temperature predominates especially in the month of May when the maximum heat waves are found. The maximum frequency of heat waves are seen in the month of May followed by June and July and the least is seen in April. The number of heat waves are fluctuating with suddenly experiencing very high values and then suddenly heat waves are highly variable and inconsistent.

Hence, from the study, it is seen that there is a development of the Urban Heat Island in Kota, Rajasthan. It is evident from the fact that maximum Land Surface Temperature has increased from 2000 to 2010 by 0.5 °C and minimum Land Surface Temperature has increased from 2000 to 2010 by 1 °C. Moreover, the Urban Heat Island effect is spreading which can be seen by comparing the LST maps of 2000 and 2010, where the extent of high temperature can be seen increasing. In fact, the increase in the minimum Land Surface Temperature by 1 °C also suggests the inclusion of the low-temperature regions to higher ones. Apart from this, Kota is also seen to a heat wave prone area of India. Though the number of heat waves are highly variable from the frequency of their occurrence it can be said their occurrence is highly probable in Kota, though the total number may vary. Again, among the heat wave types, the predominant one is the severe type of heat only. Thus, this establishes that Kota where Urban Heat Island effect is developing due to urbanisation, the situation is further worsened by the fact that it is also a heat wave prone region. Urban Heat Island Effect combined with the heat waves will create a very discomforting situation for the urban population resulting in health-related problems and even mortality. Hence, in Kota which is an education institution hub as well as industrial hub the development of built ups will increase the UHI effect and the heat waves will further worsen the thermal conditions of the urban places more in the future.

References

De US, Dube RK, Rao GS Prakasa (2005) Extreme weather events over India in the last 100 years. J Indian Geophys Union 9(3):173–187

Giannini MB, Belfiore OR, Parente C, Santamaria R (2015) Land surface temperature from landsat 5 TM images: comparison of different methods using airborne thermal data. J Eng Sci Technol Rev 8(3):83–90

Karnieli A, Qin Z, Berliner P (2001) A mono-window algorithm for retrieving land surface temperature from landsat TM data and its application to the Israel-Egypt border region. Int J Remote Sens 22(18):3719–3746

Klein Tank AMG, Zwiers Francis W, Zhang X (2009) Guidelines on analysis of extremes in a changing climate in support of informed decisions for adaptation. Climate Data Monitoring, WCDMP No. 72, WMO

Landsat 7 Science Data Users Handbook, National Aeronautics and Space Administration

Mallick J, Kant Y, Bharath BD (2008) Estimation of land surface temperature over Delhi using Landsat-7 ETM+. J Indian Geophys Union, July 2008, 12, 3:131–140

Nairn John R, Fawcett Robert J B (2015) The excess heat factor: a metric for heatwave intensity and its use in classifying heatwave severity. Int J Environ Res Public Health 12:227–253

Ngie A, Abutaleb K, Ahmed F, Taiwo OJ, Darwish AA, Ahmed M (2015) An estimation of land surface temperatures from landsat ETM + images for Durban. South Africa. Geo, Technology of Rwanda

Ramaswami C (1941) On the heat wave over Northern India in April–May

Robinson Peter J (2001) On the definition of heat wave. J Appl Meteorol 40:762–775

Shahraiyni Hamid Taheri, Sodoudi Sahar, El-Zafarany Abbas, Seoud El, Abou Tarek, Ashraf Hesham, Krone Kristin (2016) A comprehensive statistical study on daytime surface urban heat Island during summer in urban areas. Cairo and Its New Towns, Remote Sensing, Case Study

Sharma JK, Srivastava K, Bhatnagar NP (2009) A study of heat wave/ severe heatwave over Bihar and Jharkhand. Mausam 60:89–106

Souch C, Grimmond CS (2004) Applied Climatology: heat waves. Prog Phys Geogr 599–606

Spinoni J, Lakatos M, Szentimrey T, Bihari Z, Szalai S, Vogt J, Antofie T (2015) Heat and cold waves trends in the Carpathian Region from 1961 to 2010. Int J Climatol

Tan J, Zheng Y, Tang X, Guo C, Li L, Song G, Zhen X, Yuan D, Kalkstein AJ, Li F, Chen H (2010) The urban heat Island and its impact on heat waves and human health in Shanghai. Int J Biometeorol: 75–84

Walawender Jakub P, Szymanowski Mariusz, Hajto Monika J, Bokwa Anita (2014) Land Surface Temperature Patterns in the Urban Agglomeration of Krakow (Poland) Derived from Landsat-7/ETM + Data. Pure Applied Geophysics 171:913–940

Zaharaddeen I, Baba II, Zachariah A (2016) Estimation of land surface temperature of Kaduna Metropolis, Nigeria using Landsat Images. Sci World J 11(3):37–42

Zhang Z, Ji M, Shu J, Deng Z, Wu Y (2008) Surface Urban Heat Island in Shanghai, China: Examining the Relationship between Land Surface Temperature and Impervious Surface Fractions Derived from Landsat Etm + Imagery. The International Archives of the Photogrammetry, Remote Sensing and Spatial Information Sciences, Vol XXXVII., Part B8, Beijing

Part II
Urban Health and Wellbeing

Chapter 8
Growing Urbanization, Health Infrastructure, and Vector-Borne Diseases: A Study in Khammam Municipal Corporation, Telangana State

A. Rambabu, G. S. Srinivasa Gopinath and B. Srinivas

Abstract Growing urbanization, even at slow pace demands housing, safe drinking water, drainage outlets, solid waste management, public toilets, and other health infrastructure facilities for a comfortable living. Urban centers and metropolitan cities have been struggling to cope up with the growing requirements. Though, Indian culture harps upon the saying that "Health is wealth" it has not been attained to prioritize community health.

Keywords Urbanization · Health · Infrastructure · Diseases Khammam

8.1 Introduction

Urbanization in the Western countries is relatively higher and has resulted due to the expansion of industries and economic activities. It is accepted as one of the indicators of economic development in the western world. In India, urbanization has increased at a slow pace, over years from 17.3% in 1951 to over 31.2% in 2011. Still, it is very low when compared to the developed countries.

Indian urbanization has not been prompted by industrial and economic expansion as in developed countries, but it is mainly due to distressed migration from the rural to urban in search of livelihood and in the hope of comfortable living, education and hospital facilities. Against this background, this research paper makes an attempt to analyze the growth of population, health infrastructure and the incidence of vector-borne diseases in Khammam Municipal Corporation of Telangana State in south India. The basic objective of this paper is to examine the incidence of vector-borne diseases in the Khammam Municipal Corporation area mainly malaria, dengue, and Chikungunya and other trends in the growth of human population and health-related infrastructure in the study area.

A. Rambabu · G. S. Srinivasa Gopinath (✉) · B. Srinivas
VLITS, Vadlamudi, Guntur, AP, India
e-mail: gopinathgarre@gmail.com

R. B. Singh et al. (eds.), *Urban Health Risk and Resilience in Asian Cities*, Advances in Geographical and Environmental Sciences,
https://doi.org/10.1007/978-981-15-1205-6_8

8.2 Materials and Methods

The study is based on secondary data collected from the authorities of the municipal corporation offices of the District Malaria Officer; profile of the Khammam Municipal Corporation has also been used to collect information on health facilities. Simple averages and percentages are used as tools of analysis, results, and discussion. Khammam municipalities have been upgraded as corporation by including a few villages in the outskirts. Population and the three decades population figures are presented in Table 8.1 (Handbook of Statistics Khammam District 2011; 2015).

It can be noted that the growth rates of population are higher than the State/National averages in Khammam Urban area. Population of the corporation at present is close to four lakhs and the growth when compared to 1991 is 2.67 times. Rapid expansion of population in the municipal corporation necessitates extra health infrastructure to meet the growing needs. The State Government of Telangana has also accepted the challenge of "SWACHH BHARAT" which entails extra manpower and equipment. These are 50 wards, 98,548 households, and 71 slums of which 28 only are notified. Slums and slum dwellers have been increasing and the slum population is 66,918. The slum dwellers are highly exposed to the health hazards due to poor sanitary conditions. The corporation authorities are constrained to take care of 93.45 Sq. Kms of area in their efforts to ensure health and hygienic conditions of the population (Profile of the Khammam Municipal Corporation 2011) (Table 8.2).

Data shows that the corporation has initiated different measures to improve the health conditions of the people. However, majority of the slums around the town lack even the basic amenities. When compared to the area and size of population the requirements need a lot of improvement. Garbage disposal, clearance of the drainages, DDT spraying activities, cleanliness of the roads, and community toilets have been taken up in the right spirits. As Khammam town is the district headquarter and is close to Andhra Pradesh, Chattisgarh, and Odisha states the floating population is to be considered (Table 8.3).

Urban centers are prone to vector-borne diseases and Khammam District is endemic due to the concentration of tribal and forest areas. Data shows that malaria has been controlled to a large extent and chickungunya is insignificant.

The other vector-borne disease JE has not been found in the town. It implies that the health-care measures of the corporation are effective. However, dengue, the most talked about in Khammam district in 2016 decreased between 2012 and 2014. But its

Table 8.1 Population of Khammam Municipal Corporation

Years	Population	% of growth in the decade
1991	1,49,007	–
2001	1,89,469	27.0
2011	2,84,648	50.2
Present	3,98,000	40.1

Source Handbook of Statistics, Khammam District, 2011

Table 8.2 Health infrastructure in Khammam Municipal Corporation

Health aspect	Facility available
1.Drinking Water:	15-Water supply per day 100 LPCD
a.No of reservoirs	1320
b.Hand pumps	32,543% of Households once in two days (30.3)
c.Tap connections	315 KMS
d.Water release	60 KMS (Total 375 KMS)
2.Drains:	Tractors 32 + Tricycles 60 + push carts 30
a.Pucca	209 MT
b.Kuchcha	Regular 76 + outsourcing 580
3.Solid waste management:	Total = 656
a.Vehicles	One within an area of 85% of the area 38 acres in the corporation
b.Total garbage	
c.Public health workers	1
d.Dump yard	3
e.Door to door collection	46
4.Hospital:	9
Government	3
Urban health centers	3
Private	8
5.Burial grounds	18
Hindus	8-length 45Kms
Muslims	4-Identified 64
Christians	636
6.Community toilets	1717
7.Parks and plan Grounds	
8.Litter free roads	
9.Places covered with green fencing	
10.Open defecation free	
Toilets completed	
To be completed	

Source Profile of the Khammam Municipal Corporation 2011

Table 8.3 Incidence of vector-borne diseases in Khammam Municipal Corporation

Disease	Incidence in						
	2011	2012	2013	2014	2015	2016	2017
Dengue	3	29	3	15	61	116	3
Malaria	–	–	19	9	13	2	3
Chickungunya	–	2	–	4	3	3	–
Japanese encephalitis	–	–	–	–	–	–	–

Source Statistical Data Book of District Malaria Office-2017

incidence increased to 61 in 2015 and to 116 in 2016. Already three cases of dengue have been reported in 2017. People in the town are literates with higher levels of health awareness along with financial viability to some extent. Even the increase in the incidence of dengue in Khammam town is in increasing trend revealing the urgency of precautionary measures. Majority of the dengue cases have been noted in the slums where the residents are illiterate/semi-literate, poor and ignorant.

8.3 Major Findings

Population in the Khammam Municipal Corporation has increased by 40–50% in the recent decades testifying the growing urbanization. Municipal authorities have expanded the health infrastructure. But the expansion has not been commensurate with the growth of population. Shortage of regular health workers is evident in the corporation. Slum areas have been growing and they are the points of unhygienic environment where mosquitoes create the menace of vector-borne diseases.

Malaria has been under control while chickungunya and Japanese encephalitis are negligible due to the interventions of the health authorities of the corporation in coordination with the district malaria and medical and health authorities. Dengue has been in the increasing trend since 2014 in view of the poor community health scenario, though efforts for SWACHH BHARAT are encouraging.

8.4 Conclusion

It is appreciable to note that the Government of India has taken up "Smart City" programme to ensure healthy atmosphere in the selected centers. But more funds are to be allocated to the state governments to enhance the health infrastructure in the other urban areas where population growth has been swelling in view of different factors. Slum clearance drive has to be taken up by rehabilitating the slum dwellers with the much needed housing and other facilities; the number of health workers and the related equipment are to be increased as per the norms and research on urban problems is to be encouraged. Above all, awareness about community cleanliness is to be created to prevent the mosquito menace by involving all the stakeholders by designing short-term and medium-term plans.

References

Handbook of Statistics Khammam District, (2011)
Handbook of Statistics Khammam District, (2015)
Profile of the Khammam Municipal Corporation, (2011)
Reports of the District Malaria Officer and Medical Officer, (2017)

Chapter 9
Maternal Reproductive Health: A Comparison Between India and Empowered Action Group States

Deepali Chanu Sanasam

Abstract Maternal healthcare services are very important due to its multifaceted effects on mother and child's health. The present study is an attempt to assess the reproductive health of a mother through the lenses of maternal health care provided by the state in Empowered Action Group (EAG) states and rest of the Indian states. Objectives of the study are to (i) assess the interstate variants in maternal reproductive health facilities available in the EAG states along with other Indian states in the pre and postnatal period (ii) assess the outcome of maternal healthcare facilities in the EAG states. National Sample Survey Organisation (NSSO) 71st data on Social Consumption and Health has been extracted and analyzed using Stata, SPSS, and GIS software. Variables like prenatal care, institutional deliveries, postnatal care, etc. are taken. Methodology adopted includes Principal Component Analysis (PCA) for making composite index and correlation for testing the hypotheses. Interstate variants in healthcare performance have been shown in Choropleth map. Results of the analysis show that states of Uttar Pradesh and Bihar rank lowest among the EAG as compared to states like Odisha, Uttaranchal, Chhattisgarh, etc. Negative correlation between maternal mortality and maternal healthcare facilities emphasizes the need to strengthen reproductive healthcare facilities in the EAG states. Good maternal healthcare facilities will lead to good reproductive health of a mother.

Keywords Maternal reproductive health · Empowered action group states · Maternal healthcare facilities · Prenatal care · Postnatal care · Institutional delivery

9.1 Introduction

Maternal health services are very important due to its multifaceted effects on both child and mother's health. Maternal healthcare services can be used as indicator for reproductive health outcome. Reproductive health is defined as "state of complete physical, mental and social well-being in all matters related to the reproductive

D. C. Sanasam (✉)
Jawaharlal Nehru University, New Delhi, India
e-mail: sanasamdeepali@gmail.com

R. B. Singh et al. (eds.), *Urban Health Risk and Resilience in Asian Cities*, Advances in Geographical and Environmental Sciences,
https://doi.org/10.1007/978-981-15-1205-6_9

system and to its functions and processes" (United Nations Population Fund 2004, p. 45). The concept includes access to contraceptives and being able to exercise reproductive choices. It also includes the right of women to "safely go through pregnancy and provide couples with the best chance of having a healthy infant" (UNPF 2004, p. 45).

Maternal reproductive health can be assessed with the help of indicators of health outcomes such as maternal mortality ratio, neonatal mortality, and postneonatal mortality. The study tries to assess the reproductive health of a mother through the maternal health care provided by the doctors and various institutions during pregnancy and post-delivery. The basis of the study lies in the fact that reproductive health of mothers does not only mean access to contraceptives and sterilization-oriented reproductive health programs but also the providence of adequate and quality healthcare services at prenatal and postnatal period which will lead to good reproductive health outcomes of the mothers. Thus, the study tries to explore the reproductive health of a mother through the lenses of maternal healthcare services available in the Empowered Action Group of States (EAG states) and the rest of Indian states.

Accordingly, the maternal healthcare indicators like antenatal care visits, tetanus shot, Iron Folic Acid (IFA) tablet consumption and other prenatal care, Institutional delivery, etc. are taken along with postnatal care. While outcome indicators such as maternal mortality rate of EAG states are taken and correlated to find any link between the maternal reproductive healthcare services and the maternal health outcome. Also, the study tries to establish a link between Maternal Mortality Ratio (MMR) and percentage of institutional delivery. The idea behind it is that greater usage of institutional mechanism of delivery will lead to reduce reproductive morbidities due to access to trained birth attendant. However, it cannot be denied that there are various factors which affect a mother's health ranging from diet to socio-cultural practices like the practice of home delivery by untrained nurse, excessive works hours of the pregnant women without appropriate nutrition.

In India, the importance of maternal health service was realized early on from the second 5-year plan onwards and as such government initiated many schemes for looking after the need of the expecting mother and lactating mother (Reproductive and Child Health, Janani Suraksha Yojana, etc. to name a few). However, high maternal mortality ratio still persists in the country with MMR of 167 per 100,000 live births (SRS 2011–13) and the target to achieve Millennium Development Goal (MDG) for reducing maternal mortality ratio of 109 by 2015 (MOSPI 2015) had been achieved in few states like Kerala, Tamil Nadu, Maharashtra, etc. This has pointed out the need to study maternal reproductive healthcare services available in Indian states with emphasis on Empowered Action Group states which have very high maternal mortality ratio.

9.2 Objective

The objective of the study is to assess the maternal reproductive healthcare facilities available in Indian states with special focus on Empowered Action Group (EAG) states in prenatal and postnatal period. Second objective is to assess the outcome of maternal reproductive healthcare facilities in the EAG states. The third objective is to assess the relation between institutional delivery and outcome of maternal reproductive healthcare facilities in the EAG states.

9.3 Hypotheses

i. Higher the availability of maternal healthcare facilities better the outcome of maternal reproductive health in terms of maternal mortality ratio of the EAG states.
ii. The more the rate of institutional deliveries among the EAG states better the outcome of maternal mortality ratio in the EAG states.

9.4 Database and Methodology

National Sample Survey Organization (NSSO) 71st Round (2014) data on Social Consumption in India: Health has been taken, extracted, labeled and merged using Stata software. Variables like percentages of perinatal coverage, live births in the women of age group 15–49 years, other prenatal cares, etc. are taken. Perinatal coverage includes coverage of IFA tablets and Tetanus vaccine immunization which are very important for pregnant women. Meanwhile, variables like "other prenatal care" include care received from Accredited Social Health Activist (ASHA), Auxiliary Nurse and Midwife (ANM), etc., which are also taken in light of the importance of the care provided by them. Variables like institutional deliveries are also considered because of the importance stressed by Janani Suraksha Yojana (JSY) scheme to reduce maternal mortality and morbidity in India with special focus on EAG states.

Principal Component Analysis (PCA) method is used for developing composite index of all the health indicators taken to derive an index which will make the maternal healthcare facilities comparable among the states. After calculation of PCA only five variables are found to be correlated which are given in Table 9.1. The formula for calculating composite index (CI) is given in the following lines.

$$\mathrm{CI} = \mathrm{Xi} \cdot \mathrm{W}$$

where Xi = Standardize values (Xi/μ) and W = Weightage (based on PCA).

Table 9.1 Weight of the variables based on PCA

Variables	Weights given
IFA tablets	0.909
Tetanus vaccination	0.884
Other prenatal care	0.851
Postnatal care	0.854
Institutional delivery	0.835

Source Table computed from unit-level NSSO data (71st round 2014)

Table 9.2 Category of composite index of maternal reproductive healthcare facilities

Class	Value
Very High	>4.59
High	4.33–4.59
Medium	4.07–4.33
Low	<4.07

Source Table computed from unit-level NSSO data (71st round 2014)

After giving weights to the entire variables, the states are then grouped into four categories which are very high, high, medium, and low (refer to Table 9.2) using mean-standard deviation method based composite health index.

Based upon the Composite Index value and the class intervals, the maternal healthcare facilities of the states are shown in a Choropleth map using Arc GIS 10. Also, correlation is used to test the relation between MMR and Institutional delivery and MMR with maternal healthcare facilities of EAG.

9.5 Literature Review

Reproductive health as a concept has been confined to the use of contraceptives and family planning exercise. Thus, there lies the need to look beyond the normative concept of reproductive health of a mother and Jejeebhoy (1999) remarked of the need to "re-orient India's traditional population programme to go beyond demographic targets, contraceptive prevalence, and female sterilization to a more comprehensive focus on reproductive health needs and services" (p. 3075).

Jejeebhoy and Rao (1998) stated that mortality and morbidity along with problems of family planning are true reflection of reproductive health. Also, socio-economic factors like little awareness on part of women about healthcare practices together with culture of strong seclusion which prohibits women from availing healthcare facilities, frequent and un-spaced birth and poor diet at the time of pregnancy affects the reproductive health of a mother and child.

In a study conducted by Shah et al. (1984) in a prospective community-based study in Maharashtra, found that out of 3173 babies born, 90 were stillborn which accumulated to stillbirth rate of 28.4/1000 births. This may be because of various causes like asphyxia (53%), unexplained stillbirth (17.8%), preterm birth (15.6%), intrauterine growth retardation (4.4%), congenital anomaly (4.4%), other causes (4.4%), etc. It also highlighted the importance of seeking antenatal care as the causes of stillbirths were found to be low ANC visits and poor institutional delivery. Bloom et al. (1999) in a study conducted in Varanasi among the poor to middle-income groups found that women who had availed the services of antenatal care tend to use safe delivery care as compared to those women who did not avail antenatal care facilities.

Reproductive health of a mother is affected by the level of nutrition. Nutrition of a mother is the key to the safety of the mother and the child during perinatal and postnatal period. It accounts for maternal and perinatal mortality as well as occurrence of low birth weight and preterm babies. However, a study by Rush (2001) found that any increase in birth weight from low weight before pregnancy may not necessarily lead to better child survival. It is because sudden increased in weight can cause increase in perinatal mortality.

Maternal reproductive health may also be plagued by conditions of morbidity. Maternal morbidity is found to be associated with pregnancy both during perinatal and postnatal period. The morbidities occur during labor, during puerperium period, etc. Bang et al. (2004) found that during labor maternal morbidities like "prolonged labor (10.1%), prolonged rupture of membranes (5.7%), abnormal presentation (4.0%), and primary postpartum hemorrhage (3.2%). The postpartum morbidities included breast problems (18.4%), secondary postpartum hemorrhage (15.2%), and puerperal genital infections (10.2%), and insomnia (7.4%) prolonged rupture of membranes" were most common (p. 231). While nearly 15% women who did not opt for institutional delivery needed emergency obstetric care for safety to health of the mother.

According to Ravindran and Mishra (2001) maternal reproductive health also includes unmet need for contraceptives, lack of access to abortion, unmet need for sterilizations, etc. This unmet need is associated with women in India who has more or less been able to realize reproductive needs. It is based on couples preference and exhibited varied estimate of unmet need emerging from varied sources. He proposed a measure for reproductive intention which was known as "Helping Individual Achieve their Reproductive Intentions (HARI) index" (p. 105) which was for assessing whether people can plan their reproductive schedule in healthy manner. It includes proportion of women not being able to realize any form of family planning, safe abortion, preferred birth spacing, and safety from associated morbidity.

Abortion as a topic is still a taboo in India. It is not reported by most women at the time of interview. Mostly people go for illegal abortion due to complicated legal procedure. A study done by Varkey et al. (2000) in rural Rajasthan about the prevalence of abortion revealed that "of the 195 women studied, 28% had induced abortion, 5.6% considered abortion but eventually did not have an abortion and a termination was attempted unsuccessfully by 4%" (p. 86). Also, the study highlighted that only 40% of the women were aware about complications related to abortion. Some complications mentioned by them were an "infection, excessive bleeding, infertility and death" which was caused by adoption unregulated abortion method by unskilled health professional and "intra-amniotic injection of saline presents a worrisome trend" (p. 86).

Maternal reproductive health is also affected by economic factors which decide the healthcare seeking behavior of the mothers affecting their health. A study by Agarwal (2011) found that among the lowest quartile of mother using maternity care "only 54% of pregnant women had at least three antenatal care visits contrary to 83% for the rest of the urban population" (p. 18). Also, attendance by health personnel in birth cases was very low for poorest quintile group where only half of the births were assisted by health personnel in 2005–06. This was particularly so in the states of Uttar Pradesh, Delhi, Bihar, and Rajasthan's urban population.

Iyengar and Iyengar (2009) suggested that qualified nurse-midwives may lead to considerable increase in access to skilled maternal and neonatal care in rural areas. They can also deal with maternal complications on their own without the need to refer. He also pointed out the need to understand the act that different families may not follow the advice of doctors referred. In this case, initial care provided by nurse-midwives can help prevent certain complications dismissing the need for any referral.

A good state of maternal reproductive health is reflected by maternal mortality, neonatal mortality, rate of miscarriage, and still births. In India, maternal mortality is mainly caused by hemorrhage mostly postpartum hemorrhage, i.e., 38% of maternal deaths (SRS as cited by Vora et al. 2009, p. 186). While anemia was found to be plaguing nearly 60% of pregnant women who were reported as anemic (NFHS as cited by Vora et al. 2009).

9.6 Introduction to the Study Area

The Empowered Grouped States comprises of Bihar, Chhattisgarh, Jharkhand, Madhya Pradesh, Uttaranchal, and Uttar Pradesh as shown in Fig. 9.1. The region has been specially demarcated so as to facilitate execution and implementation of area-specific programs as the region has failed to curb its population growth (Government of India 2001).

The population of EAG states in 2011 is 555 million as per Census of India (2011a, b). Sex ratio in the EAG states is very low with Bihar and Rajasthan having lowest sex ratio of 916 and 926, respectively. While Chhattisgarh and Orissa records the highest sex ratio of 991 and 978 among EAG states (Census of India 2011a, b). The EAG states have very high maternal mortality ratio of 246 per 100,000 live births including Assam as compared to all India level of 167 in 2011 (SRS 2011–13). These trends in maternal mortality ratio may be due to low per capita expenditure of the government on health and family welfare which has been found to be the lowest except for Rajasthan (Arokiasamy and Gautam 2007). Social indicators like effective

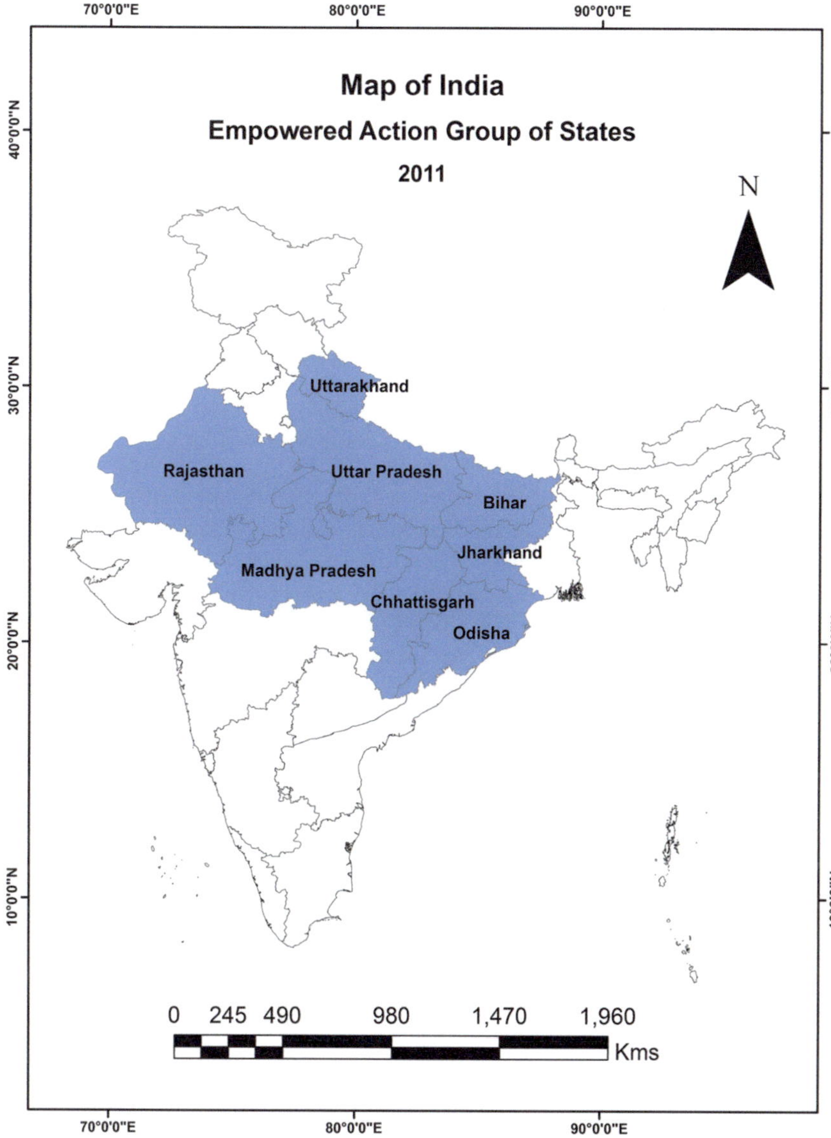

Fig. 9.1 India and EAG states

literacy rates of the EAG states (68.86%) also fare low compared to non-EAG states (84.76%) and all India level (74.04%) in 2011. The male–female gap in literacy rate is also high in EAG states which is 20.97% which is just 13.34% for non-EAG states and 16.68% at all India level (GOI 2011).

9.7 Findings

9.7.1 Indicators Taken

Indicators taken for assessing the maternal reproductive health include the prenatal care facilities like IFA tablets, tetanus vaccination, other prenatal care, and postnatal care which has been considered by Ministry of Health and Family Welfare (MoHFW 2013) as among the important indicators of maternal health. Among these, the most important indicators are tetanus vaccination, IFA tablet dosage and institutional delivery.

9.7.1.1 IFA Tablet Consumption

India has high consumption of IFA tablets with 93.83% recorded in 2014. This is one of the most important indicators of maternal health as deficiency of iron in an expecting mother may lead to anemia affecting the health of both mother and unborn child. The state with highest consumption of IFA tablet is Goa with 100% usage of IFA tablet followed by Telangana with 98.68% usage. Himachal Pradesh also has high consumption of IFA tablet (98.38%). Arunachal Pradesh ranks low in this category with only 83.65% consumption of IFA tablet out of the total pregnant women. Uttar Pradesh (87.76%) and Nagaland (88.13%) also ranks low in this indicator.

9.7.1.2 Tetanus Vaccination

As per NSSO 2014 data, tetanus vaccination seems to be one of the indicators with highest coverage with an all India value of 95.92%. While in terms of state, Goa has the most extensive coverage of the vaccine with 100% coverage. It is followed by Telangana, Chhattisgarh with more than 98% coverage. The lowest coverage is found in the state of Arunachal Pradesh with coverage of only 87.42% followed by Nagaland with 90.63% coverage and Bihar with 91.79% coverage.

9.7.1.3 Other Prenatal Care

Other prenatal care received from ANM has 92.75% coverage in India as a whole. Among the Indian states, Goa has 100% coverage of other prenatal care followed by Tamil Nadu (99.43%) and Kerala (99.14%). Arunachal Pradesh (84.91%) and Mizoram (87.25%) are the states with lowest utilization of other prenatal care facilities.

9.7.1.4 Institutional Delivery

This is very important indicator affecting the maternal health of the mother. India has all together 83.79% of the pregnant mother opting for institutional delivery in all forms. Institutional delivery includes delivery in private hospital, public hospital, primary health centre, community health centre, etc. However, compared to other indicators, this has low performance. Among the states which have low percentage in institutional delivery are Uttar Pradesh (74.6%), Uttaranchal (79.7%), Meghalaya (73.19%), etc. While the states like Goa (97.73%), Tamil Nadu (91.13%), Kerala (97.13%), etc. have high rate of institutional delivery.

9.7.2 Interstate Variants in Maternal Reproductive Healthcare Facilities Among Indian States

States such as Goa (CI 4.78), Kerala (CI 4.70), Andhra Pradesh (CI 4.67), Tamil Nadu (CI 4.66), Telangana (CI 4.63), and Maharashtra (CI 4.62) come under very high category in terms of reproductive maternal health facilities (refer Fig. 9.2). Goa ranks the highest among them because it has 100% coverage of prenatal care services like tetanus vaccine, IFA tablet, Institutional deliveries, and other indicators like postnatal care also crossed 95% mark. While Kerala has very high coverage of prenatal care services like Tetanus vaccine (94.63%), IFA tablets (97.29%), other prenatal care (99.14%), etc. The same goes for Andhra Pradesh, Tamil Nadu, Telangana, Maharashtra, etc. These states are in the very high category in terms of maternal healthcare services. States like Karnataka (CI 4.59), Gujarat (CI 4.52), Odisha (CI 4.48) fall under the high category of maternal reproductive healthcare facilities with composite index ranging from 4.59 to 4.34. These states have high percentage of consumption of in IFA tablets (above 95%) and other prenatal care services (above 95%). However, they have comparatively lower usage of postnatal care with less than 90% women availing postnatal care facilities.

States like Mizoram (CI 4.31), Manipur (CI 4.30), Madhya Pradesh (CI 4.29), Tripura (CI 4.29) Rajasthan (CI 4.14), and Chhattisgarh (CI 4.14) come in the medium category of usage of maternal reproductive healthcare facilities with composite index ranging from 4.31 to 4.14 (refer Fig. 9.2). These states have low percentage of postnatal care (less than 85%) and institutional delivery (ranging between 67 and 92%).

The lowest rank state in terms of maternal health care goes to Arunachal Pradesh with CI of 3.79 which is a huge declined as compared to Goa. This state has low institutional delivery (69.5%) and postnatal care (71.74%) compare to other states. Other states which fall under the low performance bracket are Uttar Pradesh (CI 3.90), Bihar (CI 4.04), Nagaland (CI 3.79), etc. Nagaland has very low institutional delivery with just 60.39% mothers availing this facility. This might have pulled down

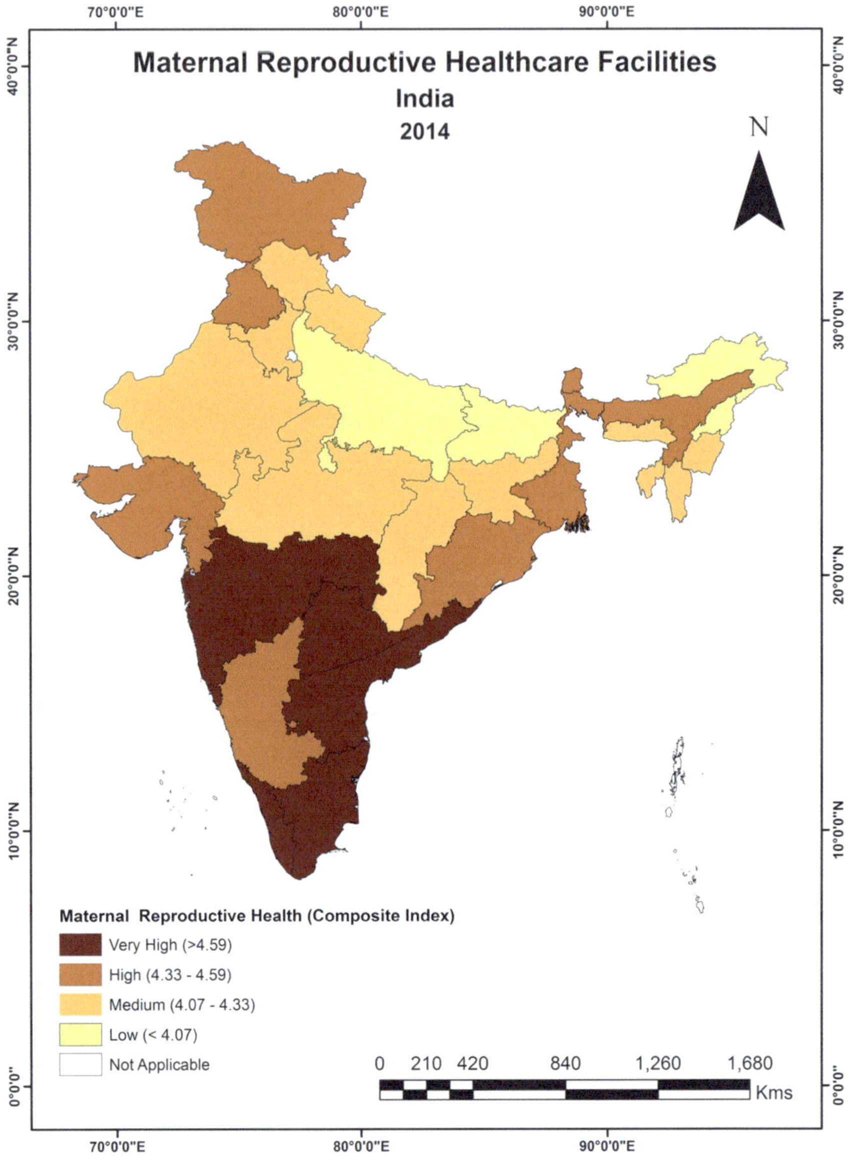

Fig. 9.2 Maternal reproductive healthcare facilities in India

the performance of the state putting it in lowest rank states. Bihar and Uttar Pradesh also have low usage of postnatal care facilities and institutional delivery (less than 80%).

9.7.3 *Interstate Variants in Maternal Reproductive Healthcare Facilities Among Indian States And EAG States*

Among the EAG states, Odisha ranks the highest falling under the high category (refer Fig. 9.2) in terms of availing maternal healthcare services. It has fairly good acceptance of prenatal care services like IFA tablet consumption (94.13%) and tetanus vaccination (95.52%). While it has comparatively lower percentage of institutional delivery with 92.21%. Other states like Uttaranchal (CI 4.15), Jharkhand (CI 4.14), Chhattisgarh (CI 4.14), Rajasthan (CI 4.14) ranks in the medium category of states. Uttaranchal ranks second among the EAG states with CI of 4.15 with high percentage of usage of IFA tablets (93.07%), tetanus vaccination (92.1%) and other prenatal care (95.54%), while postnatal care (71.74%) and institutional delivery (79.7%) are comparatively lower. Jharkhand and Chhattisgarh have high usage of prenatal care like IFA tablet usage and other prenatal care (above 85%) but low institutional delivery (less than 80%).

One contrasting feature that can be seen in Rajasthan is that it has high performance in institutional delivery (90.78%) while compared to all the other states of India, but its performance in receiving other prenatal care (87.19%) and postnatal care (71.74%) is relatively low.

Uttar Pradesh (CI 3.90) and Bihar (CI 4.04) ranks the lowest among the EAG states and among other states of India except for Arunachal Pradesh and Nagaland in terms of utilization of maternal reproductive healthcare facilities (refer Fig. 9.2). Uttar Pradesh has comparatively lower percentage of utilization of maternal reproductive healthcare facilities such as other prenatal care (less than 90%) and postnatal (less than 80%), institutional delivery (less than 80%).

9.7.4 *Maternal Mortality and Maternal Healthcare Facilities: A Performance Appraisal of the EAG States*

For testing the relationship between MMR and maternal healthcare facilities, PCA among maternal health facilities is calculated. From the PCA results, it has been found that there is a highly positive correlation among the indicators taken which is shown in Table 9.3.

The analysis from Table 9.3 reveals that variables "Other prenatal care" and "IFA tablets" are highly correlated ($r = 0.749$) at 0.01 level of significance with p-value of 0.000. The same variable "other prenatal care" has high positive correlation with tetanus vaccination with $r = 0.691$ significant at 0.01 confidence level and p-value of 0.000 (as shown in Table 9.3). It also has high positive correlation with "postnatal care" with r-value of 0.649 at 0.01 significance level. The same variable "other prenatal care" has positive correlation with "institutional delivery" with r-value = 0.600 significant at p-value of 0.000.

Table 9.3 Correlation among maternal reproductive healthcare facilities

	Other prenatal care	IFA tablets	Tetanus vaccination	Postnatal care	Institutional delivery
Other prenatal care	1	0.749**	0.691**	0.649**	0.600**
IFA tablets		1	0.872**	0.687**	0.616**
Tetanus vaccination			1	0.605**	0.649**
Postnatal care				1	0.770**
Institutional delivery					1

Source Table computed from unit-level NSSO data (71st round 2014). (**Significant at 0.01 level of significance)

The variable IFA tablets has high positive correlation with "tetanus vaccination" with value of r = 0.872 at 0.01 level of significance and p-value of 0.01 as per Table 9.3. The correlation between "IFA tablets" and "postnatal care" is positively correlated (Pearson's correlation value of 0.687) at 0.01 level of significance as seen in Table 9.3. "IFA tablets" also have positive correlation with "institutional delivery" with r-value of 0.616 (refer to Table 9.3).

From Table 9.3 it is also seen that the variable "tetanus vaccination" is also positively correlated with "postnatal care" and "institutional delivery" with r-value of 0.605 and 0.601, respectively.

The variable "postnatal care" is also highly positively correlated with "institutional delivery" at 0.001 level of significance and Pearson correlation of coefficient value of 0.770 (refer to Table 9.3).

The graph from Fig. 9.3 reveals that there is a negative correlation between maternal mortality ratio and maternal healthcare facilities found in Empowered Action Group States. This shows that for reducing maternal mortality rate, maternal healthcare facilities must be improved in the EAG states. Thus, the hypothesis that with higher availability of maternal healthcare facilities better will be the outcome of maternal reproductive health in terms of MMR is true. Previous analysis (in Fig. 9.2) has shown that these EAG states like Uttar Pradesh and Bihar rank low (with CI less than 4.07) in the availability of maternal healthcare facilities, while the maternal mortality ratio is as high as 285 per 100,000 in Uttar Pradesh and 208 per 100,000 in Bihar/Jharkhand (as per SRS Bulletin 2011–13). At the same time, states like Kerala and Tamil Nadu have very high maternal healthcare index of greater than CI 4.59 as shown in Fig. 9.2 with very low MMR of 61 and 79 per 100,000, respectively. Government should take initiatives to improve the maternal healthcare facilities of the EAG states especially with focus on postnatal care.

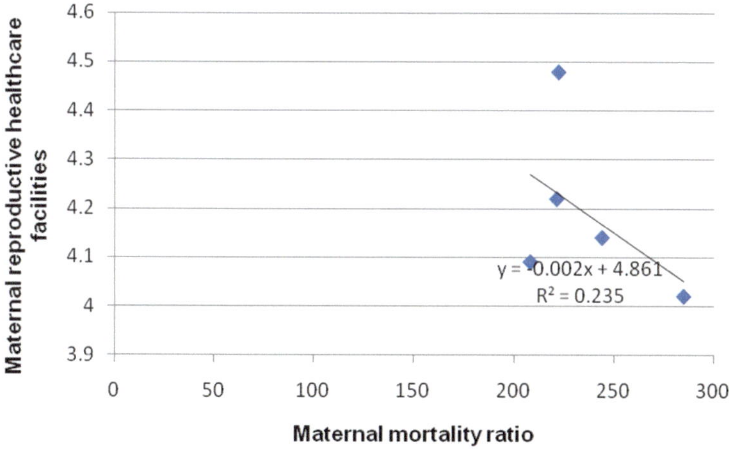

Fig. 9.3 Scattered diagram of maternal mortality rate (MMR) and maternal reproductive healthcare facilities. *Source* Table computed from unit-level NSSO data (71st round 2014) & SRS (2011–13)

9.7.5 Maternal Mortality Ratio and Institutional Delivery in EAG States a Performance Appraisal

The scatter diagram in Fig. 9.4 shows that there is negative correlation between maternal mortality ratio and percentage of institutional delivery in EAG states. This shows that so as to reduce the MMR in the region, rate of institutional delivery must be increased. The hypothesis that with higher institutional delivery, there will be

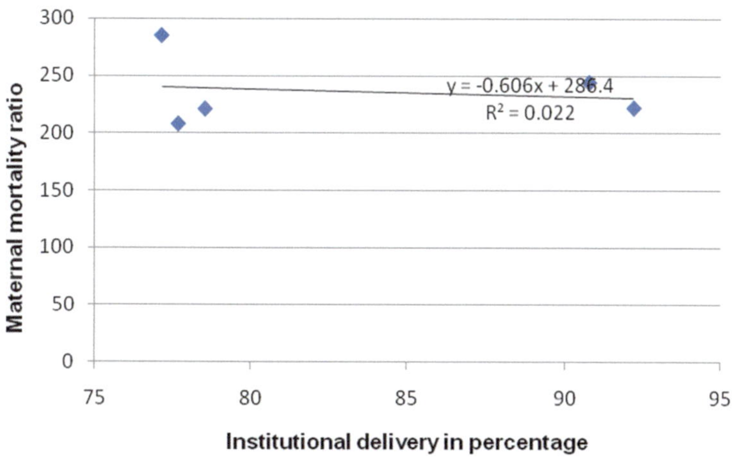

Fig. 9.4 Scattered diagram of maternal mortality rate and institutional delivery in EAG states. *Source* Table computed from unit-level NSSO data (71st round 2014) & SRS (2011–13)

lower mortality rate is also confirmed. However, since the correlation is not strong, the role of other factors such as other healthcare facilities and socioeconomic factors cannot be ignored.

9.8 Conclusion

Indicators like IFA tablets, Tetanus vaccination, and other prenatal care were found to have high utilization. While indicators like institutional delivery and postnatal care have low percentage of utilization. In terms of States, Goa, Kerala, and Tamil Nadu have very high utilization of maternal care facilities in almost all the indicators taken. Other states need to improve in order to catch up with the highly performing states. This is especially true for the EAG states like Uttar Pradesh and Bihar. The whole trend of low percentage in utilization of postnatal care and institutional delivery presents a worrisome trend for India. This shows that maternal healthcare facilities are not be followed up till the time of delivery and post-delivery which is a crucial time for preventing post-delivery complications and deaths.

It has been established that high maternal mortality in the EAG states have been influenced by low maternal reproductive healthcare facilities. However, the relation between maternal mortality ratio institutional deliveries is vaguely pointing out the importance of maternal healthcare facilities and other socioeconomic determinants in reducing MMR. Hence, government should ensure that there is utilization of healthcare facilities by the pregnant and lactating mothers through awareness, education of the masses and also making the healthcare facilities accessible to all. Steps should be taken to ensure more institutional deliveries and also lactating mothers should receive the adequate health care. Focus should be on strengthening the services of ASHA, ANM, Multipurpose Health Worker especially in EAG states. Reducing maternal mortality ratio of EAG states will go a long way in reducing the maternal mortality ratio of India. Providing adequate maternal healthcare facilities and utilization of the facilities may lead to good reproductive health of mother which may lead to reduction in maternal mortality ratio in EAG states. However, further research is needed to study the role of socioeconomic factors in increasing "Institutional delivery".

References

Agarwal S (2011) The state of urban health in India; comparing the poorest quartile to the rest of the urban population in selected states and cities. Environ Urban 23(1):13–28. https://doi.org/10.1177/0956247811398589

Arokiasamy P, Gautam A (2007) Neonatal mortality in the empowered action group states of India: trends and determinants. J Biosoc Sci 40(02):183–201. https://doi.org/10.1017/s0021932007002623

Bang R, Bang A, Reddy M, Deshmukh M, Baitule S, Filippi V (2004) Maternal morbidity during labour and the puerperium in rural homes and the need for medical attention: a prospective observational study in Gadchiroli, India. BJOG: Int J Obstet Gynaecol 111(3):231–238. http://dx.doi.org/10.1111/j.1471-0528.2004.00063.x
Bloom SS, Lippeveld T, Wypij D (1999) Does antenatal care make a difference to safe delivery? A study in urban Uttar Pradesh, India. Health Policy Plan 14(1):38–48. Oxford University Press, Oxford
Census of India (2011a) Primary census abstract total. The Registrar General and Census Commissioner, India. http://www.censusindia.gov.in/DigitalLibrary/MapsCategory.aspx
Census of India (2011b) Gender composition of the population. http://censusindia.gov.in/2011-prov-results/data_files/india/Final_PPT_2011_chapter5.pdf
Government of India (2001) Empowered action group on population stabilization to focus on Bihar and U.P. http://pib.nic.in/archieve/lreleng/lyr2001/rjun2001/20062001/r200620011.html
Government of India (2011) State of Literacy. http://www.educationforallinindia.com/chapter6-state-of-literacy-2011-census.pdf
Iyengar K, Iyengar S (2009) Emergency obstetric care and referral: experience of two midwife-led health centres in rural Rajasthan, India. Reprod Health Matters 17(33):9–20. https://doi.org/10.1016/s0968-8080(09)33459-x
Jejeebhoy SJ (1999) Reproductive health information in India: what are the gaps? Econ Polit Wkly 34(42/43):3075–3080
Jejeebhoy SJ, Rao SR (1998) Unsafe motherhood: a review of reproductive health, women's health in India: risk and vulnerability. In: Gupta M, Lincoln C, Krishnan T (eds) Women's health in India: risk and vulnerability. Oxford India Paperbacks, New Delhi, pp 122–152
Ministry of Health and Family Welfare (2013) A strategic approach to reproductive, newborn, child and adolescent health (RMNCH+A) in India. Government of India, pp 1–68. http://nhm.gov.in/images/pdf/programmes/rmncha-strategy.pdf
Ministry of Statistics & Programme Implementation (2014) Social consumption in India: health, national sample survey 71st round. Government of India
Ministry of Statistics & Programme Implementation (2015) Millennium development goals India country report 2015. Government of India. http://mospi.nic.in/sites/default/files/publication_reports/mdg_2july15_1.pdf
Ravindran T, Mishra U (2001) Unmet need for reproductive health in India. Reprod Health Matters 9(18):105–113. https://doi.org/10.1016/s0968-8080(01)90097-7
Rush D (2001) Maternal nutrition and perinatal survival. J Health Popul Nutr 19(3):S217–S264
Sample Registration System Bulletin (2011–13) Maternal mortality rate and life time risk: India, EAG & Assam, South and Other States. Registrar General of India. http://www.censusindia.gov.in/vital_statistics/mmr_bulletin_2011-13.pdf
Shah U, Pratinidhi A, Bhatlawande P (1984) Perinatal mortality in rural India: a strategy for reduction through primary care. I Stillbirths. J Epidemiol Community Health 38(2):134–137. http://dx.doi.org/10.1136/jech.38.2.134
UNFPA (2004) Reproductive rights and reproductive health. Programme of action, pp 45–60. https://www.unfpa.org/sites/default/files/event-pdf/PoA_en.pdf
Varkey P, Balakrishna P, Prasad J, Abraham S, Joseph A (2000) The reality of unsafe abortion in a rural community in South India. Reprod Health Matters 8(16):83–91. https://doi.org/10.1016/s0968-8080(00)90190-3
Vora KS, Mavalankar DV, Ramani KV, Upadhyaya M, Sharma B, Iyengar S, Gupta V, Iyengar K (2009) Maternal health situation in India: a case study. J Health Popul Nutr 27(2):184–201. https://www.ncbi.nlm.nih.gov/pmc/articles/PMC2761784/pdf/jhpn0027-0184.pdf

Chapter 10
Research Frontiers in Water, Environment, and Human Health

Dilek Eren Akyuz

Abstract Cleaner water, better environment, and healthier life are aimed in the Millennium Development Goals (MDGs) by securing safe water, adequate sanitation, and clean environment, which are essential for human well-being. Moreover, water is accepted as a basic human right in January 2003, with the UN Committee on Economic, Social, and Cultural Rights' comment No. 15. Therefore, everyone should have access to clean water for their physical health and for a healthy environment. Unfortunately, every day around one thousand children die due to preventable diseases caused by lack of water and sanitation. A key to achieving a cleaner environment is possible through an optimum sustainable integrated water resources management (OSIWRM), because water, environment, and health are interrelated and they must be taken into account holistically. Deterioration of one of these components might lead to decline of others. For example, when the environment is polluted, it increases the total nitrogen concentration in water and plants start to produce more pollen. In return, this leads to allergic diseases such as asthma, and hence, become more widespread. Moreover, climate change has clear effects on human health as well as on water availability and the environment due to increase of toxic algae in favorable climate conditions. The concentration of chemicals in water and the environment and its effects on organisms have been researched extensively. Nowadays, the relationship between environmental conditions and actions of organisms is being researched and becoming clearer. For example, investigation of the terrestrial locomotion of mummichogs and their method of navigation toward water in a terrestrial environment shows that they primarily use visual cues, specifically reflected light, to orient toward water. More research on the relationship of water, environment, and health is needed. Results of such research can help us understand the relationship and take measures of protection before the point of no return. In addition, there is some confusion in the determination of causality. If correct indicators (for example, total number, proportion, variety of grains, etc.) are not selected, results may not reflect reality and cannot demonstrate a causal relationship. Sufficient time for research projects is another crucial factor in reaching concrete results. Unfortunately, budgets define the duration of

D. E. Akyuz (✉)
Civil Engineering Department, Istanbul University - Cerrahpasa, Avcilar, Istanbul, Turkey
e-mail: dilek.akyuz@istanbul.edu.tr

R. B. Singh et al. (eds.), *Urban Health Risk and Resilience in Asian Cities*,
Advances in Geographical and Environmental Sciences,
https://doi.org/10.1007/978-981-15-1205-6_10

most research studies and bring a project to an early closure although it may be in the middle of observed event, or even at the beginning. On the other hand, one should use research findings to improve standards of living in the world and fight waterborne diseases such as preventable cholera, which has caused 1,678 deaths in Yemen with 0.6% fatality rate (October 2016–July 2017) and 1,098 deaths in Somalia with 1.5% fatality rate (January–July 2017). These outbreaks can be controlled and the number of deaths minimized with our existing knowledge combined with political will.

Keywords Water · Health · Sanitation · Climate change

10.1 Introduction

Water is an essential element for the existence of life without any doubt. It is also crucial for human health and a healthy environment. Therefore, civilization has started near water sources, in other words, clean water usage has always improved civilization, wealth, and health. Due to these advantages of water usage, humans have developed a tendency to overuse and abuse it with the assumption that there is unlimited fresh clean water whenever one demands and water cleaning process is a simple duty of nature carried out by itself. Unfortunately, the reality is completely different from this false assumption and one should wake up to this, hopefully sooner than later.

Available water is limited and on the decrease due to increase both in overall population and in the water consumption amount per person. Moreover, urbanization and industrialization raise the impermeable area percentage. As a result of this, the runoff and infiltration rate of the precipitation change as does the water cycle. Therefore, with an increased awareness of natural cycle at a personal level, one should try to reduce the negative effects of human beings on nature and cut down on own consumption of resources. Among these natural resources, water in particular, with its quantity and quality, is increasingly becoming a permanent problem all over the world. More research is required to find out sustainable solutions to this pressing problem.

The precipitation as rain, snow, or hailstones is only water resource in the world and contrary to the common knowledge lakes, rivers, or even groundwater are not water resources. These are simply water storages because they cannot survive without precipitation. Therefore, the management of precipitation is essential. Atmospheric movements and hydrological cycles, which are very complex and stochastic in their behaviors, are main factors of precipitation forms and they make precipitation management a difficult issue. The aim of precipitation management is to get maximum gain and minimum damage from precipitation according to our needs; time, location, and amount of the precipitation arrangement. More research is required especially to reveal effects of precipitation management.

The effective application of precipitation management includes irrigation with the accumulated precipitation and several best management practices (BMPs). BMPs aim to improve water quality, to increase groundwater level, to reduce the runoff, to reduce the need of irrigation, to detain the peak discharge time, and finally to reduce the negative effects of urbanization. Some precipitation management applications have local effects such as reusing the accumulated precipitation with rain catchers, some of them have global effects such as an artificial mountain construction which is considered by the US-based National Center for Atmospheric Research (NCAR) to increase rain in the United Arab Emirates where US$558,000 was spent for cloud-seeding only in 2015 (Arabian business 2017).

The problem of water quantity may occur as two extremes, drought and flood, both of which claim lives and money. Some parts of world suffer from drought (UNICEF 2017) for a list of countries to an extent that access to clean drinking water is limited and people have to use muddy water as the only alternative. On the other hand, some other parts of the world try to reduce runoff for the prevention of flood damages (International flood network 2017 for a list of recent floods). Floods and droughts reduce agricultural production and cause lack of food and drinking water. Moreover, they give damage to environment and destroy suitable land for habitation. As a result of these damages, life conditions become more difficult. In addition, floods and droughts have become more frequent with higher peak values and bigger day by day as outcomes of the climate change and urbanization.

The problem of water quality due to degradation in its chemical, physical, biological, and radiological properties is severe and exists almost in every country (Somlyody 1995). The quality of water decreases with the increase in the concentration of dissolved materials in water resources due to discharged wastewater, industrial waste, sewage, and polluted water, which contain more nutrients, heavy metals, oil, or sediments. It is an important issue because these pollutants, especially radioactive waste, may make the whole water resource unusable. Moreover, these materials cause concerns in soil quality when used in irrigation, and health concerns when food produced with polluted water is consumed or dust from polluted soil is inhaled (Marfe and Stefano 2016).

Eutrophication is another water quality problem, which occurs because of excess amount of nutrients in the water, especially total nitrogen and total phosphorus concentration. Although there are several studies on this problem, there is still need for more research. For a long time, the debate in the relevant research literature was whether for an effective solution to the problem of harmful algal blooms (HABs) nitrogen (Chaffin and Bridgeman 2014) or phosphorous should be controlled (Schindler et al. 2008). Recent research points out that both of them need to be controlled through reduction of their concentration to prevent algal blooms in freshwater and seawater (Paerl et al. 2016).

10.2 Safe Water and Adequate Sanitation Right

Safe water, adequate sanitation, and clean environment are essential to human health and well-being. Therefore, these three important issues are also aimed in the Millennium Development Goals (MDGs). Some of MDGs aims are to

(i) Eradicate extreme poverty and hunger,
(ii) Reduce child mortality and improve maternal health,
(iii) Combat diseases,
(iv) Ensure environmental sustainability (UN 2017a).

Unfortunately nowadays, 2.4 billion people lack access to basic sanitation services, and 1.8 billion people have to use contaminated water as a source of drinking water. As a result of these unsuitable conditions, approximately 1,000 children per day die from preventable water and sanitation-related diarrhoeal diseases. Therefore, there is a great need to improve efforts to fulfill these aims all over the world. Water is a basic human right and it is under guarantee with comment No. 15 of the UN Committee on Economic, Social, and Cultural Rights stating that "*The human right to water entitles everyone to sufficient, safe, acceptable, physically accessible and affordable water for personal and domestic uses.*" So, everyone deserves access to clean water to establish a healthy environment and to have a healthy body. Moreover, water demand and usage differ between habitants upstream and downstream (especially in transboundary rivers such as the Nile). This difference may cause serious conflicts whereas some scientists believe that water should flow in the world without disturbance to the water cycle like blood flows in the body (Falkenmark 1999).

10.3 Relationship Between Water and Environment

Environment means all surroundings of a living organism with nonliving materials and several natural forces such as wind, sunshine, and gravity forces. It provides an area for humans to live and civilize. We change our habitat and adjust our lives to our surroundings. Thus, we change the place we live and it changes us. Moreover, the fact that the surroundings, which we change, especially through urbanization cause changes on the ecosystem of other living beings such as the lives and activities of animals, as clearly shown in the recent research (Dahirel et al. 2017).

The rate of urbanization raises the amount of impervious surfaces as well as the quantity of precipitation runoff water. Another effect of urbanization is the drop in the quality of water. In order to reduce effects of natural disasters, we build some infrastructures. As an example, for stormwater management there are two types of infrastructures used. These are named according to use of methods adopted such as "gray" for classical infrastructure based on piping systems and "green" infrastructure based on natural alternatives including soil, vegetation, and woody plants. Newer green infrastructure brings along many social, economic, and ecological benefits.

These benefits remind us of the importance of nature and that its preservation is of paramount importance. Using green infrastructure has more economic benefits than using the gray, but it also has significant contribution to the environment as it cleans the air, water, and soil, and provides habitat for living organisms. In addition, woody plants are more effective than vegetation in the "green" infrastructure due to their stronger influence on the hydrological cycle. Some effects of trees on stormwater are intercepting incoming precipitation, removing water from soil through transpiration, reducing runoff and enhancing infiltration (Berland et al. 2017) As a result of these benefits of the woody plants, hydrological cycle improves, and hydrological cycle is an essential natural process to clean water. In other words, planting more trees is a natural and economical water cleaning method.

10.4 Relationship Between Water and Health

It is a well-known fact that living beings contain water; therefore, bodies of humans, animals, and plants are composed of a great quantity of water. Water is not only needed for initial existence, but also for sustaining it. Drinking clean water is essential for healthy life, and consuming unclean water has been proven to lead to many diseases and is detrimental to human health. Around 23.7% of total world population, in other words every fifth person has to use fecally contaminated water (UN 2017a). Unclean water usage causes waterborne diseases such as cholera, typhoid, hepatitis, dysentery, giardiasis, guinea worm, and schistosomiasis. Moreover, climate change and global warming may increase heavy rainfall events, and may lead to consequent advance in the risk of these diseases and their spread (Levy et al. 2016).

Unclean water may cause negative effects on health, especially for children and there is a need for a frame to observe the relationship between them (Taranu et al. 2017). Unfortunately, children are more sensitive and less resistant to pollutants. For example, a recent study in Spain on bone tumors revealed that toxic pollutants cause a higher risk in children (Garcia-Perez et al. 2017).

10.5 Relationship Between Environment and Health

The impact of surroundings on human health is an observable fact that we actually experience. One may become ill due to polluted water as well as polluted air, or unhealthy food. Pollution in the water, air, and soil may transfer to humans through inhalation, physical contact, and ingestion. Clean water, sanitation, and hygiene are effective means to improve human health especially by preventing associated deaths from diseases such as diarrhea, which unfortunately remains to be the third biggest killer of under-fives worldwide (UN 2017a). Ascariasis, diarrhea, dracunculiasis, hookworm infection, schistosomiasis, and trachoma are widespread conditions which are more influential on children under 5 years old (Dangour et al. 2013). However, they are preventable with sufficient sanitation, hygiene, and clean water.

Moreover, living in clean surroundings increases human health and well-being. For example, in a study conducted in Toronto, Canada differences between people living in two areas, one with more greenery and another with less were compared. Data including environmental conditions, number of trees, general health perception of the inhabitants, occurrences of cardio-metabolic and mental illnesses were used in the study. The results of this study showed that living in a green neighborhood (with more than ten trees) has a positive effect on health, especially cardio-metabolic conditions comparable to those living in less green neighborhoods who were seven ages younger (Kardan et al. 2015). Having one more tree in the neighborhood has a positive influence on the cardio-metabolic conditions comparable to those living in neighborhoods with one less tree and 1.4 years younger on average.

Another similar study focused on the influence of community-based forest management on the well-being of local stakeholders in eight Tanzanian villages with comparison to non-managed forests over a 10-year period, 2005–2015 and found that there was better water access but greater food insecurity in areas with forest management (Gross-Camp 2017).

Moreover, the relationship between environmental properties and organism activities has become clearer thanks to the research studies conducted on this matter. For example, an investigation of the terrestrial locomotion of mummichogs and their method of navigation toward water in a terrestrial environment showed that they primarily use visual cues, specifically reflected light, to orient toward water (Bressman et al. 2016).

10.6 Relationships Among Water, Environment, and Health

The relationships between water and environment; health and environment; and water and health have been discussed above. Health-, environment-, and water-related issues affect each other and they are all interrelated. The relationships among water, health, and environment have some points that need further research. What is already known is that deterioration of one of them, causes the others to worsen as well. For example, when environment gets polluted, such as through the increase of total nitrogen concentration in water, trees produce more pollen. As a result of this, allergenic illnesses such as asthma become more widespread (Kozlowski and Pallardy 2002; Reinmuth-Selzle et al. 2017).

Water storage and delivery are other important issues for preservation of good health because water storage under unsuitable conditions causes some health problems. Because water can easily become host for algae when it stays under sunlight for a duration varying from hours to a day depending on the strength of the sunshine. Both transfer and storage of the freshwater need due care and must be considered as important part of the process of preserving clean water. This means a holistic approach is necessary at different stages of dealing with clean water.

Among water, health, and environment, there are also differences in terms of data collection methods and tools available. For example, water and environmental issues can be researched with experiments and numerical data but as non-conscious agents cannot provide details on their experience. On the other hand, human subjects can provide detailed information verbally on the effects of water and environment on their health. Moreover, tolerance levels and reaction times of humans are different to deterioration of water, environment, and health. Although people can tolerate some worsening of water or environment for a longer period of time, slight worsening of health has more immediate and significant impact.

10.7 The Importance of Knowledge for Issues of Water, Environment, and Health

Some of the common serious threats to water, environment, and human health are droughts, floods, climate change, and fires. These are natural disasters which cannot be prevented but they may affect human life, health, and infrastructures. Severity of these threats shows evidently that we have to live in harmony with the environment, not in discord. In order to succeed in this aim, first we need to know the properties of nature, its cycles (such as water, nutrients and so on), its actions, and its reactions. Although these threats cannot be annulled, with sufficient knowledge they can be managed more successfully.

For sustainable management of the above threats, we need knowledge and we cannot attain it without due measurements. Therefore, measurements are of essential quality. Nowadays, new technologies and improvements in the science enable scientists to make calculations for a study area using numerical models or satellite images according to stations' measurements and significantly reduce the budget spent on measurements. Therefore, some scientific discussions in this field focus on how to reduce the number of measurement stations or measurement parameters. Indeed, modeling techniques help to simulate whole study area in acceptable precision but calibrating, testing, and validating the results with limited measurements cannot provide the desired level of certainty. Therefore, modeling results cannot replace actual measurements especially in the case of peak values. Moreover, measurement equipment has become cheaper, more precise and more common, and establishing a station is easier than before. There has been a lot of useful research which brought researchers to current level of knowledge and the research community should look at ways of sustaining and maybe increasing the measurements. Continuity is an indispensable quality to have a clear and comparable record of measurements and once time has passed, it is not possible to go back and carry out the missing measurements.

Measurements are needed to provide evidence in describing events but there are also several types of errors in measurements. Firstly, as with any type of measurement, there are random errors which happen naturally. Second type of errors is the systematic ones, which occur due to a mis-calibrated instrument or a design error.

Systematic errors affect all measurements using the same system. Another crucial error is in the selection of parameters and indicators such as types, total number, and proportion of grains. When there is a wrong parameter or indicator in the measurements, they cannot reflect reality and one cannot acquire clear results because of their natural complexity. These errors should be taken into account when evaluating and interpreting measurements.

Moreover, the sufficient time for research projects is another significant factor in reaching concrete results. Unfortunately, budgets define the duration of most research studies and bring a project to an early closure although it may be in the middle of an observed event, or even at the beginning. Continuous research and sharing the measurements through publications may help us to clarify the relationships. Moreover, some results are highly dependent on their location because each continent, country, basin, and catchment has its own characteristics and properties. These unique qualities make it difficult to find identical or very similar results obtained from different research studies by different institutes and researchers. These can cause some confusion on reliability of the data or even the researcher. To maximize benefit from the research studies and help advance scientific knowledge, one should make continuous measurements at several stations and make the results easily available for inspection.

10.8 Climate Change

Since the industrial revolution average world surface temperature has surpassed the emergency increase limit of 2 °C. If there are no preventative measures taken, climate change will lead to increased droughts and floods. The subsequent yield reduction will be >50% in 2050 and almost 90% in 2100 for the major crops especially in Africa (Li et al. 2009).

Clearly, climate change is effective on human health as well as water availability and environment (US Global Change 2017). A study in Pakistan shows that the number of P. falciparum has a trend to increase with favorable climate conditions (Bouma et al. 1996). This is a clear warning sign that climate change and the associated temperature increase threatens human health. For example, some studies highlight that for twenty-first century global warming is the most important health risk (Costello et al. 2009; Huang et al. 2013) and the risk is highest among developing countries' children (Hanna and Oliva 2016). These studies commonly point to the magnitude of the threat climate change causes for humanity.

Recent research has shown that climate change is progressively increasing severe drought events especially in the Northern Hemisphere and causing regional tree die-off events (Carnicer et al. 2011). Trees contribute to the water cycle and the global reduction of the carbon sink. Therefore, climate change effects accelerate due to reduction of tree numbers as well as an increase in forest fires.

Climate change has negative effects on citizens of developing countries with the biggest impact, although they are not the main agents in causing climate change

(Patz et al. 2007). Moreover, these countries have more sensitive economies and health systems and are more dependent on agriculture.

The negative effects of the climate change are not limited with water and environmental issues, but on happiness according to a study report including 67 countries (Rehdanz and Maddison 2005). This study found that high latitude countries have some benefit from climate change compared to low latitude ones. For example, increasing temperature will reduce happiness between 2010–2039 and 2040–2069 in Turkey. An example for increasing is Russia. In addition, it will cause major levels of migration. Before 2050 more than 250 million people will leave their homeland because of climate change (McMichael et al. 2012). Climate change effects such as North Atlantic Oscillation (NAO) may also cause droughts. In history similar droughts during the period of Western Roman Empire led to mass migration and incidents causing the empire to fall (Drake 2017). Consequently, it is necessary to have more studies to measure various effects of climate change on humans at personal and societal levels.

10.9 Is There Any Solution?

As pointed out throughout this chapter, water, health, and environment are closely related to each other. For a successful solution to problems related to one of these three, one needs to consider all of them together and develop a holistic approach. A key to achieve this is Optimum Sustainable Integrated Water Resources Management (OSIWRM) despite critical discussions about Integrated Water Resources Management (IWRM) success (Merrey 2008; Biswas 2008). The key role lies in the great influence of political and economic contexts on water, environment, and health issues. For example, a notorious preventable waterborne disease, cholera, caused 1,678 deaths in Yemen with 0.6% fatality rate and 1,098 deaths in Somalia with 1.5% fatality rate (WHO 2017). These outbreaks can be controlled and the number of the deaths can be reduced significantly with our current level of knowledge, but political issues prevent us from effective action. Therefore, successful OSIWRM should contain political and economic elements. Moreover, OSIWRM helps to establish sustainability and minimizes human effects on nature. An example of successful implementation is the zero discharge concept (Gronwall and Jonsson 2017; Visvanathan and Hufemia 1997).

Geographic Information System (GIS) and Remote Sensing (RS) are two main tools to manage water, environment, and health issues holistically (Aspinall and Pearson 2000). These tools are also very useful to find out more on the relationships among water, health, and environment (Hanari-Bojd et al. 2012; Kaminska and Turski 2004; Jaiswal et al. 2003). GIS provides a system to record data, to search data according to location, to present data, and results visually. The most important properties of GIS are calculation and query for the purpose of testing theories and finding out results according to location. These qualities of GIS and RS help to manage water, health, and environment through improved communication. Both these tools are helpful to manage water and to protect forests (Teich and Bebi 2009; Teresneu et al. 2016; Carver et al. 2006; Lant et al. 2005). The more widespread use of these tools in management will enable graphic representation of the data and lead to more accurate decisions. Indeed, lack of good management is the main cause of reduction in available clean water and amount of forests and consequently impacts on human health.

10.10 What You Can Do at Personal Level

Jonas Salk, who developed Polio Vaccine, said "*If all the insects were to disappear from the earth, within 50 years all life on earth would end. If all human beings disappeared from the earth, within 50 years all forms of life would flourish.*" These words are a strong warning to human beings to reduce their footprint. The best and easiest way to reduce footprint is to establish untouched forests, where there is no human activity including roads. Unfortunately, 46–58 thousand square miles of forest every year, equivalent to 48 football fields every minute, is lost. In addition, approximately 15% of all greenhouse gas emissions are the result of deforestation (World wildlife 2017). The questions of how to reduce effects of climate change and pollution, how to improve water cycle, how to protect environment or establish a healthy environment for a healthy life have a single clear answer: **increasing the number of trees** (Calder 2007). Furthermore, it is clear that water, environment, and health are proportionate with the number of trees. Therefore, if one wants to live in a better environment with good health and to drink natural freshwater, one should plant trees. A new, but a constant duty for scientists is to encourage managers to improve number of trees, because of their positive effects on nature such as stormwater (Berland et al. 2017), health (Kardan et al. 2015), air quality (Abhijith and Gokhale 2015), climate change (Bonan 2008), evaporation (Baldocchi and Meyers 1991), and erosion.

Please plant one more tree for your future/child!

In our houses, we should separate plumbing system according to usage aims. At home using various types of water can be a viable solution to reduce water withdraw from natural resources. The cleanest freshwater resources can be used for cooking in the kitchen, recycled water from saltwater, or wastewater for cleaning purposes and water filtered by a basic system for restroom use.

Current world population is around 7.6 billion (UN 2017b) and if each person saves an amount of water every day, it will make a massive contribution at the global level. Therefore, governments and international political organizations such as the European Union or United Nations need to concentrate their efforts on how to increase individual awareness for water and provide information to improve individual action. Increased awareness and less water consumption are described as "building artificial dams" as they have similar outcomes. It may also be likened to obtaining new artificial water resources. Another positive outcome of recycling water is that there will be less wastewater to discharge. Moreover, wiser consumption of any product including food, clothes, computers, cars is another way to save water. In a personal anecdote, the author went shopping for a silk scarf in China, when a Chinese colleague said "*One is enough*". This highlights a cultural difference in consumption. Another cultural example is from the West "*One man's trash is another man's treasure*" and emphasizes the importance of recycling. In the Turkish culture, there is a popular saying, "*Many a mickle makes a muckle*". The culture of saving and recycling needs to be embedded in education systems and spread through public awareness campaigns. Shopping, use, waste, and recycling are different stages which need more attention. Every product that we use is produced using water and increased consumption, shorter lifespan of products, lack of recycling, and production of more waste mean a greater demand for water. One person may not be seen to have a clear effect, but remember that globally there are 7.6 billion times one persons. Through individual efforts on reducing consumption and waste with more careful use of resources, there will be a significant contribution to water, environment, and health at a global level.

10.11 Conclusion

In this chapter, the interrelated nature of water, environment, and health was demonstrated with some examples. Therefore, these three cannot be considered separately but must be holistically approached and managed. The threats to water, environment, and health are real and imminent and require immediate action. OSIWRM is suggested as a method of managing natural resources and dealing with the dangers associated with them. It can be considered at various levels including personal, social, institutional, and political.

References

Abhijith KV, Gokhale S (2015) Passive control potentials of trees and on-street parked cars in reduction of air pollution exposure in urban street canyons. Environ Pollut 204:99–108

Arabian business (2017). http://www.arabianbusiness.com/exclusive-uae-mulls-man-made-mountain-in-bid-improve-rainfall-630079.html. Accessed 10 Oct 2017

Aspinall R, Pearson D (2000) Integrated geographical assessment of environmental condition in water catchments: Linking landscape ecology, environmental modelling and GIS. J Environ Manag 59(4):299–319

Baldocchi DD, Meyers TP (1991) Trace gas-exchange above the floor of a deciduous forest:1. Evaporation and Co2 efflux. J Geophys Res-Atmos 96(D4):7271–7285

Berland A et al (2017) The role of trees in urban stormwater management. Landsc Urban Plan 162:167–177

Biswas AK (2008) Integrated water resources management: is it working? Int J Water Resour Dev 24(1):5–22

Bonan GB (2008) Forests and climate change: forcings, feedbacks, and the climate benefits of forests. Science 320(5882):1444–1449

Bouma MJ, Dye C, VanderKaay HJ (1996) Falciparum malaria and climate change in the northwest frontier province of Pakistan. Am J Trop Med Hyg 55(2):131–137
Bressman NR, Farina SC, Gibb AC (2016) Look before you leap: visual navigation and terrestrial locomotion of the intertidal killifish Fundulus heteroclitus. J Exp Zool Part A-Ecol Genet Physiol 325(1):57–64
Calder IR (2007) Forests and water-ensuring forest benefits outweigh water costs. For Ecol Manage 251(1–2):110–120
Carnicer J et al (2011) Widespread crown condition decline, food web disruption, and amplified tree mortality with increased climate change-type drought. Proc Natl Acad Sci USA 108(4):1474–1478
Carver AD et al (2006) Applying spatial analysis to forest policy evaluation: case study of the Illinois forestry development act. Environ Sci Policy 9(3):253–260
Chaffin JD, Bridgeman TB (2014) Organic and inorganic nitrogen utilization by nitrogen-stressed cyanobacteria during bloom conditions. J Appl Phycol 26(1):299–309
Costello A, Abbas M, Allen A (2009) Managing the health effects of climate change (vol 373, 1693 (2009)). Lancet 373(9682):2200–2200
Dahirel M et al (2017) Intraspecific variation shapes community-level behavioral responses to urbanization in spiders. Ecology 98(9):2379–2390
Dangour AD et al (2013) Interventions to improve water quality and supply, sanitation and hygiene practices, and their effects on the nutritional status of children. Cochrane Database Syst Rev (8)
Drake BL (2017) Changes in North Atlantic oscillation drove population migrations and the collapse of the Western Roman Empire. Sci Rep 7
Falkenmark M (1999) Forward to the future: a conceptual framework for water dependence. Ambio 28(4):356–361
Garcia-Perez J et al (2017) Risk of bone tumors in children and residential proximity to industrial and urban areas: new findings from a case-control study. Sci Total Environ 579:1333–1342
Gronwall J, Jonsson AC (2017) The impact of 'zero' coming into fashion: zero liquid discharge uptake and socio-technical transitions in Tirupur. Water Altern- Interdiscip J Water Polit Dev 10(2):602–624
Gross-Camp N (2017) Tanzania's community forests: their impact on human well-being and persistence in spite of the lack of benefit. Ecology and Society. **22**(1)
Hanari-Bojd AA et al (2012) Larval habitats and biodiversity of anopheline mosquitoes (Diptera: Culicidae) in a malarious area of southern Iran. J Vector Borne Dis 49(2):91–100
Hanna R, Oliva P (2016) Implications of climate change for children in developing countries. Future Child 26
Huang CR et al (2013) Managing the health effects of temperature in response to climate change: challenges ahead. Environ Health Perspect 121(4):415–419
International flood network (2017). http://www.internationalfloodnetwork.org/. Accessed 10 Oct 2017
Jaiswal RK et al (2003) Role of remote sensing and GIS techniques for generation of groundwater prospect zones towards rural development - an approach. Int J Remote Sens 24(5):993–1008
Kaminska IA, Turski OA (2004) Geographical information system (GIS) as a tool for monitoring and analysing pesticide pollution and its impact on public health. Ann Agric Environ Med 11(2):181–184
Kardan O et al (2015) Neighborhood greenspace and health in a large urban center. Sci Rep 5
Kozlowski TT, Pallardy SG (2002) Acclimation and adaptive responses of woody plants to environmental stresses. Bot Rev 68(2):270–334
Lant CL et al (2005) Using GIS-based ecological-economic modeling to evaluate policies affecting agricultural watersheds. Ecol Econ 55(4):467–484
Levy K et al (2016) Untangling the impacts of climate change on waterborne diseases: a systematic review of relationships between diarrheal diseases and temperature, rainfall, flooding, and drought. Environ Sci Technol 50(10):4905–4922

Li YP et al (2009) Climate change and drought: a risk assessment of crop-yield impacts. Clim Res 39(1):31–46
Marfe G, Stefano CD (2016) The evidence of toxic wastes dumping in Campania, Italy. Crit Rev Oncol Hematol 105:84–91
McMichael C, Barnett J, McMichael AJ (2012) An ill wind? Climate change, migration, and health. Environ Health Perspect 120(5):646–654
Merrey DJ (2008) Is normative integrated water resources management implementable? Charting a practical course with lessons from Southern Africa. Phys Chem Earth 33(8–13):899–905
Paerl HW et al (2016) It takes two to tango: when and where dual nutrient (N & P) reductions are needed to protect lakes and downstream ecosystems. Environ Sci Technol 50(20):10805–10813
Patz JA et al (2007) Climate change and global health: quantifying a growing ethical crisis. EcoHealth 4(4):397–405
Rehdanz K, Maddison D (2005) Climate and happiness. Ecol Econ 52(1):111–125
Reinmuth-Selzle K et al (2017) Air pollution and climate change effects on allergies in the anthropocene: abundance, interaction, and modification of allergens and adjuvants. Environ Sci Technol 51(8):4119–4141
Schindler DW et al (2008) Eutrophication of lakes cannot be controlled by reducing nitrogen input: results of a 37-year whole-ecosystem experiment. Proc Natl Acad Sci USA 105(32):11254–11258
Somlyody L (1995) Water-quality management - can we improve integration to face future-problems. Water Sci Technol 31(8):249–259
Taranu ZE et al (2017) Predicting microcystin concentrations in lakes and reservoirs at a continental scale: a new framework for modelling an important health risk factor. Glob Ecol Biogeogr 26(6):625–637
Teich M, Bebi P (2009) Evaluating the benefit of avalanche protection forest with GIS-based risk analyses-a case study in Switzerland. For Ecol Manag 257(9):1910–1919
Teresneu CC et al (2016) Using the GIS tools for a sustainable forest management. Environ Eng Manag J 15(2):461–472
UN (2017a). http://www.un.org/sustainabledevelopment/water-and-sanitation/. Accessed 10 Oct 2017
UN (2017b). https://www.un.org/development/desa/publications/world-population-prospects-the-2017-revision.html. Accessed 10 Oct 2017
UNICEF (2017). https://www.unicef.org/drought/drought-countries.htm. Accessed 10 Oct 2017
US Global Change (2017). https://health2016.globalchange.gov/climate-change-and-human-health. Accessed 10 Oct 2017
Visvanathan C, Hufemia AMM (1997) Exploring zero discharge potentials for the sustainability of a bottle washing plant. Water Sci Technol 35(9):181–190
WHO (2017). http://www.emro.who.int/health-topics/cholera-outbreak/weekly-cholera-updates.html. Accessed 10 Oct 2017
World wildlife (2017). https://www.worldwildlife.org/threats/deforestation

Chapter 11
Impacts of Resource Consumption and Waste Generation on Environment and Subsequent Effects on Human Health: A Study Based on Ecological Footprint Analysis

Tanmoy Roy Tusher, Tonmoy Pondit, Moinul Hasan, Mahmuda Binte Latif and Md. Binyamin

Abstract In this study, the carbon emission and ecological footprint (EF) from resource consumption and waste generation in Tangail municipal area (TMA) of Bangladesh were estimated, and the ecological sustainability status of the area was also assessed by calculating biocapacity. Data were collected from 180 households covering 18 wards of the TMA. Households, 10 from each ward, were selected considering different income groups and a questionnaire survey was conducted to gather information about their monthly consumption patterns of energy, food, water, goods, and services along with waste generation for the year 2016. The calculated total carbon emission from different carbon uptake factor was 433,235.302 tons CO_2, while the highest (271,451.471 tons CO_2) carbon emission was derived from natural gas consumption and the lowest (380.083 tons CO_2) was observed from water consumption. Carbon uptake land was the major combined sector contributing to total EF (51.79%), whereas cropland was found as the single sector contributing to high footprint figure (47.13%) in TMA. The study found that the EF increases not only with resource consumption and waste generation, but also with household size ($p < 0.01$) and income ($p < 0.01$). Total EF and biocapacity of the area were found as 2.21 and 0.08 gha/capita, respectively, while the ecological deficit was estimated as 2.13 gha/capita which may have both direct and indirect impacts on human health. The study concluded that EF of TMA exceeds its biocapacity by almost twenty-eight (28) times. The study recommended to set annual targets, to reduce EF at household levels by awaking people about how much they are consuming, its environmental impacts and subsequent effects on their health.

T. R. Tusher (✉) · T. Pondit · M. Hasan · M. B. Latif
Department of Environmental Science and Resource Management,
Mawlana Bhashani Science and Technology University, Tangail 1902, Bangladesh
e-mail: tusrroy@yahoo.com

Md. Binyamin
Department of Statistics, Mawlana Bhashani Science and Technology University, Tangail 1902, Bangladesh

R. B. Singh et al. (eds.), *Urban Health Risk and Resilience in Asian Cities*,
Advances in Geographical and Environmental Sciences,
https://doi.org/10.1007/978-981-15-1205-6_11

Keywords Resource · Consumption · Waste · Ecological footprint · Human health · Bangladesh

11.1 Introduction

Human life and all human activities depend on nature and ecosystem products and services including resources, waste absorptive capacity, and space to host urban infrastructure (Ewing et al. 2010a). Since human populations and their consumption continue to expand rapidly, the pressure on natural resources is also increasing at an alarming rate resulting in environmental unsustainability because of limited supplies of resources while many of them are finite (Pimentel et al. 2006; Shakil et al. 2014). Along with increasing pressure on natural resources, human activities such as urbanization, industrialization, transportation, energy consumption, food production, and waste disposal account for most of the greenhouse gas (GHG) emissions, mainly carbon emission which primarily contributes to global climate change (IPCC 2007) and subsequently affects the physical, social, and psychological health of humans (Portier et al. 2010). The systematic accounting tool is thus needed for tracking the combined effects of human pressures on the planet and also for determining management strategies to ensure healthy life and future prosperity (Gali et al. 2012; Portier et al. 2010). Ecological footprint (EF) is a scientifically reviewed tool which measures how much life-supporting natural capital, expressed in biologically productive area, is necessary to meet the resource demand, including goods and services, and waste absorption requirements of a given population using prevailing technology (Wackernagel et al. 2004; Kitzes et al. 2007). It is used as a core indicator of environmental sustainability and for tracking unsustainable use of biological resources (Bala et al. 2012; Ewing et al. 2010a). It was developed as a means of making our ecological constraints clear and sustainability strategies more effective and livable (Shakil et al. 2014). Several studies have already been reported on applications of EF during the past decade to address environmental sustainability (Monfreda et al. 2004; Medved 2006; Chen and Chen 2006; Bagliani et al. 2008; Niccolucci et al. 2008; Bala and Hossain 2010).

The accounting of EF is of utmost importance for the context of Bangladesh especially in municipal areas where the unplanned development with excessive resource consumption and waste generation is familiar which results in unsustainable environmental situation (Shakil et al. 2014), while the bio-productive land and water body are reducing day by day (Rahman et al. 2012; Byomkeshet al. 2009) on which more than one-third of the country's population depends for livelihood (Asaduzzaman 2002). Such conditions may disrupt regional environmental stability and pose additional load on rest of the world. It is thus essential to measure EF if municipalities want to manage their demand pattern, to reduce their carbon emissions, and to minimize succeeding health problems of EF (Sprangers 2011; Quam et al. 2017). Moreover, the EF is must to generate an aware citizenry which is required to make wise decisions regarding policies and practices aimed at reducing GHG emissions and the human impact on the earth's resources (Shakil et al. 2014; Cordero et al. 2008). In this study, Tangail municipal area (TMA) was selected as one of the representatives of cities or municipal areas of Bangladesh as well as other developing countries, which has also been experienced speedy urban growth as it is recognized for its educational institutions and handloom industries. The enlarged urbanization has resulted in higher population density, increased demand for food and transportation, increased waste generation, disappearing green and open spaces in the TMA (Sarker et al. 2015). Considering these, the study was conducted to measure the carbon emission and EF from household resource consumption and waste generation in TMA focusing on interaction effects of household size and income, and also to calculate and compare biocapacity with the EF for assessing the environmental sustainability status of the area.

11.2 Materials and Methods

11.2.1 Study Area

Tangail municipal area (TMA), lies on 24°15′ N latitude and 89°55′ E longitude, is located at the south-eastern part of Tangail Sadar Upazila and occupies an area of 29.43 km^2 covering 18 wards (Fig. 11.1). The total number of population and household of TMA is 167,412 and 26,700, respectively, with a population density of 4,376 per km^2. The average household size in TMA is 5.01 person. The household number and population in TMA with respective ward no. is shown in Table 11.1. Average monthly income of the people is BDT. 10,796 and 9% families are living below the poverty level (below BDT. 4000.00) which is lower than the country's national average (28%) (Master Plan for Tangail Paurashava 2013).

11.2.2 Selection of Ecological Footprint Parameters

According to Calculation Methodology of National Footprint Accounts (2010) obtained from Global Footprint Network, EF consists of six components: carbon uptake land, cropland, grazing land, fishing ground land, forest land, and built-up land (Ewing et al. 2010b). Among these six components, grazing land is absent in TMA (Master Plan for Tangail Paurashava 2013) and, thus, was not included in the calculation index. Forest land cannot also be included in the calculation index since TMA does not get any timber supply for fuel and other uses from existing forest area or trees within the study area. So, in this study the footprint accounts of TMA comprise four components, i.e., carbon uptake land, cropland, fishing grounds land, and built-up land. Biocapacity of any area consists of all the land use type considered in footprint account except the carbon uptake land. Since grazing land is absent in

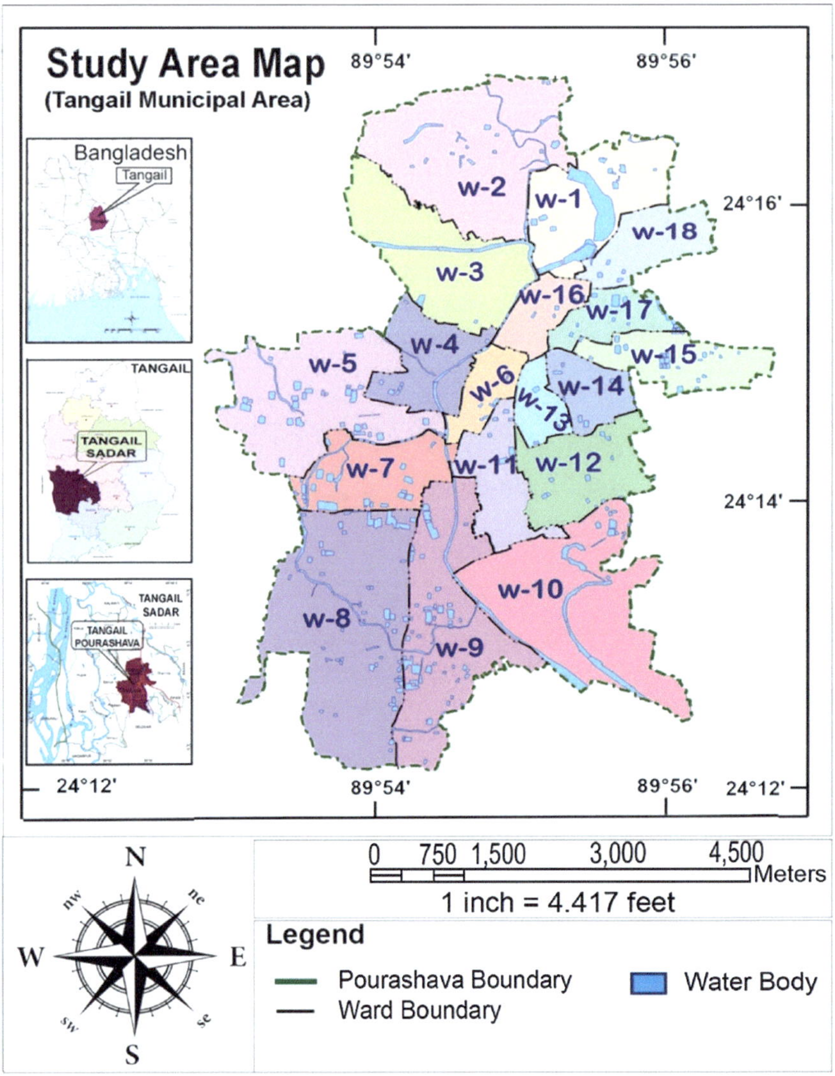

Fig. 11.1 Study area in Tangail municipal area (TMA) of Bangladesh. Modified from Master Plan for Tangail Paurashava (2013), with reuse permission from Tangail municipal authority, Bangladesh

TMA, four components were included in the calculation on biocapacity index of the present study. The components considered in the present study for accounting EF and biocapacity of TMA are given in Table 11.2.

Table 11.1 The numbers of households and population with respective ward no. and comprising area in Tangail municipal area (TMA), Bangladesh

Ward No.	Comprising area	Number of population	Number of households	Household/Family size
1	Akur Takur Para-north, Dewla, District headquarters	10,388	1518	6.89
2	Enayetpur	8760	1455	5.15
3	West Akur Takur Para, North Kagmara, South Kagmara	13,903	1662	4.85
4	Bepari Para, Fakir Para, Bera Doma, Dighulia, Char Dighulia	9208	1421	4.73
5	Kalipur, Lakshimpur, Sarutia, Shakrail	7235	1279	4.79
6	College Para, Paradise Para, Par Dighulia-part	6712	1122	5.17
7	Baluk Kandi, Bagbari, Patuli Bhabani Bagbari, Uttar Santosh-part, Santosh Palpara	7325	1413	4.51
8	Aloa Bhabani Pahim, Aloa Pahim, Dakshin Santosh-part	8026	1453	4.72
9	Aloa Baratia, Aloa Paikasta, Aloa Tarini, Baluk Kandi-part, Char Patuli, Purba Aloa	8575	1496	4.33
10	Bajitpur, Berai, Basrakandi, Kazipur, Patal	6189	1188	4.88
11	Berabuchna, Kachua Para, Kanda Para	7536	1972	3.78
12	Adi Tangail, Bepari Para, Bil Para	7160	1295	4.85

(continued)

Table 11.1 (continued)

Ward No.	Comprising area	Number of population	Number of households	Household/Family size
13	Tangail Mahalla, Chayanir Bazar, Pachanir Bazar, Thana Para, Uttar Thana Para	7572	1512	5.36
14	Purba Adalat Para, Adalat Para, Biswas Betka-part S.W corn, Shaha Para	11,786	1566	4.90
15	Ashekpur, Biswas Betka-part	12,118	1744	4.82
16	Akur Takur Para-part, Par Dighulia-part	10,670	1679	5.16
17	Kumudini College Para, Munshi Para, Registry Para, Biswas Betka-west	11,081	1396	5.45
18	Kodalia, Sabalia	13,168	1529	5.78
Total		167,412	26,700	–

Data source Master Plan for Tangail Paurashava (2013), with reuse permission from Tangail municipal authority, Bangladesh

Table 11.2 Components of ecological footprint (EF) and biocapacity applicable for Tangail municipal area (TMA), Bangladesh

Component	Unit (ha)	Existence in TMA	
		Ecological footprint	Biocapacity
Carbon uptake land	Area	Present	Not applicable
Cropland	Area	Present	Present
Grazing land	Area	Absent	Absent
Fishing ground land	Area	Present	Present
Forest land	Area	Absent	Present
Built-up land	Area	Present	Present

11.2.3 Data Collection

Ecological footprint (EF) and biocapacity calculation is a data intensive research work. Data regarding crop, fishing ground, forest and built-up lands of TMA were collected from land use survey (2009) done for preparing TMA Master Plan (2013), while data regarding the household consumption patterns (energy, waste, water, transportation, food, goods, and services) of the inhabitants to determine carbon uptake land were collected through household's questionnaire survey. Since the total number of population in TMA is 160,412 and the total number of households is 26,700 (Table 11.1), the sample size was determined 379 households at 95% confidence level and confidence interval of 5 (Creative Research System 2010). The final survey was carried out on 180 households of TMA for time constraints and difficulties in getting accessibility for survey purpose.

The questionnaire was composed of 25 questions covering both the closed- and open-ended questions to gain the maximum possible results. However, ecological footprint is influenced by income status (Ewing et al. 2010a). Thus, the data were collected from all income groups such as low (BDT <10,000), middle (BDT 10,000–30,000), and high (BDT >30,000) income households in order to make an overall estimate of EF account of TMA and also to investigate the effects of income on households' EF. During the household survey housewives of each household were interviewed because they are mainly responsible for the daily/weekly/monthly grocery shopping. The households were also categorized into three groups as small (<5 persons), medium (5–8 persons), and extended (>8 persons) families in order to assess the relationship between the family/household size and their EF.

The conversion factors (i.e., sequestration factor, equivalence factor, yield factor) used in the study were collected from various international research agencies such as Stockholm Environment Institute (SEI), Berkeley Institute on the Environment (BIE), Intergovernmental Panel on Climate Change (IPCC), United States Environmental Protection Agency (USEPA), and Global Footprint Network, etc.

11.2.4 Estimation of Ecological Footprint and Biocapacity

The EF and biocapacity of the TMA were estimated using the standard equations given by National Footprint Accounts (NFA) accounting framework.

11.2.5 Calculation of Total Footprint

The total EF was calculated by calculating each of its components, i.e., carbon uptake land, cropland, fishing ground land, and built-up land using the formulae shown in Fig. 11.2. Finally, the total footprint was estimated as the sum of footprints of all components. Moreover, the carbon uptake land footprint was calculated following the component (electricity, water, energy, waste, transportation, foods, goods, and services)-based approach using the equations (Simmons et al. 2000) shown in Fig. 11.3, and the strategies applied to collect representative data are also briefly described below.

(a) $$\left(\text{CO}_2 \text{ emissions [t]} \,/\, \text{Sequestration [t/ha/year]} \right) = \text{Carbon uptake land [w ha]} \times \text{Forest equivalency factor} = \text{Carbon footprint [g ha]}$$

(b) $$\text{Agricultural products [t]} \,/\, \text{World crop yields [t/ha]} = \text{Cropland [w ha]} \times \text{Cropland equivalency factor} = \text{Cropland footprint [g ha]}$$

(c) $$\text{Fish products [t]} \,/\, \text{World fish yields [t/ha]} = \text{Fishing ground [w ha]} \times \text{Fishing equivalency factor} = \text{Fishing ground footprint [g ha]}$$

(d) $$\text{Built-up land [w ha]} \times \text{Cropland equivalency factor} = \text{Built-up land footprint [g ha]}$$

Fig. 11.2 Formulae used to calculate footprint of different components present in Tangail municipal area (TMA), Bangladesh: **a** carbon uptake land footprint (Simmons et al. 2000); **b** cropland footprint (Borucke et al. 2013); **c** fishing ground footprint (Borucke et al. 2013); and **d** built-up land footprint (Borucke et al. 2013)

11.2.6 Footprint from Electricity Consumption

Average monthly electricity bill was calculated for the variation of summer and winter. Then it was multiplied with 12 to get total amount of electricity bill of a year, and converted into unit of electricity consumption—MWh (MegaWatt/hour) with the help of unit price of electricity rated by Bangladesh Power Development Board (BPDB). Finally, the total amount of carbon emission from electricity consumption for each unit of MWh electricity use was determined from GHG Equivalencies Calculator-United States Environmental Protection Agency (USEPA) using the standard equation given in Fig. 11.3.

11.2.7 Footprint from Natural Gas Consumption

Respondents were asked about the type and number of hours of their gas burner use per day on an average. After that it was converted to total amount of cubic meter gas use per year. During this conversion per hour of gas consumption rate by a double burner (majority household use double burner gas cooker) was used by The Engineering Toolbox (2001). At last the total amount of carbon emission from natural gas consumption was determined from GHG Equivalencies Calculator-USEPA using the equation shown in Fig. 11.3.

$$\textbf{Electricity Consumption} = \overbrace{\left(\text{annual MWh} \times \frac{\text{tons } CO_2}{\text{MWh}}\right)}^{CO_2\text{ emission}} \div \overbrace{\left(\frac{\text{annual tons } CO_2}{\text{acre}}\right)}^{\text{Sequestration}} \times \text{ha} \times \overbrace{\left(\frac{\text{annual gha}}{\text{hactere}}\right)}^{\text{Equivalency Factor}}$$

$$\textbf{Natural Gas Consumption} = \overbrace{\left(\text{annual } m^2 \times \frac{\text{tons } CO_2}{m^3\text{gass}}\right)}^{CO_2\text{ emission}} \div \overbrace{\left(\frac{\text{annual tons CO}}{\text{acre}}\right)}^{\text{Sequestration}} \times \text{ha} \times \overbrace{\left(\frac{\text{annual gha}}{\text{hactere}}\right)}^{\text{Equivalency Factor}}$$

$$\textbf{Waste Generation} = \overbrace{\left(\text{annual tons} \times \frac{\text{tons } CO_2}{\text{tons waste}}\right)}^{CO_2\text{ emission}} \div \overbrace{\left(\frac{\text{annual tons } CO_2}{\text{acre}}\right)}^{\text{Sequestration}} \times \text{ha} \times \overbrace{\left(\frac{\text{annual gha}}{\text{hactere}}\right)}^{\text{Equivalency Factor}}$$

$$\textbf{Water Consumption} = \overbrace{\left(\text{million liter} \times \frac{\text{tons } CO_2}{\text{million liter}}\right)}^{CO_2\text{ emission}} \div \overbrace{\left(\frac{\text{annual tons } CO_2}{\text{acre}}\right)}^{\text{Sequestration}} \times \text{ha} \times \overbrace{\left(\frac{\text{annual gha}}{\text{hactere}}\right)}^{\text{Equivalency Factor}}$$

$$\textbf{Food Consumption} = \overbrace{\left(\text{million U.S.\$} \times \frac{\text{tons } CO_2}{\text{million U.S.\$}}\right)}^{CO_2\text{ emission}} \div \overbrace{\left(\frac{\text{annual tons } CO_2}{\text{acre}}\right)}^{\text{Sequestration}} \times \text{ha} \times \overbrace{\left(\frac{\text{annual gha}}{\text{hactere}}\right)}^{\text{Equivalency Factor}}$$

$$\textbf{Goods and Services Consumption} = \overbrace{\left(\text{million U.S.\$} \times \frac{\text{tons } CO_2}{\text{million U.S.\$}}\right)}^{CO_2\text{ emission}} \div \overbrace{\left(\frac{\text{annual tons } CO_2}{\text{acre}}\right)}^{\text{Sequestration}} \times \text{ha} \times \overbrace{\left(\frac{\text{annual gha}}{\text{hactere}}\right)}^{\text{Equivalency Factor}}$$

$$\textbf{Fuel (Petrol) Consumption} = \overbrace{\left(\text{liter} \times \frac{\text{tons } CO_2}{\text{liter}}\right)}^{CO_2\text{ emission}} \div \overbrace{\left(\frac{\text{annual tons } CO_2}{\text{acre}}\right)}^{\text{Sequestration}} \times \text{ha} \times \overbrace{\left(\frac{\text{annual gha}}{\text{hactere}}\right)}^{\text{Equivalency Factor}}$$

$$\textbf{Fuel (CNG) Consumption} = \overbrace{\left(\text{cubic feet} \times \frac{\text{tons } CO_2}{\text{cubic feet}}\right)}^{CO_2\text{ emission}} \div \overbrace{\left(\frac{\text{annual tons } CO_2}{\text{acre}}\right)}^{\text{Sequestration}} \times \text{ha} \times \overbrace{\left(\frac{\text{annual gha}}{\text{hactere}}\right)}^{\text{Equivalency Factor}}$$

Fig. 11.3 Equations used to calculate the footprint of various components of carbon uptake land in Tangail municipal area (TMA) of Bangladesh (Simmons et al. 2000)

11.2.8 Footprint from Waste Generation

Households of the study area were divided into some clusters on the basis of income level. Households were then randomly selected from each stratum. By calculating how much waste is produced each day from a household and multiplying by 365, the annual waste production was determined. Then it was multiplied with the total household of a particular ward. Finally, it was multiplied by a carbon intensity to get the footprint by using the standard equation as shown in Fig. 11.3.

11.2.9 Footprint from Water Consumption

The water consumption footprint is an indicator of water use that looks at both direct and indirect water use. Water consumption impact on GHG emission has not been mentioned in the Component Method of ecological footprint assessment (Simmons et al. 2000). However, in case of some recent footprint studies (Chambers et al. 2005) it has been added as a component. Volume of daily water use from households was collected and then it was converted into annual millions of liters of water and the footprint was calculated by using the equation given in Fig. 11.3.

11.2.10 Footprint from Food Consumption

For assessing the food consumption footprint, expenditure for food consumption was collected under different category and subcategory. Categorization was determined on the basis of availability of standards for Economic Input-Output Life Cycle Assessment [EIO-LCA] (Green Design Institute 2011). The total consumption unit was converted into annual consumption unit, and then using the unit price, yearly cost in taka was calculated. Then it was transformed into yearly expenditure in Millions US$ using a standard conversion rate (1 US$ = 79.53 BDT). Finally, CO_2 emission from each consumption category of food was calculated by using the GHG emission standards and the equation as in Fig. 11.3.

11.2.11 Footprint from Goods and Services Consumption

In this study, goods and service activity expenditure was assessed under five broad categories as: education, health, entertainment, clothing, and others (Tobacco). The monthly expenditure of a household was collected for the specified categories. Then GHG emission resulted for per million-dollar expenditure in each category was calculated. After collecting the monthly expenditure CO_2 emission from each consumption category of goods and services was calculated by using the GHG emission standards following the procedure discussed in food consumption earlier using the standard equation as given in Fig. 11.3.

11.2.12 Footprint from Fuel Consumption (Transportation)

To determine the total amount of fuel use of different types, respondents were asked about the number and type of vehicle they have, and per month average fuel cost on each respective vehicle. Following the determination of total annual expenditure for each fuel type, annual quantity was estimated using the unit price of each fuel type. At last, the total amount of carbon emission from electricity consumption for each unit of MWh electricity use was determined from GHG Equivalencies Calculator-USEPA using the equation shown in Fig. 11.3.

11.2.13 Calculation of Biocapacity

Biocapacity of TMA was calculated by using the equations as shown in Table 11.3 and as described in Borucke et al. (2013).

Table 11.3 Formulae used to calculate the biocapacity of Tangail municipal area (TMA), Bangladesh

Production area (ha)	×	Yield factor	×	Equivalency factor	=	Biocapacity (gha)	
Cropland area	×	Cropland yield factor	×	Cropland equivalency factor	=	Cropland biocapacity	Total biocapacity
Grazing land area	×	Grazing yield factor	×	Grazing equivalency factor	=	Grazing biocapacity	
Fishing ground area	×	Fishing yield factor	×	Fishing equivalency factor	=	Fishing biocapacity	
Forest area	×	Forest yield factor	×	Forest equivalency factor	=	Forest biocapacity	
Built-up land area	×	Cropland yield factor	×	Cropland equivalency factor	=	Cropland biocapacity	

11.2.14 Data Analysis

The collected and estimated data were analyzed using Microsoft Office excel 2013, IBM Statistics 20.0, and R software. Various descriptive statistical measures, e.g., frequency, percentage, etc. were used for categorization and describing the variables. Bar charts, pie charts, whisker box plots, etc. were used to represent the results. Tukey multiple comparison test and ANOVA was performed to assess the interaction effects of family size and income groups on the households' EF. The ANOVA was done using three models: (i) Model 1 comprises only the family size, (ii) Model 2 includes the income groups, and (iii) Model 3 comprises both the family size and income groups to show the interactions.

11.3 Results and Discussion

11.3.1 Ecological Footprint of Tangail Municipal Area

11.3.1.1 Carbon Uptake Land Footprint

The total carbon (CO_2) emission from different carbon uptake factor was calculated as 433,235.302 tons CO_2, while the highest (271,451.471 tons CO_2) carbon emission was derived from natural gas consumption and the lowest (380.083 tons CO_2) was observed from water consumption (Fig. 11.4). The ecological footprint calculated from carbon uptake land components is shown in Table 11.4. The total carbon uptake land footprint was calculated as 1.146 gha/capita, of which natural gas (44.58%) and fuel for transportation (24.35%) were contributed to the significant portion.

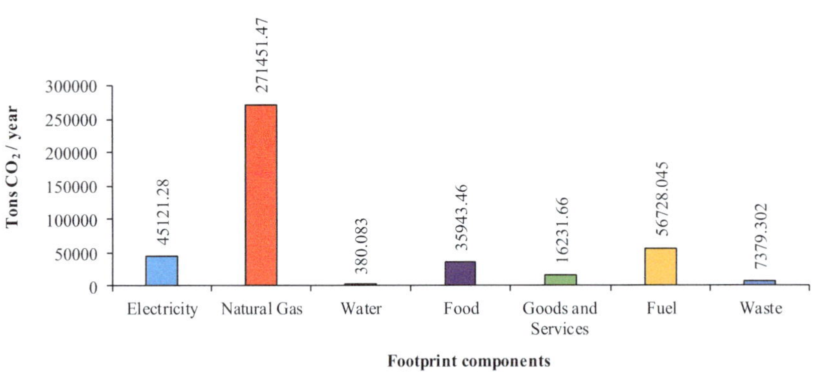

Fig. 11.4 Total carbon emission from carbon uptake land in Tangail municipal area (TMA) of Bangladesh

Table 11.4 Carbon uptake land footprint of Tangail municipal area (TMA), Bangladesh

Footprint components		Footprint (ha)	Footprint (gha)	Footprint (gha/capita)	% contribution to carbon uptake land footprint
Electricity		11,289.386	14,224.626	0.085	7.42
Natural gas		67,917.409	85,575.935	0.511	44.58
Waste		1846.308	2326.348	0.014	1.22
Water		95.097	119.822	0.001	0.09
Transportation-fuel		37,134.303	46,789.222	0.279	24.35
Food	Meat	6366.043	0.048	0.048	4.19
	Dairy	2133.094	0.016	0.016	1.40
	Fruits	3573.55	0.027	0.027	2.36
	Cereal	6683.875	0.050	0.050	4.36
	Confectionary	1095.058	0.008	0.008	0.70
	Drinks	847.429	0.006	0.006	0.52
	Others	2829.606	0.021	0.021	1.83
Goods and services	Education	5220.823	0.039	0.039	3.40
	Health	1591.892	0.012	0.012	1.05
	Clothing	1594.584	0.012	0.012	1.05
	Entertainment	469.279	0.004	0.004	0.35
	Others	1748.701	0.013	0.013	1.13
Total		152,436.4	192,069.9	1.146	100

11.3.2 Cropland and Fishing Ground Footprint

The study calculated that the total cropland footprint in TMA was 174,664.346 gha, while the per capita cropland footprint was 1.043 gha/capita (Table 11.5). Meanwhile, the total fishing ground footprint was calculated as 212.578 gha in TMA, whereas the per capita fishing ground footprint was calculated 0.001 gha/capita (Table 11.5).

11.3.3 Built-Up Land Footprint

The built-up land footprint is calculated depending on the areas of land comprised of human infrastructure— residential, commercial, institutional, mixed use, roads, and transportation. The built-up land footprint of the study area is shown in Table 11.6. The cropland equivalency factor (2.51) was considered for the calculation of built-up land footprint of the area.

Table 11.5 Cropland and fishing ground footprint of Tangail municipal area (TMA), Bangladesh

Cropland footprint								
Agricultural products (t/yr)	/	World crop yields (t/ha/yr)	=	Cropland (wha)	×	Cropland equivalency factor	=	Cropland footprint (gha)	Footprint (gha/capita)
12,875.845	/	1.852	=	69,587.389	×	2.51	=	174,664.346	1.043
Fishing ground footprint									
Fish products (t)	/	World fish yields (t/ha)	=	Fishing ground (wha)	×	Fishing equivalency factor	=	Fishing ground footprint (gha)	Footprint (gha/capita)
574.536	/	1.000	=	575.536	×	0.370	=	212.578	0.001

Table 11.6 Built-up land footprint in Tangail municipal area (TMA), Bangladesh

Land type	Area (ha)		Equivalency factor (of cropland)	Footprint (gha)	Footprint (gha/capita)
Residential	1310.68	×	2.51	3289.81	0.0197
Commercial	20.49	×		51.43	0.0003
Institutional	51.74	×		129.87	0.0008
Roads and transportation	104.55	×		262.42	0.0016
Mixed use	33.60	×		84.34	0.0005
Total	1521.06	×		3817.86	0.0229

The result of the study depicted that the total built-up land footprint in TMA was 3817.86 gha, while the per capita built-up footprint was 0.0229 gha/capita. Shakil et al. (2014) stated that the built-up footprint in Dhanmondi residential area, Dhaka, Bangladesh was 434.644 gha/hectare. The results of the study found that the built-up footprint in TMA was 8.74 times higher than Dhanmondi residential area of Bangladesh.

11.3.4 Total Ecological Footprint

The total EF of TMA considering different components is presented in Table 11.7. The result of the study explicated that carbon uptake land was found as the main

Table 11.7 Total ecological footprint (EF) of Tangail Municipal Area, Bangladesh

Footprint components		Footprint (ha)	Footprint (gha)	Footprint (gha/capita)	% contribution to total footprint
Carbon uptake land footprint		152,436.4	192,069.9	1.146	51.79
Cropland footprint		69,587.389	174,664.346	1.043	47.13
Fishing ground footprint		575.536	212.578	0.001	0.05
Built-up land footprint	Residential	1310.68	3289.810	0.0197	0.89
	Commercial	20.49	51.430	0.0003	0.01
	Institutional	51.74	129.870	0.0008	0.04
	Roads and transportation	104.55	262.420	0.0016	0.07
	Mixed use	33.60	84.340	0.0005	0.02
Total footprint		224,120.422	370,764.705	2.2129	100

incentive behind the rising EF (51.79%) since it includes number of different sub-components, whereas the cropland was singly contributing to the highest amount of footprint (47.13%) in TMA, wherein the lowest amount of footprint (0.05%) was contributed by the fishing ground sector. Among the carbon uptake land, the study revealed that the energy consumption (electricity and natural gas consumption) was mainly responsible for rising EF of a household followed by transportation-fuel, food, goods and services, waste and water. However, the fishing ground and built-up land were found as limited contributing sector to EF in TMA.

11.3.5 Ecological Footprints in Response to Family Size and Income Groups

The ecological footprint (EF) in response to family sizes with their corresponding lower, middle, and upper quartiles is shown in Fig. 11.5. The upper and lower quartiles showed the highest and lowest EF values for a particular family size, while the median (middle quartile) clearly depicted that the EF of small families was much lower than the EF of extended families followed by medium families. The boxes generated for different family sizes showed that the EFs of 50% households of each family size lie within these ranges. The study, therefore, revealed that the household's EF increases with increasing household/family size. In the meantime, the study also explicated that the EF of a household increases with family income. In Fig. 11.6, the highest and lowest ecological footprint values for a specific income group was represented by the upper and lower quartiles, respectively, whereas the median clearly showed that

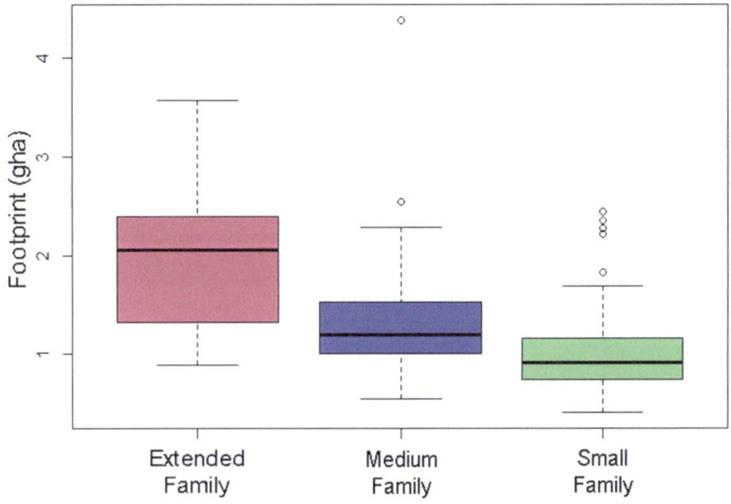

Fig. 11.5 Box and whisker plots showing the ecological footprint in response to family sizes

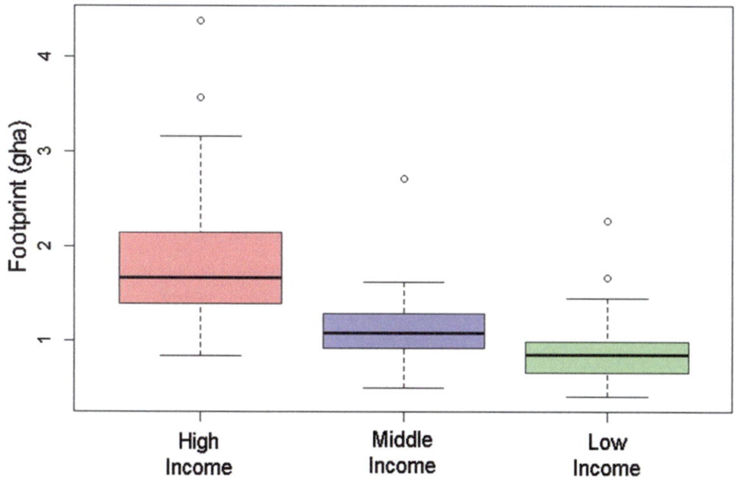

Fig. 11.6 Box and whisker plots showing the ecological footprints in response to income

the EF of high income groups was much higher than the EF of small income groups followed by middle income groups. The mean EFs are significantly different for different family sizes ($p < 0.01$) and income groups ($p < 0.01$), while no significant difference was observed in case of the interaction effects of family size and income groups (Table 11.8).

The multiple comparison of means of EFs in response to family size and income group is shown in Table 11.9. The 95% confidence interval for the difference between the means of different pair groups showed whether their effects on EFs were statistically significant or not, where the difference is statistically significant if the range of confidence interval did not include zero. The study found a number of pair groups which was statistically significant at the 0.01 and 0.05 levels. The confidence intervals for the remaining pairs of means all include zero, which indicated that the differences are not statistically significant (Table 11.9).

The grand mean of the surveyed household was calculated as 1.208 gha. The study analyzed that a total of 15 extended (>8 persons), 74 medium (5–8 persons), and 91

Table 11.8 ANOVA of ecological footprints in response to family size and income groups

Model	SV	Df	Sum Sq.	Mean Sq.	F value	*p*-value
ANOVA1	Family size	2	13.34	6.668	26.66	0.000*
ANOVA2	Income	2	24.24	12.120	64.28	0.000*
ANOVA3	Family size	2	13.337	6.668	40.909	0.000*
	Income	2	16.16	8.080	49.568	0.000*
	Family size: Income	4	0.239	0.060	0.366	0.833

*Comparison is significant at the 0.01 level

Table 11.9 Multiple comparison of means of ecological footprints in response to family size and income group

Model	Difference	95% confidence interval	*p*-value
ANOVA 1 (Family size)			
Small–Medium	**–0.307**	**(–0.492, –0.122)**	**0.000****
Small–Extended	**–0.968**	**(–1.298, –0.639)**	**0.000****
Medium–Extended	**–0.661**	**(–0.661, –0.996)**	**0.000****
ANOVA 2 (Income)			
Low–High	**–0.962**	**(–1.167, –0.758)**	**0.000****
Middle–High	**–0.721**	**(–0.919, –0.523)**	**0.000****
Middle–Low	**0.241**	**(0.065, 0.416)**	**0.003***
ANOVA 3 (Family size: Income)			
Family size			
Small–Medium	**–0.307**	**(–0.457, –0.158)**	**0.000****
Small–Extended	**–0.968**	**(–1.234, –0.702)**	**0.000****
Medium–Extended	**–0.661**	**(–0.932, –0.391)**	**0.000****
Income			
Low–High	**–0.749**	**(–0.939, –0.559)**	**0.000****
Middle–High	**–0.566**	**(–0.750, –0.382)**	**0.000****
Middle–Low	**0.183**	**(0.020, 0.346)**	**0.024***
Family size: Income			
Medium: High–Extended: High	−0.456	(–0.971, 0.056)	0.122
Small: High–Extended: High	**−0.747**	**(–1.289, –0.205)**	**0.001****
Extended: Low–Extended: High	**−1.121**	**(–2.112, –0.129)**	**0.014***
Medium: Low–Extended: High	**−1.320**	**(–1.826, –0.815)**	**0.000****
Small: Low–Extended: High	**−1.485**	**(–1.953, –1.017)**	**0.000****
Extended: Middle–Extended: High	−0.684	(–1.447, 0.078)	0.117
Medium: Middle–Extended: High	**−1.090**	**(–1.566, –0.615)**	**0.000****
Small: Middle–Extended: High	**−1.311**	**(–1.782, –0.839)**	**0.000****
Small: High–Medium: High	−0.289	(–0.736, 0.158)	0.523
Extended: Low–Medium: High	−0.663	(–1.606, 0.280)	0.405
Medium: Low–Medium: High	**−0.862**	**(–1.264, –0.461)**	**0.000****
Small: Low–Medium: High	**−1.027**	**(–1.380, –0.673)**	**0.000****
Extended: Middle–Medium: High	−0.227	(–0.924, 0.471)	0.984
Medium: Middle–Medium: High	**−0.633**	**(–0.996, –0.269)**	**0.000****
Small: Middle–Medium: High	**−0.853**	**(–1.212, –0.495)**	**0.000****
Extended: Low–Small: High	−0.374	(–1.333, 0.585)	0.950
Medium: Low–Small: High	**−0.574**	**(–1.011, –0.136)**	**0.002****

(continued)

Table 11.9 (continued)

Model	Difference	95% confidence interval	*p*-value
Small: Low–Small: High	**−0.738**	**(−1.132, −0.344)**	**0.000****
Extended: Middle–Small: High	0.063	(−0.657, 0.782)	0.999
Medium: Middle–Small: High	−0.344	(−0.746, 0.059)	0.163
Small: Middle–Small: High	**−0.564**	**(−0.962, −0.166)**	**0.000****
Medium: Low–Extended: Low	−0.199	(−1.139, 0.738)	0.999
Small: Low–Extended: Low	−0.364	(−1.283, 0.555)	0.945
Extended: Middle–Extended: Low	0.436	(−0.662, 1.535)	0.944
Medium: Middle–Extended: Low	0.030	(−0.893, 0.953)	1.00
Small: Middle–Extended: Low	−0.190	(−1.11, 0.731)	0.999
Small: Low–Medium: Low	−0.164	(−0.506, 0.178)	0.850
Extended: Middle–Medium: Low	0.636	(−0.560, 1.328)	0.990
Medium: Middle–Medium: Low	0.230	(−0.122, 0.582)	0.510
Small: Middle–Medium: Low	0.098	(−0.337, 0.356)	1.000
Extended: Middle–Small: Low	**0.800**	**(0.135, 1.465)**	**0.007****
Medium: Middle–Small: Low	**0.394**	**(0.098, 0.690)**	**0.001****
Small: Middle–Small: Low	0.174	(−0.115, 0.463)	0.622
Medium: Middle–Extended: Middle	−0.406	(−1.077, 0.264)	0.612
Small: Middle–Extended: Middle	−0.626	(−1.294, 0.0414)	0.085
Small: Middle–Medium: Middle	−0.220	(−0.522, 0.081)	0.351

**Comparison is significant at the 0.01 level
*Comparison is significant at the 0.05 level

small (<5 persons) families were surveyed, wherein the mean of EF of extended, medium, and small families was 1.97, 1.308, and 1.001 gha, respectively. The mean of EFs for different interactions between family size and income group with their percentages is given in Table 11.10. The highest mean of EF was found for extended family: high income pair group (2.30 gha), whereas the lowest mean was observed for small family: low income pair group (0.82 gha). Moreover, the mean of low,

Table 11.10 Mean of ecological footprints for different interactions between family size (FS) and income group (IG)

FS	IG		
	High income (%)	Middle income (%)	Low income (%)
Extended family	2.30 (5)	1.62 (2.22)	1.18 (1.11)
Medium family	1.84 (10.56)	1.21 (18.89)	0.98 (11.67)
Small family	1.55 (7.78)	0.99 (20.55)	0.82 (22.22)

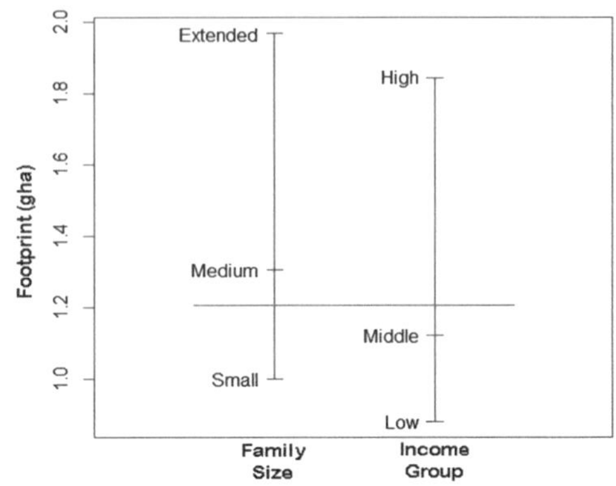

Fig. 11.7 Effects of family size and income on mean ecological footprints

middle, and high income families' EF was calculated 0.883, 1.124, and 1.846 gha, respectively (Fig. 11.7).

11.3.6 Biocapacity of Tangail Municipal Area

In the land use map of TMA, a total of 1214.99 ha of land was found which was used for agricultural practices and was considered as cropland. No layer or database for greenery existed within the area. So, the area of open space was considered as the area of greenery or forest land in this study. The total open area of TMA which is considered as forest land was 51.865 ha. Water bodies like river, pond, and ditch encompass 299.487 ha or 9.36% of the study area. Louhajang River, which branches out from Jamuna River, passes through the middle of the TMA. The built-up land area of TMA was 1521.06 ha (Master Plan for Tangail Paurashava 2013). In case of built-up land, its productivity or the infrastructure yield is set equal cropland yield in the Global Footprint Network, National Accounts Methodology (Borucke et al. 2013). Built-up land yield is assumed to be the same as that for cropland because urban areas are typically built on or near the most productive agricultural lands. Equivalency factors for all types of land uses were derived from Ecological Footprint Atlas (Ewing et al. 2010a). The result of the study revealed that the total amount of biocapacity in TMA was 12,852.365 gha, while the per capita biocapacity was estimated as 0.0767 gha/capita. The study also expressed that the highest amount of biocapacity (5647.938 gha) was observed in cropland, whereas the lower amount of biocapacity (22.938 gha) was observed in forest land (Table 11.11).

Table 11.11 Biocapacity of Tangail municipal area (TMA), Bangladesh

Land use	[a]Area (ha)		[b]Yield Factor for Bangladesh		[c]Equivalency Factor (gha/ha)		Biocapacity (gha)	Biocapacity (gha/capita)
Cropland	1214.997	×	1.852	×	2.51	=	5647.938	0.0337
Grazing land	0.00	×	1.898	×	0.46	=	0.000	0.0000
Forest land	51.865	×	0.351	×	1.26	=	22.938	0.0001
Fishing ground	299.487	×	1.000	×	0.37	=	110.810	0.0007
Built-up land	1521.06	×	1.852	×	2.51	=	7070.678	0.0422
Total						=	12,852.364	0.0767

Data source [a]Master Plan for Tangail Paurashava (2013); [b]Shakil et al. (2014), [c]Ewing et al. (2010a)

11.3.7 Status of Environmental Sustainability: Ecological Footprint Versus Biocapacity

The EF of TMA was calculated 2.21 gha/capita (370,764.705/167,412 person) for the year 2016. On the other hand, biocapacity for the same year was determined 0.08 gha/capita (12,852.364/167,412 person). It means that each person of TMA needs 2.21 ha of global average productive land in the year 2017 to meet their various ecological demands. On the other side, only 0.08 ha of global average productive land is available for each person to support the demand. So, from the comparison TMA's ecological demand (footprint) exceeds its ecological capacity (biocapacity) by almost twenty-eight (28) times. The actual extent of TMA (3198.635 ha) was compared with its footprint (224,120.422 ha) and found that it needs about seventy (70) "TMA" to support the present one in terms of resource supply and waste absorption. From the above outcomes, it can be derived that TMA is very unsustainable in terms of ecological resource consumption and waste production. If this condition prevails and could not be planned properly, TMA can be compared with the high carbon cities in the long-time duration. The spatial variations of ecological footprint, biocapacity, and ecological deficit are shown in Fig. 11.8.

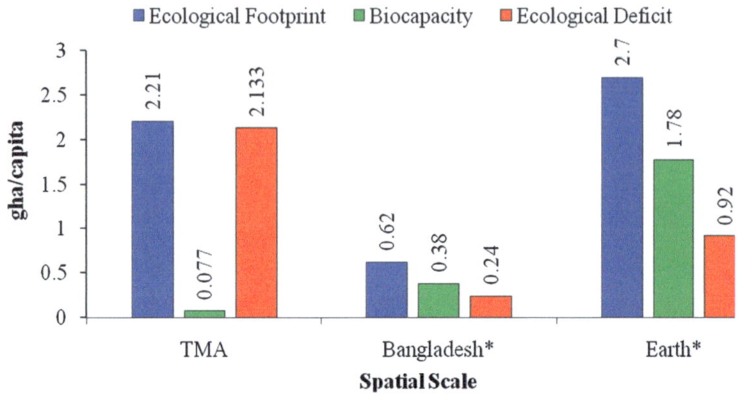

Fig. 11.8 Spatial variations of ecological footprint, biocapacity and ecological deficit (**Data source* Global Footprint Network 2010)

11.3.8 Human Health Effects of Ecological Footprint

Human health is essentially dependent upon the natural environment. With rapidly growing global population and per capita consumption of natural resources, global carbon emission and ecological footprint (EF) is increasing, resulting in the alteration of the structures and functions of Earth's natural systems like climatic systems, hydrological systems, atmospheric dynamics, and diverse biosphere (Myers et al. 2013). That means the way we choose to house, cloth, shelter, and meet the needs for resources such as food, energy and water, not only affecting long-term availability of our finite resources but also disrupting the natural balance of the environment. Since the natural environment is viewed as a principal determinant of human health, questions arise about how and at what extent our EF is transforming our natural systems and also is affecting the global human health. Climate change is the most visual consequence of such change which is impacting the health of people more than the mean and in greater number. Environmental consequences of climate change, such as temperature extremes, sea level rising, changes in precipitation resulting in flooding and droughts, and frequent natural calamities, directly and indirectly affect the physical and psychosocial health of humans including morbidity, mortality, and transmission of vector- and food-borne diseases (Ebi et al. 2006; Campbell-Lendrum et al. 2015). Apart from climate change-induced health impacts, study found that air pollution resulting from carbon emission might cause thousand additional deaths and more cases of respiratory illness and asthma in the United States (Jacobson 2008). Thus, calculating and reducing carbon emission and EF at household level will definitely help to mitigate the direct and indirect human health impacts.

11.4 Conclusion

The study calculated the ecological footprint (EF) considering the maximum possible components which provides an understanding of the demands of Tangail municipal area (TMA) from nature and its distance from achieving environmental sustainability. The study revealed that footprint account increases with increasing resource consumption and waste generation, while per capita footprint account increases with decreasing total population if similar patterns of resource consumption and waste generation prevail. The study also found highly significant relationship between EF and household size as well as income. Biocapacity was found as a very small amount because of unplanned urban cluster. The study revealed that TMA's ecological demand exceeds its ecological capacity by almost twenty-eight times which will ultimately affect the human life and health of the area. In substance, this additional land acquisition has been collected over the years, but in effect, it belongs to other inhabitants elsewhere on the planet. This study has great scope in future and can be used as a persuasive tool during urban policy formulation to enhance environmental sustainability through utilizing our resources more judiciously. The study recommended to set annual targets, to reduce ecological footprint at household levels by awaking people about how much they are consuming, its environmental impacts and subsequent effects on their health.

Acknowledgements Authors are thankful to the officials of Tangail municipal authority, Bangladesh, for providing us necessary data and extending their cooperation throughout the study period.

References

Asaduzzaman M (2002) Frontiers of change in rural Bangladesh: natural resources and sustainable livelihoods. In: Hands not land: how livelihoods are changing in rural Bangladesh, Department for International Development (DFID), Dhaka, Bangladesh, pp 65–70

Bagliani M, Galli A, Niccolucci V, Marchettini N (2008) Ecological footprint analysis applied to a sub-national area: the case of the Province of Siena Italy. Environ Manag 86:354–364

Bala BK, Hossain MA (2010) Food security and ecological foot print of the coastal zone of Bangladesh. Environ Dev Sustain 12:531–545

Bala BK, Hossain MA, Haque MA, Majumder S (2012) Ecological footprint of the hill tracts of Chittagong in Bangladesh. Oecol Mont 21:36–41

Borucke M, Moore D, Cranston G, Gracey K, Iha K, Larson J, Lazarus E, Morales JC, Wackernagel M, Galli A (2013) Accounting for demand and supply of the biosphere's regenerative capacity: the national footprint accounts' underlying methodology and framework. Ecol Ind 24:518–533

BPDB (Bangladesh Power Development Board) (2015) Tariff rate. Bangladesh Power Development Board. http://bd.bpdb.gov.bd/download/tariff/Tariff%20order.retail_9.1.2015.pdf. Accessed 10 Jan 2017

Byomkesh T, Nakagoshi N, Shahedur RM (2009) State and management of wetlands in Bangladesh. Landsc Ecol Eng 5:81–90
Campbell-Lendrum D, Manga L, Bagayoko M, Sommerfeld J (2015) Climate change and vector-borne diseases: what are the implications for public health research and policy? Philos Trans R Soc Lond B: Biol Sci 370(1665):20130552. https://doi.org/10.1098/rstb.2013.0552
Chambers N, Child R, Jenkin N, Lewis K, Vergoulas G, Whiteley M (2005) Stepping forward: a resource flow and ecological footprint analysis of the south west of England. Technical report, The Stepping Forward Report Series, Best Food Forward Limited, UK
Chen B, Chen GQ (2006) Ecological footprint ac-counting based on energy—a case study of the Chinese society. Ecol Model 198:101–114
Cordero E, Todd AM, Abellerra D (2008) Climate change education and the ecological footprint. Bull Am Meteorol Soc 865–872
Creative Research System (2010) Sample size calculator. The survey system. http://www.surveysystem.com/sscalc.htm. Accessed 23 Oct 2016
Ebi KL, Mills DM, Smith JB, Grambsch A (2006) Climate change and human health impacts in the United States: an update on the results of the US national assessment. Environ Health Perspect 114(9):1318–1324. https://doi.org/10.1289/ehp.8880
Ewing B, Moore D, Goldfinger A, Oursler A, Reed A, Wackernagel M (2010a) Ecological footprint atlas 2010. Global Footprint Network, Oakland
Ewing B, Reed A, Galli A, Kitzes J, Wackernagel M (2010b) Calculation methodology for the national footprint accounts, 2010th edn. Glob Footpr Netw, Oakland
Gali A, Wiedmann T, Ercin E, Knoblauch D, Ewing B, Giljum S (2012) Integrating ecological, carbon, and water footprint into a "Footprint family" of indicators: definition and role in tracking human pressure on the planet. Ecol Ind 16:100–112
Global Footprint Network (2010) Ecological footprint and biocapacity, 2007. Based on National Footprint Accounts, 2010 edn
Green Design Institute (2011) GHG emissions: the economic input-output life cycle assessment (EIO-LCA). Carnegie Mellon University, Pittsburgh
IPCC (Intergovernmental Panel on Climate Change) (2007) Climate change 2007: The physical science basis. Contribution to Working group I to the Fourth Assessment Report of the Intergovernmental Panel on Climate Change, Cambridge University Press, Cambridge, United Kingdom and New York, USA
Jacobson MZ (2008) On the causal link between carbon dioxide and air pollution mortality. Geophys Res Lett 35(3):L038809. https://doi.org/10.1029/2007gl031101
Kitzes J, Peller A, Goldfinger S, Wackernagel M (2007) Current methods for calculating national ecological footprint accounts. Shiga Univ Environ Res Cent Annu Rep 4(1):1–9
Master Plan for Tangail Paurashava (2013) Preparation of master plan for Tangail paurashava: structural plan, urban area plan and ward action plan (2013–2033), Government of People's Republic of Bangladesh
Medved S (2006) Present and future ecological footprint of Slovenia—the influence of energy demand scenarios. Ecol Model 192:25–36
Monfreda C, Wackernagel M, Deumling D (2004) Establishing national natural capital accounts based on detailed ecological footprint and biological capacity assessment. Land Use Policy 21:231–246
Myers SS, Gaffikin L, Golden CD, Ostfeld RS, Redford KH, Ricketts TH, Turner WR, Osofsky SA (2013) Human health impacts of ecosystem alteration. Proc Natl Acad Sci USA 110(47):18753–18760. https://doi.org/10.1073/pnas.1218656110
Niccolucci V, Gall A, Kitzes J, Pulselli RM, Bor-sa S, Marchettini N (2008) Ecological footprint analysis applied to the production of two Italian wines. Agric Ecosyst Environ 128:162–166
Pimentel D, Huang X, Cordova A, Pimentel M (2006) Impact of a growing population on natural resources: The challenge for environmental management. Front: Interdiscipl J Study Abroad 13:1–12

Portier CJ, Thigpen TK, Carter SR, Dilworth CH, Grambsch AE, Gohlke J, Hess J, Howard SN, Luber G, Lutz JT, Maslak T, Prudent N, Radtke M, Rosenthal JP, Rowles T, Sandifer PA, Scheraga J, Schramm PJ, Strickman D, Trtanj JM, Whung PY (2010) A human health perspective on climate change: a report outlining the research needs on the human health effects of climate change. Environmental Health Perspectives/National Institute of Environmental Health Sciences, Research Triangle Park, NC. https://doi.org/10.1289/ehp.1002272, www.niehs.nih.gov/climatereport

Quam VGM, Rocklov J, Quam MBM, Lucas RAI (2017) Assessing greenhouse gas emissions and health co-benefits: a structured review of lifestyle-related climate change mitigation strategies. Int J Environ Res Public Health 14:468. https://doi.org/10.3390/ijerph14050468

Rahman MM, Rahman MM, Tanu TA, Rahman MM (2012) Spatial environmental impact on land degradation in Bangladesh. Asian J Appl Sci Eng 1(2):84–90

Sarker BC, Shutradhar SC, Khan A, Saifullah ASM, Ruma AB (2015) Urban growth and its impact on Tangail municipal area. J Environ Sci Nat Resour 8(2):163–166

Shakil SH, Kuhu NN, Rahman Z, Islam I (2014) Carbon emission from domestic level consumption: ecological footprint account of Dhanmondi residential area, Dhaka, Bangladesh—a case study. Aust J Basic Appl Sci 8(7):265–276

Simmons C, Lewis K, Barrett J (2000) Two feet-two approaches: a component based model of ecological footprinting. Ecol Econ 32(3):375–380

Sprangers S (2011) Calculating the carbon footprint of universities. Doctoral dissertation, Erasmus School of Economics, Erasmus University, the Netherlands

The Engineering Toolbox (2001) Natural gas consumption. http://www.engineeringtoolbox.com/natural-gas-consumption-d_172.html. Accessed 08 Nov 2016

USEPA (United States Environmental Protection Agency) (2011) Greenhouse gas equivalencies calculator. United States Environmental Protection Agency. http://www.epa.gov/cleanenergy/energyresources/calculator.html. Accessed 12 Nov 2016

Wackernagel M, Monfreda C, Schulz NB, Erb KH, Haberl H, Krausmann F (2004) Calculating national and global ecological footprint time series: resolving conceptual challenges. Land Use Policy 21:271–278

Chapter 12
The Impact of Climate Change on Human Eyes

Murali Krishna V. Iyyanki and Prisilla Jayanthi

Abstract The era has come to recognize the changes in environment, people are living in hazardous climate because of man-made atmosphere, and it cannot be reversed but definitely can be reduced. The climate change on its way ensures the risk to understand and aims at improving the healthcare and emergency services to implement disaster planning relevant to preventing and educating risk communication. The assessment can improve understanding of the relations between climate and health and provides an opportunity for the cloud networking.

Keywords Cloud computing · Carbon dioxide · Diseases · Ultraviolet radiation

12.1 Introduction

The urbanization is the key cause of climate change. This predictable change causes for less rainfall and otherwise leads to more flooding; extreme and unusual weather events are common with sea levels increasing. The atmospheric global change increases the UV radiation on the earth that contributes to chronic diseases by exposure of the human skin and eye tissues. This paper describes the climate changes, chronic diseases related to human eye and cloud computing which makes the network faster in certain few remote areas.

M. K. V. Iyyanki (✉)
DRDO, Hyderabad, India
e-mail: iyyanki@gmail.com

P. Jayanthi
CSE department, K G Reddy College of Engineering and Technology, Moinabad, Hyderabad, India
e-mail: prisillaj28@gmail.com

R. B. Singh et al. (eds.), *Urban Health Risk and Resilience in Asian Cities*,
Advances in Geographical and Environmental Sciences,
https://doi.org/10.1007/978-981-15-1205-6_12

12.2 Do Volcanoes Harm?

The carbon dioxide emitted by volcanoes and fossil fuel is shown in Fig. 12.1 for the year 2015, it states that through deforestation, burning of coals and other fossil fuel emits 40 billion metric tons over the volcanoes is depicted in the graph (Scott and Lindsey 2016). The human carbon dioxide emissions measure to be 90 times greater than global volcanic carbon dioxide emissions in 2011. The man-made pollution which causes changes in the atmosphere and creates dangerous health hazards. The global warming is a widespread of pollution that traps hotness around the earth. This effect is distressing the innocent people living in remote rural and undeveloped regions.

Cities play a vital role in economic development of the nation as many enterprises are located in the urban areas. Hence, cities are referred to as the primary cause of the climate change. The growing industries pollute the atmosphere with the deadly gases like particulate, Carbon Monoxide, Hydroxide, Oxide of Sulfur, and Nitrogen shown in Fig. 12.2. The effect depends on the location and adaptive capacity of the cities. The impact of the climate change affects the poor communities at health risk the most because of their lack of access to food and information resources and the more urbanization mean the more pressure on the food production and water.

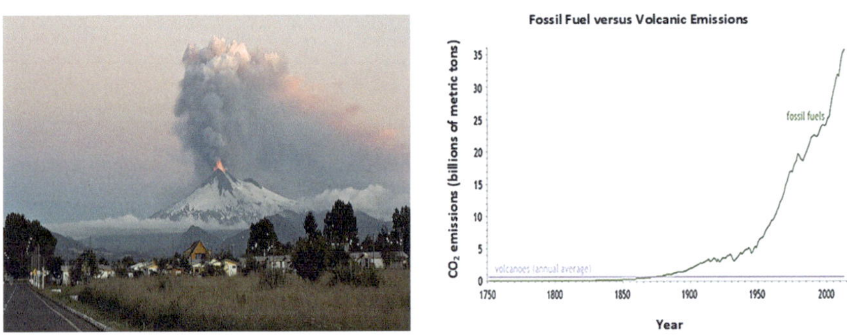

Fig. 12.1 The CO_2 emitted by volcanoes and to the right is the graph indicating fossil fuel versus volcanic emissions

Fig. 12.2 The CO_2 emitted from industries

Fig. 12.3 Sun in UltraViolet light

Science that addresses the various diseases due to higher temperature, heat waves, and extreme weather changes in air quality affect the human eye health. Economic loss increases globally owing to extreme weather events with influence of climate change. Lot of research is being done for identifying key measures to help ensure the availability of appropriate health care. Research must study the relationship between human development and adaptation to climate change. The climate change is linked with shortage of food, malnutrition and contamination, seafood from chemical reactions, crops by pesticides, and how water will mix with sewage in surface and the supplies of underground water distribution system (Portier et al. 2010) (Fig. 12.3).

12.3 Human Eyes and Climate

A researcher at Wilmer Eye Institute states that the three sources that climate change can have impact on the eyes of human. Firstly, as the areas of the earth get hotter and lesser moist, people are prone to suffer from dry eyes. The dry climate increases watery eyes resulting in dryness and redness of eye. Secondly, the increase in airborne particulates from fire seasons and clearing of crops due to climate change can affect the eyes. The smoke emitted by cooking fires leads to increased scarring of the eyelids and cornea (West et al. 2013). The third point relates to UV exposure bringing risk to cataract or other eye diseases.

Climate change contributes diseases through the direct effect of global warming and extreme weather events are likely to increase diarrheal disease and malnutrition. Changes in fauna and flora have larger impact on the ecosystem, agriculture, and foremost food security giving way to the migration of people away from threatening environment into low-affected environment. All of these may impact eye health: blindness in children due the deficiency of vitamin A. And research reveals that trachoma is associated with the high temperatures and low rainfall, more concern about *Musca sorbens* flies for transmission of trachoma. Anitha et al. (2016) showed in Fig. 12.4 prevalence of active trachoma in Asia, Africa, and few Pacific Islands. Climate variables: temperature, rainfall, humidity, flooding are the main causes for

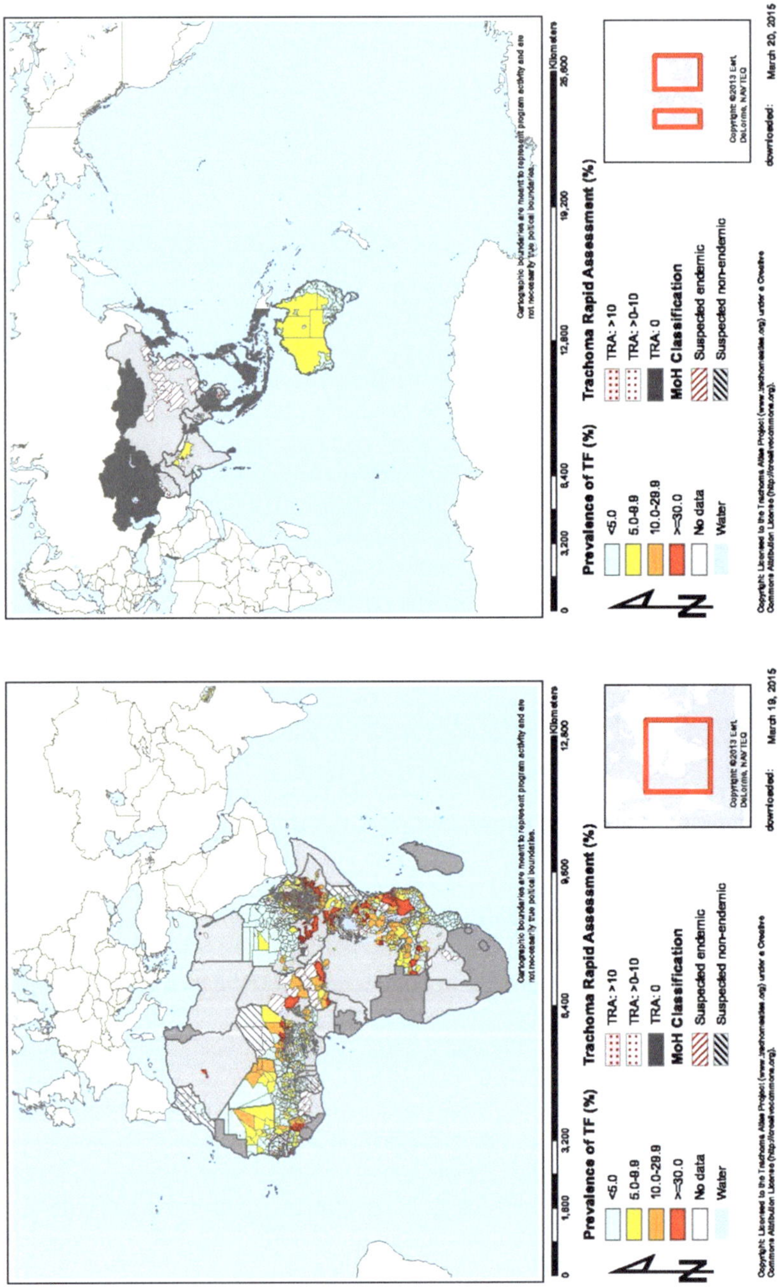

Fig. 12.4 Prevalence of active Trachoma in Africa, Asia, and the Pacific Islands

Trachoma, *Musca sorbens* flies feed on human ocular and nasal emissions to attain the nutrients for egg production (http://iceh.lshtm.ac.uk/climate-change-eye-health/).

12.4 Ultra Violet Radiations

Scientists proved that the serious eye disorders arise due to increased levels of ultraviolet radiation reaching the surface of the earth and toxic mix of air pollutants as shown in Fig. 12.3. Experts quantify that the effect of temperature changes (ultraviolet–B radiation), suspended particles, pollutants on the prevalence of eye diseases. The ozone layer depletion decreases the sun's harmful ultraviolet radiation (http://www.ncbi.nlm.nih.gov/pmc/articles/PMC1298; http://www.bcairquality.ca/101/ozone-depletion-impacts.html; McKenzie et al. 2011; Abisko).

WHO states that 18 million people are blind due to cataract and 5% of all cataract-related disease is directly related to ultraviolet radiation exposure. UVR levels increase by about 5% with every 1000 m altitude. Ozone absorbs UVR from the sun and at higher UVR level, a person is said to be closer to equator (http://www.ncbi.nlm.nih.gov/pmc/articles/PMC1298; http://www.bcairquality.ca/101/ozone-depletion-impacts.html; McKenzie et al. 2011; Abisko). UV light is divided into three parts, UV-C (100–280 nm), UV-B (280–315 nm) and UV-A (315–400 nm). The exposure to UV-B radiation after age 15 and cortical cataract formation was found related recently (http://www.eco-action.org/dt/ozone.html). The analysis from INSAT-3D at below 700 hPa of number of winds, and speed bias and mean observation is displayed in Fig. 12.5 (http://www.ncmrwf.gov.in).

The analysis and forecasts of ensemble mean wind and spread of wind speed at 500 h Pa is displayed in Fig. 12.6 (http://www.ncmrwf.gov.in; Frederik and van Kuijk 1991). The easterly winds associated with the passage of western disturbance over North Arabian Sea near west coast of India, though weaker and with more uncertainty is seen in the Day-10 forecast. However, the mean wind speed/direction is indicated by arrows and spread in wind speed is indicated by the shades. Spread in wind forecast increases with increase in forecast days. The regions showing more spread coincide with the dynamically active regions having higher wind magnitude suggest more uncertainty compared to calmer wind places. This wind causes climate change and spread eye infectious diseases.

Aura satellite, one of the devices that detect Nitrogen dioxide gases happens to be common emission from cars, power plants, and industrial activity. Nitrogen dioxide is a respiratory pollutant in urban smog used as an indicator of air quality. The year-to-year trends in Nitrogen dioxide levels have increased tremendously in the world, was analyzed by the science team.

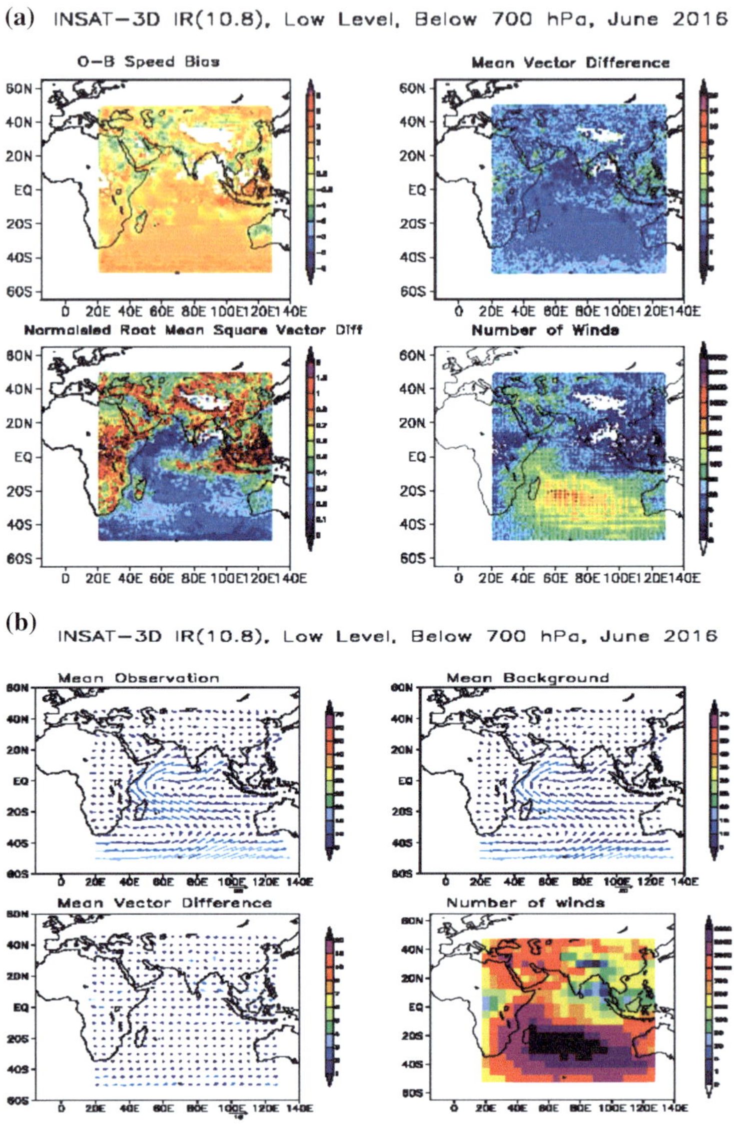

Fig. 12.5 **a** Indicates the number of winds, speed bias. **b** Indicates the number of winds, speed bias and mean observation Abisko

12.5 The Climate Prediction

The datasets collected from aircraft, ships and satellite put together helps to provide a comprehensive explanation of the global system. The measurements of the temperature of the different components of the urban surroundings such as the river,

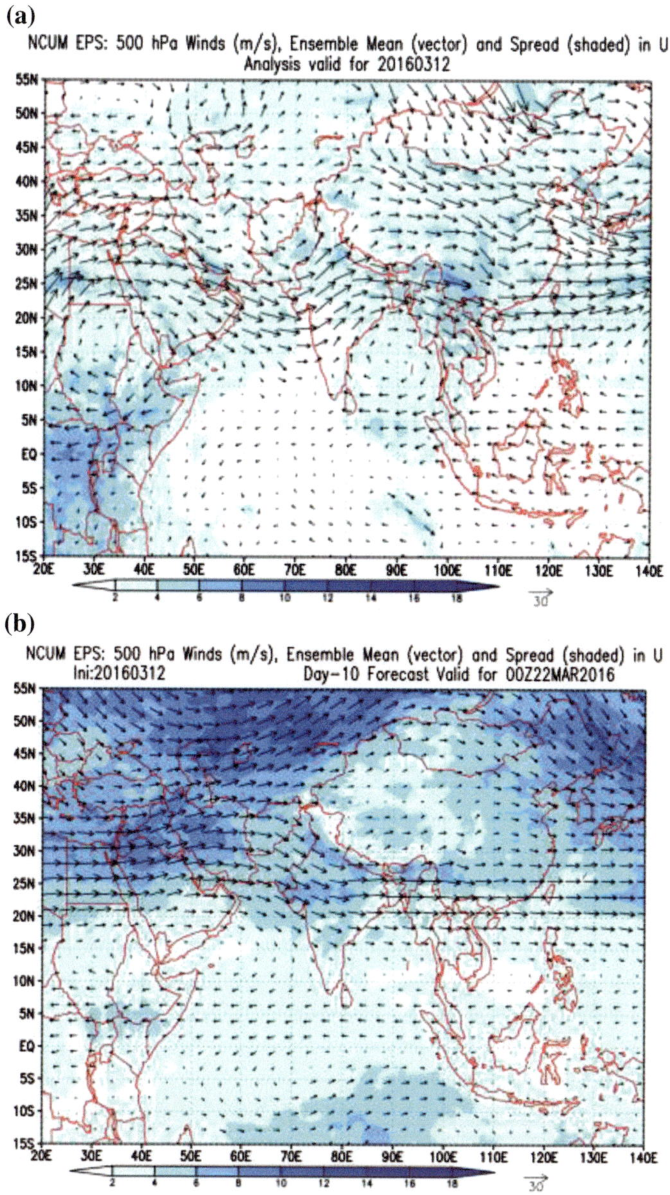

Fig. 12.6 **a** Analysis of forecast of winds. **b** Day-10 forecast of winds and ensemble spread of wind speed (shaded) at 500 hPa

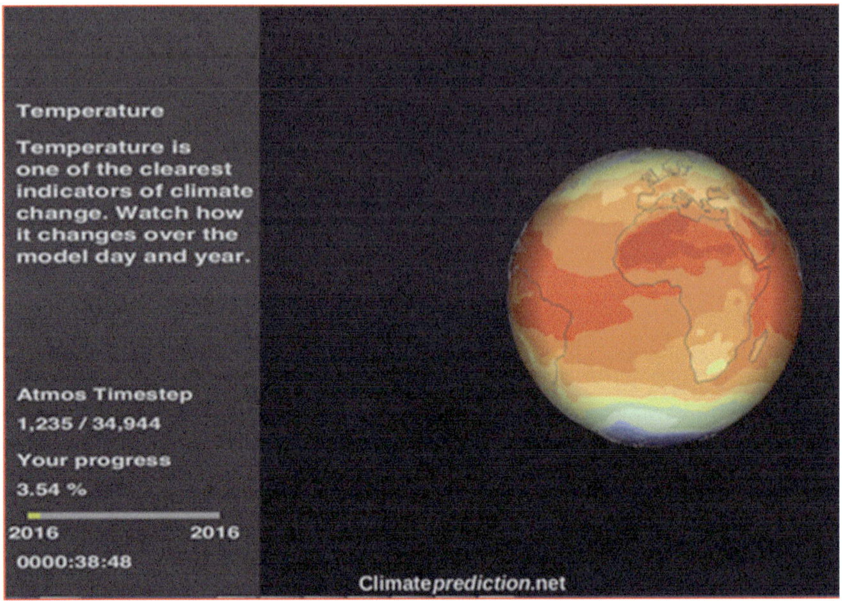

Fig. 12.7 The screen shot 1 of the climate model indicating temperature at 3.54%

the vegetation, the concrete, and tar are the other measurements that are logistically challenging and expensive.

Comparing Figs. 12.7, 12.8 and 12.9, it clearly suggests that the temperature is the main indicator of climate change, notice the variation in color shades in certain areas it changes over every second, minute, hours all together. These variations are captured by running the Climateprediction.net of BOINC manager on windows 7 64-bit. The BOINC is an open-source middleware system supporting grid computing and distributed applications like mathematics, climatology, environmental science, and medicine. As a high-performance distributed computing platform, it was developed at Space Sciences Laboratory at university of California. The small changes to the model can represent huge effects on the predictions of the climate. In Figs. 12.8 and 12.9 a white color spot on the model indicates the low temperature in few areas whereas red color indicates very extremely high temperature.

12.6 Cloud Networking

Cloud computing supports sharing of resources and carries out tasks remotely by connecting to a remote data center provided by an IT services. It provides a way for migration of data from in-built data centers. Cloud computing with its own concerns of security and reliability has become a big contributor for carbon emission because

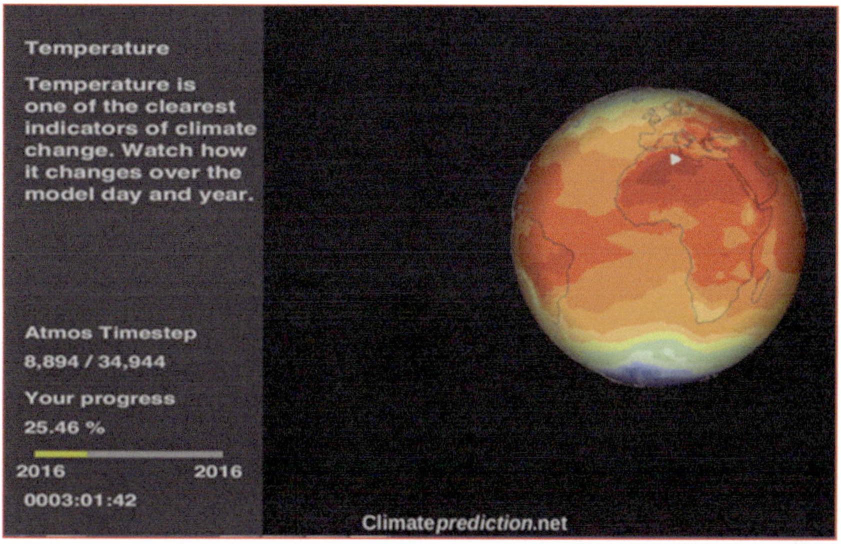

Fig. 12.8 The screen shot 2 of the climate model indicating temperature at 25.46%

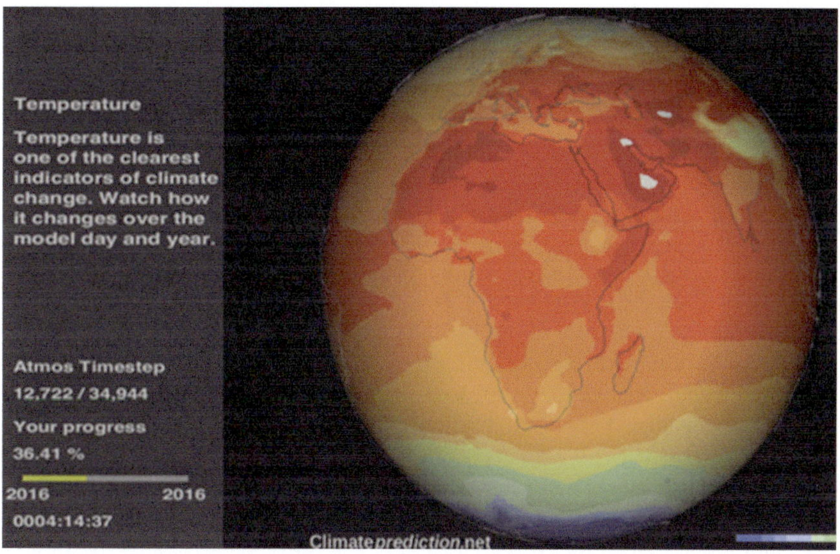

Fig. 12.9 The screen shot 3 of the climate model indicating white spots

of dirty energy. The cloud is expanding when climate changes and reducing carbon dioxide emissions. A recent survey states that cloud computing reduces energy consumption of IT departments and save monetary on energy bills. A demand shifting in data centers is significantly increasing; data backup are extending cloud's potentiality and can be distributed to countries on the other side of globe. It is intellect to shift data on the cloud where electricity supply is ideal or where wind power is produced maximum but consumption is low. Hence researchers are designing the cloud network at sites with exposure to high winds which can increase capacity and performance. Cloud computing not only save billions of dollars in energy costs but can reduce carbon emissions by millions of tons. Cloud computing has proved itself in maintaining and keeping updated information in climate change and hereby playing a significant role in the human health (Cloud Computing).

The government has introduced several schemes to restore equality, to eradicate poverty and hunger from the rural undeveloped places. To fund basic requirements and improve the building infrastructure and as a researcher, the most imminent need is communicating with the developed nations. The communication happens to be essential to improve the medical facilities; it is very common practice for doctors to refer to the health websites. The medical practitioner may require a knowledge-based sharing service to get connected to the network communities that has an interaction between doctors-to-doctor, patient-to-doctor. Thus, the cloud computing is an open, global communication networking facilitating data exchanges in the field of medical care (Shawish and Salama 2014; Choubisa 2012; Validation of INSAT 2016; http://www.esrl.noaa.gov/csd/assessments/ozone/1998/faq9.html).

12.7 Conclusions

The climate change on its way ensures the aim at improving the healthcare and emergency services to implement disaster planning relevant to preventing and educating risk communication. Responding to climate change the people living the remote areas can be addressed about the awareness of eye diseases and other chronic diseases. Climate change has social impact and intensifies the seasonal changes in the area, as well as endangering the lives of its populations. Increasing doses of UV-B cause human eye cataract and even extreme temperature variation, ultraviolet radiation, and speed of wind are core cause for global climate change affecting human eyes. Migrating population to lesser risk area can mitigate the human health hazard. Hence, the ozone depletion, UV radiation, climate change are source of cause related to human eye cataract. And cloud computing offers an excellent data centers to access a huge remote resource in an efficient and swift way.

Conflicts of Interest The authors declare that there are no conflicts of interest regarding the publication of this paper.

References

Abisko, Climate change, ozone depletion, and ultraviolet radiation. Climate Change 158–169

Choubisa K (2012) Cloud computing & rural development. Int J Informat Technol Knowl Manag 6(1):98–100

Cloud computing—a green opportunity or climate change risk? Greenpeace Contributor

Does Ozone Depletion cause Climate Change. http://www.esrl.noaa.gov/csd/assessments/ozone/1998/faq9.html

Frederik JGM, van Kuijk (1991) Effects of ultraviolet light on the eye: role of protective glasses. Environ Health Perspect 96:177–184

McKenzie RL, Aucamp PJ, Bais AF, Björn LO, Ilyas M, Madronich S (2011) Ozone depletion and climate change: impacts on UV radiation, pp 182–198. https://doi.org/10.1039/c0pp90034f. Epub 2011 Jan 20

Ozone depletion will bring big rise in number of cataracts. http://www.ncbi.nlm.nih.gov/pmc/articles/PMC1298891/

Portier CJ, Tart KT (2010) A human health perspective on climate change. A Report Outlining the Research Needs on the Human Health Effects of Climate Change, April 2010

Ramesh A, Bristow J, Kovats S, Lindsay SW, Haslam D, Schmidt E, Gilbert C (2016) The impact of climate on the abundance of *Musca sorbens*, the vector of trachoma. https://doi.org/10.1186/s13071-016-1330-y

Scott M, Lindsey R (2016) Which emits more carbon dioxide: volcanoes or human activities? June 15, 2016

Shawish A, Salama M (2014) Cloud computing: paradigms and technologies. Springer, Berlin Heidelberg. https://doi.org/10.1007/978-3-642-35016-0_2

Trachoma and climate. http://iceh.lshtm.ac.uk/climate-change-eye-health/

The impacts of ozone depletion. http://www.bcairquality.ca/101/ozone-depletion-impacts.html

The ozone layer, depletion and UV radiation—eco-action.org. http://www.eco-action.org/dt/ozone.html

Validation of INSAT-3D Atmospheric Motion vectors, June 2016

West SK, Bates MN, Lee JS, Schaumberg DA, Lee DJ, Adair-Rohani H, Chen DF, Araj H (2013) Is household air pollution a risk factor for eye disease? Int J Environ Res Public Health 10(11): 5378–5398. Published online 2013 Oct 25. https://doi.org/10.3390/ijerph10115378

http://www.ncmrwf.gov.in

Chapter 13
Health and Well-Being of Ageing Population in India: A Case Study of Kolkata

Lakshmi Sivaramakrishnan and Ambika Roy Bardhan

Abstract Ageing of population, the most inevitable consequence of demographic transition has been one of the most significant social transformations of the twenty-first century witnessed worldwide. Worldwide, there has been a 48% increase in 2015 from the 607 million aged persons in 2000. India located in southern Asia shelters approximately 1.16 million aged persons ranking second in the world after China and is projected to surpass China by 2050. The city of Kolkata, one of the oldest cities of India has recorded fast increase in the number of its aged population. Rapid urbanisation, globalisation and modernisation have left the aged in the city isolated, lonely, insecure and helpless, thus pushing them to old age homes. Living in old age homes is quite different than living in own homes and so understanding the well-being of these aged persons becomes important. Hence, the present study aims to find out the well-being of the aged persons residing in selected five old age homes of Kolkata based on age, gender, marital status, education and living arrangement. Well-being of two hundred aged respondents living in the five chosen old age homes was studied with respect to three dimensions, namely—health, economic and social relationships and engagements. For each dimension, an index has been constructed based on the composite scores calculated for all respondents. Based on the three indexes, the well-being index was computed. Finally, the obtained well-being index scores were categorised as high, moderate and low. Findings of the study clearly show that well-being of the respondents who were younger aged, male, married, living with spouse and those with higher education were better off than those who were very old, females, unmarried/widows/widowers, living single and illiterate.

Keywords Ageing · Health · Social relationships · Social engagements · Well-being

---

L. Sivaramakrishnan (✉) · A. Roy Bardhan
Department of Geography, Jadavpur University, Kolkata, India
e-mail: lakshmi.bu@gmail.com

A. Roy Bardhan
e-mail: ambika.roybardhan@gmail.com

R. B. Singh et al. (eds.), *Urban Health Risk and Resilience in Asian Cities*, Advances in Geographical and Environmental Sciences,
https://doi.org/10.1007/978-981-15-1205-6_13

13.1 Introduction

Ageing of population is the most inevitable consequence of demographic transition. Declining fertility rate on one hand and increasing life expectancy on the other hand have been the major causes for the rapid increase in the total number of the aged population worldwide. According to United Nations Report (2015a) the number of global population of 60 years and over has been approximately 901 million in 2015, an increase of 48% over the 607 million aged persons in 2000. By 2030, the number of aged persons in the world is projected to grow by 56%, to 1.4 billion, and by 2050, it is projected to more than double its size of that of 2015, reaching nearly 2.1 billion. Percentage share of aged population worldwide was 8.6% in 1980 which rose to 12.3% in 2015 and is projected to increase to 16.5% in 2030. Further, it has been estimated that for the first time in human history in 2050 aged persons will outnumber the population of children (0–14 years). In 2012 one out of every nine persons in the world has been aged being 60 years and above. By 2050, one out of every five persons is projected to be in that age group (UN 2012).

Figure 13.1 clearly depicts how over the years the share of male, female and total aged population have increased continuously and are projected to increase further in the coming years requiring attention of the planners, government, NGOs, researchers, etc. The 80+ population has recorded a share of 0.9% in 1980 which increased to 1.7% in 2013 and is projected to increase almost four times to 4.1% in 2050. For 60+ as well as 80+ populations females recorded higher percentages for all the years and are also projected to increase greater compared to male aged. This is because of feminisation of ageing population.

The distribution of the aged population in the world if taken together shows that just five countries—China, India, the United States, Japan and the Russian Federation—accounted for half of the world's aged population 60 years and above in 2015. The world's aged population 80 years and above was similarly concentrated in the small number of countries. The five countries with the largest number of

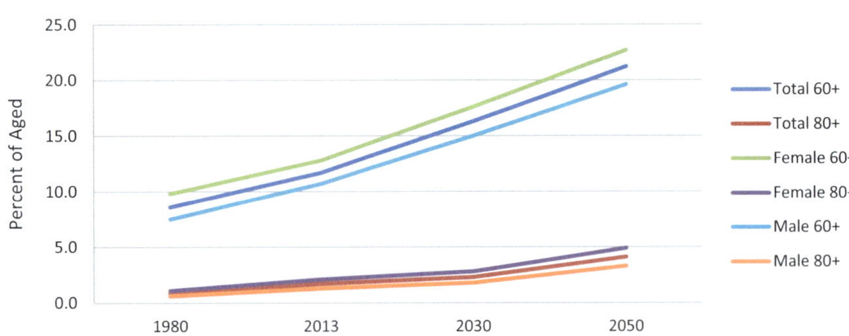

Fig. 13.1 Share of aged persons by age and sex. *Source* United Nations (2013)

oldest-old persons—China, the United States, India, Japan and Germany—collectively accounted for 48% of the world's oldest-old in 2015 and 19 countries held three-quarters of the global population aged 80 years or over (United Nations 2015c).

Thus from Fig. 13.2, it is very much evident that India accommodates a considerable number of aged population in the world. It ranks second after China by sheltering 1,16,553 aged above 60 years of age in 2015 which is projected to rise to 1,90,730 in 2030.

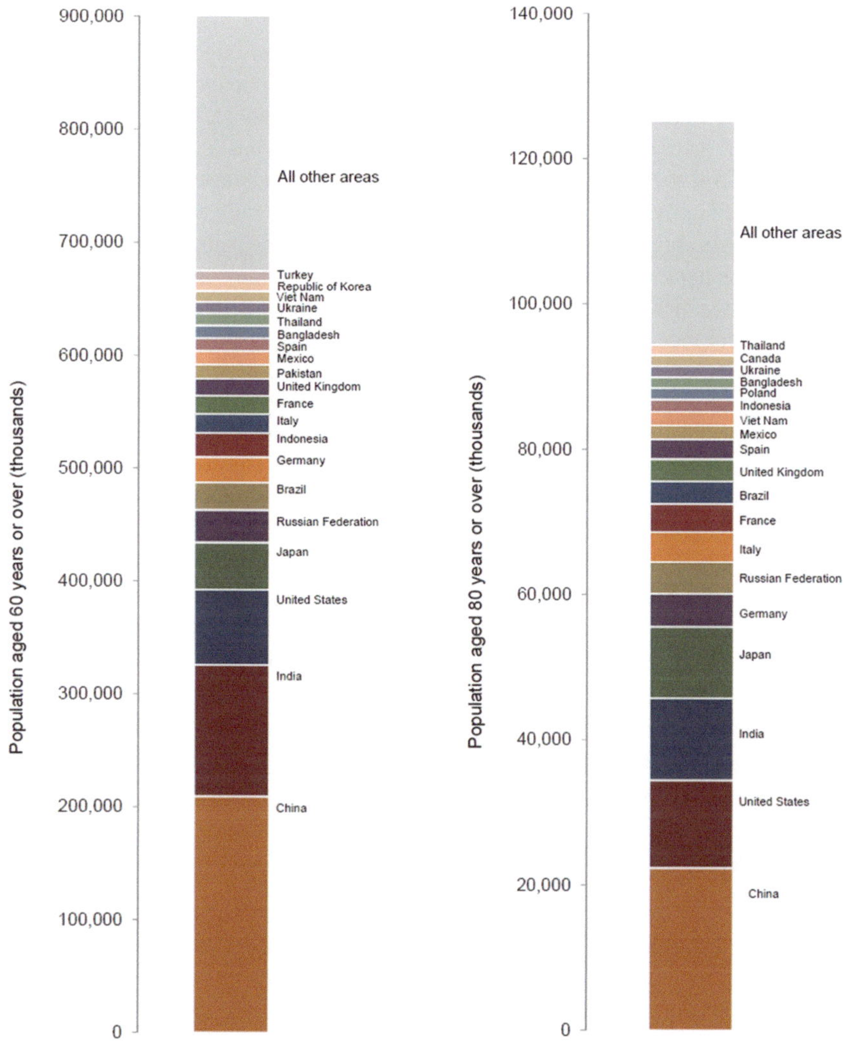

Fig. 13.2 Population Aged 60 years and above and aged 80 years and above by Country, 2015

Table 13.1 Percentage of aged in some metropolitan cities of India

City	20–24 Years (in %)	25–29 Years (in %)	30–59 Years (in %)	60 Years And Above (in %)
Kolkata	**8.9**	9.03	42.67	**11.76**
Chennai	9.3	10.2	41.12	9.9
Mumbai	10.7	10.53	38.88	8.9
Delhi	10.33	9.9	36.8	7.8
Bangalore	11.12	12.4	38	7.7

Source Sen (2015)

From Table 13.1, it can be noted that among the most important metropolitan cities of India, Kolkata records lowest number of younger population of 20–24 years while highest number of aged population that accounts to 11.76%. Rapid rate of urbanisation, modernisation and globalisation in the city has led to changes in the social structure, weakening of social ties and values and erosion of social institutions like the joint family system, changing economic structure with both of a couple are found to be working having less time for their family and most importantly migration of younger generations abroad or in other parts of the country for better job opportunities have somehow led the aged isolated, insecure and helpless in the city. This has led to very recently widespread growth of old age homes in the city for the aged who are either forcefully or willingly moving to such alternative accommodations.

In India, there has been a steady rise in the number of old age homes from 728 old age homes in 1998 to 1176 old age homes in 2009 (HelpAge India 1998, 2009). For the maintenance of old homes rupees, 34 crores were provided by the central government over the past four years out of which 71% of the fund went to five leading states of India accommodating 32% of India's aged population. Out of which West Bengal is one to have received rupees 2.14 crores (Bharti 2016). West Bengal ranks second with 164 old age homes just after Kerala in 2009 (HelpAge India 2009). Out of the 164 enlisted old age homes in West Bengal 77 are in Kolkata (around 47%).

Thus, in this respect studying the well-being of the aged becomes important and necessary as living in old age homes is quite different than living in own homes. Hence, the present study has been conducted in order to perceive the well-being of the aged population residing in five different old age homes in the city.

13.2 Objective

To find out the well-being of the aged residing in five different old age homes of Kolkata in terms of social, health and economic conditions and social relationships and engagements.

13.3 Database and Methodology

In West Bengal different types of old age homes are found. There are free of cost old age homes while there are old age homes on pay and stay basis. Certain old age homes have both free as well as pay and stay facilities. Some old age homes are meant exclusively for women while some for men only, others accommodate both men and women.

Considering this in mind the data were collected from five old age homes of similar types located within the Kolkata Municipal Corporation Area (KMC). The chosen old age homes were similar in terms of the facilities and services that they provided to the respondents like availability of doctors and nurses, medical facilities, ambulance service, entertainment facilities, quality of food, outing facilities, nature of accommodation, assets in the old age homes like open space for walking, yoga room, library, etc. accommodating both male and female aged and also in terms of the same monthly fee of rupees 15000/- that all the old age homes charged.

A sample of 200 respondents above 60 years of age residing in the five chosen old age homes for more than 6 months was randomly selected. Forty respondents were chosen from each old age homes. The locations of the five chosen old age homes chosen for the study have been shown in Fig. 13.3.

Data for the present study was collected through a primary survey using questionnaire prepared to find out the well-being of the respondents relating to their health, economic and social conditions in the five old age homes. Beside the questionnaire on social, economic and health conditions, a general information schedule was also used to collect information relating personal details about the respondent, i.e. name, gender, age, marital status, living arrangement and education received (Thanakwang and Soonthorndhada 2006). The data were collected between January and May 2016.

The details of the old age homes and the general information about the respondents are given in Table 13.2.

The three dimensions, namely, health consisting of six indicators, economic consisting of six indicators and lastly social relationships and engagements consisting of four indicators. Indicators relating to health conditions of the respondents included suffering from any chronic illness, health status, psychological well-being (Thanakwang and Soonthorndhada 2006), availability of medical facilities and access to medical facilities (Ramesh 2013). Indicators related to social relationships and engagements conditions included relation with the family members (Lalan 2014), relation with friends and relatives, frequency of visit by family members and participation in social activities. Economic indicators included income level of the respondents, source of monthly income (Bhatt et al. 2014), ownership of asset (Thanakwang and Soonthorndhada 2006), engagement gainfully in any economic activity.

Details of the broad three dimensions and the indictors are given in Table 13.3.

A weighted score was given to each of the indicators. Since there were variations in the range of possible answers to the question for each indicator the formula given by McGrahan and others in 1986 was used in order to bring the obtained weighted scores for each indicator into equilibrium with each other. These obtained scores were

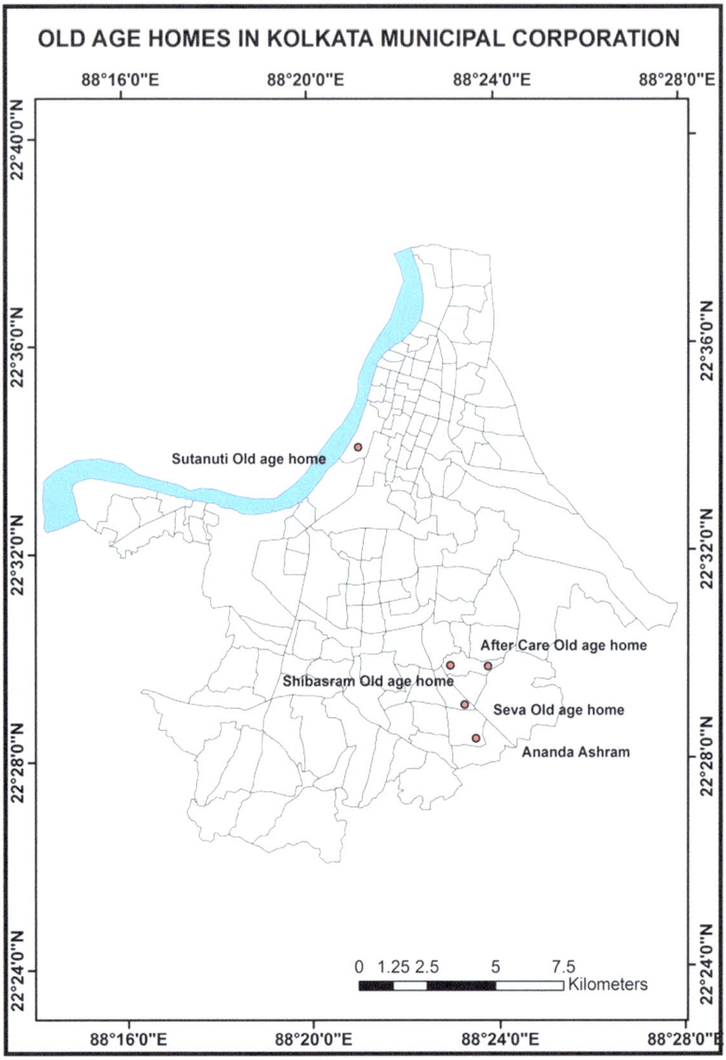

Fig. 13.3 Location of the old age homes chosen for the study in KMC area

then summed up to get the composite score for each dimension (Thanakwang and Soonthorndhada 2006). For example, the composite score on the health dimension is composed of six indicators (X1–X6). This was done mathematically as shown in the formula below:

Composite score = X1/MxT + X2/MxT + X3/Mxt + X4/MxT + X5/MxT + X6/MxT

where X = the score of the each indicator

M = the maximum answer value of each indicator

Table 13.2 General information about the respondents residing in the five Old Age Homes of Kolkata

Name of the Old Age Homes	**Location**	Age Groups			Gender		Marital Status			Living Arrangement		Education			
		60–70	70–80	80+	Male	Female	Married	Unmarried	Widows	Alone	With Family	Illiterate	Up to Primary	Secondary	Above Secondary
Shibasram	Santoshpur	11	11	18	18	22	27	3	10	11	29	1	17	1	21
Ananda Ashram	Naktala	11	17	12	17	23	22	3	15	16	24	1	20	1	18
Sutanuti Old Age Home	B.B.D Bagh	14	16	10	15	25	26	1	13	14	26	1	22	1	16
Seba Old Age Home	Baghajatin	11	18	11	15	25	28	1	11	12	28	1	22	1	16
After Care Old Age Home	Ajaynagar	11	18	11	15	25	23	6	11	17	23	1	17	2	20

Source Primary Survey (January–May, 2017)

Table 13.3 Details of the indicators under the broad three dimensions

Dimensions	Indicators	Measurement Scale
Health	Suffering from Chronic disease/illness	0 = Yes 1 = No
	Self-assessed health status	1 = Very bad 2 = Bad 3 = Fair 4 = Good 5 = Very Good
	Psychological Well-being	1 = Low 2 = Moderate 3 = High
	Ability to perform activities of daily living	1 = Yes 0 = No
	Access to health care facilities	1 = Yes 0 = No
	Availability of health care facilities	1 = Yes 0 = No
Economic	Income level	1 = Low 2 = Moderate 3 = High
	Adequacy of income	1 = Yes 0 = No
	Source of Income	1 = One Source 2 = Two or More Source
	Ownership of any asset	1 = Yes 0 = No
	Engagement in gainful economic activity	2 = Fully 1 = Partially 0 = No Engagement
Social	Relationship with family	3 = Good 2 = Fair 1 = Bad 0 = No Relation
	Relationship with friends/relatives	3 = Good 2 = Fair 1 = Bad 0 = No Relation
	Frequency of visit of family members	4 = Very Frequently 3 = Frequently 2 = Occasionally 1 = Rare 0 = Never
	Participation in social activities	3 = Frequent Participation 2 = Occasionally 1 = Rare 0 = Never

T = the total number of indicators of a dimension

Composite score of health = X1/5x6 + X2/3x6 + X3/1 × 6 + X4/1 × 6 + X5/1 × 6 + X6/1 × 6.

Based on the composite scores an index of each dimension (health, economic and social relationships and engagements) was constructed. To calculate these dimension indices, minimum and maximum values (goalposts) were chosen for each underlying indicator. Performance in each dimension is expressed as the minimum and maximum value between 0 and 1 in accordance with the construction method of the Human Development Index developed by the United Nations Development Programme (UNDP 2005) as follows:

Dimension index = Actual value − Minimum value

Maximum value − Minimum value

Simple average of the obtained three indices was computed to get the well-being index as shown in the formula below:

Well-being index = 1/3(health index) + 1/3(economic index) + 1/3(social relationships and engagements index).

Each index was classified into three levels based on the UNDP criteria of human development level, which constitutes an indicator of the quality of life, as follows:

(1) Index score less than 0.5 is low level;
(2) Index score between 0.5 and 0.79 is moderate level;
(3) Index score equal or higher than 0.8 is high level.

13.4 Findings and Discussions

From Table 13.4 it is evident that the health index of the male respondents was far better than that of females. 21.1% of the males recorded higher health index compared to 4.5% of females while 28.4% females recorded lower health index compared to only 7.3% of males. Poor economic condition among the female aged could be the reason which prevented them from taking time to time required medical treatments. Health index of the younger aged was far better than those of older aged. This is because of the commonly accepted fact that ageing is a biological process and it is associated with deterioration in health condition. With respect to marital status, married aged recorded better health index where 69.4% had higher health index while 56.2% widowed aged had lower health index. 68.9% of the unmarried aged had moderate health index. This can be because as noted while surveying the unmarried respondents were mostly mentally strong, independent and were more acceptable in their attitude about their present condition and mentally prepared for their future days to come. The health indexes of the respondents living with their family member that is their spouse were far better than those living alone. This can be because as observed while surveying being together, both the couple looked after their partners, both were mental and physical support for each other. Finally, the health indexes of the higher educated respondents were better. This can be because

Table 13.4 Conditions of the aged residing in Old Age Homes of Kolkata with respect to their health index

General characteristics		Health index		
		Low (in percent)	Moderate (in percent)	Higher (in percent)
Gender	Male	7.3	71.6	21.1
	Female	28.4	67.1	4.5
Age groups (in years)	60–70	14.3	55.8	29.9
	70–80	45.6	37.6	16.8
	80+	69.6	28.3	2.1
Marital status	Unmarried	11.8	68.9	19.3
	Married	7.3	23.3	69.4
	Widowed/Divorced/separated	56.2	28.7	15.1
Living arrangement	With family	30.2	33.5	36.3
	Single	33.1	38.9	28.0
Level of education received	Illiterate	88.9	9.1	2.0
	Up to Primary	78.1	19.5	2.4
	Secondary	66.5	22.3	11.2
	H.S	58.8	21.9	19.3
	Graduate	29.1	48.9	22.0
	Above graduate	6.1	42.0	51.9

Source Primary Survey (January–May, 2017)

most of the higher educated respondents were males, whose economic conditions were good. So, they could afford proper health care facilities. Thus, it can be said that the health indexes were better among males, younger aged, married, those living with their family and higher educated respondents.

Table 13.5 depicting the economic indexes clearly portray that male aged recorded a higher economic index than females. This can be because greater percentage of the male respondents reported to have worked when they were young and have retired with more than one source of income. Most of them received pension and also enjoyed interest from their savings. The economic index among the younger aged was higher because most of them reported that they were engaged in some kind of economic work on a part-time basis. Aged who were married and living with their family recorded higher economic index. This can be due to the fact that the couples in most cases had more than one source of income and ownership of assets. Finally, aged who were more educated had a higher economic index because they reported that they earned higher income during their working ages and enjoy in most cases pension and interest from their savings. Some of them have been found to be working on a part-time basis. Thus, it can be seen that the economic index was higher among male, younger aged, married, those living with family and higher educated aged.

Table 13.5 Conditions of the aged residing in Old Age Homes of Kolkata with respect to their Economic Index

General characteristics		Economic Index		
		Low (in percent)	Moderate (in percent)	Higher (in percent)
Gender	Male	3.7	30.5	65.8
	Female	38.2	39.3	22.5
Age Groups (in years)	60–70	28.2	33.4	38.4
	70–80	26.5	38.8	34.7
	80+	39.4	35.9	24.7
Marital Status	Unmarried	58.7	25.8	15.5
	Married	12.8	38.6	48.6
	Widowed/Divorced/Separated	50.8	37.3	11.9
Living Arrangement	With family	9.3	32.1	58.6
	Single	49.2	48.9	1.9
Level of Education Received	Illiterate	81.2	13.9	4.9
	Up to Primary	61.2	28.8	10.0
	Secondary	27.6	33.2	39.2
	H.S	17.5	30.4	52.1
	Graduate	12.2	21.2	66.6
	Above Graduate	8.1	19.8	72.1

Source Primary Survey (January–May, 2017)

From Table 13.6 it can be said that the social index of female aged has been higher than that of males. 58.7% of the females recorded high social index compared to just 12% among males. This can be due to the fact as many studies have revealed that females socialise more than that of males. Moreover, the social index among the younger aged has been better with 69.2% of age group 60–70 years than the older aged with just only 0.7% with age group 80 years and above. This has been because the younger aged reported to be much more physically healthy, psychologically stronger with more ability to accept the harsh realities of life that has led them to socialise more than the aged in the other older age groups. Surprisingly, the social index among the unmarried aged was higher with 68.4%. This has been because most of the unmarried aged surveyed were from 60 to 70 years age group, many respondents reported that they met their friends most often and had cordial relationship with their family members. Those aged living with their family recorded higher social index (48.8%) compared to those living single. This is because those staying with their family revealed more willingness to participate in social activities with their partners, many reported that they willingly came to the old age home as they wanted to live an independent life and so had good relations with their children, relatives, etc. Among the higher educated aged respondents the social index was higher. This can be due to the fact that the higher educated aged were more economically independent which gave them more opportunity to meet their friends, relatives often, take part

Table 13.6 Conditions of the aged residing in Old Age Homes of Kolkata with respect to their social index

General characteristics		Social Index		
		Low (in percent)	Moderate (in percent)	Higher (in percent)
Gender	Male	38.1	49.9	12.0
	Female	8.9	32.4	58.7
Age groups (in years)	60–70	1.7	29.1	69.2
	70–80	14.7	39.2	46.1
	80+	78.2	21.1	0.7
Marital status	Unmarried	3.5	28.1	68.4
	Married	51.1	32.6	16.3
	Widowed/Divorced/separated	62.8	21.1	16.1
Living arrangement	With family	17.9	33.3	48.8
	Single	31.3	33.4	35.3
Level of education received	Illiterate	12.1	32.9	55.0
	Up to Primary	11.9	33.5	54.6
	Secondary	10.1	31.6	58.3
	H.S	7.1	28.7	64.2
	Graduate	6.2	30.1	63.7
	Above Graduate	4.3	31.2	64.5

Source Primary Survey (January–May, 2017)

in different tours, occasion, etc. Thus, it can be said that the social index among the females, younger aged, unmarried, those living with their family and higher educated aged accounted better social indexes.

The overall well-being index has been computed based on the three indexes—health, economic and social. It can be said from Table 13.7 that the overall index of male aged is higher compared to that of females. This is because male aged recorded higher economic and health index compared to that of female aged. The better economic condition of the male aged has helped them in accessing better health care and medical facilities. On the other hand the overall well-being indexes of the younger aged were better compared to the very old respondents. This was because the younger aged respondents were much more economically stable, socialised more and were healthier than the older aged respondents. Married aged and those living with their family also accounted higher overall well-being index. This would be because the respondents accounted higher health, social and economic indexes. Lastly, the higher educated aged respondents accounted higher overall well-being index. This had been because the higher educated aged respondents were economically better off that had helped them to avail all the better health care and medical facilities, and also afford to participate in all social activities and functions, thus recording higher economic, health and social indexes.

Table 13.7 Conditions of the aged residing in Old Age Homes of Kolkata with respect to their overall well-being index

General characteristics		Overall index		
		Low (in percent)	Moderate (in percent)	Higher (in percent)
Gender	Male	16.4	50.7	32.9
	Female	25.2	46.3	28.5
Age groups (in years)	60–70	14.7	39.4	45.8
	70–80	28.9	38.5	32.5
	80+	62.4	28.4	9.2
Marital status	Unmarried	24.7	40.9	34.4
	Married	23.7	31.5	44.8
	Widowed/Divorced/separated	56.6	29.0	14.4
Living arrangement	With family	19.1	33.0	47.9
	Single	37.9	40.4	21.7
Level of education received	Illiterate	60.7	18.6	20.6
	Up to Primary	50.4	27.3	22.3
	Secondary	34.7	29.0	36.2
	H.S	27.8	27.0	45.2
	Graduate	15.8	33.4	50.8
	Above Graduate	6.2	31.0	62.8

Source Primary Survey (January–May, 2017)

13.5 Conclusion

Ageing of population has been a major social issue of the twenty-first century which has left its impact in the city of Kolkata. The reason for more concern is that rapid urbanisation, modernisation and globalisation have left the aged isolated, helpless, insecure and alone in the city pushing them to the old age homes. Better well-being of the aged is needful for the overall development of the ancient city of Kolkata in true sense of the term. Detailed study has clearly portrayed that there exist differences in the well-being of the aged with respect to their gender, age, marital status, living arrangement and education. Aged who are male, younger aged, married, living with their family and higher educated enjoy better well-being compared to those who are female, very old, unmarried or divorced or separated, living alone, and have received no or less education. The conditions of the aged with lower well-being are miserable and need attention. Both it is the responsibility of the Government as well as the society to safeguard the interest of the aged for making an inclusive society in its all way in the city. The aged should be considered as an asset and not a burden on the society whose experience if harnessed can be directed in the betterment of the society.

Moreover, development in the city can only take place when a higher well-being of its entire population can be achieved.

References

Bharti N (2016) India is ageing, but it's not spreading central funds evenly among the aged. Accessed from https://scroll.in/article/809500/india-is-ageing-but-its-not-spreading-central-funds-evenly-among-the-aged

Bhatt BM, Vyas V, Joshi JP (2014) Ageing and health: a health profile of inmates of old age home. Nat J Commun Med 5(1):2

HelpAge India (1998) Directory of Old Age Homes in India 1998. Policy Research and Development Department HelpAge India. New Delhi

Help Age India (2009) Directory of old age homes in India Revised Edition 2009. Policy Res Development Department Help Age India, New Delhi

Lalan Y (2014) A sociological study of old persons residing in an old age home of Delhi, India. Int Res J Soc Sci 3(4):22

McGahan PL et al (1986) Composite score manual. University of Pennsylvania/Veterans Administration Center for Studies of Addiction, Philadelphia, United States

Ramesh (2013) Health status of elderly living in the old age homes: an overview. J Media Social Develop 139

Sen S (2015) Kolkata is ageing faster than other metros. Times of India

Thanakwang K, Soonthorndhada K (2006) Attributes of active ageing among older persons in Thailand: evidence from the 2002 survey. Asia-Pacific Population J 113

United Nations Development Programme (2005) Human development report 2005. United Nations Development Programme, New York

United Nations (2012) Population Ageing and Development (No. ST/ESA/SER.A/323). Department of Economic and Social Affairs, Population Division, New York: United Nations

United Nations (2013) Profiles of Ageing 2013. United Nations, Department of Economic and Social Affairs, Population Division, New York

United Nations (2015a) World Population Ageing (No.ST/ESA/SER.A/390). Department of Economic and Social Affairs, Population Division, New York: United Nations, 9

United Nations (2015b) World Population Ageing (No.ST/ESA/SER.A/390). Department of Economic and Social Affairs, Population Division, New York: United Nations, 17

United Nations (2015c) World population ageing (No.ST/ESA/SER.A/390). Department of Economic and Social Affairs, Population Division, New York: United Nations, 16, 17

Chapter 14
Noise Pollution and Its Consequences on Urban Health in Sylhet City

Tulshi Kumar Das and Rony Basak

Abstract Sylhet is one of the fast growing cities in Bangladesh. It has witnessed unprecedented population growth in the recent years. Migration of people from different parts of the country to Sylhet city especially for education and employment has substantially increased over the last 20 years. The trend of urbanization has engulfed adjacent rural areas of the city which has been attracting investments in multiple sectors like education, health, industry, commerce, and trade. As a result, phenomenal growth of transports of varied types, converted natural gas (CNG) refueling stations, engineering and welding workshops of different sorts is seen throughout the city. Moreover, a power generating station has been located very close to the residential areas, just around five kilometers away from the centre of the city. The roads inside the city always remain busy with vehicles-jam resulting in a severe chaotic situation that causes harm to human physiological and psychological health. The main sources of intolerable noise in the city are huge numbers of transports, CNG stations, engineering and welding workshops and a power generating station that all are generally located along with the major thoroughfare. The aims of the current study are to measure and realize the amount of noise pollution that occurs at industrial, commercial, residential, and silence zones of the city. It investigates physiological and psychological sufferings of the affected people living in different areas and experiencing continuous noise from varied sources. The study tries to perceive the level of awareness among the people who regularly face unbearable noise, and it also explores the kinds of treatment people might have taken from the physicians because of noise pollution. The nature of the study is basically qualitative. But quantitative measurement of noise is carried out meticulously so that the rate of noise pollution at different areas could be understood better. Purposive sampling technique is used to select 20 persons from four different zones–industrial, commercial, residential and silence—in order to understand the physiological and psychological sufferings

---

T. K. Das (✉)
Department of Social Work, Shahjalal University of Science and Technology, Sylhet, Bangladesh
e-mail: tulshikumardas@gmail.com

R. Basak
Department of Geography and Environment, Shahjalal University of Science and Technology, Sylhet, Bangladesh

R. B. Singh et al. (eds.), *Urban Health Risk and Resilience in Asian Cities*, Advances in Geographical and Environmental Sciences,
https://doi.org/10.1007/978-981-15-1205-6_14

of the people. Another 20 persons are selected using accidental sampling technique from the same four zones in the same way to realize people's awareness about noise pollution that they encounter in their day-to-day life and also their suggestions to reduce its rate. Two interview guides are adopted for the study; one is for collecting qualitative data relating to physiological and psychological sufferings of the people and the treatment sought because of noise pollution, and the other one is for collecting qualitative data that helps understand the level of people's awareness with regard to noise pollution. The findings of the study show that noise pollution at different zones goes beyond the acceptable limit, and people have multiple physiological and psychological sufferings due to noise pollution, which have sometimes forced them to seek medical treatment. It also shows that people's awareness about noise pollution seems to be gradually growing high in Sylhet city.

Keyword Noise · Pollution · Urban · Health · Sylhet · People's awareness

14.1 Introduction

Noise pollution is considered as one of the important sources of environmental degradation in the fast growing urban ambience across the globe (Hunashal and Patil 2012; Phan et al. 2010). Though its adverse effects on human health have been recognized widely, it still does not get adequate attention like many other pollutions, especially in developing and least developed countries (Baloye and Palamuleni 2015; Jamrah and Al-OmariA 2006; Zannin et al. 2002). Rapid growth of urbanization, industrialization, means of communication, transports, traffic, construction works, business and commerce, etc. has caused excessive noise that often crosses normal limit of tolerance for the urban people (Hunashal and Patil 2012; Pathak et al. 2007, 2008). This situation creates an impact on physical and mental well-being of those who are regularly exposed to it (Piccolo et al. 2005: Phan et al. 2010; Mato and Mufuruki 1999). Studies (Alberola et al. 2005; Baloye and Palamuleni 2015; Georgiadou et al. 2004; Hunashal and Patil 2012; Pathak et al. 2008; Phan et al. 2010; Vidyasagar and Rao 2006) conducted on noise pollution in different countries throughout the world clearly show its negative consequences on public health. Noise is always undesirable, unpleasing and disturbing and it affects physical and mental health of the exposed people with so many psychological and physiological impacts. Urban people facing unremitting excessive noise in their day-to-day life generally suffer from different health problems like irritability, hearing loss, hypertension, blood pressure, palpitation, headaches, stomach ulcers, sleeping disturbance, stress, less productivity, feeling of fatigue, vertigo, etc. (Allen et al. 2009; Baloye and Palamuleni 2015; Brainard et al. 2004; Hunashal and Patil 2012; Morrell et al. 1997; Oni et al. 2016; Pandya 2003; Piccolo et al. 2005; WHO 2011; Zannin et al. 2002). Normal tolerable noise level has been considered as between 45 dB (A) and 60 dB (A) while measured by A-weighted decibel scale with 0 dB (A). People with noise level more than 80 dB (A) for a long time can cause hearing loss or becoming deaf, and on the other hand, people facing noise level between 130 and 140 dB (A) could be extremely painful for them (Baloye and Palamuleni 2015; Berglund and Lindvall 1995). Like any other South Asian countries Bangladesh is also affected due to growing noise pollution in its different cities. Dhaka, the capital city of Bangladesh, is worst affected because of noise pollution, which has become an issue of serious concern. The unprecedented growth of Dhaka city with rapid urbanization, industrialization and motorization has attracted millions of people from all over the country, which in turn, has made the city as one of the noisiest cities in the world (Ahmed and Rahman 2015; Chowdhury et al. 2010; Haq et al. 2012). Other big cities like Chittagong, Khulna and Sylhet are equally affected due to noise pollution. Noise pollution in Sylhet city has been increasing at an alarming rate because of its extraordinary growth in terms of urbanization, educational institutions, business and commerce establishments along with many other government and non-government activities over the years. The number of motorized vehicles has increased substantially in the Metropolitan area of the city, and many people from all over the country, especially people from the northeast region have chosen Sylhet city as their final destination. This development has made the city extremely noisy and crowded (Alam et al. 2006; Amin et al. 2014; Farzana et al. 2014). Though the effects of noise pollution are well recognized across the globe, the Sylhet city inhabitants feel helpless as there is nobody looking into the matter despite their immense sufferings from the growing noise pollution in the city. This study makes a modest effort to understand the level of noise pollution in Sylhet city, and investigates the sufferings of the people because of noise-induced pollution that they face in their everyday life.

14.2 Conceptual Framework

Noise pollution and its negative consequences on the quality of human life and human health have been recognized as serious problem in most of the cities across the world (Fiedler and Zannin 2015; Georgiadou et al. 2004; Goudreau et al. 2014; Jamrah and Al-OmariA 2006; Mehdi et al. 2011; Murthy et al. 2007; Pathak et al. 2008; Ramazani et al.2017; Skanberg and Ohrstrom 2002). It generally causes annoyance, hypertension, hearing loss, blood pressure, cardiovascular effects, ischemic stroke, sleep disturbance, stress, anxiety, nuisance, fatigue, bad performance, depression, etc. for regularly exposed people (Agarwal and Swami 2011; King and Davis 2003; Murphy et al. 2009; Ramazani et al. 2017; Singh and Davar 2004). The problem of noise pollution is much more in developing and least developed countries compared to developed ones, and it has been increasing at an alarming rate in many of the developing and least developed countries (Al-Harthy and Tamura 1999; Baloye and Palamuleni 2015; Banerjee 2012; Jamrah and Al-OmariA 2006; Mehdi et al. 2011; Murthy et al. 2007; Shastri et al. 1996). Thus the people living in the cities of third world countries have been suffering from numerous physiological and psychological problems (Abdel et al. 2000; Agarwal and Swami 2011; Pathak et al. 2008; Phan et al. 2010; Vidyasagar and Rao 2006; Zannin et al. 2002). Bangladesh is still a least developed country, and all the major cities in the country, including Dhaka—the capital city, have currently been going through an escalated rate of urbanization and industrialization resulting in an exodus of rural people arriving and permanently settling in different cities. As a result, Dhaka has become most crowded and nosiest city in the country. The inhabitants of Dhaka city are worst affected as they have been suffering from numerous physiological and psychological illness (Alam et al. 2001; Chowdhury et al. 2010; Haq et al. 2012; Hoque et al. 2013; The Daily Sun 2017). Other major cities like Chittagong, Khulna, and Sylhet also have started receiving people from rural and semi-urban areas of the country. Thus, these cities have gradually become crowded, and noise pollution in all these major cities has gone beyond the acceptable limit. Sufferings of these city dwellers have also been pointed out in a few studies (Alam et al. 2006; Al Mamun et al. 2017; Amin et al. 2014; Haq et al. 2012; Hoque et al. 2013; The Daily Star 2013). Sylhet, one of the major cities in Bangladesh, has witnessed unprecedented growth and expansion of educational institutions, medical colleges and hospitals, business and commerce, motorized vehicles, roads and transports, civil construction works, government and non-government offices and their activities in recent times. A few studies (Alam et al. 2006; Amin et al. 2014; Farzana et al. 2016; Rahman et al. 2005) have indicated the level of noise pollution and its effects on the people who regularly experience noise in Sylhet city. But noise pollution and its consequences in Sylhet Metropolitan city area is still not much known. Even the awareness about noise pollution among the city dwellers is not investigated yet. Therefore, effective policy and programs to reduce the level of noise pollution in Sylhet Metropolitan city area do not seem to be made in the agenda of the concerned authorities.

14.3 Objectives of the Study

The overall objective of the study is to explore the level of noise pollution and its consequences among the people of Sylhet Metropolitan city area. Specific objectives of the study are

(a) To measure and understand the level of noise pollution at different zones in Sylhet city;
(b) To investigate the physiological and psychological consequences of noise pollution among the exposed people;
(c) To explore the level of awareness among the people about the noise pollution;
(d) To know about the treatments affected people seek because of noise pollution

14.4 Methodology

14.4.1 Area of Study and Measurement of Noise

Sylhet is located in the northeastern part of Bangladesh (Fig. 14.1), and it is one of the major cities in the country. The total population of Sylhet district is 3,434,188. The geographical area of Sylhet district is 3,452.07 km^2 (1332.85 $miles^2$), and the density of population is 995/km^2 (2600/mi^2) (BPC 2011). The total population of Sylhet Metropolitan city is 531,663, and the area of the city is 26.50 km^2 (10.23 mi^2) (BBS 2014) (Fig. 14.2). The study is conducted in Sylhet Metropolitan city area from July to September 2017. In some cases, adjacent places of Sylhet Metropolitan city area are also chosen to measure the level of noise as the places are considered important with regard to their industrial and educational value. Different areas of the city have been identified as silence, industrial, commercial, and residential zones according to the characteristics of a particular area. The area which has hospitals, schools, colleges, and universities is considered as silence zone, and the area which has small and medium-size industries, factories, power plant, engineering, and welding workshop, etc. is considered as industrial zone. On the other hand, Railway station, bus stands and some of the business centers located in different parts of the city have been identified as commercial zone; and different locations of residence of city dwellers have been identified as residential zone. Noise level of different zones from nine different points of the city has been measured for 15 days during morning (9.00–11.30 am), evening (4.00–6.00 pm), and night (8.00–11.00 pm). High and low level of noise during morning, evening and night in each zone has been recorded for 15 days; and finally measurements of noise (high and low) during different times have been made average to understand the real scenario of noise pollution in the city. A portable digital noise levels meter has been used to measure the noise levels in different zones of the city. Table 14.1 shows the measurement of noise level at different zones in Sylhet Metropolitan area:

The above table clearly shows the current situation of noise level at different zones in Sylhet Metropolitan city area. The measurement of noise level indicates that noise has gone beyond the acceptable limit in most of the important areas inside the city.

14.4.2 Sampling and Sample Size

The sample size for this study is 40. Of 40, 26 are male and 14 are female, selected as encountered in different zones of the city. Purposive sampling method is used to select 20 persons from industrial, commercial, silence and residential zones of the city to investigate their psychological and physiological sufferings caused by noise-induced pollution. Another 20 persons are selected using accidental sampling technique under purposive sampling method aiming to explore the level of their awareness about growing noise pollution in the city. Two interview guides are used;

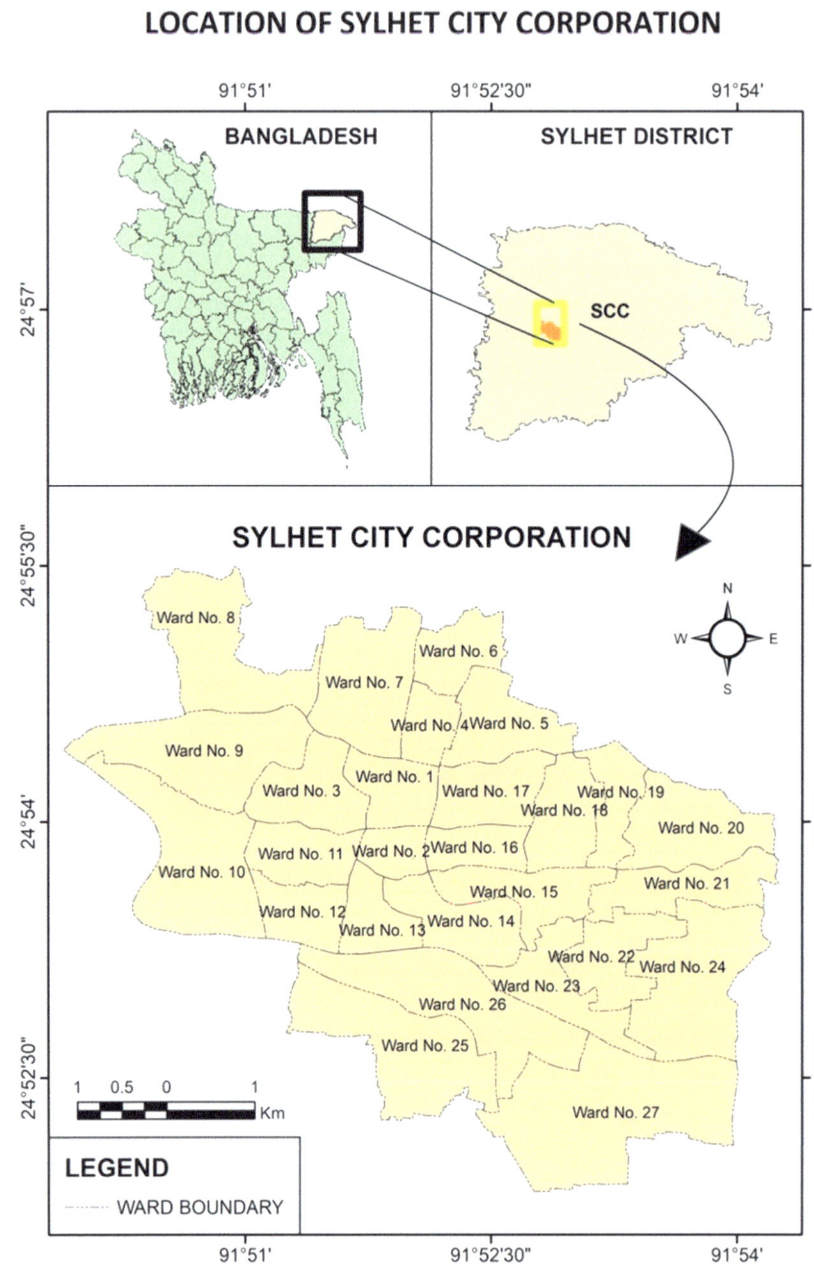

Fig. 14.1 Location of Sylhet City Corporation (*Source* Own made)

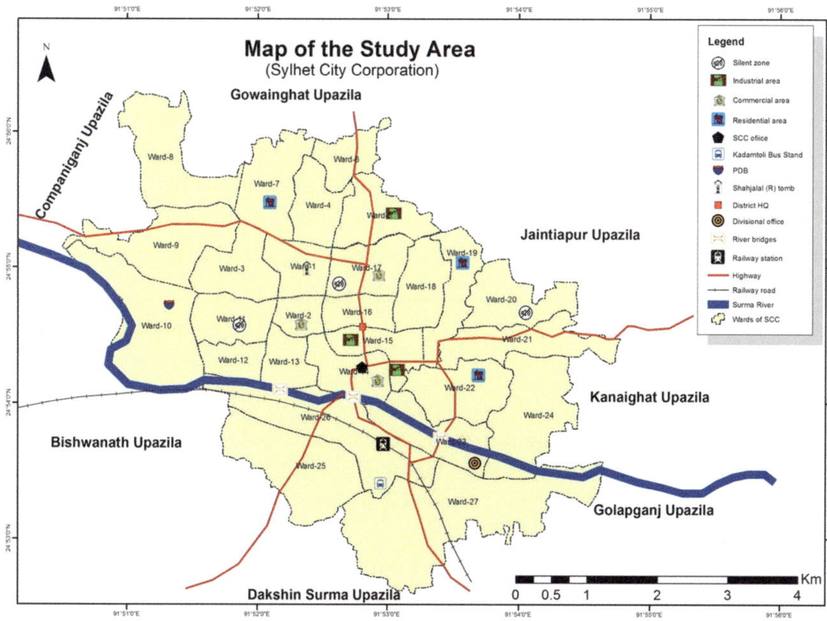

Fig. 14.2 Map of the study area (*Source* Own made)

Table 14.1 Measurement of noise level at different zones in Sylhet metropolitan area

Zone	Name of the Location	Morning	9–11:30 am	Evening	4–6:00 pm	Night	8–11:00 pm
		High	Low	High	Low	High	Low
Silence	Kajijalaluddin school	62.3	49.6	71.6	52	70.5	55.2
Silence	Scholarshome TV gate	89.3	72.6	87.5	66.5	80.2	69.2
Industrial	Chalibondor	56.2	43.5	76.7	66	70.5	59.4
Commercial	Bondor court point	89.1	75.1	90.4	76.7	87.3	72.1
Silence	Sylhet Osmani medical college and hospital	76.9	61.4	76.2	63.8	69.9	59.5
Commercial	Sylhet railway station	69	52.7	68.7	57.6	83.1	59.6
Commercial	Kadamtoli bus stand	88	73.5	80.8	66.8	87.1	68.4
Commercial	Zindabazar	85.2	78.7	92.2	77.8	89.5	75.3
Residential	Subidbazar Kolapara	65.5	49.1	72.2	54.2	71.3	57.5

Data Source Field Survey

one for collecting qualitative data about physiological and psychological sufferings of the people and the treatments they felt forced to take from the doctor because of noise-induced problem, and the other one is also for qualitative data about the level of participants' awareness regarding growing trend of noise pollution and their suggestions to address it in the context of Sylhet city. Moreover, quantitative data have been collected in relation to the current level of noise measurement in different zones of the city to understand the overall situation of noise pollution in Sylhet Metropolitan city area.

14.4.3 Data Collection

Both the researchers of this study have jointly collected all data from 40 selected research participants. The researchers themselves visited different zones namely, industrial, commercial, silence and residential within Sylhet Metropolitan area from August 16 to September 17, 2017. They met and built up rapport with potential research participants in all the zones and collected data from them through 3–4 times interactions. The objectives of the research were first explained to all the potential participants, one by one whenever someone met. Not everybody agreed to be interacted with the researchers. At least 115 persons were contacted, but at last 40 persons were fully interacted with. Others mostly did not have much time to spend with the researchers and therefore could not be interviewed. The researchers' first ensured permission from all 40 research participants and jointly signed informed consent form for full length interview. These participants agreed to be interacted with several times on several days. They were given freedom not to answer particular questions if wished, and also assured their freedom to stop answering questions midway through if felt so. Participation of the respondents in the interaction with the researchers was completely voluntary, and the participants had refused to take any money or gift for spending time. Each interaction was held in native Bangla language and was audio-recorded with their permission.

14.4.4 Data Analysis and Interpretation

Both qualitative and quantitative data are collected. All the data are transcribed first and then translated into English. Translated data are arranged and put in different categories formulated in relation to the objectives of the study. Then they are presented in different charts and tables. In some cases, qualitative data are quantified and presented in charts. Different charts and tables reflect the thoughts and ideas of research participants about the situation of noise pollution and its consequences on urban health in Sylhet Metropolitan area. Each chart and table is explained and analyzed under different categories to catch up the real scenario of noise pollution in the city. Some of the narratives extracted from the research participants have also been

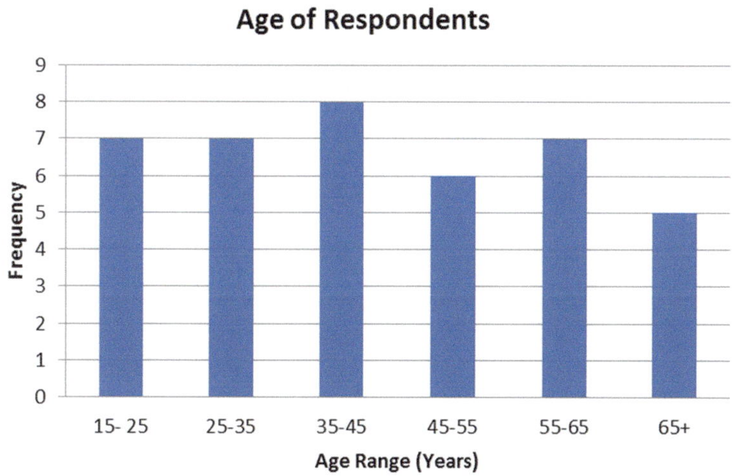

Fig. 14.3 Age of respondents (*Source* Own made)

presented along with charts and tables taking the objectives of the study into consideration. Pseudonyms are used in order to present some narratives of the research participants.

14.5 Findings of the Study

14.5.1 Background Information

14.5.1.1 Age Structure

Of 40 research participants, 26 are male and 14 are female. The age of the participants' ranges from 19 years lowest to 80 years highest. Each age group has almost similar number of participants excepting one participant who is of more than 80 years old (Fig. 14.3).

14.5.1.2 Educational Status

Since we have conducted our study in the Metropolitan city area, there is nobody illiterate among our research participants. Of 40, 33 participants are found graduate or more than that, and four participants have the education of higher secondary level, and 2 of secondary level, and 1 of primary level education (Fig. 14.4).

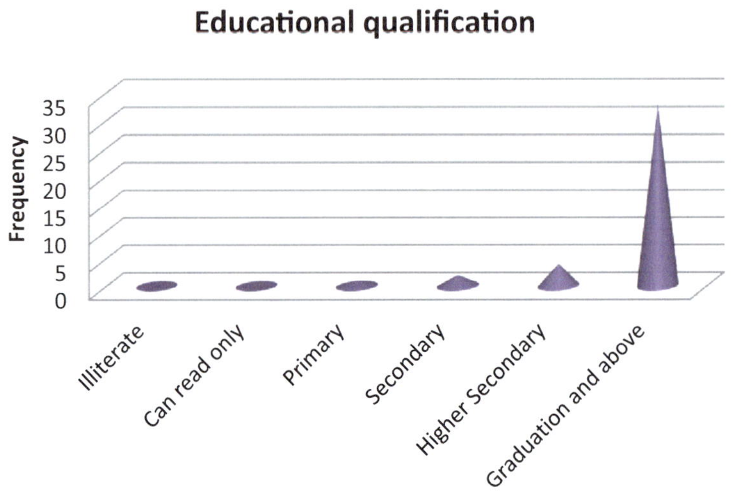

Fig. 14.4 Educational qualification (*Source* Own made)

14.5.1.3 Occupational Diversification

Among the participants, 11 are lawyer, 8 are student, 6 are teacher and 5 are NGO worker. The rest belongs to some other occupation. Occupational diversification of the research participants may be understood through Fig. 14.5.

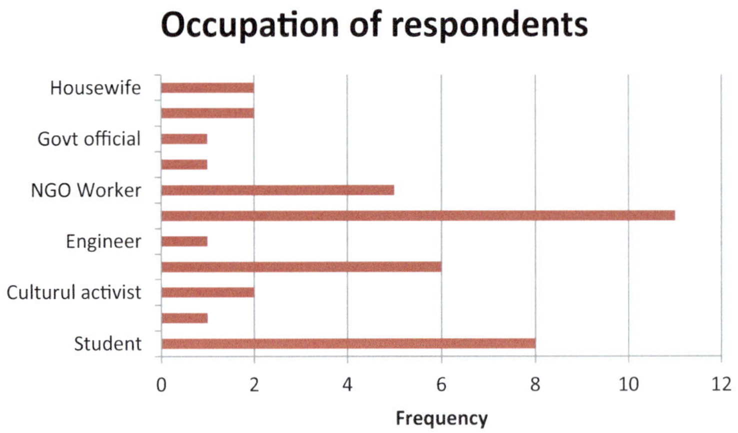

Fig. 14.5 Occupational diversification (*Source* Own made)

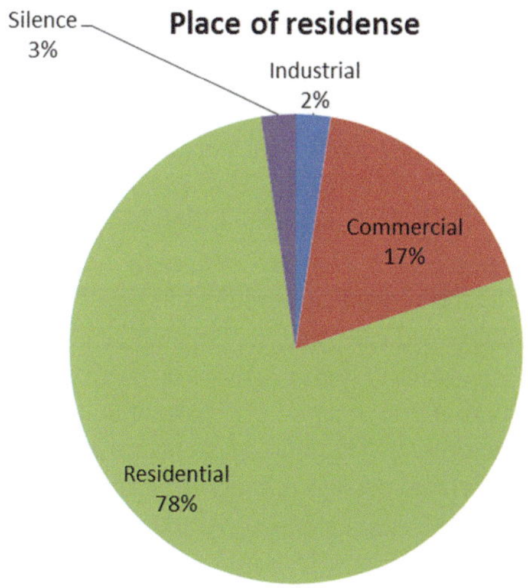

Fig. 14.6 Place of residence (*Source* Own made)

14.5.1.4 Place of Residence

Most of the research participants (31 out of 40) live in the residential zone of Sylhet Metropolitan City Area. Seven participants live in the commercial zone, and other 2 live in industrial and silence zone, respectively. The place of residence of the research participants is shown in Fig. 14.6.

14.5.1.5 Zones of Workplace

Seventeen research participants work in the commercial zone, 14 work in the silence zone, and 5 and 4 research participants work in the residential and industrial zone of Sylhet Metropolitan City Area respectively (Fig. 14.7).

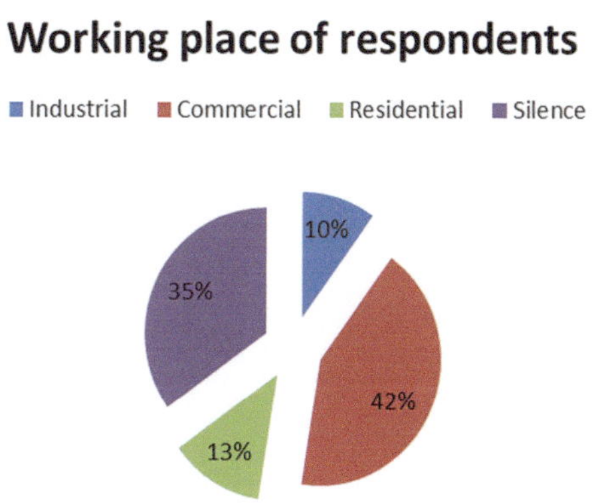

Fig. 14.7 Types of workplace (*Source* Own made)

14.5.2 Causes of Noise Pollution

There are multiple causes identified for noise pollution by the research participants. Most of the participants had numerous responses to identify causes of noise pollution across the city. Thirty-nine participants hold huge traffic jam and unbearable horn sounds of different vehicles plying on the road responsible for causing noise pollution in the city. Sounds with high intensity coming from horns of vehicles seriously jeopardize physical and mental health of human beings, especially children, elderly people, patients, and people with heart disease are affected most. Around 25 respondents identify that frequent use of mike for different purposes causes noise pollution. There are as many as 11 participants who have argued that religious gathering of Muslims and Hindus, and construction works at different places all over the city also create noise pollution. Moreover, some of the research participants blame political gathering, engineering and welding workshop, use of loudspeaker and musical instruments, power plant, etc. for causing noise pollution (Fig. 14.8).

Rashid (40), a resident of residential zone of Sylhet city, argues…

Level of noise caused by motorized vehicles and traffic jam has gone beyond acceptable limit. It is really unbearable now. Unfortunately, nobody cares for it.

Habib (36), a resident of a silence zone of the city, speaks up…

Vehicle traffic creates unbearable noise on the road. I take my son to school every day. My son generally covers both of his ears with his hand to avoid noise while going to the school alongside the road. He does it spontaneously on his own.

Anis (48), a resident of commercial zone, blames…

Use of mike in Muslim and Hindu religious gathering causes huge noise that results in annoyance and headache

Most of the research participants had different opinions regarding the causes of noise pollution.

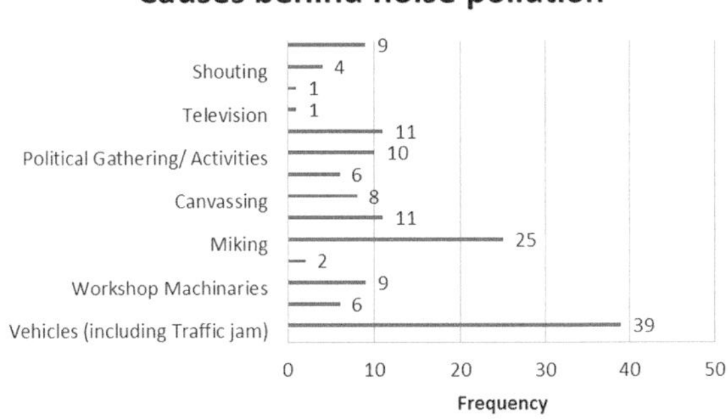

Fig. 14.8 Causes of noise pollution (*Source* Own made)

14.5.3 The People Who Suffer from Noise Pollution

The research participants argue that school going children are worst affected by noise pollution. Around 37 responses identify children who suffer most because of noise pollution. On the other hand, 21 participants think that patients are most affected, and 20 participants argue that it is the elderly and aged people who suffer most. About 10 participants think that everybody is in fact affected by noise pollution. Multiple responses may be seen in Fig. 14.9 about the people who suffer because of noise pollution:

Alam (60), a resident of silence zone, argues…

Since vehicles of different kinds are the main source of noise pollution, school going children generally suffer most as they daily experience huge noise caused by the horns of vehicles while going to school.

Roma (35), a resident of industrial zone, expresses her concern…

I think patients and aged people are worst affected because of increasing noise pollution in the city. It is high time to take some measures, especially for patients and the aged who are unable to adapt with growing noise pollution in the city.

Each of the participants is found concerned but has expressed helplessness for the growing trend of noise pollution in the city. Most of them argued that everybody more or less suffers from noise pollution.

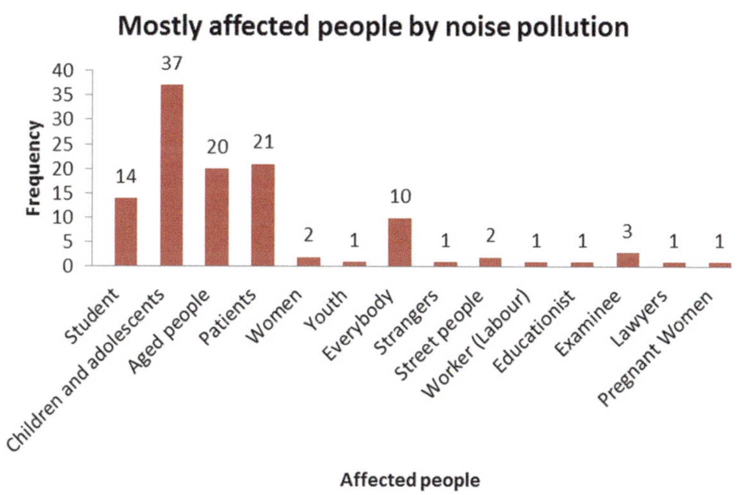

Fig. 14.9 Sufferers of noise pollution (*Source* Own made)

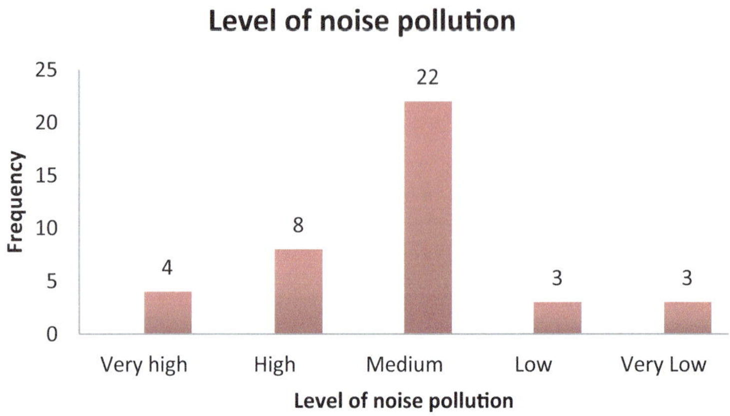

Fig. 14.10 Level of noise pollution around residence (*Source* Own made)

14.5.4 Noise Pollution Around the Residence

Most of the research participants generally think that noise pollution is very much felt around their residence. About 22 research participants admit that noise pollution is average around their residence. But 8 participants rate the noise pollution around their residence as high, and 4 participants rate it as very high. Only 3 participants describe noise pollution around their residence as very low (Fig. 14.10).

Khaleque (55), a resident of residential zone, says…

Though my home is located in the residential zone, still we (all family members) experience noise of medium level around my residence as construction works nearby go on round the year and also high sound of vehicles' horn enter there from the main road which afflicts the quality of our living.

Aziz, (52), a resident of commercial zone, shares…

My residence is located nearby the commercial zone, and therefore, noise around my residence is always high. We have become habituated to it, but strongly feel that we are getting affected because of continuous noise.

Most of the research participants experience medium to high and very high level of noise pollution around their residence.

14.5.5 Noise Pollution Causes Health Hazard

Many of the research participants think that they do not have direct health problem because of noise pollution, but they are definitely affected indirectly in the long run. Most of the participants argue that noise pollution affects almost everybody directly or indirectly. Of 40 participants, 18 did not feel anything adverse directly on their physical and mental health. But 12 participants have experienced physical adversity

Table 14.2 Health hazard caused by noise pollution

Response	Frequency
No	18
Yes (Didn't specified)	4
Physical	12
Mental	7
Impact not understood	1
No response	1
Total	43

Data Source Field Survey

due to unbearable noise. Even 7 participants do complain of mental hazards caused by noise pollution. Another 4 participants are disturbed one way or the other because of noise pollution (Table 14.2).

Sunil (60), a resident of industrial zone, expresses…

I have been suffering from irritation inside the ear over the past couple of years. The doctor has detected that I have partial hearing loss. I also understand it and feel that my ears are affected because of my regular exposure to excessive noise. The doctor has asked me not to expose myself to high level of noise.

Zahid (49), a resident of silence zone, shares…

I extremely feel irritated while I experience high level of noise outside my home, especially when I go to the market or use vehicles to go from one place to another. I sometimes get headache like anything because of incessant high level of noise.

Some of the research participants do not have direct impact on their physical and psychological well-being. But many of them had psychological and physiological botheration due to excessive noise they face in their daily life.

14.5.6 Types of Physical and Mental Adversity Faced

Multiple responses are received from the research participants when asked about types of physical and mental adversity they face because of noise pollution. Of 40 research participants, 8 think that excessive noise often causes irritation; 6 complain of rising blood pressure, another 6 argue that noise pollution causes insomnia. Other responses are like sudden awakening, mental stress, inattention, complexities inside the ear, headache, palpitation, hearing loss, memory loss, feeling of discomfort, exhaustion, depression, etc. Figure 14.11 shows different types of physical and mental adversity that are caused by noise pollution.

Majid (62), a resident of commercial zone, feels…

Excessive noise may cause many dangerous diseases. I myself have been suffering from blood pressure because of perhaps my daily exposure to unbearable noise.

Sultan (60), a resident of industrial zone, says…

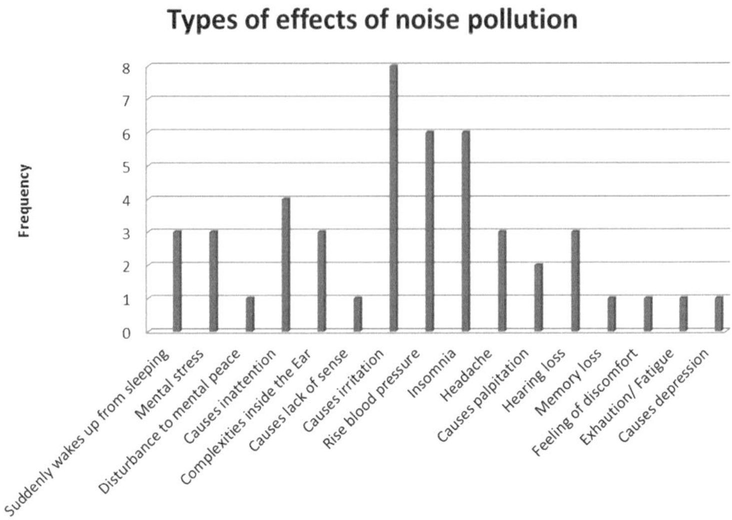

Fig. 14.11 Physical and mental adversity (*Source* Own made)

Noise causes huge irritation for me. I just can't tolerate it as I feel headache and sometimes feel like vomiting.

Research participants had multiple responses in relation to physical and mental adversity caused by noise pollution. Each of the participants admits that current level of noise pollution in the city causes bad impact on the physiological and psychological health of the city dwellers.

14.5.7 Health Hazard Forces the Participants Consult a Doctor

It is understood that people generally do not visit a doctor for a health issue that might be caused by hearing excessive noise. They are used to it. In many cases, people have no idea about noise pollution causing direct and indirect health hazard for many people coming across manifold noise frequently. Of 40 participants, 34 have never consulted a doctor because of facing some health hazard that might be caused by noise pollution. Only 6 participants admit that they had to consult a doctor as they faced health hazard due to noise pollution.

Moloy (58), a resident of silence zone, argues…

The city is no more livable as it has become extremely noisy these days. Too much of noise caused by motorized vehicles has affected most of the people, particularly the patients like me. I have heart problem and often suffers from palpitation. Whenever I go out I feel very bad, and my heart beats also go up while experiencing even medium level of noise. I consulted the doctor about it, and the doctor has suggested me not to face excessive noise.

14.6 Recommendations and Suggestions Given by the Research Participants to Reduce or Control Noise Pollution

There are so many suggestions put forward from the part of the research participants in order to reduce or control noise pollution in and around the city of Sylhet. It is clearly understood that most of the people living in the city are more or less concerned about the increasing trend of noise pollution, and want the government or city corporation authorities to take concrete steps toward addressing the problem. A few participants, however, found very pessimistic as they do not think that noise pollution could be controlled by any means. The recommendations and suggestions given by the research participants are noted below:

- Use of horn by the drivers of different vehicles must be controlled
- Existing law needs to be applied to punish those not complying with it
- Use of mike with high intensity for different purposes should be controlled
- Use of hydraulic horn must be prohibited
- Public awareness about the increasing trend of noise pollution and its consequences needs to be raised
- Train up the drivers of vehicles for appropriate use of horns
- Power plant should not be allowed around the residential area
- Industrial area should be separated and away from the residential area
- Traffic on the road should be effectively controlled
- New law may be enacted to control noise pollution strictly
- Private car plying on the road needs to be controlled and restricted
- Engineering and welding workshop should be relocated outside the city
- Making any kind of noise near residential area needs to be prohibited
- Maximum sound level at different parts of the city needs to be announced by the competent authorities
- Use of "noise or sound controlling device" in all the vehicles plying on the road should be made mandatory
- Controlling noise at the construction sites, especially near residential area/silence zone, is a must
- Public functions during night near residential area should be prohibited
- Announcing time schedule for religious and cultural functions near residential area may be effective
- Alternative to using horn may be thought of
- Expired vehicles plying on the road should be withdrawn
- Checking the fitness of vehicles regularly
- Heavy vehicles need to be redirected and bypassed the main city thoroughfare

- Construction of factories near the residential area should not be allowed
- Minimum age to drive motor bikes needs to be announced
- Religious and political leaders need to be made aware about using mike in public gathering
- Tree plantation around the city needs to be encouraged
- Noise-reducing device may be provided to those who work in industries/factories
- Use of sign language for those working in the industries may be introduced
- Sufficient public transports need to be arranged instead of private ones
- Members of law enforcing agency should be made aware and trained to appropriately control excessive noise caused by different sources
- Mike should not be allowed in the street and religious gathering
- Speaker may be used instead of mike
- Noise of engine boat plying on the river should be kept under control.

14.7 Discussion

The findings of the study prove that noise pollution in Sylhet, an important city located on the northeastern part of Bangladesh, has been increasing in an alarming rate because of mostly rapid growth of urbanization in the region. Though the sample size of the research participants drawn from the city is small, people in general seem to be well aware about the increasing trend of noise pollution across the city. It may be noted that Sylhet city has been receiving rural, urban as well as semi-urban migrants from all over the country, particularly from the greater northeastern region, as the city has witnessed a quick growth of different types of educational institutions, governmental, non-governmental and international organizations, business and commercial activities, and small and medium-size industries over the last three decades. As a result, the population of the city has substantially increased. Consequently, the mobility of the population within the city has also increased, and therefore the number of different types of vehicles plying on the city has gone ten times higher than what was two and half decades before. It is thus understandable as to why the research participants identify 'horns' used in numerous vehicles and the 'daily traffic' as major sources of noise pollution in the city, and that is what causes physical and mental health hazard for the city inhabitants. The same explanation for noise pollution and its effects has been presented by some of the scholars in their studies (Agarwal and Swami 2011; Doygun and Gurun 2008; Fiedler and Zannin 2015; Georgiadou et al. 2004; Jamrah and Al-OmariA 2006; Li et al. 2002; Mehdi et al. 2011; Murthy et al. 2007; Pathak et al. 2008; Phan et al. 2010; Ramazani et al. 2017; Skanberg and Ohrstrom 2002; Tang and Wang 2007). The participants blame indiscriminate use of mike with loud sound by advertisers of different commodities, and others like religious and political leaders delivering a speech in public gatherings with the use of mike often cause noise pollution which generally remains undisclosed. Increasing number of people, construction works at different sites, engineering and welding workshop, CNG refueling station and power plant near the residential area have caused the people living in Sylhet city to suffer from excessive noise. This scenario is more or less similar in most of the developing and least developed countries (Alam 2011; Amin et al. 2014; Hunashal and Patil 2012; Mato and Mufuruki 1999; Zannin et al. 2002). The findings of the study show no one is spared from the harmful impacts of noise pollution, though the children, aged and the patients are worst affected. Some other studies on noise pollution also find children, elderly people and patients mostly suffer from noisy environment (Alam et al. 2006; Al Mamun et al. 2017; Hoque et al. 2013). Research participants of this study argue that noise pollution has direct as well as indirect negative impacts on physical and mental health of the people. There are participants who have already experienced physical as well as mental health hazard because of excessive noisy atmosphere around their residence or near the workplace. This finding may also be substantiated by other studies on noise pollution and its impact (Allen et al. 2009; Al Mamun et al. 2017; Amin et al. 2014; Baloye and Palamuleni 2015; Doygun and Gurun 2008; Goudreau et al. 2014; Jamrah and Al-OmariA 2006; Mehdi et al. 2011; Murthy et al. 2007; Pathak et al. 2008; Phan et al. 2010; Piccolo et al. 2005; Ramazani et al. 2017; Skanberg and Ohrstrom 2002). According to the experience of the research participants, noise pollution generally causes the people to feel irritated and it raises blood pressure that leads to hypertension of the people who frequently come across it. Some of the participants complain of insomnia because of facing excessive noise in their everyday life. Moreover, participants have experienced disturbance during sleeping, mental stress, feeling of discomfort, inattention, pain inside the ear, loss of hearing, loss of memory, feeling of fatigue, depression, headache, palpitation, etc. These findings may well be corroborated by many other studies on noise pollution and its effects undertaken by different scholars (Agarwal and Swami 2011; Alam 2011; Allen et al. 2009; Brainard et al. 2004; Hunashal and Patil 2012; Mato and Mufuruki 1999; Oni et al. 2016; Piccolo et al. 2005; Ramazani et al. 2017). It is a fact that people generally do not visit a doctor because of health hazard caused due to noise pollution. They adapt with it; feel used to it, but sometimes they also visit a doctor if the problem caused by noise is felt severe. A few of the participants admit that they had to seek treatment from a qualified doctor because of health hazard they faced from the excessive noise pollution. The research participants strongly feel that the concerned authorities must take stringent steps to reduce and control the increasing trend of noise pollution in Sylhet city before it deteriorates further and goes out of hand.

14.8 Conclusion

Along with the process of urbanization and industrialization, the problem of increasing noise pollution has been a matter of serious concern all over the world, and Bangladesh is no exception to it. Sylhet city has been expanding in terms of its population growth, growing number of manifold educational institutions, development of commercial and business establishments, small and medium-size industries. etc. over the last couple of decades. The city has, therefore, become a destination of many rural, urban and semi-urban migrant youth who come and stay in this city for different purposes. The rapid growth of vehicles plying inside the city is simply unprecedented, which has been resulted from the same process of urbanization fast taking place in this northeastern region of the country. Traffic vehicles and horns used in the vehicles cause major noise pollution in the city. The measurement of noise level in industrial, commercial, residential and silence zone of the city is taken at different times, mostly during morning, evening, and night, which shows that the level of noise in many cases in different areas has crossed the normal limit. Commercial and industrial areas are affected most, and sometimes residential and silence zones are also affected. Apart from traffic and horns used in the vehicles, indiscriminate use of mike, building construction works, roadside engineering and welding workshop, religious and political gathering, power plant, etc. are the major sources of excessive noise in Sylhet city. School going children, aged population and the patients with different diseases are mostly affected from noise pollution in the city. The city inhabitants generally do not visit a doctor for being affected through noise pollution, but they consult a doctor if the suffering gets worse. The research participants often experience irritation, rise of blood pressure, hypertension, insomnia, sleeping disturbance, loss of hearing, mental stress, headache, discomfort inside the ear, feeling of fatigue, etc. because of their daily exposure to excessive noise around the city. Noise pollution has caused physical and mental health hazard for the city inhabitants. It is high time for the concerned authorities to take appropriate measures to address growing level of noise pollution in Sylhet city and thereby help the inhabitants attain a noise free and environmentally sound–a healthy city life. It is of utmost importance for our next generation who are right now at the risk of facing severe environmental disaster.

References

Abdel IR, Zeid Q, She M (2000) Noise measurements in community of Nablus in Palestine. Acoustical 86:578–580

Agarwal S, Swami BL (2011) Road traffic noise, annoyance and community health survey: a case study for an Indian city. Noise Health 13(53):272–276

Ahmed T, Rahman T (2015) Non-auditory health hazard vulnerability to noise pollution: assessing public awareness gap. Am J Eng Res 4(4):143–147

Al Mamun MM, Shil A, Paul A, Paul P (2017) Analysis of noise pollution impacting educational institutes near busy traffic nodes in Chittagong city. International conference on green architecture (ICGrA 2017). (https://www.researchgate.net/publication/319879633 Accessed 25 Oct 2017)

Alam W (2011) GIS based Assessment of Noise Pollution in Guwahati City of Assam. Int J Environ Sci 2(2):743–751

Alam MJB, Rauf AFMA, Ahmed MF (2001) Traffic induced noise pollution in Dhaka city. J Civil Eng CE 29(1):55–63

Alam JB, Alam MJ, Rahman B, Dikshit MMAK, Khan SK (2006) Study on traffic noise level of Sylhet by multiple regression analysis associated with health hazard. Iran J Environ Health Sci Eng 3(2):71–78

Alberola J, Flindell H, Bullmore J (2005) Variability in road traffic noise levels. European Commission, Environmental Noise Directive 2002/49/EC; Off J European Communities. L189, 12–25

Al-Harthy I, Tamura A (1999) Sound environment evaluation and categorization of audible sounds: the first survey of the human response to the sound environment in Muscat city (Oman). J Acoust Soc Jpn (E) 20(5):353–564

Allen RW, Davies H, Cohen MA, Mallach G, Kaufman JD, Adar SD (2009) The spatial relationship between traffic-generated air pollution and noise in 2 US cities. Environ Res 109:334–342

Amin N, Sikder I, Zafor MA, Chowdhury MAI (2014) Assessment of noise pollution of two vulnerable sites of Sylhet city, Bangladesh. Int J Water Resour Environ Eng 6(1):112–120

Baloye BO, Palamuleni LG (2015) A comparative land use-based analysis of noise pollution levels in selected urban centers of Nigeria. Int J Environ Res Public Health 12:12225–12246. https://doi.org/10.3390/ijerph121012225

Banerjee D (2012) Research on road traffic noise and human health in India: Review of literature from 1991 to current. Noise Health 14:113

BBS (Bangladesh Bureau of Statistics) (2014) Population and housing census, pp 25–48

Berglund B, Lindvall T (1995) Community noise. Center for sensory research. Stockholm University and Karolinska Institute, Stockholm

BPC (Bangladesh Population Census 2011) (2011)

Brainard JS, Jones AP, Bateman IJ, Lovett AA (2004) Exposure to environmental urban noise pollution in Birmingham, UK. Urban Stud 41(13):2581–2600

Chowdhury SC, Razzaque MM, Helali MM, Bodén (2010) Assessment of noise pollution in Dhaka city. The 17th international congress on sound and vibration, Cairo. 18–22 July 2010

Daily Sun (2017) Noise pollution turns for the worse in city. Sun Online Desk. http://www.daily-sun.com/post/234293/Noise-pollution-turns-for-the-worse-in-city. Accessed 26 Oct 2017

Doygun HD, Gurun DK (2008) Analysing and mapping spatial and temporal dynamics of urban traffic noise pollution: a case study in Kahramanmaraş, Turkey. Environ Monit Assess 142:65–72. https://doi.org/10.1007/s10661-007-9908-7

Farzana SZ, Nury AH, Biswas B, Das A (2014) A study on noise pollution of stone crusher machine at Jaflong, Sylhet. Proceedings of 5th International Conference on Environmental Aspects of Bangladesh (ICEAB 2014). http://benjapan.org/ICEAB/proceedingsICEAB14/i7%20p20.pdf. Accessed 25 Oct 2017

Farzana SZ, Biswas B, Das A, Zafor A (2016) A study on cause and effects of noise pollution due to stone crusher machine at Jaflong, Sylhet. Int Res J Eng Technol 3(8):32–35

Fiedler PEK, Zannin PHT (2015) Evaluation of noise pollution in urban traffic hubs—noise maps and measurements. Environ Impact Assess Rev 51:1–9

Georgiadou E, Kourtidis K, Ziomas I (2004) Exploratory traffic noise measurements at five main streets of Thessaloniki. Greece Glob Nest: Int J 6(1):53–61

Goudreau S, Plante C, Fournier M, Brand A, Roche Y, Smargiassi A (2014) Estimation of spatial variations in urban noise levels with a land use regression model. Environ Pollut 3(40):48–58

Haq MA, Islam MM, Ali MS, Haque MF, Akhand MR (2012) Status of noise pollution in mixed areas of Dhaka city: a GIS approach. J Environ Sci Nat Resour 5(1):09–17

Hoque MMM, Basak LK, Rokanuzzaman M, Roy S (2013) Level of noise pollution at different locations in Tangail Municipal area, Bangladesh. Bangladesh J Sci Res 26(1, 2):29–36

Hunashal RB, Patil YB (2012) Assessment of noise pollution indices in the city of Kolhapur, India. Procedia—Soc Behav Sci 37(2012):448–457

Jamrah A, Al-Omari A, Sharabi R (2006) Evaluation of traffic noise pollution in Amman, Jordan. Environ Monit Assess 120:499–525. https://doi.org/10.1007/s10661-005-9077-5

King RP, Davis JR (2003) Community noise: health effects and management. Int J Hyg Environ Health 206:123–131

Li B, Tao S, Dawson RW (2002) Evaluation and analysis of traffic noise from the main urban roads in Beijing. Appl Acoust 63(2002):1137–1142

Mato RRAM, Mufuruki TS (1999) Noise pollution associated with the operation of the Dares Salaam international airport. Transp Res Part D 4(1999):81–89

Mehdi MR, Kim M, Seong JC, Arsalan MH (2011) Spatio-temporal patterns of road traffic noise pollution in Karachi, Pakistan. Environ Int 37:97–104

Morrell S, Taylor R, Lyle D (1997) A review of health effects of aircraft noise. Aust N Z J Public Health 21:221–236

Murphy E, King EA, Rice HJ (2009) Estimating human exposure to transport noise in central Dublin, Ireland. Environ Int 35:298–302

Murthy VK, Majumder AK, Khanal SN, Subedi DP (2007) Assessment of traffic noise pollution in Banepa, a semi-urban town of Nepal. Kathmandu Univ J Sci, Eng Technol 6:1–9

Oni OO, Adetimirin OI, Ajibade SA, Kolade KR, Ilesanmi OI, Akinluyi FO (2016) Noise vulnerability mapping using geospatial technique in part of Ibadan north east local government area. World Sci News 52:143–161

Pandya GH (2003) Assessment of traffic noise and its impact on the community. Int J Environ Stud 60(6):595–602

Pathak V, Tripathi BD, Mishra VK (2007) Dynamics of traffic noise in a tropical city Varanasi and its abatement through vegetation. Environ Monit Assess. https://doi.org/10.1007/s10661-007-0060-1

Pathak V, Tripathi BD, Mishra VK (2008) Evaluation of traffic noise pollution and attitudes of exposed individuals in working place. Atmos Environ 42:3892–3898

Phan HYT, Yano T, Sato T, Nishimura T (2010) Characteristics of road traffic noise in Hanoi and Ho Chi Minh City, Vietnam. Appl Acoust 71(2010):479–485

Piccolo A, Plutino D, Cannistraro G (2005) Evaluation and analysis of the environmental noise of Messina, Italy. Appl Acoust 66(2005):447–465

Rahman MT, Dikshit AK, Khan SK (2005) Traffic induced noise pollution in Sylhet city, Bangladesh. https://www.researchgate.net/publication/298564177_Traffic_induced_noise_pollution_in_Sylhet_City_Bangladesh. Accessed 27 Oct 2017)

Ramazani ME, Mosaferi M, Rasoulzadeh Y, Pourakbar M, Jafarabadi MA, Amini H (2017) Temporal and spatial evaluation of environmental noise in urban area: a case study in Iran. Int J Environ Sci Technol. https://doi.org/10.1007/s13762-017-1462-4

Shastri SC, Bakre PP, Khan TI (1996) Industry, environment and the law. Jaipur, I-Publication, RBSA Publishers

Singh N, Davar S (2004) Noise pollution-sources, effects and control. J Hum Ecol 16:181–187

Skanberg A, Ohrstrom E (2002) Adverse health effects in relation to urban residential soundscapes. J Sound Vib 250(1):151–155. https://doi.org/10.1006/jsvi.2001.3894

Tang UW, Wang ZS (2007) Influences of urban forms on traffic-induced noise and air pollution: results from a modelling system. Environ Model Softw 22:1750–1764

The Daily Star (2013) Increasing noise puts Chittagong people's health in danger, Dhaka. http://www.thedailystar.net/news/increasing-noise-puts-ctg-peoples-health-in-danger.html. Accessed 15 Sept 2017)

Vidyasagar T, Rao N (2006) Noise pollution levels in Visakhapatnam city (India). J Environ Sci Eng 48(2):139–142

WHO (World Health Organization) (2011) Burden of disease from environmental noise: quantification of healthy life years lost in Europe. WHO Regional Office for Europe and JRC European Commission. Denmark: Author

Zannin PHT, Diniz FB, Barbosa WA (2002) Environmental noise pollution in the city of Curitiba, Brazil. Appl Acoust 63:351–358

Chapter 15
Neighbourhood Deprivation and Health; A Study of Low-Income Neighbourhoods in Azamgarh City

Saleha Jamal and Uzma Ajmal

Abstract The term neighbourhood is generally used to describe the subdivisions of any urban or rural settlement. Neighbourhood is a simply a locality in which people. Neighbourhoods can adversely affect health of the residents by its poor water quality, air quality, inadequate drainage system, water logging conditions, accumulation of solid and liquid waste, overcrowding conditions, adverse traffic conditions, medical and health services, etc. Present study is an attempt to examine physical environmental conditions and associated health problems in low-income neighbourhoods in Azamgarh City. In this study, deprived or low-income neighbourhoods were identified on the basis of income, population density and household density. This type of neighbourhoods is mainly occupied by low and middle-income groups of population. The study is based on primary source of data collected through questionnaire, interviews and observation and finds that deprived or low-income neighbourhoods are generally found in environmentally degraded and unhealthy conditions. Most of the environmental problems are confined in these neighbourhoods including inadequate drainage, poor waste collection, erratic water supply, water logging and waste accumulation. Many people especially children, elderly and women, spend much of their time close to home. Thus their health is directly affected by these problems and they are more prone to diseases like malaria, diarrhoea, jaundice, typhoid, worm infestation, cholera, chickenpox, etc.

Keywords Low-income neighbourhoods · Deprived neighbourhoods · Neighbourhood environment · Health · Diseases

---

S. Jamal (✉) · U. Ajmal
Department of Geography, Aligarh Muslim University, Aligarh 202002, Uttar Pradesh, India
e-mail: salehajm@gmail.com

U. Ajmal
e-mail: uzmakhan667@gmail.com

R. B. Singh et al. (eds.), *Urban Health Risk and Resilience in Asian Cities*, Advances in Geographical and Environmental Sciences,
https://doi.org/10.1007/978-981-15-1205-6_15

15.1 Introduction

Just like the conditions in our homes have important effects on our health, conditions in the neighbourhoods around our homes also can affect our health (Cubbin et al. 2008). Neighbourhoods can influence our health in many ways. First and perhaps most important is through the physical characteristics of neighbourhood (Macintyre and Ellaway 2003). A number of previous studies have established that neighbourhood's poverty and economic deprivation are associated with poor health condition of the residents (Wheaton et al. 2015). Neighbourhood environmental conditions like poor drainage system, water logging conditions, accumulation of solid and liquid waste, overcrowding conditions, adverse traffic conditions, lack of access to safe places to exercise, medical and health services, etc. can have major implication on the health of the residents (Jamal and Ajmal 2018a). Many cities are currently burdened by a number of infectious diseases like typhoid, malaria, and diarrhoea, etc., intensified by poor living conditions. Unfair differences have been recorded among the health of the city dwellers with most disadvantaged areas having the highest rate of poor health outcomes (World Health Organization 2010). Environmental threats have a greater tendency to influence the poor and most vulnerable groups in the developing countries. Various studies have pointed out that most of the environmental problems of the city are confined in the most deprived parts of it. Poor people that are both economically and politically deprived and living in the most deprived areas, bear the brunt of underdevelopment and suffer most from the environmental problems (Mushir and Khan 2007). Major environmental problems related to water supply, sanitation, air pollution, solid waste and housing, etc. are the serious problems that are found in almost all categories of population but are particularly abundant in poor neighbourhoods (Songsore and McGranahan 1993). It was observed that in high-density populated areas, where largely poor people reside, conservancy services for clearing wastes by municipality are inadequate in comparison with high-income areas. This situation is responsible for unclean surroundings in low-income areas with dumped wastes that are breeding grounds of various insects and pests resulting in spread of infectious (Das and Angshuman 2007). Most of the environmental problems are confined in deprived areas of the city including inadequate drainage, poor waste collection, erratic water supply, water logging and dilapidated housing conditions. Since many people especially women, children and elderly, spend much of their time close to home, thus their health is directly affected by these problems. Considering all these aspects the present study attempts to examine major environmental issues in deprived/low-income neighbourhoods and their impact on the health of the residents, in Azamgarh City.

15.2 Study Area

Azamgarh city is a small-sized city located in eastern part of Uttar Pradesh in the fertile Gangetic plain (Fig. 15.1). The city lies between 26° 4′ North latitude and 83° 1′ East longitude. It comprises 25 wards. The total population of Azamgarh city is 110,892, out of which 52.3% are males and 47.7% are females. Azamgarh City has 12.91% SC Population and 918 Sex Ratio (District Census Handbook and Azamgarh 2011).

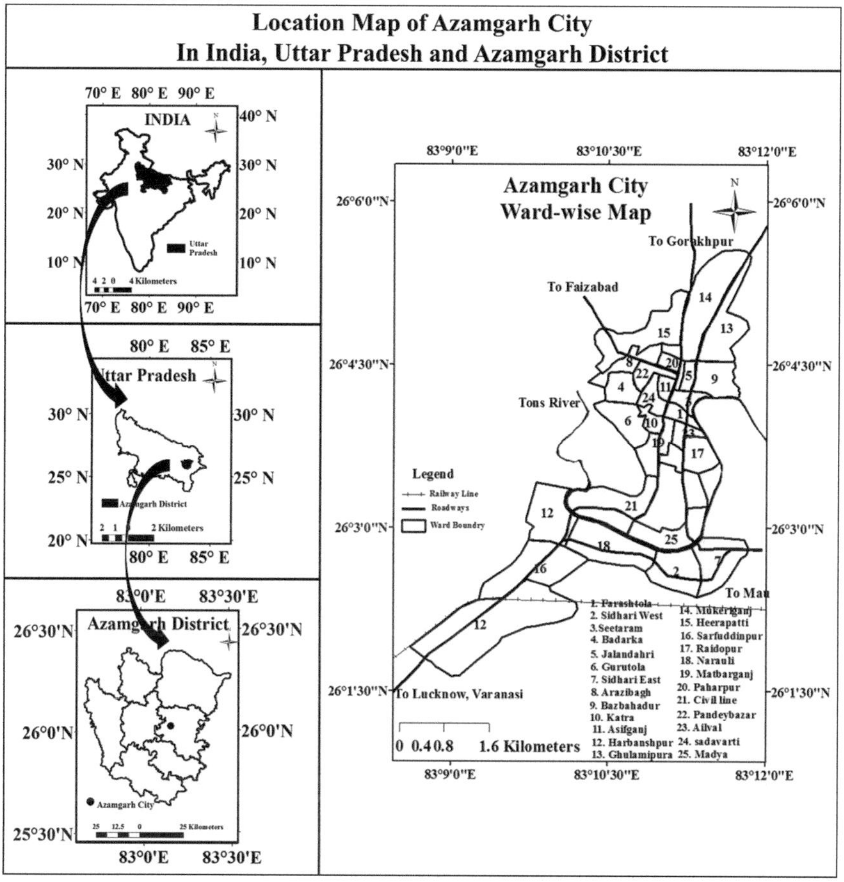

Fig. 15.1 Location map of Azamgarh city. *Source* Municipal Office, Azamgarh (2017–2018)

15.3 Data Base and Methodology

The study is mainly based on primary sources of data which were collected with the help of questionnaire interviews. Field work was conducted in the year 2016–2017. The following methodology has been employed.

- A questionnaire for the present study was prepared with the help of questionnaire used in similar studies (Das and Angshuman 2007; District Census Handbook and Azamgarh 2011; Rahman 1998; Karn et al. 2003; Pokhrel and Viraraghavan 2004; Singh and Baba 2014) and by observing the population and socio-economic condition of the neighbourhoods in Azamgarh City.
- For selecting the sample, multistage stratified random sampling technique was adopted.
- In the first stage, all 25 wards of the city were grouped into 8 different neighbourhoods on the basis of (i) income, (ii) population density and (iii) household density (Table 15.1 and Fig. 15.2). A neighbourhood was considered as low-income when more than 30% families of the neighbourhood belong to the low-income category (Lemstra et al. 2006).
- In the second stage, 3 low-income/deprived neighbourhoods (having dominant income group of below Rs 5000 per month) were selected, namely, Low-income with High Density (LI/HD), Low-income with Medium Density (LI/MD) and Low-income with Low Density (LI/LD) (Table 15.2 and Fig. 15.3).
- In the third stage, from each low-income neighbourhood 10% households have been sampled for the analysis. Total sample size consists of 588 households (Table 15.3).
- Residents were interviewed regarding environmental conditions in their neighbourhoods and focussed group discussions were also conducted for the same.
- Data regarding the health issues associated with various environmental problems were collected through (i) personal interview with the help of questionnaire and (ii) from the records of Azamgarh District hospital and various private doctors' clinics in Azamgarh City.
- Karl Pearson's correlation coefficient method was applied to examine the relationship between neighbourhood environmental risk factors and associated diseases.

$$r = \frac{n\Sigma x_i y_i - \Sigma x_i y_i}{\sqrt{n\Sigma x_i^2 - (\Sigma x_i)^2} \times \sqrt{n\Sigma y_i^2 - (\Sigma y_i)^2}}$$

where

r coefficient of correlation
X, Y the two given variables.

Table 15.1 Ward-wise categorization of Azamgarh City on the basis of

Income-wise dominance			
Income group category	Income group[a] (Rs/month)	Ward number	No of wards
High	Above 25,000	2, 21, 25, 17, 14, 15, 11, 8, 9, 17	10
Medium	10,000–25000	10, 4, 12, 16	4
Low	Below 10,000	3, 9, 5, 20, 22, 13, 24, 1, 6, 13, 18	11
Population density			
Population density category	*Range (persons per sq.km)*	*Ward Number*	*No of wards*
High	Above 20,000	12, 14, 18, 21, 25, 16, 15, 2, 17	9
Medium	10,000–20,000	13, 8, 7, 24, 1, 22, 6, 19, 23	9
Low	Below 10,000	20, 10, 4, 5, 3, 11, 9	7
Household density			
Household density category	*Range (households per sq.km)*	*Ward Number*	*No of wards*
High	Above 2700	12, 14, 18, 21, 16, 15, 2, 17	8
Medium	1400–2700	13, 8, 7, 24, 1, 22, 6, 19, 23, 25, 20	11
Low	Below 1400	10, 4, 5, 3, 11, 9	6

[a]Dominant income group of the neighbourhood
Source (i) District census handbook, Azamgarh, 2011. (ii) Municipal office, Azamgarh (2017–2018). (iii) Based on city survey (2017–2018)

Student t test has been chosen to identify the significant relationship between the variables at 1% and 5% level of significance.

$$t = r\sqrt{\frac{n-2}{1-r^2}}$$

where

t calculated value of "t" in the test of significance
n no of observations
r computed value of coefficient of correlation.

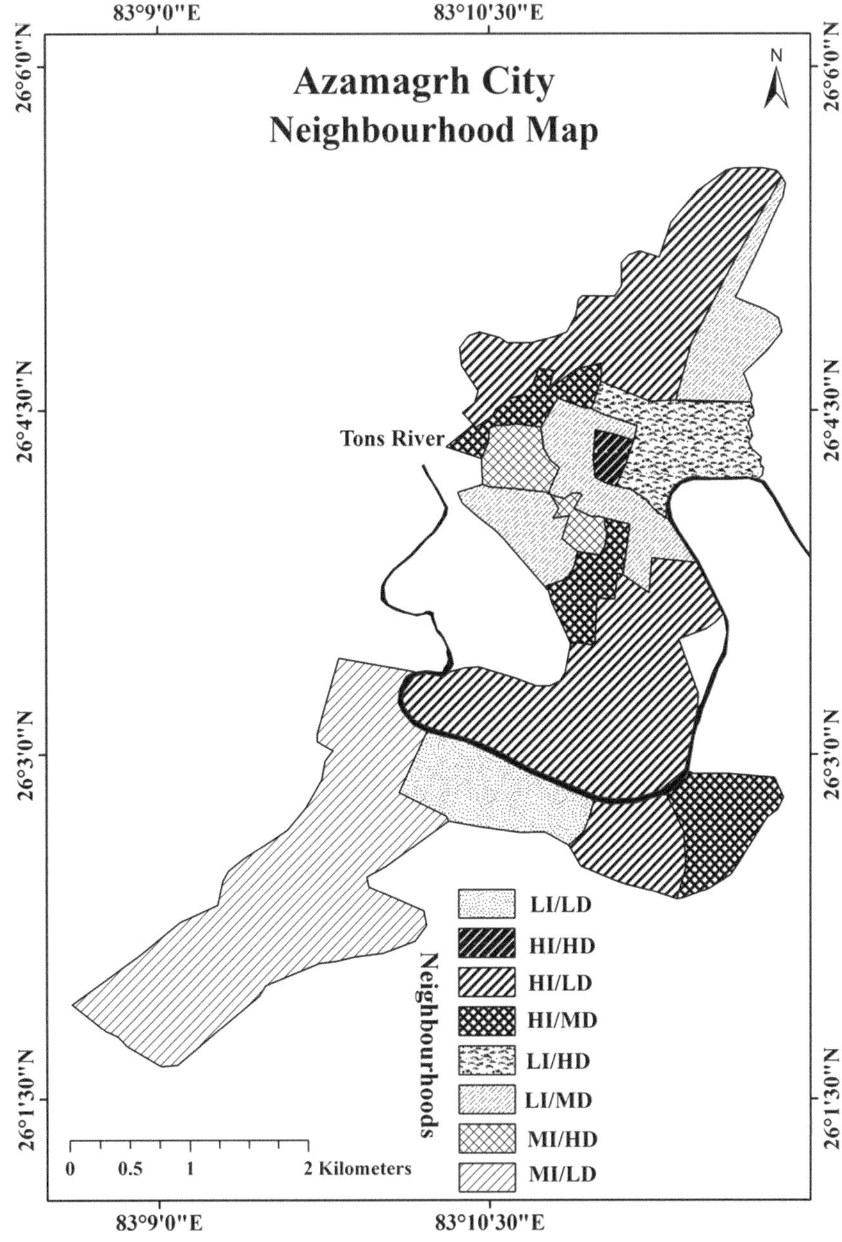

Fig. 15.2 Neighbourhood map. *Source* (i) District census handbook, Azamgarh, 2011. (ii) Municipal office, Azamgarh (2017–2018). (iii) Based on city survey (2017–2018)

Table 15.2 Low-income neighbourhoods of Azamgarh city

S. No.	Sampled neighbourhoods	Acronyms	Ward numbers	Sampled households (10%)
1	Low-income with high density	LI/HD	20, 5, 9, 3	240
2	Low-income with medium density	LI/MD	1, 24, 6, 13, 23, 22	289
3	Low-income with Low Density	LI/LD	18	59
Total	3	3	11	588

Source (i) District Census Handbook, Azamgarh, 2011. (ii) Municipal Office, Azamgarh (2017–2018). (iii) Based on city survey (2017–2018)

15.4 Results and Discussion

15.4.1 General Profile of Sampled Households

A perusal of Table 15.3 shows the general profile of the sampled households, which exhibits that about 75% people belong to medium to low-income groups (27.41% in Upper middle, 28.05% in Lower middle and 27.3% in Lower income group). About 70% belong to age group of 25–44 (29.26% in 25–34 and 44.67 in 35–44), 61.94% are females, 55.84 are Hindu, 90% of the respondents are married and 40% people belong to General Caste and rest come under Other Backward Caste and Schedule Caste (OBC 37.03% and SC 22.97%). Occupation status revealed that most of the respondents work as labourers (27.9), vendors (18.43), driver (13.36), businessman (11.23), etc. Education status revealed that 10.85% respondents are uneducated while remaining are educated up to Primary (38.91), High School (21.13) and Intermediate (12.9).

15.4.2 Neighbourhood Environmental Conditions

Survey of the households revealed that neighbourhood environmental conditions are affected by a variety of factors, for example, overcrowding, poor water and air quality, inadequate sanitation, etc. can increase concentration of air and waterborne pathogens and facilitate the concentration of communicable and infectious diseases. On the basis of household surveys environmental conditions which contributed most to poor health were identified (i.e. overcrowding, inadequate drainage, poor waste collection, erratic water supply, water logging and waste accumulation) which play catalytic role and leave adverse effects on neighbourhood residents.

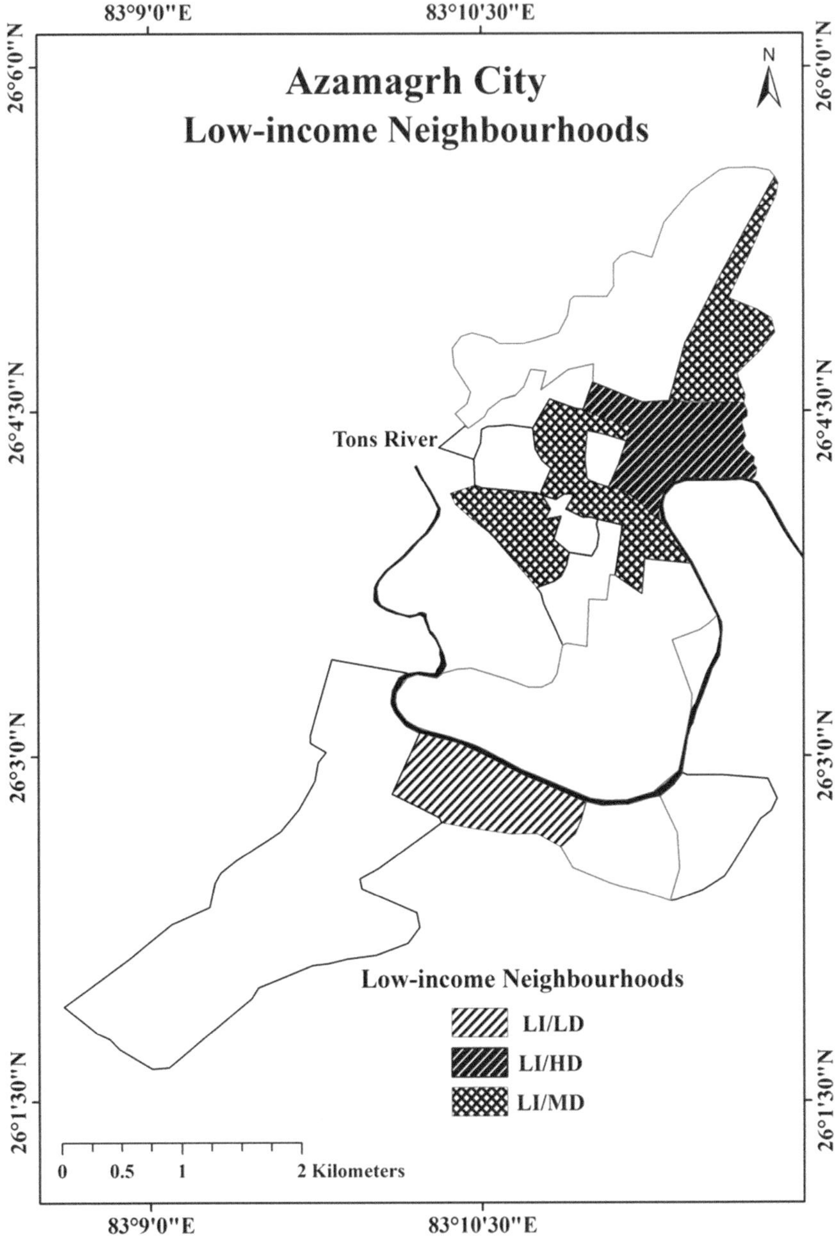

Fig. 15.3 Low-income neighbourhoods of Azamgarh city. *Source* (i) District census handbook, Azamgarh, 2011. (ii) Municipal office, Azamgarh (2017–2018). (iii) Based on city survey (2017–2018)

Table 15.3 General profile of the sampled household

Sex-wise distribution		Education-wise distribution	
Gender	Percentage	Education	Percentage
Female	61.94	Uneducated	10.85
Males	38.06	Primary/Middle	38.91
Religion-wise distribution		High school	21.13
Religion	Percentage	Intermediate	12.9
Hindus	55.84	Graduate and above	16.21
Muslims	43.62	**Income-wise distribution**	
Christians	0.55	Income level	Percentage
Caste-wise distribution		High income	17.21
Caste	Percentage	Upper middle	27.41
General	40	Lower middle	28.05
OBC	37.03	Low-income	27.3
SC	22.97	**Occupation-wise distribution**	
Age-wise distribution		Labour	27.9
Age groups	Percentage	Vendor	18.43
15–24	7.3	Driver	13.36
25–34	34.35	Service	6.44
35–44	20.35	Business	11.23
45 and above	18	Other	22.64

Source Based on field survey 2017–2018

15.4.3 Waste/Garbage Dumps

Field survey revealed that waste dump is a major neighbourhood environmental issue faced by the residents. Household survey has revealed that about 32.95% households throw the household waste directly from the house (29.57% in LI/HD, 31.97% in LI/MD, 37.33% in LI/LD neighbourhood) while 45.52% people keep it in houses in closed container and 21.52% in open container before disposal (Table 15.4). About 49.56% dispose their garbage at roadside collection points (51.46% in LI/HD, 44.03% in LI/MD and 53.20% in LI/LD neighbourhood) while remaining throw the garbage in open field (20.02%), water body (14.30%) and in backyards (16.11%). It has been observed that waste collection practice from roadside collection point is not being performed on a daily basis resulting into heaps of garbage spilled over in surrounding and adjoining areas. Nearly 32% (34.06% of LI/HD, 25% of LI/MD and 37% of LI/LD) reported a daily collection of solid waste, 32.6% (25% of LI/HD, 34.83% of LI/MD and 38% of LI/LD) biweekly and rest have reported weekly (30.51) and monthly (7.29) collection of waste. About 65% households are unsatisfied with waste collection services in their neighbourhood and 84.66% in LI/MD, 79.71% in LI/HD and 63.7% in LI/LD neighbourhoods have reported accumulation of garbage

Table 15.4 Azamgarh city: neighbourhood-wise percentage distribution of waste collection conditions

S. No.	Neighbourhoods		LI/HD	LI/MD	LI/LD	Total
1	Waste storage in house	Throwing out	29.57	31.97	37.33	32.95
		Closed container	49.89	47.00	39.67	45.52
		Open container	20.54	21.03	23.00	21.52
2	Disposal of household waste	Roadside collection point	51.46	44.03	53.2	49.56
		Open field	16.96	23.63	19.47	20.02
		Back yard	13.07	21.27	14.00	16.11
		Water body	18.51	11.07	13.33	14.30
3	Frequency of waste collection	Daily/Adequate	34.06	25.00	37.00	32.02
		Biweekly	25.00	34.83	38.00	32.61
		Weekly	34.69	31.84	25.00	30.51
		Monthly	6.25	8.33	–	7.29
		Inadequate/not daily	65.94	75.00	63.00	67.98
4	Satisfaction with waste collection services	Satisfactory	28.25	25.00	51.38	34.88
		Not satisfactory	71.75	75.00	48.62	65.12
5	Condition of Waste in the neighbourhoods	Spilled on the road	57.33	51.47	39.67	49.49
		In open space	23.10	37.33	35.50	31.97
		Collects in the drain	19.57	11.20	24.89	18.54
6	Waste accumulation	Yes	79.71	84.66	63.7	76.02
		No satisfaction	20.29	15.34	36.3	23.98

Source Based on field survey 2017–2018

in their neighbourhoods. As far as condition of garbage in neighbourhood is concerned about 50% have reported garbage spilled on roads and roadside (57.33% in LI/HD, 51.47% in LI/MD and 39.67% in LI/LD neighbourhood), 31% in open space (23.1% in LI/HD, 37.33 in LI/MD and 35.5 in LI/LD) while about 19% reported collected in drain (Table 15.4). Since the waste remains uncollected for weeks, the risk of health from existence of solid waste in the neighbourhood is potentially high. Waste accumulation is a major environmental threat and can result in soil, air and water pollution and other problems like bad odour, germs and bacteria as well as various pathogens resulting in various waste-related diseases like typhoid, cholera, jaundice, etc.

Population increase and urbanisation have had their shared roles of various disadvantages and one of the main concerns is waste and garbage dumps. With the increase in population and rising demand of food and other essentials, there is a rise in waste being generated by households. According to the Central Pollution Control Board (CPCB), per capita waste generation is 0.2–0.6 kg per day (Ministry of Urban Development, Govt. of India 2016). These wastes are thrown into waste collection points and then collected by municipality and further disposed to landfill sites. However either due to inefficient municipal services, infrastructure, resource crunch and mismanagement, not all of the waste get collected properly on routine basis resulting into spilled over heaps of garbage which can cause serious health hazards and problems in the surrounding areas. Although the amount of waste generated by a rich family is far more than middle class and poor family but during the field survey dumping of generated waste was also observed from high-income areas to low-income areas.

15.4.4 Water Supply and Sanitation

An adequate supply of easily accessible, potable water is central to the survival and growth of any city. However in case of low-income neighbourhoods of the city, Table 15.5 revealed that only 45% households avail regular supply of water (41.56% in LI/HD, 35.09% in LI/MD and 61% in LI/LD neighbourhood). A high value of 61% in LI/LD neighbourhood is due to absence of municipal water supply which compels all the residents to get their own supply of water, i.e. own hand pump/boring while some poor people still fetch water either from roadside hand pumps or from their neighbours. As far as duration of water supply is concerned, it is found that 26.37% get water supply only for 1–6 h, 22.03% for 7–12 h and 5.71% for 13–18 h while remaining (45.89) avail water supply for 24 h. Regarding water quality condition (assessed through resident's perception) it is found that about 33% respondents have reported problem regarding water quality like bad odour, dirt in water as well as bad taste of water, (37.37 in LI/HD, 44.39 in LI/MD and 19.85 in LI/LD neighbourhood). By field survey, it has been found that 56% households consume water without any kind of treatment while 44% households treat water before drinking in which treating by Reverse Osmosis (RO) is more common i.e. 37.79%, 5.78% households boil the water and 1.04% households add chlorine before consumption.

City survey has revealed that the inadequate sanitation and drainage are very common problem in low-income neighbourhoods. A perusal of Table 15.5 exhibits that only about 67% of households have proper toilet facility in their houses while 24.59% go for open defecation, (24.17% in LI/HD, 27.84% in LI/MD and 21.78% in LI/LD), which can result in foul smell in neighbourhood, soil and water pollution as well as contamination of food. Open defecation often called as mother of all infections and morbidity, is the major cause of diarrhoeal and other diseases in developing countries.

Regarding the state of drains, only 38.37% respondents have reported closed drain in their neighbourhood (37% in LI/HD, 31% in LI/MD and 47.11 in LI/LD neighbourhood) while remaining reported of open drains. Most of low-income households do not have proper drainage connection and their waste water goes to the main drain via small open/kachcha drains while in some areas households deliberately do not let the main drains close, due to fear of chocking and difficulty in proper cleaning, resulting in open drain sites which create nuisance as well as breeding site to a myriads of

Table 15.5 Azamgarh City: Neighbourhood-wise percentage distribution of water supply and sanitation condition of sampled households

S. No.	Neighbourhood		LI/HD	LI/MD	LI/LD	Total
1	State of water supply	Regular	41.56	35.09	61.00	45.88
		Irregular	58.44	64.91	39.00	54.12
	Duration	1–6 h.	16.14	23.96	39.00	26.37
		7–12 h.	31.74	34.37	–	22.03
		13–18 h.	10.56	6.58	–	5.71
		24 h.	41.56	35.09	61.00	45.89
	Problem with water quality	Water quality problem[a]	37.37	44.39	19.85	33.87
		No problem	62.63	55.61	80.15	66.13
	Treating of Water	Yes	43.41	44.04	46.39	44.61
		No	56.59	55.96	53.61	55.39
	If Yes	RO	35.51	35.65	42.22	37.79
		Boil	6.72	6.45	4.17	5.78
		Adding Chlorine	1.18	1.94	0.00	1.04
2	Toilet facility	Yes	68.80	63.04	70.22	67.35
		No	31.20	36.96	29.78	32.65
	If No	Open defecation	24.17	27.84	21.78	24.59
		Community latrine	5.54	7.01	8.00	6.85
		Other	1.49	2.11	0.00	1.20
3	State of drains	Open	63.00	69.00	52.89	61.63
		Closed	37.00	31.00	47.11	38.37
	Cleaning of drains	Yes	17.3	15.66	41.33	24.76
		No	82.7	84.34	58.67	75.24
4	Water logging	Yes	59.25	67.33	57.7	61.43
		No	40.75	32.67	42.3	38.57
	Type of water logging	Rainwater	67.50	63.00	50.00	60.16
		Sullage	32.50	32.00	50.00	39.84
	Place of water logging	Around the houses	17.33	26.05	23.18	22.18
		On the roads	43.56	36.98	66.44	48.99
		Open spaces	39.11	36.97	10.38	28.82

[a]Based on resident's perception
Source Based on field survey 2017–2018

health problems. Periodical cleaning of drains is also found rare as only 24.76% have reported proper cleaning (17.3% in LI/HD, 15.66 in LI/MD and 41.33% in LI/LD).

While examining water logging condition in neighbourhoods, it was found that about 61.43% have reported water logging in the neighbourhood (59.25% in LI/HD, 67.33% in LI/MD and 57.7% in LI/LD neighbourhood) for which rain water (60.16%) and sullage (39.84%) are the main source. Regarding place of water logging 22.18% reported of around houses, 48.99% on roads and 28.82% in open spaces. Water-logging is said to be open invitation for waterborne diseases (especially due to mosquitos). These water logging sites are ideal breeding ground for various vectors, resulting in diseases particularly malaria and dengue. Improper drainage is also responsible for pollution of water supplies resulting in diseases like viral, diarrhoea and typhoid, etc.

15.4.5 Over Crowding, Urban Green Spaces, Air and Noise Pollution

Neighbourhood overcrowding is equally important as housing overcrowding, and it is measured as one of the important indicators to influence its resident's health status. Neighbourhood-level crowding is defined as the percentage of households within a census tract with more than one person per room (Gove et al. 1979). Percentage of housing per unit area decides ration of open space to the all available land of a neighbourhood and proper ventilation to the streets and houses. Overcrowding is responsible for rapid and continued spread of airborne diseases. Neighbourhood-level overcrowding in Azamgarh city has been measured by resident's perception. Household survey reveals that about 67% people have reported overcrowding condition in their neighbourhood, i.e. 87.6% in LI/HD, 64% in LI/MD and 59% in LI/LD neighbourhood.

Urban Green spaces are an important part of any urban area and their significance is very well recognised for conserving the environmental quality and sustainability. Urban green spaces include parks, gardens and recreational venues, informal green spaces like river fronts and indigenous vegetation types, etc. Green space in urban areas offers a sense of better place for living, as it arranges for a place of recreation for all groups of society (Gupta et al. 2012). City survey revealed that only 33% households have reported Urban Green spaces (parks, gardens, playgrounds, river-front and informal open spaces, etc.) in their neighbourhoods (Table 15.6). Open spaces are significantly deficient in LI/HD and LI/MD neighbourhoods, i.e. 8.25% and 29.95%, respectively.

Azamgarh city is a small-sized city which not much urbanised and industrialised and hence less polluted in comparison to other megacities of India. However, as far as resident's perception is concerned about 50% people of LI/LD neighbourhood have reported problem of air and noise pollution (Table 15.6). LI/LD neighbourhood

Table 15.6 Azamgarh city: neighbourhood-wise percentage distribution of overcrowding, air and noise pollution (Based on resident's perception)

Neighbourhoods		LI/HD	LI/MD	LI/LD	Total
Overcrowding	Yes	77.60	64.33	59.41	67.11
	No	22.40	35.67	40.59	32.89
Urban green spaces	Yes	8.25	29.95	61.33	33.17
	No	91.75	70.05	38.77	66.83
Air pollution	Yes	12.50	10.00	50.00	24.17
	No	87.50	90.00	50.00	75.83
Noise pollution	Yes	12.50	10.00	50.00	24.17
	No	87.50	90.00	50.00	75.83

Source Based on field survey 2017–2018

is basically located in southern extension of the city in which air and noise pollution are higher due to constant traffic movement.

15.4.6 Incidence of Diseases in Low-Income/Deprived Neighbourhoods

Environmental factors are major cause of a bulk of environment related diseases (Jamal and Ajmal 2018b). Many of the neighbourhood environmental factors discussed above may pose health risks to the residents. Among the frequently occurring diseases as reported by residents, within the previous two years, associated with poor neighbourhood environmental conditions were Typhoid (54.8%), Malaria (53.33%), Diarrhoea (51.88%), Cold and Fever (51.74%), Worm Infestation (51.51%), Dengue (42.07%), Cholera (38.64%), Jaundice (32.88%) Chicken pox (26.21%). The high frequency of these diseases was also reported by Health Department of Azamgarh City as most frequently occurring diseases (Table 15.7).

15.4.7 Relationship Between Neighbourhood Environmental Problems and Associated Diseases

An attempt has been made to establish an association of causal relationship between the neighbourhood environmental risk factors and different types of diseases that are probably caused by these risk factors. Karl Pearson's Correlation coefficient technique has been applied. Here environmental risk factors are taken as independent variables and symbolised as X1, X2, X3…… so on and diseases are considered as

Table 15.7 Azamgarh city: neighbourhood-wise percentage distribution of most frequently occurring diseases during the last two years in low-income neighbourhoods

Neighbourhoods	LI/HD	LI/MD	LI/LD	Total
Typhoid	54.06	59.54	50.81	54.8
Malaria	52.54	58.91	48.54	53.33
Dengue	41.31	47.67	37.25	42.07
Diarrhoea	57.86	63.56	34.23	51.88
Worm infestation	56.05	64.65	33.84	51.51
Cholera	38.3	44.42	33.21	38.64
Jaundice	35.8	46.8	16.03	32.88
Chickenpox	31.43	25.36	21.85	26.21
Cold and fever	50.66	57.75	46.82	51.74

Source Based on field survey 2017–2018

dependent variables symbolised as Y1, Y2, Y3…….so on. A positive value of correlation coefficient and ranging from 0.9 to 1.00 shows a strong correlation between independent and dependent variable. The student 't' test has been applied to find out the level of significance.

Neighbourhood Environmental Problems/Independent variables

X1 Inadequate Drainage/Open drains
X2 Poor cleaning of drains
X3 Water logging
X4 Irregular Supply of water
X5 Water Quality Problem
X6 Waste accumulation in neighbourhood
X7 Improper waste disposal
X8 Open defecation
X9 Overcrowding

Associated Diseases/Dependent variables

Y1 Typhoid
Y2 Malaria
Y3 Dengue
Y4 Diarrhoea
Y5 Worm Infestation
Y6 Cholera
Y7 Jaundice
Y8 Chicken pox
Y9 Fever and Common Cold

The correlation Coefficient (r) given in Table 15.8 shows that all the selected disease, namely, Typhoid, Malaria, Dengue, Diarrhoea, Worm Infestation, Cholera,

Table 15.8 Correlation coefficient between dependent variable (associated diseases) and in dependent variable (risk factors in the neighbourhood environment)

	Y1	Y2	Y3	Y4	Y5	Y6	Y7	Y8	Y9
X1	0.957	0.962	0.963	0.981	0.995	0.980	1.000[a]	0.494	0.950
X2	0.818	0.827	0.829	0.992	0.977	0.868	0.954	0.743	0.804
X3	0.974	0.971	0.970	0.757	0.812	0.949	0.860	−0.003	0.979
X4	0.910	0.916	0.918	0.998[a]	0.999[a]	0.945	0.993	0.607	0.900
X5	0.925	0.931	0.933	0.995	1.000[b]	0.957	0.997	0.575	0.916
X6	0.904	0.910	0.912	0.999[a]	0.999[a]	0.940	0.991	0.618	0.893
X7	0.990	0.988	0.987	0.811	0.860	0.973	0.901	0.084	0.993
X8	1.000[a]	1.000[b]	1.000[b]	0.896	0.932	0.998[a]	0.960	0.247	0.999[a]
X9	0.249	0.264	0.268	0.672	0.603	0.338	0.532	1.000[a]	0.226

[a]Correlation is significant at the 0.05 level,[b]Correlation is significant at the 0.01 level. *Source* Based on field survey 2017–2018

Jaundice, Chicken pox, Fever and Common Cold are positively correlated with the neighbourhood environmental problems (Inadequate Drainage/Open drains, Poor cleaning of drains, Water logging, Irregular Supply of water, Water Quality Problem, Waste accumulation in neighbourhood, Improper waste disposal, Open defecation and Overcrowding). However a very strong positivecorrelation has been found between Inadequate drainage (X1) and Typhoid (Y1) ($r = 0.957$), Malaria(Y2) ($r = 0.962$), dengue (Y3) ($r = 0.963$), diarrhoea (Y4) ($r = 0.981$), worm infestation (Y5) ($r = 0.995$), Cholera (Y6) ($r = 0.980$), Jaundice (Y7) ($r = 1.000$), and Fever and common cold (Y9) ($r = 0.950$), Water logging (X3) and Typhoid (Y1) ($r = 0.974$), Malaria (Y2) ($r = 0.971$), dengue (Y3) ($r = 0.970$), Cholera (Y6) ($r = 0.949$), and Fever and common cold (Y9) ($r = 0.979$), Irregular supply of water (X4) and Typhoid (Y1) ($r = 0.910$), diarrhoea (Y4) ($r = 0.998^*$), worm infestation (Y5) ($r = 0.999^*$), Cholera (Y6) ($r = 0.945$), Jaundice (Y7) ($r = .993$), Water quality problem (X5) and Typhoid (Y1) ($r = 0.925$), diarrhoea (Y4) ($r = 0.995$), worm infestation (Y5) ($r = 1.000^{**}$), Cholera (Y6) ($r = 0.957$), Jaundice (Y7) ($r = .997$), Waste accumulation in the neighbourhood (X6) and diarrhoea (Y4) ($r = 0.999^*$), worm infestation (Y5) ($r = 0.999^*$), Open defecation (X8) and Typhoid (Y1) ($r = 1.000^{**}$), Malaria (Y2) ($r = 1.000^{**}$), dengue (Y3) ($r = 1.000^{**}$), Cholera (Y6) ($r = 0.998^*$), and Fever and common cold (Y9) ($r = 0.999^*$) and Overcrowding (X9) and Chicken pox (Y8) ($r = 1.000^{**}$). Air and noise pollution are though not strongly correlated with these diseases; however, significance of these problems in deteriorating health cannot be denied.

15.5 Conclusion

The study concludes that deprived or low-income neighbourhoods in the Azamgarh city are mostly located in old, central, congested part of the city, along the river Tons. These neighbourhoods are generally found in environmentally degraded and unhealthy conditions and face a lot of environmental problems such as inadequate drainage and sanitation, improper cleaning of drains, poor waste collection services, accumulation of waste, erratic water supply, water quality problem, water logging, etc. A strong positive correlation between environmental problems and most frequent diseases affirms that these problems are directly or indirectly responsible for occurrence of various diseases and deterioration of health conditions of the residents. Since deprived neighbourhoods are predominantly inhabited by deprived or low-income groups of the society, most vulnerable towards diseases related to environmental deprivation. Thus the study arrives at a conclusion that there is an urgent requirement of stringent environmental legislation with active government and community participation to encounter ever increasing environmental pollution and degradation which would subsequently lead to improvement in the surrounding as well as health condition of the residents.

References

Cubbin C, Egerter S, Braveman P, Pedregon V (2008) Where we live matters for our health: Neighborhoods and health
Das D, Angshuman G (2007) Some issues associated with the municipal conservancy services to dispose off solid waste in a small town of Birbhum District, W.B. Ed. Ghosh A. Environment drinking water and public health problems and future goals. Daya Publishing House Delhi, pp 48–71
District Census Handbook, Azamgarh, 2011
Gove WR, Hughes M, Galle OR (1979) Overcrowding in the home: an empirical investigation of its possible pathological consequences. Am Sociol Rev 59–80
Gupta K, Kumar P, Pathan SK, Sharma KP (2012) Urban neighborhood green index–a measure of green spaces in urban areas. Landsc Urban Plan 105(3):325–335
Jamal S, Ajmal U (2018a) Neighbourhood Environment: Assessing and Mapping of vulnerable neighbourhoods of Azamgarh City fro Planning, Annals of the National Association of Geographers, India, 38(1), 156–174
Jamal S, Ajmal U (2018b). Neighbourhood Disorder and its Influence on Resident's Health A Case Study of Azamgarh City. The Geographer, 65(1), 76–90.
Karn SK, Shikura S,Harada H (2003) Living environment and health of urban poor: a study in Mumbai. Econ Polit Wkly
Lemstra M, Neudorf C, Opondo J (2006) Health disparity by neighbourhood income. Can J Public Health/Rev Canadienne de Sante'ePublique 435–439
Macintyre S, Ellaway A (2003) Neighborhoods and health: an overview. In: Kawachi I, Berkman LF (eds) Neighborhoods and health. Oxford University Press, New York, pp 20–42
Ministry of Urban Development, Government of India (2016) Swachch Bharat Mission, Municipal Solid waste Manual
Mushir S, Khan MF (2007) Quality of urban environment. Idarah-i-Adabiyat-i-Delhi Delhi
Pokhrel D, Viraraghavan T (2004) Diarrhoeal diseases in Nepal vis-a'-viswater supply and sanitation status. J Water Health
Rahman A (1998) Household environment and health Delhi B R Publishing Corporation
Singh AL, Baba SA (2014) Understanding water supply and sanitation conditions in the neighbourhoods in Srinagar City (J&K). Natl Geogr J India 60(3):197–210
Songsore J, McGranahan G (1993) Environment, wealth and health: towards an analysis of intra-urban differentials within the greater accra metropolitan area. Ghana Environ Urban 5(2):10–34
Wheaton B, Nisenbaum R, Glazier RH, Dunn JR, Chambers C (2015) The neighbourhood effects on health and well-being (NEHW) study. Health Place 31:65–74
World Health Organization (2010) Why urban health matters WHO

Chapter 16
Quality of Living, Health and Well-Being of Slum Dwelling Women Domestic Workers in Kolkata

Sudeshna Roy

Abstract Urban sustainability is a major concern in the rapidly urbanising developing world. With escalation in city-ward in-migration and resultant dense population concentrations in informal settlements coupled with higher consumption patterns; smart, efficient and sustainable management of environmental and human resources becomes imperative. In this background, the paper focuses on the intra-urban spatial inequality in standard of living, availability and access to basic services across selected slum pockets distributed across Kolkata city. Using field survey data of slum dwelling women participating in live-out paid domestic work, an analysis is attempted regarding the level of inequality in quality of living and associated health and well-being status of the households. The paper endeavours to study the distribution pattern of housing amenities across the socio-religious groups among slum households, with special focus on women domestic helpers regarding their occupational health. Housing amenities directly influence health of household occupants. In this scenario, women's health issues become vital to be addressed but unfortunately is often neglected because of her precarious status in the socio-economic hierarchy. There is conspicuous disparity among the northern and southern Kolkata slums in living conditions and consequential health situation. Women domestic workers endure tedious working hours under indecent paid working conditions for paltry income and simultaneously shoulder unpaid care-work responsibilities. Time-use data reveals lack of sufficient sleep time, personal care and leisure time among them, leading to psychosomatic and other diseases adversely affecting their work productivity. General caste, Hindu women and cooks perform better in health relative to scheduled categories, Muslims, house-cleaners, babysitters.

Keywords Slum · Housing amenities · Well-being · Occupational health · Time-use

---

S. Roy (✉)
Jawaharlal Nehru University, New Delhi, India
e-mail: sudeshna.jnu@gmail.com

R. B. Singh et al. (eds.), *Urban Health Risk and Resilience in Asian Cities*,
Advances in Geographical and Environmental Sciences,
https://doi.org/10.1007/978-981-15-1205-6_16

16.1 Introduction

Sustainability approach brought about through efficient management of environmental and human resources is the most appropriate way forward for the urban growth and development in the backdrop of accelerating pace of urbanisation in the world. Sustainable Development Goals (SDG) 2015 which improved upon the Millennium Development Goals (MDG) is expected to facilitate in integrating the subsystems of economic, social, infrastructural, environmental and demographical sectors with a broader view to achieve reduction in inequalities among different social groups, economic communities, gender and age structure. According to Census of India, 31.16% of the country's population resides in urban areas in 2011 in about 53 million plus cities, which rose from 27.81% and 35 cities in 2001 respectively, with dense concentrations in metropolises and its immediate suburbs, which produces 60% of Gross Domestic Product (GDP) of the country. (NBO 2013) The level of urbanisation is expected to escalate along with growth of new towns brought about by natural increase and an augmented rural to urban migration and migration from lower class towns to Class I cities and metropolises. This puts additional overload on the existing infrastructure and thus calls for increased financial investments in creation and expansion of technologically proficient public utility services that would be made equitably available to all sections of urban population.

A decent living necessitates the presence of minimum and quality house and housing amenities and that in turn affects the health and well-being stature of the occupants. House structures and availability and access to civic amenities such as safe drinking water, sanitation and sewerage disposal, solid waste management, nature of cooking fuel, power connection determine the basic health status of the population. In this regard, as is evident from secondary data sources most deprived are the disadvantageous sections, comprising primarily poor, economically weaker and socially backward slum and squatter settlement dwellers. Poverty creates some adverse situations that expose these people to severe health risks and vulnerabilities due to poor quality of living. Thus one of the primary aims of SDG is inclusiveness of the marginalised groups of Scheduled Castes (SC), Scheduled Tribes (ST), disabled, women, aged and other minorities in urban setting. SDG Goals 11 and 8 are vital in this regard which seek to promote empowering, resilient, productive cities with an aim of sustainable economic growth achieved through decent work trajectory. SDG Goals 3, 6, 7 pertaining to good health, safe water and clean energy combined under the reduction of inequality agenda of goal 10 are important constituents in the course. Urban areas of the developing countries are at more risk owing to the increasing levels of 'health differentials' across population subsections. (ICSU group 2011) Likewise Chandrasekhar and Mukhopadhyay (2008) discusses the existence of 'social fragmentation' in urban areas of the developing world which obstruct in the active creation and functioning of 'community-based health risk-sharing mechanisms.'

16.2 Hypothesis

Quality of living has a two-directional interrelation with housing amenities and the latter in turn determines the health status. Health status is culmination of the interplay of multitude of factors which have inter-linkages among each other and are constituent elements in the overall health system. The variables are related to the immediate physical environment of the individual in which he or she lives and those are largely dependent on nature of civic amenities available and accessed, biological configuration of the individual which is also influenced by nutrition and medicinal intake, the work activity related environment and socio-cultural environment such as familial, recreational, societal cohesion influencing the mental health. Occupational health is an important area of concern as it affects the worker productivity and in consequence the income earning capacity and in the long run the 'burden of care' (O'Neill and Ross 1991) at household and societal level. But labour and health literature have not given much attention to this aspect particularly the health conditions of women earning livelihoods from informal, irregular and insecure low paying jobs. The paper hypothesises that the subtle interconnections between the physical and working environment mutually affect the general health status of women domestic workers and there are noticeable spatial deviation and social heterogeneity within the domestic worker group.

16.2.1 Objectives

The paper attempts to analyse the location differentials among the slum households in Kolkata Municipal Corporation (KMC) area in terms of availability and access to essential housing amenities and the interconnection of the standard of living of the households with their health status. The study will also focus on identification of existent inter-communal inequality in terms of quality of living among slum dwellers in order to hypothesise that intra-slum disparity exists in different dimensions. The initial section will be devoted to study the distribution and pattern of housing amenities across the socio-religious groups among selected slum households in Kolkata city. Thereafter, the spatial as well as socio-economic disparity in degree of asset ownership, standard of living and health status of the slum households will be analysed in order to delve deeper into the aspect of inequality thus highlighting the need for inclusive urban development. The third objective deals with the assessment of level of health and well-being of the women engaged in paid domestic work across the geographical zones of the city. Women health research has been principally concentrated in the realm of reproductive and maternity-related health. The general health and job-related health and well-being component have been chiefly missing especially for the informal sector workers. Paid domestic workers also suffer from the lack of unionised collective representations, statistical invisibility and absence of legal and social protection, thereby making it all the more difficult to address their health and well-being issues. Thus the final aim of this paper is to investigate the effect of working conditions on occupational health pattern among the categories of women domestic workers and across city geographical zones. Denial of decent work is also seen to culminate into violating the human rights of the women workers and it is in these vulnerabilities that they earn their livelihood (Negi 2014).

16.3 Study Area

Kolkata located on the eastern bank of river Hooghly and the capital city of West Bengal is one of the four major mega-cities of India functions as the administrative, financial, manufacturing, commercial and educational nucleus for not only the state but for whole of the eastern and north-eastern region of the country and thus attracts scores of inter-state and intra-state migrants. It is a port and focal centre of the third largest Urban Agglomeration (UA) in India; Kolkata Urban Agglomeration (KUA) or Greater Kolkata spread across 6 neighbouring districts sheltering 14.1 million population, having a municipal corporation as the Urban Local Body (ULB) delineating 141 wards under its administrative jurisdiction spread across 185 square kilometres with a population of 4.5 million as per 2011 Census of India. The eastern and southern margin of KMC is bounded by North and South 24 Paraganas district and Bay of Bengal, while the west and north merge into districts of Howrah and North 24 Parganas. Slum households distributed within the KMC area are canvassed for the present study.

16.4 Database and Methodology

The paper is based on a primary field survey conducted during December 2015–March 2016 within the municipal administrative limits of Kolkata city across the selected slum households wherein women participating as paid domestic workers resided. About 360 slum households were canvassed distributed uniformly across four geographical zones in the city namely north, west, south and central. Stratified random sampling technique has been applied to interview only women who are 18 years of age and above and who are employed as paid domestic labourers at the time of survey. The snowballing method is also used to identify the women workers and further they were interviewed on their socio-economic background, family details, house and housing amenities, assets possessed, paid and unpaid care-work structure, income-expenditure pattern and other economic details along with health conditions. Simultaneously a time-use questionnaire schedule also allowed collection of the different activities performed by these women workers per day, thus assisting in understanding the work and leisure balance in their lives. A comparative approach is adopted to analyse the data across the spatial locations within the city and among the work categories of domestic workers using descriptive analysis, principal component analysis, binomial logistic regression analysis, Pearson's correlation and computed weighted indexes.

16.5 Findings and Discussions

Kolkata city has a '*bustee*' or '*thika*' tenancy form of slum settlement whereby the tenants pay rent to the landowners or the slum house owners. Older slum quarters typical of north and central Kolkata like Paikpara, Shyambazar and Park Circus are aligned in *thika* system of accommodation, while many newer slums have come up in southern and western parts of city especially along railway network such as Dhakuria and Tollygunge, Brace Bridge, where the lodgers are owners themselves. Availability and access to safe drinking water and functional clean toilets are the two basic essential housing infrastructure for safeguarding general health of the population. KMC is helmed with provision of tap water supply, tube-wells, hand-pumps and construction and maintenance of public toilets that service within the slums. There were instances where households living within a slum house pooled collective lump sum to build and maintain pour latrines and bathroom facilities. All the components are discussed in detail further on in the sections below.

16.5.1 Drinking Water and Toilet Facilities

Mere availability does not guarantee access and quality of service, rather per capita availability depends on location of the service and distance from the residence, duration of tap water supply, digging and serviceability of bore-wells and hand-pumps, functional operation and upkeep of sufficient number of latrines through cleaning and repairing services by municipality or the users themselves. Satapathy (2014) points out that making provisions of safe drinking water requires not only laying down physical infrastructure but also needs focus on integrated system of public health and behavioural transformation of community.

Almost all slum households (98.1%) draw water primarily from tap source point, the rest 2% rely on tube-well as the primary source, with 40.3% households having water source within premise. 86% of the taps or tube-wells that are located within the household premises are shared among more than one household while 63.4% of households that do not have water source within precincts have to share community facilities located at the roadsides. It is especially difficult for the women of the households, the working maids especially who work hard at employer's home to earn money and again endure long hours at household chores with additional chore-time collecting drinking water. The distribution across social groups shows the caste and religion-wise inequality which has been discussed earlier. Only 33%, 30.3% and 21.4% of SC, OBC and Muslim households have access to water within premises, respectively, while a much higher proportion of general castes (53.1%) and Hindus

(41.2%) are classified with the same. On similar note, 11% and 7% of general caste and Hindu households enjoy individual water usage while a much less incidence at 3.7% and 3% is observed among the SC and OBC households in that order. No Muslim households have individual water facility. Water quality is affected by not only presence of superior source point of drinking water, but also its duration of availability and ease of access, distribution and per capita quantity availability along with the facets like cleanliness, turbidity, odour, taste, handling and hygiene of water collection.

Duration of supply is mostly limited to few hours in the morning or afternoon. Barring 8.4% households who get 24 h water supply, most households rely on supply which lasts 4–8 h (46.7%) or even less than 4 h (26.4%) a day. Tap density defined as number of households sharing one source point of tap water is found to be high demonstrating population pressure on the public service. (Table 16.1) Around 35% of the slum households share water with 6–10 houses, while 34% share with 11–30 houses. At the lowest and highest extremes of scale, 17.5% and 8.1% of the sampled slum households share one tap source with 2–5 houses and 31–50 households in that order. Only 5.8% households have individual private water source and do not share with other households. The zonal variation in tap density points out overcrowding in northern and western slum clusters with 16.1% households in north sharing water with 31–50 households and 25% of households in west sharing water with 21–30 households. Slum households in the northern Kolkata (21.5%) like Shyambazar,

Table 16.1 Distribution of slum households according to drinking water and toilet facilities, Kolkata

Water quality			Tap density			Toilet density		
Duration of supply/day (h)	N	%	No. of households/tap source	N	%	No. of households/toilet seat	N	%
Less than equal to 4	95	26.4	2–5	63	17.5	Individual household	12	3.3
4.01–8.00	168	46.7	6–10	126	35.0	2–5	83	23.1
8.01–12.00	24	6.7	11–15	56	15.6	6–10	91	25.3
12.01–16.00	9	2.5	16–20	31	8.6	11–20	100	27.8
16.01–20.00	28	7.8	21–30	34	9.4	21–40	36	10.0
Throughout day (24)	30	8.3	31–50	29	8.1	41–60	19	5.3
No tap water	6	1.7	51–100	0	0.0	61 and above	1	0.3
Total	360	100.0	101–250	0	0.0	Defecate in open	18	5.0
			Individual household	21	5.8	Total	360	100.0
			Total	360	100.0			

Source Computed from primary survey, 2015–2016

Paikpara, DumDum, and Beleghata have good water quality, with availability for almost 24 h. Southern (11%) and central (14.3%) Kolkata slums get 8–12 h of treated water supply per day. Most households approximately 47% in southern and 41% in western regions are the worst with supply duration of less than 4 h. The households in Dhakuria, a slum which evolved along the suburban railway tracks in the southern Kolkata, depend on a two-hour window in the mid-afternoon for collecting palatable sweet drinking water that is procured from the distributing water tankers. It puts the women under severe stress to schedule their household chores and paid work accordingly and set aside ample time for awaiting their turn standing in the queue for collection and storage of water.

Likewise, the toilet facility is primarily pouring latrine and 90% of households having the facility within premises share it with community and only 10% have individual usage. Out of the canvassed households, only a small fraction (4.2%) has separate toilets only for themselves and 42% of the toilets are located within premises. The general castes (49.2%) and Hindu (42%) households are in a better position than the SC (36.1%), OBC (36.4%) and Muslim (36%) households in terms of toilets within premises. Toilets are cleaned either by the households themselves or they collectively pay the corporation sweepers to clean the latrines. Unfortunately, there is also 5% incidence of households reporting to defecate in the open especially among 11.3% and 6.3% households in central and western regions of the city, respectively. Toilet density which is number of households using one toilet seat is more than 40 households (5.6%) and 38% households share one toilet and one bathing facility with 11–40 houses (Table 16.1). Quality of toilets is poorer in some northern and western zones of Kolkata where sharing of one toilet seat with up to 20 households or more is accounted by 19.4% and 34.4% of slum households, respectively. Though the recent Jawaharlal Nehru National Urban Renewal Mission (JNNURM) and KMC efforts have improved the sanitation facilities across the slums but there is much scope for improvement for the maintenance and upgrade of the existing toilets especially in the western and eastern fringes of the city. Households without toilet within premises use corporation toilets (72.3%), *Sulabh* pay and use facility (13.5%). It is a fascinating and perplexing occurrence that even though penetration of mobile telephony and satellite cable television within the slum habitats has seen rapid acceleration and wider coverage in recent years, but still the practice of open defecation is highly prevalent. This can be attributed to the habitual preferences, attitudinal quandary and dearth of health knowledge about hygiene sensibility among the population. Inequality and variability in the availability of two essential municipal services across the slum settlements in Kolkata city can be ascertained statistically. Distribution of slum households in terms of availability of water facility within premises showed the highest variation across the city slums, with a coefficient of variation (CV) value of 68%, followed by toilet facility within premises (52.4%). Simultaneously among households with water and toilet facilities within premises, the degrees of variation for the individual usage were very low with CV of 29% and 30.2%, respectively. It is noteworthy that there is a strong positive correlation between the presence of water and toilet facilities within household premises with a Pearson's correlation value of 0.99 at a high level of significance.

16.5.2 Cooking Fuel, Kitchen Space and House Structure

The other amenities that affect household health substantially are type of cooking fuel and location of kitchen along with house structure and number of rooms. Kerosene is the primary source of cooking fuel (51.7%), followed by liquefied petroleum gas (LPG) (40.3%) and wood (7.8%). Inter-slum and inter-community difference in affluence level is clearly evident from the use of inferior sources of fuel such as wood and coal. About 12.5% slum households in western part of the city and 12% in southern areas use wood as primary combustible substance, while 51% of northern city slums use LPG but the lowest is recorded by central zone at 17.1%. Central Kolkata slums have the highest proportion of casual labour households belonging to Bihari and Muslim community where members along with household head participate in occupations which are characterised by marginal and piece rate daily wages such as masonry, cobblers, shoe making, freight-loading and construction workers. Thus most households in this zone use cheap and air-polluting cooking fuels such as dung cakes (4.3%), wood (4.3%) and coal (3%) which is highest among all other slum zones. Proportion of general castes consuming LPG is slightly higher at 41.4% than SC (36.1%) and ST (37.5%) households. Wood is used more among the Muslims (14.3%) than the Hindu (7.5%) households.

Location of kitchen can determine levels of air quality within the house along with household size and number of occupied rooms. In the slums, where the hutments are built very close to each other with almost passable narrow alleys and congested living arrangements of people within limited space, the kitchens are mostly located within the living room (42.8%). Practice of cooking activities is also done in the narrow *verandah* (43.6%) and open courtyard space (8.3%) which can cut down the indoor pollution but might seriously affect the food hygiene. Interestingly 50% SC households have their kitchen set up within living room, which is higher in occurrence relative to rest of the castes, lowest recorded by the general castes (33.6%), while none of the Muslim households reported to practice cooking in a separate room.

House structure and construction material indicate economic level of the owner but also affect the quality of life at large. Temporary houses usually termed '*kacçha*' are prepared of mud, with thatched roofs or use of tiles and polythene shacks for roofing and are prone to easy physical and environmental damage. The proportion of '*pacca*' house is high at 57.2%, where bricks, tiles, and cement mortar are the building materials and *semi-pacca* or mixed structure houses amount to 40.6%. Slums, namely, Beleghata, Tangra, Topsia in the central zone (6%) has the highest proportion of households living in *kaccha* houses. Again economic inequality can be seen from the figures, 62.5% of general caste and 58% of Hindu households live in *pacca* house as against only 51.5%, 53.4% and 35.7% of OBC, SC and Muslim households correspondingly. Room density not only affects the indoor pollution and physical environment of the living space but also determines the psycho-social living conditions of the inhabitants. Lack of floor space, crowding and small sized rooms without appropriate ventilation and lighting are principal features that differentiate slums from non-slum residential structures. Room density in Kolkata slum residences is an average 3–6 person per room with 69% of households account for such distribution, while 9.2% are single member households and 14.2% being occupied by 2 persons; usually spouses or children. Central zone slums are overcrowded with 18% households sheltering 7–11 people per room followed by western slums (8.3%).

16.5.3 Asset Ownership and Standard of Living

Durable assets such as television, refrigerator and personal vehicle such as bicycles and two-wheelers are proxy indicators for the household wealth and economic condition. Mobile telephony has seen vast coverage in India in recent years, and slum households also show better ownership level in Kolkata, with 76% of slum households reporting to own a mobile handset. Bicycles (21.1%) are seen to be one of the chief possessions of the households as is the satellite television connections (44.7%). Television is owned by 73.6% and is the chief source of entertainment, followed by radio (14%). Furniture objects such as cot, bed and *almirah* (cupboard) and refrigerator are in possession by 24.4%, 47.5%, 54.4% and 15.8% households, respectively. Land is owned by only 8% households either at their native place from which some receive annual income from quantities of agro-produce or as monetary share of the agricultural sale or land is owned in form of house built up on the land by the households themselves. Average size of the landholdings is less than 0.5 acres owned by 75% of the landowning households. Around 10.7% households primarily Hindus concentrated in southern slum areas of Kolkata own land in the 0.5–1 acre range and 14.3% own 1–2 hectares of land.

A better comparative comprehension about the economic standing of the slum households can be estimated from the computed asset ownership index and weighted standard of living index. The former is calculated using possession of 12 objects of durable assets such as television, mobile, bed or cot, own vehicle (bike, cycle, autorickshaw and rickshaw), land, *almirah*, fan, light, fridge, sewing machine and LPG cylinder and based on the composite score that households are categorised into low, medium and high and no assets index classes which amount to 20.8%, 59%, 16.9% and 3.3%, respectively. Spatial variation can be ascertained from the ward-wise asset index scores. Proportion of households with high asset index tag is highest in slum north (20.4%) and slum west (18.8%), while 7.1% of central Kolkata slums have no specified assets and are relatively poorer (Table 16.2). Ward numbers 57, 58 and 59 comprising slums in Tangra, 50, 54 and 38 comprising Boubazar, 38 in KC Sen Street in the central zone and 131,141 in Brace Bridge in the west have most households under lowest asset index. While the slums of Chetla, Panditiya in the south, Tollygunge and Ranikuthi in west and Wellington, Shyambazar in the central and north zones have high index. The proportion of high asset ownership is greater for literates (21.1%) than illiterates (13.9%), Disparity among socio-religious communities is observed whereby 14.3% of Muslim households own no assets as against only 2.9% of Hindus, and about 6.1% of OBC, 4.7% of SC households with no assets as against mere 0.8% of general caste households. Many widowed and

Table 16.2 Distribution of slum households according to asset ownership index, Kolkata

Asset ownership index	Slum households across city zones			
	North	West	South	Central
Low	18	24	19	14
	19.4%	25.0%	18.8%	20.0%
Moderate	56	49	65	42
	60.2%	51.0%	64.4%	60.0%
High	19	18	15	9
	20.4%	18.8%	14.9%	12.9%
No assets owned	0	5	2	5
	0.0%	5.2%	2.0%	7.1%
Total	93	96	101	70
	100.0%	100.0%	100.0%	100.0%

Source Computed from primary survey, 2015–2016

separated domestic workers in the respondents list are head of their households who in most cases were compelled to participate in paid labour market to earn livelihood after death and desertion of their spouses. Thus, their poor economic condition is conspicuous with only 9.1% of women-headed households being accounted in high assets index class while approximately 22.1% of male headed households in the same. Likewise, women living alone in single family structure have generally low index (56%) or no assets (8%) unlike other categories. Households with heads having higher educational attainments and employment in regular-salaried jobs in private (33.3%) and public sector (20%) mostly come under high index categories owing to income security that education brings forth and financial stronghold of regular jobs. Among the work categories, domestic workers who perform tasks of a house-cleaner have lower asset index (23.4%) and about 6.3% have no assets while 29% of the cooks and 15.4% of babysitters fall in the high index category. This can be explained in conformance of the differential wage structure across the work categories based on nature of tasks and skills required to perform them.

On the other hand, the weighted standard of living index scores ranked as 'fair', 'bad' and 'worse' in the descending order of living standards is also generated using normalised values of 19 selected indicators grouped under three broad heads namely asset ownership, living condition, and socio-economic condition which are given weights in decreasing order, respectively. Living condition is calculated using presence of indicators such as households practising open defecation, kitchen inside living room, use of unclean cooking fuel and non-tap source of drinking water. Level of socio-economy is measured using illiteracy, percentage of women-headed household and incidences of underemployment. Underemployment is gauged from women domestics who work more than 6 h per day but who would like to work more number of employer houses if available. Asset ownership variable here sums up the non-ownership of specified assets by the households. Higher index scores reflect poorer standard of living and lower index values indicate good living standards. The general poor standard of living which characterises slum settlements justifies the rationale behind computing degrees of variation in poor standard of living. Majority of the slum households about 76% live under 'bad' conditions, while 11% have 'worse' index scores and 14% households are classified under 'fair' index scores. Standard of living index also helps to categorise the women groups according to spatial distribution and by employment categories. There exists conspicuous inequality in general among the households classified based on geographical zones in which they are located within the city. The slum households of the north (17.2%) and south (15.8%) Kolkata have 'fair' standard of living, whereas 20% households in central and 15.6% in western part of Kolkata are classified under 'worse' standard of living.

For understanding the intra-group inequality tabulation is made within the individual zones and worker categories. Among the work categories, the domestic helpers (17%) rank poorest having the worst standard of living relative to the cooks (2.4%) and ayahs (5.1%). The domestic workers employed as babysitters, ayahs-cum-cooks perform better in standard of living index, with 18% of their households each enjoy 'fair' index. The women employed in multiple tasks of cooking, babysitting along with cleaning and washing have better living standards. This can be exemplified from the task-based unequal wage structure which permeates down the economic position of the households and keeps them intact in their level of standard of living. Similarly, the economically backward communities such as Muslims (14.3%), OBC (12.1%), SC (13.1%) and women-headed households (21.4%) have higher share of households who live under 'worse' standards, unlike that of Hindus (11%), general castes (8.6%) and male-headed households (4.6%) (Table 16.3).

It is most likely that widowed and separated women maids are single-earning members of their households and they exist under dire financial situations. Literacy status and standard of living are found to have a positive relation which can be seen

Table 16.3 Distribution of slum households by castes and standard of living index, Kolkata

Castes	Standard of living index			
	Fair	Bad	Worse	Total
SC	20	146	25	191
	10.5%	76.4%	13.1%	100.0%
ST	1	7	0	8
	12.5%	87.5%	0.0%	100.0%
OBC	3	26	4	33
	9.1%	78.8%	12.1%	100.0%
Non-SC-ST	25	92	11	128
	19.5%	71.9%	8.6%	100.0%
Total	49	271	40	360
	13.6%	75.3%	11.1%	100.0%

Source Computed from primary survey, 2015–2016

from the fact that about 17.3% of illiterate maids live in worse living conditions as against on meagre 2.6% of the literate maids. Monthly household expenditure which represents economic condition across the city zones also depicts presence of economic inequality. Slum households in southern and western Kolkata are relatively affluent than their counterparts in other zones with about 16% and 5% of the households, respectively, which spend in the range of Rs 5000 to more than Rs 20,000 per month, which is much higher than the mean monthly expenditure of Rs 2644 for slum households in totality. The Gini coefficient computed for monthly household expenditure also points out that western city slums have the highest economic inequality (0.53) while the northern slums have the lowest with Gini value of 0.30.

The ward-wise classification generated across the slum provides a better comprehension of the spatial concentrations of households with ordered levels of standard of living. The slums such as Tangra in east, Baghajatin in south, Bracebridge railway bustee in the west and K. C. Sen Street in the north come under worse category. The slums of Paikpara in north, central Kolkata slums and southern zone slums like Ballygunje, Tollygunge, Behala are better performing in terms of living standards. Location economics does influence the intra-urban cost of living, land value, wage rate and income earnings, rental prices of housing which are manifested in discrepancy in wealth condition and economic position among the slum settlements. The better off socio-cultural subgroups within the slum category move into relatively advantageous slum pockets situated in prime spaces within the city having suitable living cost which they can afford, while the inferior land and housing locations remain to be colonised by the lowest stratum of economically weakest section (EWS).

16.6 Health Condition of the Women Domestic Workers

Health status of the women workers is strictly tied to the housing and housing amenities surrounding them and the nature of work and working conditions they are exposed to. Ghai (2002) refers to the four constituents that form the backbone of decent work, namely ‘employment, social security, worker’s rights and social dialogue’. But informal workers and in this context the women domestic workers work precariously with no job tenure security and are non-unionised into labour unions and collectives. The proportion of part-time domestic workers has risen substantially in India relative to full time workers (Neetha 2013). But owing to the fact that a domestic worker works in multiple employer houses on daily basis, the amount of work-time stretches to long hours of paid work coupled with the burden of unpaid work responsibilities at own households and it takes a toll on their health. On the other hand, in the crowded, unhygienic and insanitary environmental surroundings of the slums where these workers reside, in general one may note that health structure of the resident population of the slums is not robust. Disease prevalence rates such as communicable diseases like typhoid, tuberculosis, diarrhoea can be above average compared to rest of the parts of the city. This is owing to flooding, water contamination, lack of room space, higher number of heads sharing one room accommodations, household indoor pollution due to usage of polluting and cheaper unclean soot producing cooking fuels such as coal, cow dung cakes, wood, hay and organic wastes, lack of sufficient ventilation, narrow alleys and open drains amongst all.

Location of kitchen is a crucial determining factor behind level of indoor pollution and consequent health status of the household. About 43% of slum households have their kitchen located in the living room, and central Kolkata slums have the highest share (56%). While the lowest figures are found in the northern Kolkata slum houses (21.5%), which are spread out with customary courtyard space attached. The microscopic respiratory particulate matter (RPV) that remains suspended in the air and is easily inhaled more in amount when choice of cooking fuel is non-LPG. According to Duflo et al. (2008) about one-third of acute lower respiratory infections (ARI) are caused by indoor air pollution.

Health assessment of the women domestic workers envisages the degree of productivity at workplace and 'burden of care' (O'Neill and Ross 1991) they have to endure in response to disease outbreaks and other health-related dysfunctions. Concerns about the health status of the categories of women domestic workers show that more than 50% women in all classes suffer from some chronic disease. More than 13.4% reported to suffer from blood pressure-related chronic diseases, followed by diabetes (6.4%), gastric or abdominal ulcer and tumour incidences (5.6%), arthritis and joint pain (5%) and eye infections (3.1%). Thyroid deficiency, spondylitis and nerve disorders, gynaecological diseases and unitary tract infections (UTI) occurrence is also widely reported by the maids. More than 40% women in all worker categories reported to live with musculoskeletal diseases for years. Most women reported frequent occurrence of primary illnesses of gastro-intestinal problems (33.3%), fever and cold (13.6%) skin allergies, knee joint pain (18.1%), headache, sinusitis and migrane (3.3%) bronchitis and acute respiratory tract infections which affect their productivity at the workplace, often resulting in job loss on account of consequent absenteeism. About 60 respondents (16.1%) of domestic workers lost their job in the reference period of 1 year before the survey because of their illness-related leave or absenteeism from employer household and in most cases adequate compensation was not handed over to them. Usually, these women reported to have neglected their health problems and become accustomed to live and work with these over the course of time. Lack of financial stronghold, lack of employer cooperation in arrangements of temporary substitute workers and fear of job loss due to frequent work leaves, non-cooperation from family members, absence of work and unemployment of male head of the household, attitude of nonchalance toward own health priorities can be listed among many reasons that prohibit these women to seek timely and regular treatment for their ailments. Poverty and low affordability in these low-income households make adequate medical interventions beyond reach.

The situation for women's health in Indian society is all the more agonising because limited household monetary resources are generally entirely expended on males and children of the household in seeking treatment, buying medicines and consumption of balanced diet and women are overlooked. Low-income households are no exception; rather the low educational attainments make them all the more ignorant about diverse health issues that women must address to. Moreover, tobacco and betel leaf intake and addiction habit of the domestic workers was found to be quite conspicuous with 25.3% of women maids being habituated to chew and smoke tobacco products, indicative of bad health. The higher incidence of tobacco consumption is recorded by domestic helpers, (31%), the lowest among cooks (19%).

Around 10% and 11.2% women seek no treatment and continue with their self-medication methods, respectively, for treating diseases like asthma, diabetes and gynaecological health problems. Amongst the women who seek treatment, allopathy stands out as the primary treatment procedure followed by homoeopathy with 56.17% and 10.3% women reporting, respectively. It is usually the government facility or hospital (20%) that they visit for curative or operative care. Private clinics are easy to access and so it is the primary place of treatment for 30% of women and their households, who can afford consultation fees for private medical practitioners, while around 14.5% go to the medical shop and local dispensaries to buy medicines according to self-medication or as suggested by others. Southern slum households are in financially better position to afford higher amounts of private clinic visit fees (45%) and other medical expenditures compared to the poorer households of slum west (19%) and north (22.6%).

16.6.1 Occupational Health

Work-related diseases and medical problems can also be listed but most respondents seem to be clueless about how their rudimentary work activities that they also are accustomed to perform daily in their own homes can create medical situations. Almost all of the respondents (95.6%) believe that they do not suffer from any medical condition that may arise or be exasperated by their nature of work performed. Likewise, it is also difficult to isolate work environment and investigate the diseases that may emanate from work profile of a person. Nevertheless, some occupational hazards were listed by few of the maids that they assume to stem up from the tasks they perform. Hazards are categorised into physical, chemical and biological. Few women (3.3% of the respondents) who were aware of such medical risks, claimed that dust and gasses (25%) emitted from cooking and sweeping activities and harsh chemicals in floor cleaners and detergents (25%) used for mopping and washing tasks can expose them to respiratory and dermatological diseases like bronchitis, eczema, allergies. Long hours of standing (10%), lifting heavy objects (15%) and repetitive motions like sweeping can result in musculoskeletal and nerve-related deformities such as rheumatoid arthritis, tennis elbow syndrome, frozen shoulder, muscle dystrophy, auto-immune diseases and osteoporosis. Following the Pareto rule, these four above mentioned hazard cumulatively compose 80% of the health problems which are seen to have direct linkage with their profession.

Ala-Mursula et al. (2006) opine that work stress is also determined by the perceived control over work schedules and it further influences the health issues that originate from balancing the demands of paid and unpaid work. Similarly, when probed into the work-related psychological conditions which have usually been experienced in spells of short-lived events and affected these women, an interesting picture came forward. The interviews portray the absence of self-awareness and indifference of the domestic workers about their mental health issues. About 68.3% respondents report about their lack of knowledge or non-existence of any day to day psychological health concerns such as phases of irritability, tensions and work-related exhaustion. Monotony (8.3%), anxiety (14.2%) and depressive mood conditions (6.4%) are common occurrences among those who acknowledged to have suffered bouts of such mental behavioural issues and these three diseases cumulatively make up 76% of the mental health problems among the working domestic workers. (Pareto rule) Work satisfaction can be determined by combination of efficient utilisation of physical, mental, emotional and spiritual capabilities which are linked to biological health, cognitive faculty stimulation, happiness and purpose the work serves in the larger existence.

The activities that are generally performed by a person in his or her daily life are a key to understand their quality of life. (Camporese et al. 1998) Working women grapple with long working hours of managing own homes, sifting through to the multiple employer workplaces and perform monotonous manual work, which provide them with no sense of pride or contentment in lives. This often neglected resultant facet of work participation must be addressed so that frameworks can be made to enhance worker productivity and well-being. It is clearly discerned from the time-use data gathered alongside that domestic workers devote less than 30 min (18.1%) to personal care activities and sleep time is often less than 5 h (17%) on account of performing manual repetitive domestic chores both at employers home and own homes. These women are at more risk of suffering from psychosomatic distresses especially in the environment of constant work pressure, juggling paid work and home making, worries regarding earning enough to enable savings and making day to day ends meet, bringing up their children in desired way.

Working conditions also determine the overall well-being and health. Ventilation, appropriate lighting, wholesome timely meals during work, rest period and access to hygienic toilet facilities, first aid, relation with the employer and co-workers amongst other criterion do influence physical and mental health status of a worker (Table 16.4). Availability and access to toilet facility which is one of the crucial factor for well-being at workplace is there for only 55.2% of workers, while 32% maids get no access to first aid and emergency medical attention, 47% get no time and space to rest, 94% and 93.4% women are not given protective kits like gloves, slippers and warm water for washing purposes during winter months. Another discomforting occurrence is that many times the community toilets at housing societies have no separate toilets for women and often the maids are not allowed by the security guards to access them.

16.6.2 Health and Well-Being Index

Thus to ascertain the variation in health and well-being levels across the work categories of domestic workers and the geographical zones across the city slums, a

Table 16.4 Distribution of employer households of women domestic workers according to working conditions, Kolkata

Distribution of employer households/workplaces according to working conditions provided to domestic workers

	First aid		Hot water during winter		Resting place and time		Appliances aiding work	
Availability	N	%	N	%	N	%	N	%
Yes	613	68.2	59	6.6	478	53.2	55	6.1
No	286	31.8	840	93.4	421	46.8	844	93.9
Total	899	100.0	899	100.0	899	100.0	899	100.0
Toilet at workplace	N	%	**Food/beverage**	N	%	**Lighting**	N	%
Employer's toilet	319	35.5	Yes	413	45.9	Sufficient	885	98.4
Separate toilet	178	19.8	No	403	44.8	Moderately sufficient	14	1.6
Common housing/community toilet	206	22.9	Do not eat at employer's house	71	7.9	Insufficient	0	0
No toilet	184	20.5	Do not have time to eat	12	1.3	Total	899	100.0
Do not go toilet outside home	12	1.3	Total	899	100.0	–	–	–
Total	899	100.0	–	–	–	–	–	–
Ventilation	N	%	–	–	–	–	–	–
Sufficient	836	93.0	–	–	–	–	–	–
Moderately sufficient	61	6.8	–	–	–	–	–	–
Insufficient	2	0.2	–	–	–	–	–	–
Total	899	100.0	–	–	–	–	–	–

Source Computed from primary survey, 2015–2016

composite index has been computed taking into consideration 19 indicators broadly clustered into four variables. Physical environment is comprised of room density measured as more than 4 persons sharing a room, open defecation, usage of polluting fuel and kitchen inside living room and inferior non-tap source of drinking water. Health conditions is summarised with consumption of tobacco, presence of chronic disease and musculoskeletal condition, non treatment of diseases and amount of sleep time of less than 6 h per day which can be termed as sleep deprivation form another variable. Working conditions of women maids who do not enjoy paid leave per month, no sick leave, incidence of job loss within reference period of last 1 year due to health-related absenteeism, work pressure experienced and extreme hours of paid work of more than 60 h per week are found to profoundly worsen health conditions. Lastly well-being is measured using incidence of domestic violence at home, work related psychological stress, lack of time and scope for leisure activity and performing unpaid household care-work without help from household members or what can be termed as unpaid burden of care. Incidence of wife-beating by husbands and verbal abuse by son and daughter in-law has been reported by 28% of respondents, and this has a great contribution in sense of emotional well-being of the women. Standardised values for each variable is then used to compute a weighted average for the composite health and well-being index. The higher the values for the index, the poorer will be the health condition and vice versa.

The index scores for the worker categories and zones are ranged into 'fair', 'bad' and 'worse', indicative of progressive deterioration of health. 3.3% women are categorised under 'worse' health and well-being conditions and 70.3% have 'bad' index scores with just 26.4% registering 'fair' heath status. Women of all zones except the western zone have the worst health status among all the subgroups. Central Kolkata slums, namely, Beleghata, Tangra, K. C. Sen Street have the lowest proportions of women maids with 'fair' health status (20%). Among the task-based worker classes, ayahs and the house-cleaners are at the lowest stratum of the health and well-being index having the 'worse' health condition, relative to cooks and cook-cum-house-cleaners who have 'fair' health status (Table 16.5). The nature of work of the ayahs is such that it involves minimum of 8 h, usually stretching up to 12 h or more, and this extreme work-time cause sleep deprivation, change of body clock and other health issues. About 80% of babysitters and 44% ayahs work more than normal 8 h and get less than 6 h of sleep, respectively, and likewise 72% get no time for leisure. But most of them receive help from family members in completion of own household chores, while 46.4% cooks and 40% house-cleaners perform more than 6 h of unpaid care-work without any supplementary support from family.

Among the exclusive category of cooks 29% women have fair health and only 1.2% come under 'worse' status. The house-cleaner and babysitter women groups

Table 16.5 Distribution of domestic worker categories among slum dwelling women according to health and well-being index, Kolkata

	Health well-being index			
Worker categories	Fair	Bad	Worse	Total
Ayah/Babysitters	4	33	2	39
	10.3%	84.6%	5.1%	100.0%
Cook	24	59	1	84
	28.6%	70.2%	1.2%	100.0%
House-cleaners	47	120	8	175
	26.9%	68.6%	4.6%	100.0%
Cook and house-cleaners	17	35	0	52
	32.7%	67.3%	0.0%	100.0%
Ayah and cook and house-cleaners	3	6	1	10
	30.0%	60.0%	10.0%	100.0%
Total	95	253	12	360
	26.4%	70.3%	3.3%	100.0%

Source Computed from primary survey, 2015–2016

do not have good health conditions, as seen from the distributive figures of 4.6% and 5.1% women, respectively, classified under 'worse' health category. The long duration of paid work time which is more than 8 to 12 h per day and the no-work so no-pay rule of ayahs who are generally paid on per day basis makes only about 10.3% of ayahs under 'fair' health status.

The geographical disparity in health and well-being score can be well depicted by disaggregating the index at the ward level. Among the slums, the worst performing wards are 58 in the central zone and 89, 95, 97, 101 in the southern zones. Slum women in the northern quarters around Dum Dum and Chatubaburbazar neighbourhoods have 'fair' health status, while all other wards come under 'bad' health status category. Health and well-being situation among the women workers are studied according to the literacy status, religion and caste membership, age and marital status and place of residence. There is marked distinction between literate and illiterate women, where 5.3% of the latter has 'worse' health relative to 0.7% of the former. With progressive rise in educational attainment among the literates, the proportion of women with 'fair' and good health also rises, exemplified by improved health awareness and better quality of life.

Logistic regression performed to ascertain the factors explaining the variation among the domestic worker women in terms of being healthy and not healthy, shows that age, marital status and work category of the women, occupation of the head of the household, living arrangement of the women and level of decent work index of the worker to have perceptible influence on determining state of health of the women domestics. There is strong negative relationship of decent work index with women workers being unhealthy which is highly significant at 1% level. With the rise of level of decent work the probability of domestic workers being unhealthy decreases. Among the worker categories, the cooks have less probability to be unhealthy than ayah (Tables 16.6 and 16.7). The house-cleaners have higher probability to be not healthy and all the relationships are statistically significant at 5% level. Age of the domestic worker has significant and positive relationship with unhealthy state at 5% level of significance. Both the widowed and currently married have positive relation with not being healthy and the relation is significant at 5% level. Within the criteria of living arrangement, the women who live with only their children have more probability of being healthy. Lastly there is more probability of a domestic worker to be unhealthy if head of the household is a casual labour and the relationship is significant. Other factors like household monthly expenditure, education level of the domestic worker, sex of the head do not explain any difference on the health condition.

Currently married women workers have greater responsibility for household chores and bringing up children alongside paid work and this entails them with lesser personal care and recreational time. Work-life balance becomes all the more difficult to manage for married women, especially in the middle age-group of 36–59 years, which has the highest proportion of women with worse health among all age-groups (4%). Muslim (7.1%) and OBC (6.1%) groups have greater proportions of women under worse health category, almost double as compared to the Hindu (3.2%) and general caste (2.3%) women. On comparison of health and well-being

Table 16.6 Binary logistic regression for determinants of unhealthy status of slum dwelling women domestic workers, Kolkata

Logistic regression explaining factors influencing unhealthy status of women domestic workers				
Determinants	B	S.E	Sig.	Exp (B)
Worker category (Ref. Ayah, Cook and House-Cleaner)			0.053	
Ayah/Baby-sitter	−1.166	0.482	**0.015**[b]	0.312
Cook	−0.791	0.456	**0.083**[c]	0.454
House-cleaner	−1.312	0.516	**0.011**[b]	0.269
Cook and house-cleaner	−0.061	0.867	0.944	0.941
Household head (Ref Male)				
Women headed	0.819	0.6	0.172	2.268
Age of domestic worker	0.026	0.013	**0.056**[b]	1.026
Marital status (Ref. Unmarried)			0.246	
Currently married	1.457	0.634	**0.022**[b]	4.292
Divorced	21.613	40192.97	1	2.43E + 09
Separated	1.154	0.759	0.128	3.17
Widowed	1.284	0.656	**0.05**[b]	3.61
Occupation of household head (Ref. Regular-Salaried)			0.087	
Self employed	0.114	0.401	0.777	1.12
Casual labour	1.058	0.428	**0.013**[b]	2.88
Unemployed	0.718	0.514	0.162	2.051
Dependant	0.794	0.593	0.181	2.211
Domestic worker	0.039	0.69	0.955	1.04
Total expenditure of household	0	0	0.319	1
Decent work index	−4.288	1.163	**0**[a]	0.014
Education level of domestic worker (Ref. Illiterate)			0.966	
Standard 1–5	0.112	0.283	0.694	1.118
Standard 6–10	0.113	0.404	0.78	1.119
Standard 9–10	−0.096	0.551	0.861	0.908
Living arrangement (Ref. Alone)			0.133	
With spouse	0.402	0.785	0.609	1.494
With spouse and children only	0.536	0.617	0.385	1.71

(continued)

Table 16.6 (continued)

Logistic regression explaining factors influencing unhealthy status of women domestic workers				
Determinants	B	S.E	Sig.	Exp (B)
With children only	−1.277	0.747	**0.087**[c]	0.279
With others	0.416	0.521	0.425	1.516
Constant	0.902	1.434	0.529	2.464
[a]Significant at 1% level				
[b]Significant at 5% level				
[c]Significant at 10% level				

Source Computed from primary survey, 2015–2016

Table 16.7 Binary logistic regression (Model summary) for determinants of unhealthy status of slum dwelling women domestic workers, Kolkata

Model summary			
Step	−2 Log likelihood	Cox and snell R square	Nagelkerke R square
1	434.160[a]	0.165	0.220

Source Computed from primary survey, 2015–2016

index with the pattern of paid work, it was revealed that about 80.5% and 84.2% of the women who are very dissatisfied and dislike their work profile have bad health, respectively, while 48.3% and 37% women who strongly like their occupation and are very satisfied have fairly good health in that order. Among the women who desire to leave paid domestic work as a profession, 76.2% come under bad health status. Women who want to change into other vocation enjoying a good health are 8% less than ones who have good health status but who do not want to leave.

Numerous indirect health and well-being indicators designate an inferior health condition among the sampled women. About 7% of domestic workers go untreated for their diseases and 62% have no time for leisure activity. Paid weekly leave is not received by 65% of maids, while 47% get no paid sick leave. Good working conditions that reflect one aspect of decent work as proposed by International Labour Organization (ILO) (2010), is not enjoyed by 40% of the slum maids. Other indicators have a direct bearing on paid work as health condition of the women influences their work participation. Almost 15.5% of maids do not want to work more number of employer houses and about 3% women want to leave paid domestic work due to sickness. Similarly 4% refused to work for potential employers who wanted to hire these women on account of health-related factors. Ageing related health issues and lack of stamina and poor health were frequently cited by 4% women maids to take a work break and leaving working with any employer house (10%).

16.7 Conclusion

It can be concluded that living conditions, housing infrastructure and related public utility and environmental services have overwhelming influence in the health determination of the population. Health and quality of life of the slum households becomes critical in the wake of rising level of urbanisation and migratory behaviour in India. This is true on account of socio-cultural vivacity and demographic dividends and untapped economic opportunities that exist in the densely populated slum clusters. Kolkata city has witnessed slum life since colonial times, the essence of which has spread with diversity and complexity in recent times. Under the neo-liberal urbanism, informal employment and informal work participation have seen tremendous boost in which women domestic workers form an ever-growing group.

One can identify not only dimensions of inequality between the slum and non-slum settlements but among the intra-urban slum pockets and within a slum locality itself which is evidently opposed to the idea of homogeneity of slum settlements. Social inequality permeates through various channels and the differences in distribution of basic amenities spatially and socially forms the foundation of the inter-caste and religion-based disparity which transcendences into standard of living and health sectors of the inhabitants. Women domestic workers are already lying at the bottom of economic and societal structure owing to their unremunerative and invisible work pattern. Thus, their health status especially the segment of occupation-related health remains neglected and uninformed. Living with ailments and anxiety disorders might impede job performance, daily routines and hamper social life of the women workers. Overcoming health disorders is not easy for these women due to poverty, ignorance, lack of adequate institutional, social and legal protections and governmental health coverage schemes specifically targeted towards domestic workers. There is an urgent need for giving these women domestic workers visibility through legislative and definitional acknowledgment, opportunities to voice their issues collectively through unionised associations, adequate job and social security and definitive holistic public health policy schemes. In the absence of suitable urban planning measures by the concerned authority and effective participation at the grass-roots by all population groups, the disparity and discrepancies in standards of living, well-being and quality of life in general would otherwise be aggravated.

References

Ala-Mursula L, Vahtera J, Kouvonen A et al (2006) Long hours in paid and domestic work and subsequent sickness absence: does control over daily working hours matter? Occup Environ Med 63(9):608–616

Camporese R, Freguja C, Sabbadini LL (1998) Time use by gender and quality of life. Soc Indic Res 44(1):119–144

Chandrasekhar S, Mukhopadhyay A (2008) Poverty gender and youth, multiple dimension of urban well-being: Evidence from India. Working paper No. 11, Population Council, New York, pp 1–23

Duflo E, Michael Greenstore M, Hanna R (2008) Cooking stoves, indoor air pollution and respiratory health in rural Orissa. Economic and Political Weekly, Special article, pp 71–76
Ghai D (2002) Decent work: concepts model and indicators. Discussion paper, Series 139, International Institute of Labour Studies, Geneva, pp 1–42
ICSU group (2011) Health and well-being in the changing urban environment: a systems analysis approach. In: An interdisciplinary science plan, Report of an ICSU Planning Group, pp 1–42
ILO (2010) Decent work for domestic workers. Report IV (1). In: International Labour Conference, 99th Session. Geneva
NBO (2013) Status of slums in India. A statistical compedium. Government of India, Ministry of Housing and Urban Poverty Alleviation, National Buildings Organization, New Delhi, pp 1–217
Neetha N (2013) Paid domestic work, making sense of the jigsaw puzzle. Econ Polit Wkly XLVIII(43):35–38
Negi A (2014) Social movements, governance and domestic work: the Indian experience. In: 14th EADI general conference, Bonn
O'Neill G, Ross MM (1991) Burden of care: an important concept for nurses. Health Care Women Int 12(1):111–121
Satapathy BK (2014) Safe drinking water in slums, from water coverage to water quality. Econ Polit Wkly XLIX(24):50–55

Chapter 17
Urban Growth and Environment and Health Hazards in Kathmandu Valley, Nepal

Bandana Pradhan, Puspa Sharma and Pushkar K. Pradhan

Abstract This paper deals with the environment and health hazards with respect to urban growth in the Kathmandu valley cities of Nepal. Air and water pollutions and poor sanitation are the most serious public health issues in the cities of Kathmandu valley. The urban growth in the valley is very rapid and urban expansion has taken place haphazardly encroaching upon prime agricultural land, public land, open spaces, riverbanks, and forestland. The urban environment continues to deteriorate and provision of urban utility services is quite inadequate as to the need of alarmingly rising urban population in the valley cities. The health of the people living in the valley's urban areas is affected by the pollution of air, water, and sanitation, as well as by a huge loss of greenery and watershed coverage. However, urbanization is essential for social and economic development in Nepal and so, well-being of the people, if it is well planned and the existing urban policies and acts implemented sincerely and dedicatedly.

Keywords Urbanization · Air and water quality · Sanitation · Health hazard

17.1 Introduction

Urban environment and urban utility services are major concerns of the health and well-being of the people living in the urban areas of Nepal. Urban environment includes greenery, land use, air, water bodies such as rivers, lakes and ponds, watershed, groundwater, energy sources (cooking and heating), solid and liquid wastes, and climate and weather in the urban areas, whereas urban utility services comprise drinking water, electricity, transport services, health, communications, education, etc. Both national and urban governments have facing a serious challenge to deal with the provision of these facilities concerning the health and well-being in the urban areas of Nepal, though urbanites enjoy an advantage in health relative to rural dwellers. This is mainly due to the fact that urban areas in Nepal are growing rapidly.

B. Pradhan · P. Sharma · P. K. Pradhan (✉)
Tribhuvan University, Kathmandu, Nepal
e-mail: pushkar@reachpuba.org

R. B. Singh et al. (eds.), *Urban Health Risk and Resilience in Asian Cities*,
Advances in Geographical and Environmental Sciences,
https://doi.org/10.1007/978-981-15-1205-6_17

The annual urban growth rate during the 2001–2011 census was 3.43%, to which the urban bound migration has contributed largely (CBS 2014a). But the rapidly growing urbanization has taken place haphazardly, which is said due to weak efforts of both national and municipal governments for implementation of the urban policies, local governance acts, and municipal plans. As a result, the urban environment on the whole continues to deteriorate and delivery of urban utility services is quite inadequate as to the need of alarmingly rising urban population. Several diseases related to polluted environment appear to prevail, affecting the health of urban dwellers. Yet, there is lack of disintegrated data on the attributes as mentioned above at individual urban level. So, we intend in this paper to describe environment and health hazards with respect to urban growth in the Kathmandu valley cities, based on available data.

17.2 Data and Methods

Designated urban areas refer to incorporated municipalities in Nepal. They are defined as an area comprising several agglomerated settlements having minimum population of 10,000 in the Hills and Mountains and of 20,000 in the Tarai region of Nepal (CBS 2012). The population census of Nepal is the main source for describing demographic, social, and economic attributes of the urban areas. The population census is carried out at every 10 years interval and the last census was carried out in 2011. This is a public source and anyone can access to its digital data. This paper deals with 58 urban areas on which data and information were available in the 2011 population census[1] as well as in previous census years. Urban environment and health have been described including land use change in urban areas, climate and weather (temperature and precipitation), air quality, water quality, urban wastes and sanitation and their health impacts. Secondly, the 1996 digital data have been acquired upon paying from the Survey Department, the Government of Nepal. Thirdly, climate and weather digital data have been acquired from the Department of Hydrology and Meteorology, the Government of Nepal. Lastly, the data and information related to this paper have been acquired from individual studies and reports.

Though the population census of Nepal provides data on demographic, social, and economic features and some utility facilities of the urban areas, there is extremely lacking of disaggregate data on the attributes related to major aspects such as environment, utility services, and human health at the individual urban level. Further, there are indeed available several studies on various aspects including those three

[1]There were 58 urban areas (municipalities) in 2011. After this, 235 new municipalities have been added five times as the government changed, thus making it up to a total of 293 by 2016. However, no data are available on population and other urban features of those newly inducted municipalities. Further, municipalities are designated mainly by the political decision and so it often ignores functional criteria and therefore comparisons can be misleading. Municipalities are the lowest level of administrative and political entity and there are three levels of municipalities such as municipality, sub-metropolis, and metropolis and their numbers now, respectively, are 276, 11, and 6.

major aspects in the cities of Kathmandu Valley as compared to other cities in Nepal and thus, this paper draws on data and information related to environment and utility facilities of five urban areas of Kathmandu valley, which are available in the 2011 census, whereas data and information in regard to the health status of the inhabitants and their measures of those urban areas have been acquired from the health-related documents and specific individual studies. At the national level, urban areas are characterized by diversified geographical features such as mountain, hill, and plain, which also accordingly exhibit different socioeconomic conditions.

Noted that the data on water and air pollutions are sites location based, indicating pollution level of those sites. However, these mean not only to pollution of those sites, but also their surrounding areas, which is because of the contribution or effects of surrounding areas. For instance, in case of water pollution, it refers to condition of the area lying between two sites along the river and so it does not include area outside of those two sites, as the condition of water quality may be affected-improved or polluted. Thus, the sample sites were determined covering entire length of river. In case of air pollution level, the recording site may be affected by flow direction of local wind and other obstruction factors. In the bowl-shaped Kathmandu Valley surrounded by high hills, local wind directions change generally between summer and winter seasons. During summer, it is usually from east-west and south-north, whereas during winter it is west-east and north-south. Inverse temperature condition often occurs during the winter months and so, the condition of surrounding foothills and higher slopes remains to be warmer and cleaner air than the valley's floor.

17.3 Findings

17.3.1 Kathmandu Valley: Physical Settings

Kathmandu valley with coordinates of 27°32'13"–27°49'10"N and 85°11'31"–85°31'38"E lies in the central hill region of Nepal (Fig. 17.1a). The bowl-shaped valley is a tectonic tertiary structural basin, made up of fertile lacustrine sediments and is encircled by hills on all sides with above 1,800 masl, while the valley floor has an average elevation of 1,300 masl (Fig. 17.1b). Administratively, the valley comprises three districts, viz., Bhaktapur, Kathmandu, and Lalitpur, the area of which is 899 km^2. Kathmandu and Bhaktapur districts lie completely within the valley, while some part of Lalitpur district lies outside of the valley. The valley comprising all its 20 municipalities[2] has an area of 475 km^2 (19 × 25 km). The climate of Kathmandu valley is of subtropical temperate regime with relatively a well-balanced climate. The mean annual temperature is 18.4 °C with mean maximum of 25.3 °C and mean minimum of 11.4 °C (DHM 2015). However, the temperature varies greatly between

[2]This includes existing five large municipalities, viz., Bhaktapur, Kathmandu, Kirtipur, Lalitpur, and Madhyapurthimi and other 15 municipalities, which were newly inducted after 2011 population census.

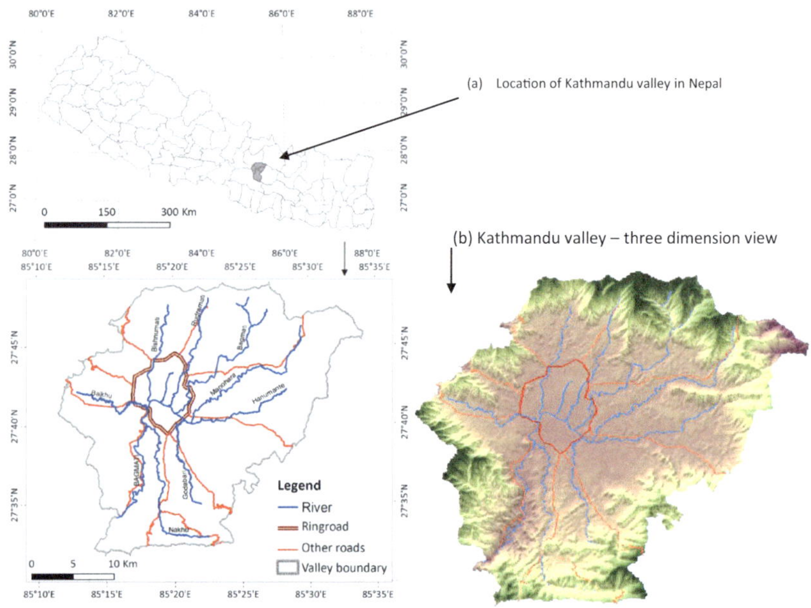

Fig. 17.1 Study region: Kathmandu valley, Nepal (Authors' construct; Sources: Survey Department's toposheets 1996 and Goggle image 2017)

summer and winter seasons. The average temperatures during the summer season range from 18 °C to 28 °C with a maximum of 32 °C in April and during the winter season the average temperatures vary from 2 °C to 20 °C with a minimum of zero or sometimes minus degree in January. The mean annual precipitation is 1,667 mm, over four-fifths of which fall during June-August (DWIDP 2009). The weather across all the year round is pleasant, which is one of the reasons to attract people from different parts of the country, as well as from other countries of the world. The forest coverage represents nearly 35% of the valley area, most of which is mainly confined to the surrounding hill slopes (Pradhan and Sharma 2016).

The Bagmati River is the main drainage system of Kathmandu valley. Its tributaries are the Bishnumati, Manahara, Hanumante, Dhobi khola, Nakkhu khola, Balkhu khola, Godavari, and Tukucha; all flow through the valley and flow out of the valley at Chobhar point to the south.

17.3.2 *Urban Growth and Distribution in Kathmandu Valley*

(i) *Historical account of urban growth up to 1941*

Early towns including all five major towns and other traditional settlements of Kathmandu Valley's are one of the oldest human settlements in Asia (Depuis 1962). Their development in the valley is believed to have taken place probably during the period ca 300–800 AD and most of the prominent early settlements appear to have enlarged and consolidated by the beginning of the thirteenth century (HMG/UN/UNESCO 1975).

It is worthwhile to explore historical account of population growth of early settlements in Kathmandu Valley. Though historical record of population census in the very beginning is not available, counting of houses was practised prior to the Lichchhavi period (300–800 AD). This can be evidenced by the fact that there was prevalent of "*Guthis*" (Cooperative Societies) in the valley, which sought help from each and every household in time and need (Kansakar 1974). Prior to the year 1769 when the Kathmandu Valley conquered by Prithvinarayan Shah, the number of houses in the towns of Kathmandu, Patan (Lalitpur), and Bhadgaun (Bhaktapur) was estimated at 22,000, 24,000, and 12,000, respectively (Bishop 1952). During the regime of Bhimsen Thapa (1806–1838), there was an enumeration of adult males, which was possibly with a view to recruiting them into the army (Giuseppe 1799). According to Hodgson (1874), the population of Kathmandu Valley at that time was estimated to be 350,000, indicating already densely populated.

In Nepal, population census was carried out for the first time in 1911, which enumerated the population for the Kathmandu Valley districts at 290,879 that increased to 306,909 in 1920 census (Kansakar 1974). The valley's population declined to 262,233 in 1930 census but again increased to 325,139 in 1941 census. During the 1911–1920 census, the annual growth rate of population in the valley was 0.6%, which though is said to be very small. But during the same year's interval, there was a negative growth rate of population at –0.13% in the country. The annual growth rate of population across all districts of Nepal continued to decline (–0.07%) during the 1920–1930 census, which in the valley's districts was much larger rate (–1.57%). There were then positive growth rates of population in both the valley and the nation at 1.95 and 1.16%, respectively, during the 1930–1941 censuses.

(ii) *Urban population sharing*

The census of 1952/1954 was the benchmark of modern scientific census in Nepal, which enumerated population for both urban and rural areas. Since then, the share of urban population to the total population of Kathmandu Valley districts varied largely by different censuses.[3] From 1952/1954 to 1981, for instance, the population living in the urban areas of the valley was below 50%, which, however, increased to over 50% in 1991 census and to almost 61% in 2001 census (Table 17.1).

[3]In Nepal, the definition of urban areas or incorporated municipalities based mainly on population size differed largely among the population censuses. In census of 1952/1954, the urban area was defined with a population of over 5,000. In 1971 census, the population size criteria for designating urban area increased to 10,000 and over, but by 1976 it was reduced to 9,000. Again in 1994, the minimum population size for the municipality remained to be 10,000. Based on this threshold population size, the number of designated urban areas reached 58 in 1996 and the number continued to remain the same until 2011 census. The "Local Self-Governance Act 1999" redefined the municipalities with a minimum population size of 10,000 for the Mountain and Hill regions and 20,000 for the Tarai region (HMG 1999).

Table 17.1 Sharing of urban population in Kathmandu valley's district population

Census year	Kathmandu valley population			National population (000)		
	District	Urban	Urban sharing (%)	Total population	Urban population	Urban sharing
1952/1954	410,871	196,800	47.90	82,570	2,383	2.9
1961	459,990	218,100	47.41	94,130	3,362	3.6
1971	618,911	249,600	40.33	115,560	4,619	4.0
1981	766,820	363,500	47.40	150,230	9,567	6.4
1991	1,105,379	598,500	54.14	184,910	16,957	9.2
2001	1,645,091	996,000	60.54	231,510	32,279	13.9
2011	2,517,023	1,465,000	58.20[a]	264,950	45,238	17.1

Source CBS (2014a)
[a]According to the 2011 census, population of the valley's all 20 municipalities combined comes to 2,420,505, sharing about 96% of Kathmandu valley district population

The Kathmandu valley is the largest urban agglomeration in Nepal. Juxtaposing population of the valley's five urban areas, viz., Bhaktapur, Kathmandu, Kirtipur, Lalitpur, and Madhyapurthimi shares nearly one-third of the national urban population and nearly three-fifths of the valley total population. With over 58%, the valley's urban population is preponderantly large over the national urban level at 17.1%. The national urbanization level is said to be one of the least urbanized countries in Asia. The valley is the hub of Nepalese urbanization and dominates the national urban system in terms of relatively greater importance of road connectivity, economic circulation, social diversity, public organizations, urban-rural linkages, etc. Despite being the largest metropolitan as well as capital city of Nepal, Kathmandu has its environmental condition being most deteriorated.

(iii) *Urban size and distribution*

The urban population of Kathmandu valley has evolved eminently over the last decades. The valley has the largest sharing of urban population to the national total urban population. Though the valley's supremacy in urban population sharing has maintained throughout all censuses, its relative share has declined consistently. For example, the valley's urban population share declined to 32.4% during the last census 2011 from 83% during the census of 1952–1954 (Table 17.2). This is due to relative growth of population of other large cities[4] in several parts across the country.

According to 2011 census, Kathmandu with over 1 million population is undoubtedly the largest city in Nepal, whereas Lalitpur with population of 227 thousand is said to be the third-largest city, after Pokhara (265,000). According to the national level, cities with population of over 100,000 can be defined as "large" and those with

[4]They include Pokhara (Western Hill), Hetauda (Central Hill) and Dharan, Itahari, Biratnagar, Janakpur, Kalaiya, Birganj, Bharatpur, Butwal, Nepalganj, Dhangadhi, and Mahendranagar (across Tarai region).

Table 17.2 Growth of size of urban population of valley's five major cities by census year

Municipalities	Set up year	Population ('000)						
		1952–1954	1961	1971	1981	1991	2001	2011
Kathmandu	1919	106.6	121.0	150.4	235.2	421.3	671.8	1,003.3
Lalitpur	1950	42.2	47.7	59.0	79.9	115.9	163.0	226.7
Bhaktapur	1997	32.3	33.9	40.1	48.5	61.4	72.5	83.7
Madhyapurthimi	1918	8.7	97.2	–	–	–	47.8	84.1
Kirtipur	1997	7.0	5.8	–	–	–	40.8	67.2
Valley urban population	Size	196.8	218.1	249.6	363.5	598.5	996.0	1,465.0
	Sharing (%)	82.6	64.9	54.0	38.0	35.3	30.9	32.4
National urban total		238.3	336.2	461.9	956.7	1695.7	3,227.9	4,523.8

Source CBS (2014a)

below 100,000 as "medium" (Pradhan 2013). Accordingly, valley's three municipalities, viz., Bhaktapur, Madhyapurthimi, and Kirtipur can be called as "medium sized cities" as they have population size between 50,000 and 100,000 (see Table 17.2).

There is a huge gap between the smallest city and the largest city in the valley. In 2011, this gap in the population size between Kirtipur and Kathmandu was nearly 15 times. Further, there exists huge gap even between population size of all four cities and Kathmandu alone. The total population size of four cities represents almost 32% as compared to 68% of Kathmandu in the valley's total urban population. In other word, the population size of Kathmandu is over two times to the population size of four cities. This gap in urban population sharing between Kathmandu city and the four cities has, however, been variable over the past decades. During the census of 1952–1954, the gap was 1.2 times that rose to 2.4 times in 1991, declined to 2.1 times in 2001 and again grew to 2.2 times in 2011.

(iv) *Urban density*

Kathmandu valley has a highest density of urban population in Nepal. It has urban population density of 15,098 persons per km^2, which is 11 times over the national urban density of 1,381 persons per km^2 (Table 17.3). Figure 17.2 shows four classes of urban density, which varies from core to outer area of the valley. Among the urban areas, the density with 4,551 persons per km^2 of Kirtipur is the lowest whereas the density with 20,289 persons per km^2 of Madhyapurthimi is the highest. Three cities—Kathmandu, Lalitpur, and Madhyapurthimi—have population density with above 10,000 persons per km^2 (CBS 2014a).

(v) *Urban growth*

There is a fluctuation in average annual urban growth rates in the Kathmandu valley as well as in the country during the past six decades. Figure 17.3 depicts that the urban growth rates in the valley remained to be less than those in the country from 1952–1954 to 2001. During the 2001–2011 census the urban growth rate in the valley

Table 17.3 Density of population of Kathmandu valley municipalities

Municipalities	Area km^2	Persons per km^2						
		052–54	1961	1971	1981	1991	2001	2011
Kathmandu	15.15	7,035	7,988	9,928	15,522	27,806	13,586	14,966
Lalitpur	6.56	6,430	7,273	9,001	12,176	17,662	10,758	12,753
Bhaktapur	11.11	2,909	3,049	3,610	4,363	5,527	11,058	7,574
Madhyapurthimi	49.45	175	197	0	0	0	4,298	20,289
Kirtipur	14.76	477	391	0	0	0	2,767	4,551
Valley urban	97.03	2,028	2,248	2,572	3,746	6,168	10,265	15,098
National urban	3,276.28	73	103	141	292	518	985	1,381

Source CBS (2014a)

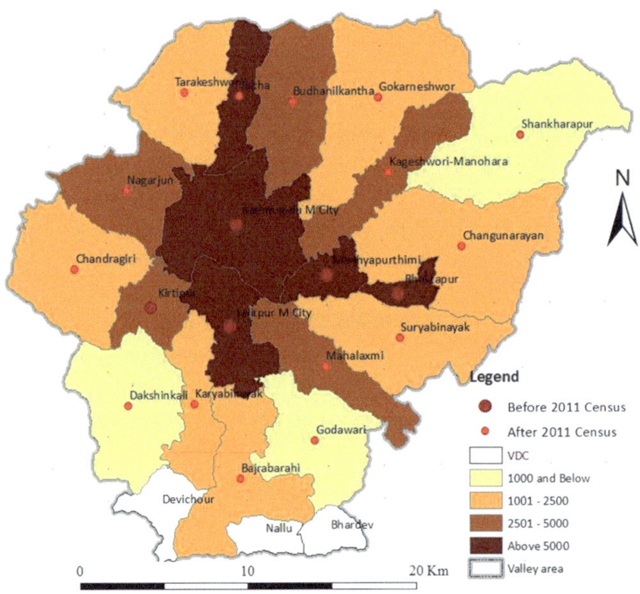

Fig. 17.2 Classification of urban population density, Kathmandu valley (*Source* CBS 2012)

with nearly 4.89% per year has surpassed the national annual urban growth rate of 3.43%. The current national urban growth is said to be one of the highest among the Asian megacities (World Bank 2013).

The growth of urban population in the valley has been rapid over the past six decades. During the 1952–1954 census, the valley's urban population was 197 thousand that grew to 1,465 thousand in 2011, showing an increase by over seven times during the past six decades (Table 17.4). However, there are varying rates of growth of population among the five cities in the valley. Of these, Madhyapurthimi, the oldest municipality being designated in 1918, has the highest growth of population

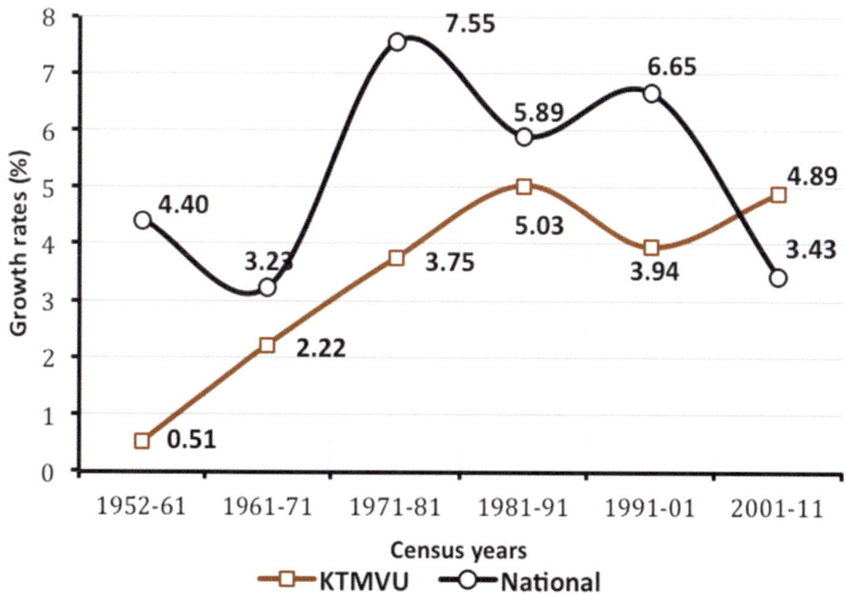

Fig. 17.3 Urban growth rates in Kathmandu Valley and country (*Source* CBS 2012)

with 9.7 times during the past six decades from 1952 to 2011 (CBS 2014a). Next to it is Kirtipur municipality with 9.6 times in growth of population between 1952 and 2011. Madhyapurthimi was relegated into non-municipality in 1971 census and again designated as municipality in 2001 census, while Kirtipur was designated into municipality only in 1997. In terms of designation of municipality by year, Kathmandu was second oldest city as it was designated as municipality in 1919. The population of Kathmandu has continued to grow and risen by 9.4 times during the past six decades. Bhaktapur city has the slowest growth of population at only 2.6 times during the 1952–2011 decades.

The growth rates of valley's urban population also vary largely among its five cities. Table 17.4 depicts that Kathmandu city has continually recorded highest urban growth rates during all census years from 1952/1954 to 2011, while Bhaktapur has shown the lowest growth rates during all those census years. Only in Kirtipur, the urban growth rate was negative during 1952/1954–1961 decade.

In Nepal, three important factors of urban growth are designation of new municipalities, expansion of municipal boundaries by amalgamating surrounding rural areas and growth and migration of population (MOUD 2015; Pradhan 2013). As stated above, after 2011 new municipalities have been added five times making it a total of 293 in the country up to 2016 and those newly inducted municipalities have been made by annexing several surrounding rural areas. In case of Kathmandu valley too, all of rural towns of Kathmandu and Bhaktapur cities within the valley have now been turned into designated municipalities, making it a total of 20. In so doing, the

Table 17.4 Annual population growth rate (%), Kathmandu valley urban areas (KTMVU)

Municipalities	1952–1961	1961–1971	1971–1981	1981–1991	1991–2001	2001–2011	1952–2011
Kathmandu	1.35	2.43	5.64	7.91	5.95	4.93	4.70
Lalitpur	1.31	2.38	3.53	4.51	4.07	3.91	3.29
Bhaktapur	0.48	1.84	2.08	2.67	1.81	1.53	1.74
Madhyapurthimi	1.23	–	–	–	–	7.62	4.43
Kirtipur	−1.81	–	–	–	–	6.45	2.32
Valley average	0.51	2.22	3.75	5.03	3.94	4.89	3.39
Nation total	4.40	3.23	7.55	5.89	6.65	3.43	5.19

Source CBS (2014a)

Kathmandu valley now has become almost an urban valley having a total population of 2,420,505, sharing about 96% of the Kathmandu valley district population.

Migration of population[5] has been an important driver of the rapid urban growth in the Kathmandu valley over the years. During the decade of 1961–1971, the population growth in the Kathmandu district alone was 171%, of which 151% was contributed by the migration. This overwhelming flow of migrants included both Nepalese and foreigners, particularly Indians (Skeldon 1985; Kansakar 1985). The valley's urban population between 1981 and 1991 grew by over 82%, in which migration contributed 59% (Pradhan 2004). During the 1990s, migration contributed 40% to the urban growth in Kathmandu valley as compared to nearly 30% in Nepal's total urban growth. Of the valley's cities, the migrant populations to the Kathmandu and Lalitpur cities were 46 and 35%, respectively. In 2001, migrant population constituted nearly two-fifths of the total population of the urban valley, the largest ever since the 1950s (CBS 2003). In 2011, the net inflow of migrants in the valley accounted for 36% of its total urban population, which is said to be the largest net inflow of migrants among all urban areas in Nepal (CBS 2012; ADB 2010).

So, Kathmandu valley is an important place for pulling more and more migrants for economic, social, and political opportunities available in the valley. Of these, employment opportunity has been the largest reason (25%) for in-migration to Kathmandu, followed by education, health, and other reasons (Pradhan 2003). For instance, there has been increased in the government expenditure on urban infrastructure development such as water pipelines, electricity, sewerage system, telephone and street network, land and housing, and the construction of ring road, which all have given ancillary employment opportunities. Other factors include growth in foreign aid, tourism, industries, hotels, and vehicular traffic and expansion of external transport lines linking the valley with the country's other parts. The valley owns over 2500 institutions of education, government, and private companies, over 173 health service institutions and over 260 standard hotels and resorts. The valley has over 950 km of road and over hundreds of thousand vehicles move every day (Thapa and Murayama 2010).

The rapid growth of the cities in the Kathmandu Valley has brought about several environmental and socioeconomic consequences as well. Urbanization in itself is not necessarily a problem, as it promotes economic activities and helps to intensify agricultural production and other local products in the hinterland areas. But haphazard and uncontrolled urban growth generally invites many environmental and health problems.

[5]According to the population census of Nepal (CBS 2003), a person of a particular place (district, municipality or village) borne elsewhere in other districts within the country and foreign places is said to be "migrant".

17.3.3 Urban Environment

Urban environment in Nepal is defined to include climate and weather, air, water bodies (rivers, lakes and ponds, groundwater), energy sources (for cooking and heating), wastes (liquid and solid), and land use in urban areas.

The environment in Kathmandu valley has been degraded over the years and is considered more severe than in any other large cities of Nepal. The environmental concerns are reaching critical levels in the valley. The problems are related to land, wastes, water, air, and noise pollutions, coupled with encroaching upon prominent agricultural land, flood plains and banks of the rivers and open spaces. Managing these problems remains difficult due mainly to the lack of urban land use zoning acts and lack of commitment of the concerned government agencies in implementing existing environmental protection acts. The urban infrastructure and utility services are invariably inadequate and poor in quality. There exists a huge gap between investment needs and financing and implementation capability of the urban governments. These all combined intensify their impact on the environment and the quality of urban life.

17.3.3.1 Urban Land Use Change

Land use is a fundamental measure of how the environment is organized in a setting (Axinn and Ghimire 2011). Over time as the population changes, as the economy grows, and as the government infrastructure spreads, land use is likely to be transformed in many ways, for instance, the conversion of agricultural land to land for housing and other non-agricultural enterprises, the reduction of greenery lands, and the intensification of agriculture system.

There has been a remarkable change in urban land uses in the valley over the past decades. The built-up area has increased remarkably, while the agriculture and forest areas declined drastically. The built-up area has consistently grew over the past decades. Two decades ago in 1996, it occupied merely 4.2% of the valley's total area (Table 17.5); it grew to 14% in 2000 (Thapa and Murayama 2009) and to 23% in 2014 (Table 17.5). Between 1996 and 2014, the agriculture land declined

Table 17.5 Change (in percent) in major urban land use categories, Kathmandu valley

Broad land use and cover categories	1996	2014
Agriculture land	62.7	44.7
Forest and shrub	32.1	31.8
Built-up	4.2	22.6
Water bodies	0.8	0.8
Others	0.2	0.1
Total	100.0	100.0

Source Survey department: toposheet 1996; Google image 2014

drastically from nearly 63% to 45%, while forest coverage shows no significant change. It signifies that the agriculture land has been badly encroached by urban built ups and the encroaching trend yet continues. Yet, agriculture and forest are two important land uses and natural resources of the valley in terms of area coverage and environmental conservation.

Figures 17.4 and 17.5 show, how the urban sprawl has encroached upon prominent agricultural land, badlands around rivers, and open spaces in the valley in recent years. Floods have become a common phenomenon during the rainy days, often inundating those settlements built around riversides and low basin (*Dol*). Haphazard building of houses in such areas has also affected the water recharge capacity of groundwater and surface water sources. Compare to the rapid and haphazard urbanization, the current forest and shrub coverage seems inadequate to recharge groundwater and surface water sources. Migration of population including urban bound migration and internal shift from the core to periphery within the urban areas is a key driving force for urban sprawl causing change in the urban land uses of Kathmandu valley. The contribution of urban bound migration is found significant to the growth of urban population in the valley. In case of internal shift of population, studies show that the

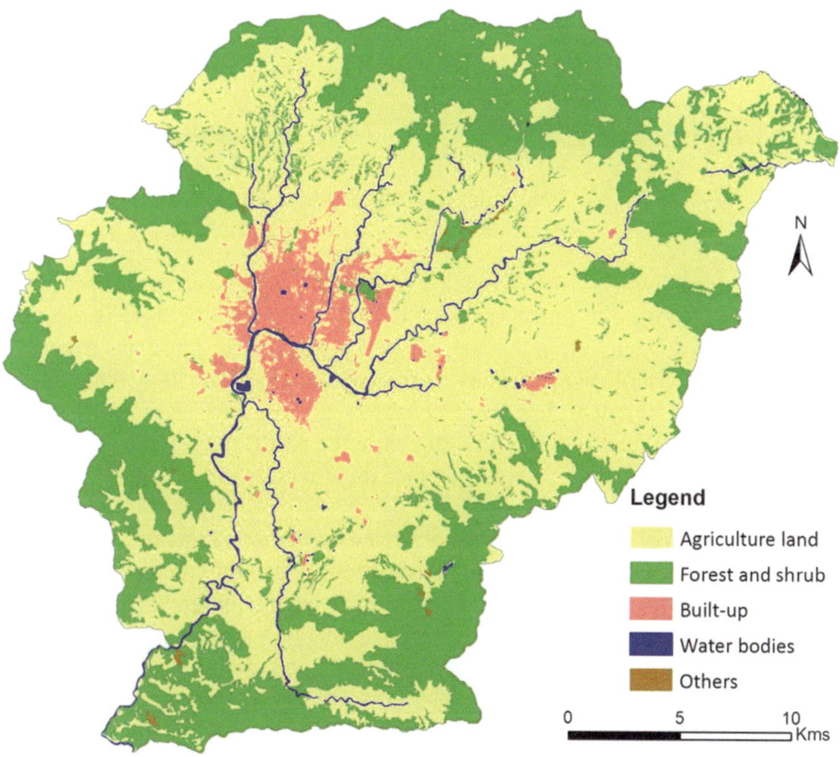

Fig. 17.4 Urban land use types, 1996 (Authors' construct, based on Survey Department's toposheet)

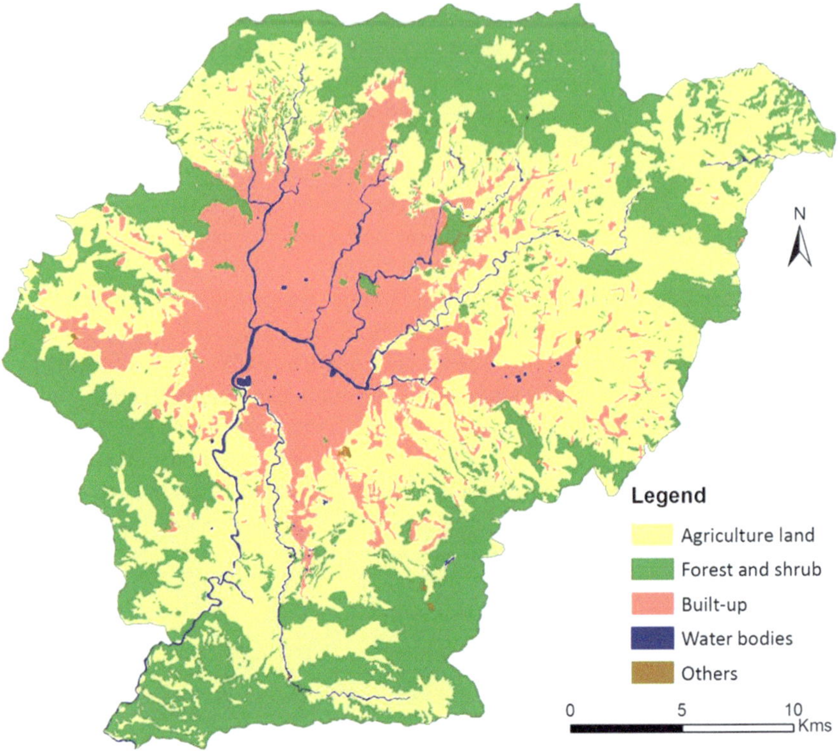

Fig. 17.5 Urban land use types, 2014 (Authors' construct, based on Google image)

built-up area grew with less than 3% per year (Pradhan 1998), whereas the periphery growth rate was over 7.5% per year (MoPE 1999). The fact that there is virtually no physical space being left in the core areas. The built-up area has thus expanded toward outward. The housing density in the core areas shows 500–1000 persons per hectare. This high densification of houses in the core areas is probably due to building height limitation, which was 80 feet in the past (Pradhan 2003). It means that the horizontal expansion has taken place more rapidly than vertical expansion of the buildings.

The urban core of Kathmandu is characterized by slums and squatter settlements. The study by Rabenau (1990) showed that number of squatter settlements increased from 17 in 1985 to 33 in 1990 and likewise their population grew from 3,000 to 15,000 during the same years. In 1990, 60% of squatting was on public land and the remaining 40% on the public buildings, temple sites, and rest houses. About 46% of the squatter settlements were on river banks (MoPE 1999). The number of squatter settlements further grew to 44 with population of 11,250 in 2002 (Baniya 2002) and 45 in 2008, occupying mainly the public land, shrine complexes, and river banks, which are ecologically sensitive, and with deficient of basic facilities (Pradhan 2008; LSGS 2008). All the squatter settlements and some of the core areas are said to be slums due to lack of basic sanitation and utility facilities. Yet, the squatting activity appears to be increasing in the valley in the recent years.

On the other hand, the valley is a legendary for its arts and culture and possesses seven protected monument zones, which have been inscribed in the UNESCO World Heritage (ETG 2012). But the aesthetic values of the temples and shrines of historical importance are now increasingly being threatened because of losing of their original traditional arts and culture.

17.3.3.2 Climate and Weather

(i) *Temperature*

Kathmandu valley has experienced change in climate over the past several decades. There has been a rising temperature thus warming climate in the valley over the decades as in other parts of the country. Figure 17.6 depicts that the trends of both annual average maximum and minimum temperatures in the valley have risen at 0.066 °C and 0.035 °C, respectively, over the past five decades from 1971 to 2014 (DHM 2017). The mean annual trend of temperature has also risen with 0.0662 °C. The rising annual average trends of temperatures in the valley show higher over the country's annual trends of average maximum and minimum temperatures at 0.056 °C and 0.002 °C, respectively.

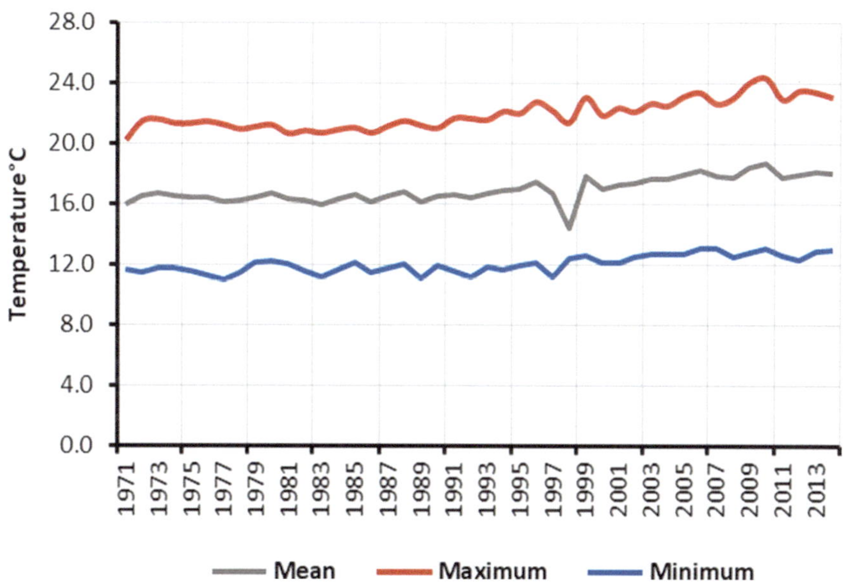

Fig. 17.6 Annual average temperature trend, Kathmandu Valley (*Source* DHM 2017)

(ii) *Precipitation*

The annual average precipitation over the last decade from 2004 to 2014 in Kathmandu valley was 1,740 mm. There has been a decreasing trend with 66 mm per year in the valley (DHM 2017). This decreasing trend of precipitation can be related to increasing temperature. Based on the trends over the past five decades, it is calculated that for each degree increase in temperature, there will be a decrease of 365 mm rainfall (DHM 2017).

As a consequence, the weather condition over the Kathmandu valley has changed considerably—causing warmer condition. One best instance is a disappearing of morning fog during the cold winter months in the valley that has happened since the recent decades. Further, the weather condition often fluctuates, unexpectedly warming in both hot summers and cold winters and appears to be unpredictable due to climate change. These phenomena of change in climate and weather can be said due to rapidly changed in urban land use coverages, rapidly growing of urbanization and population, rapidly increasing of road network and vehicles, industrial smokes, burning of plastic materials, and so on (MoPE 2017).

17.3.3.3 Air Quality

Good health and well-being of the people requires clean air. Polluted air causes health hazard. Here, air pollution includes indoor and outdoor or ambient air pollution. Type of fuel or energy used for cooking and heating is the main cause to indoor air pollution. In the valley's cities, the use of energy sources for cooking and heating by households contributes to indoor air pollution. Table 17.6 shows that burning of fuel wood, cow dung, kerosene, and others contribute to both ambient and indoor air

Table 17.6 Sharing (%) of source type of drinking water by Valley's urban households

Municipality	Piped water	Tube well	Covered well	Uncovered well	Spout tap	River	Others
Lalitpur	61.2	1.5	11.1	1.7	5.0	0.1	19.6
Madhyapurthimi	66.2	7.0	10.3	3.0	6.8	0.0	6.8
Bhaktapur	90.7	0.6	2.5	1.2	2.0	0.0	2.9
Kathmandu	64.2	7.3	4.3	0.5	1.9	0.0	21.7
Kirtipur	75.8	0.2	2.0	0.3	3.9	0.0	17.8
Valley total	65.8	5.7	5.4	0.9	2.7	0.0	19.5
Kathmandu average	70.7	3.7	5.9	1.3	3.7	0.0	14.7
National urban (%)	59.2	24.5	3.4	1.6	2.9	0.3	8.1

Source CBS (2014a)

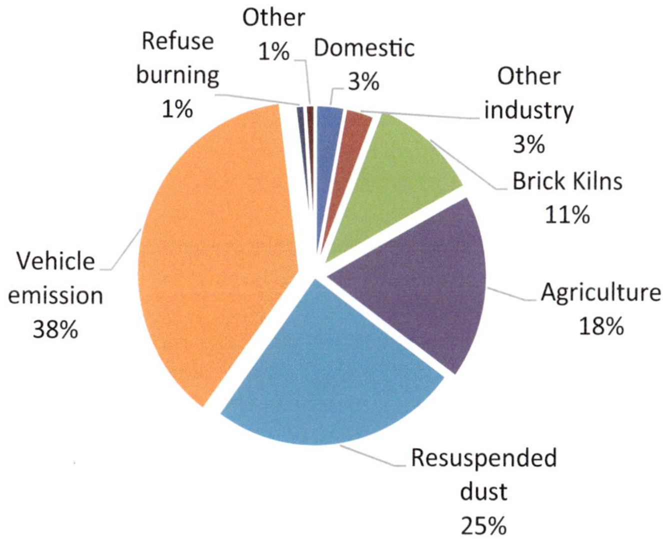

Fig. 17.7 Sources of PM10 in Kathmandu Valley (*Source* CANN/CEN 2014)

pollutions, which combined constitutes about 7% of the total energy consumption, while the remaining with more than 92% use biogas.

In the valley's cities, ambient air pollution is basically contributed by the emission from the vehicle, industrial, and burning of plastic and firewood and crop residues and dust or particles from any source (Fig. 17.7). Thus, produced pollutants can deviate the air quality from the normal composition by adding the concentration of solid, liquid, or gaseous particles in the atmosphere, which ultimately cause health hazard.

Though both natural and man-made phenomena are responsible for air pollution, the latter is more dominant than the former. In Kathmandu valley, the man-made pollutants include mainly vehicle exhaust, burning of plastic materials, brick kilns, unpaved streets, and domestic kitchens. Among the natural phenomena, the westerly wind during winter (December–February) is the major source of pollutants carried from distant lands, while in other seasons such as pre-monsoon or spring (March–May), summer or monsoon (June–September), and post-monsoon or autumn (October–November), the pollutants in the air are largely affected by the amount of precipitation across the country including Kathmandu valley (DHM 2015). Rainfall in particular is an effective agent in washing out gaseous pollutants and particulate matter present in the atmosphere (Giri et al. 2008). So, air pollution concentration in Kathmandu valley can be expected highest before beginning of monsoon and lowest after it. The ambient air quality is analyzed quantitatively by air pollution index (API), which is obtained by combining values for various air pollutants into a single measurement (UN 1997). This includes respirable particulate matter (PM) size such as $PM_{2.5}$ or PM_{10} representing particulate matter of ≤ 2.5 or ≤ 10 μm in diameter, respectively, in the air. The smaller the size of PM the greater is the risk. These particles include dust, smoke, and smog exited from burning of fuels, vehicle exhaust, and pollen plants among others used for any activities like power plants, cooking, and home heating. According to the National Ambient Air Quality Standard (NAAQS) 2012, the parameters for $PM_{2.5}$ and PM_{10} being set for Nepal are 40 $\mu g/m^3$ and 120 $\mu g/m^3$, respectively (MOPE 2017).

The valley hosts almost half of country's 1.3 million motorized vehicles in 2013 that was risen from 244 thousands in 2000. But the proportion of public transport is less than 3% (DoTM 2016). In addition, dusty roads and many unpaved and poorly maintained roads due to the snail-paced road expansion projects and delay in the underground installation of Melamchi Drinking Water pipes in Kathmandu have also contributed to high particulate loading. As indicated above in Fig. 17.7, the transport sector emitted highest proportion of PM10 (Gautam 2006; CANN/CEN 2014). About country's two-fifths industries are located in Kathmandu valley (CBS 2014a). Of these, the industries generating smoke and dust are brick kilns and stone crushers, which together contribute 14% of the total particulate matter in Kathmandu valley (CANN/CEN 2014). Kurmi et al. (2014) found that brick kilns located in the suburban areas in the valley have released toxic pollutants into the air. In the valley, bricks are the preferred material for construction of private and public buildings that causes to diminish soil quality and crop productivity (Haack and Khatiwada 2007). The bowl-shaped valley surrounded by hills restricts wind movement and retains the pollutants in the atmosphere. This is especially bad during the winter season (Nov–Feb) when thermal inversion occurs in the valley late night and early morning. Cold air flowing down from the mountains is trapped under a layer of warmer air and acts as a lid. As a result, the pollutants are trapped close to the respirable height for extended periods of time increasing the risk of air pollution to the vulnerable people (Gurung and Bell 2012).

The recent study on ambient air quality by MOPE (2017) carried out at three stations representing core (Ratnapark), middle (Putalisadak) and periphery (Pulchowk) in the valley recorded mean values of $PM_{2.5}$ at 103.5 $\mu g/m^3$, 260 $\mu g/m^3$, and 58.5 $\mu g/m^3$, respectively; all values were higher against the maximum limit of the NAAQS 40 $\mu g/m^3$. Constant yearly monitoring of ambient PM2.5 in Kathmandu Valley from 2014 to 2015 showed that the valley's ambient air exceeded the daily National Ambient Air Quality Standards (NAAQS) for the majority of days, i.e., 57%. Daily averages of PM2.5 increased 3–5 times over the national standard of 40 $\mu g/m^3$. Similarly, the concentrations of NO_2 in ambient air are also found to be higher, which was 12 times over the 24 h national standard of 80 $\mu g/m^3$ (MoPE 2012). Seasonal and monthly variations show that winter and spring months are heavily polluted with ambient PM2.5 levels (Karki et al. 2014).

Similarly, the ambient air quality measured in terms of PM10 ($\mu g/m^3$) during 24 h in dry month of May at 10 different stations[6] across the valley shows the recordings ranging from 72 $\mu g/m^3$ at Godavari, the most periphery site to 2928 $\mu g/m^3$ at Chabahil core site (MoPE 2017). All nine sites except Godavari site show the recorded particulate matter above the NAAQS 120 $\mu g/m^3$. The recorded PM_{10} concentrations varied from 1.5 to 24 times over the acceptable level throughout the valley (MOPE 2017).

A cross-sectional comparative study of air quality between workers of brick industries and grocery stores has been carried out in the Kathmandu valley (Sanjel et al. 2016). The air quality measurements included average suspended Particulate Matter (TSPM), PM10, PM 2.5, and PM1. The results show that the average concentrations of those particulate matter measurements among the brick workers were at 5.179 mg/m^3, 4.958 mg/m^3, 3.965 mg/m^3, and 3.954 mg/m^3 respectively, as compared to those average concentrations of 0.089 mg/m^3, 0.089 mg/m^3, 0.082 mg/m^3, and 0.082 mg/m^3 respectively, among the grocery workers. Thus, the concentration of particulate matter in terms of TSPM, PM10, PM 2.5, and PM1 between the exposed (brick industry) and non-exposed (grocery stores) showed 56, 56, 48, and 48 times higher in the former group than in the latter group. The study also found that the incidence of respiratory disease was significantly higher among the brickfield workers, which was due to high level of exposed airborne particulate matter in the brickfield.

According to IARC (2015), polycyclic aromatic hydrocarbons (PAHs) classified as confirmed carcinogen belongs to Group 1, which was detected in the air of Kathmandu valley with an average concentration of 16.1 ng/m^3 at 95% confidence interval (Ci = 6.4–28.6) (Pearce et al. 2015). Its high concentration was detected from coal burning especially in brick industry, vehicle emissions, and religious site with intense incense burning. The brick factory sites where extensive coal combustion was common and the market place with heavy traffic emission were associated with higher risk than other areas (Pokhrel et al. 2018).

Another PAHs study conducted by Chen et al (2015) in Kathmandu valley found that the concentration of 155 ± 130 ng/m^3 varied with seasons, with maximum average concentration during the post-monsoon season that declined consistently in other seasons such as winter, pre-monsoon, and monsoon. In the winter and pre-monsoon seasons, increase in ambient TSP and PAH concentration was due to emissions from brick kilns and use of numerous small generators. Evaluation of diagnostic molecular ratios indicated that the atmospheric PAHs in the valley originated mainly from diesel and biomass combustion. The toxic equivalent quantity (TEQ) of particle phase PAHs ranged between 2.74 and 81.5 ng TEQ/m^3, which is considerably higher with

[6]They included locality sites such as Thamel (696 $\mu g/m^3$), Chabahil (2928 $\mu g/m^3$), Kalanki (620 $\mu g/m^3$), Minbhawan (315 $\mu g/m^3$), Lagankhel (362 $\mu g/m^3$), Sanepa (545 $\mu g/m^3$), Bhaktapur (178 $\mu g/m^3$), Budhanilkantha (336 $\mu g/m^3$), and Godavari (72 $\mu g/m^3$).

2–80 times over the World Health Organization guideline (1 ng TEQ/m^3). This suggests that ambient PAH levels in Kathmandu Valley pose a serious health risk to its residents (Chen et al 2015).

(i) *Impacts of air pollution*

Air pollution is a major environmental risk to public health. Ailments relating to air pollution such as pneumonia, bronchitis, and asthma have been common in the Kathmandu valley since the last few years. About 17% of the bronchitis case was due to indoor smoke pollution in Kathmandu valley (NHRC 2009). The risk of PM10 in premature mortality per year was estimated at 212 cases, at 95% confidence interval, ranging from 127 to 338 cases (NHRC 2004). The study on the investigation of the impact of particulate matter with an aerodynamic diameter less than or equal to 10 μm (PM10) on hospitalization in Kathmandu valley by Gurung et al. (2017) was based on the individual-level daily inpatient hospitalization data from six major hospitals from 2004 to 2007. In this study, the daily PM10 concentration was measured with an average of 120 μg/m^3 and with daily maximum of 403 μg/m^3. The study found that a 10 μg/m^3 increase in PM10 level was associated with increased risks of hospitalization at different levels, such as 1.00% at 95% confidence interval: 0.62, 1.38, 1.70% at 95% confidence interval: 0.18, 3.25, and 2.29% at 95% confidence interval: 0.18, 4.43 for total, respiratory, and cardiovascular admissions, respectively.

Analysis of respiratory health effects of 11,300 inpatient records during the years 2014–2015 from 13 major hospitals in Kathmandu valley showed respiratory diseases such as COPD at 39.5%, pneumonia at 29.1% and ARI excluding pneumonia at 15.3% (Karki et al. 2016). The comparative assessment among different age groups of their study shows that children with years of age ranging from 0 to 9 and adult people with 50 and above years of age were the most vulnerable groups to respiratory disorders, with 25.5% patients being children and around 55% being aged persons.

According to the study by CEN/ENPHO (2003), major hospital records showed steadily increasing of Chronic Obstructive Pulmonary Disease (COPD) in Kathmandu valley and the number of patients with COPD admitted to hospitals remained to be highest during the winter season, when air pollution reaches at its peak. There was a rising trend in both cardiovascular and respiratory diseases among the patients in Kathmandu from 2011 to 2016 (Fig. 17.8; IOM 2017). So, air pollution is responsible to cause both communicable and non-communicable diseases. The air pollution has affected not only the health of the people, but also tourism in the valley, which is one of the major sources of economy. The survey by CANN/CEN (2012) in Thamel, a key tourist locality, found that 54% of the tourists rated air quality as fairly bad.

17.3.3.4 Water and Sanitation

(i) *Water quality*

Drinking water quality is directly related to the health of people. The purpose of drinking water quality standard is that a supply of drinking water, which complies

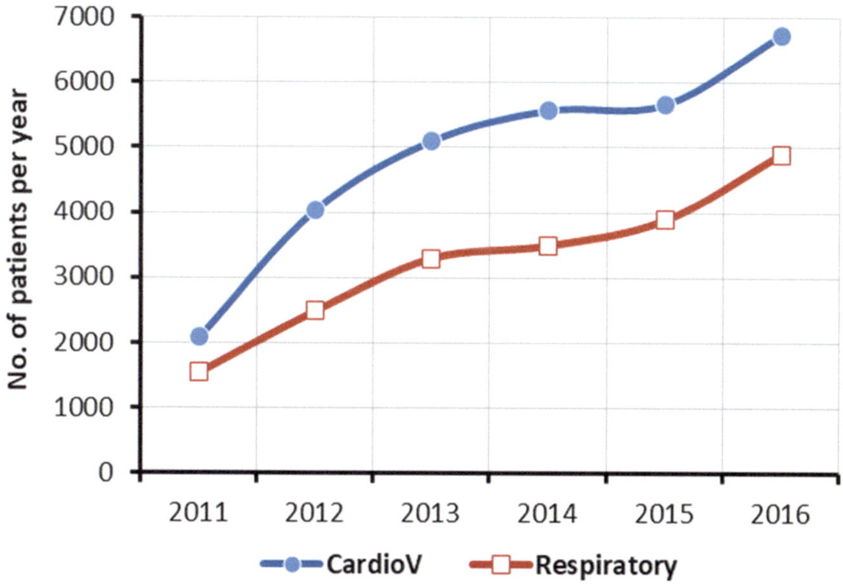

Fig. 17.8 Trend of patients' visits of tertiary hospital Kathmandu (*Source* IOM 2017)

with the standard, should not produce any adverse effect to the health of people while consuming it throughout the life (WHO 1997). The standard is always a country driven and once a country develops its standard, it should be mandatory to implement it as per its norms. Nepal has also developed its National Drinking Quality Standard (NDWQS) (DWSS 2005). However, the country is not able to implement NDWQS strictly.

The water quality of different sources such as groundwater (dug wells, deep tube wells, and stone spouts) and surface water (ponds, rivers and pipe water) in the valley is found to be contaminated at various proportions and thus not recommend for direct consumption as per the NDWQS (CBS 2015). The treatment of the water supply in the valley constituted about 80% of the total water supply (DWSS/NMIP 2014). Drinking water quality is either contaminated at the source or at the consumption point. In fact, even when the safe water at sources reaches to the consumption point it often gets contaminated due to poor maintenance of pipe network, indicating contamination during transmission or poor status of household sanitation (Pradhan et al. 2005a). Based on the analysis of water samples, the drinking water quality of Kathmandu valley in terms of basic physical and chemical parameters is found to be safe but it is found contaminated with *Escherichia coli* or coliform bacteria at consumption point or at the source (Pradhan et al. 2005b). It means the water supply is not safe to consume, but if consumed in any case, it may bring any kind of epidemics of waterborne diseases. According to the study of CBS (2015), more than four-fifths of the household members across the country including both urban and rural households were at risk of E. coli concentration $\geq$ 1 cfu/100 ml in their household water. Likewise, 71% of the household populations were at risk of E. coli due to its concentration in drinking water at source.

(ii) *Water quantity*

The quantity of water is another important component of health and hygiene. Normally, a person requires 2.5 L of water per day for consumption, so as to maintain the basic physiological processes (Tebbutt 1992). In addition to it, more water is required for other purposes such as washing, bathing, and cleaning. The Government of Nepal has classified the water need in terms of quantity per person per day into three categories such as basic, medium, and high with $\geq$ 45 L, $\geq$ 65 L, and $\geq$ 112 L, respectively (MOUD 2014).

There is variation in using drinking water by sources in the urban areas of Kathmandu valley as well as at the national level. On the whole, the access to tap water is about 71% in the valley's urban areas, which is, however, higher over the national level with below 60% (Table 17.6).

The population of Kathmandu valley grew from 1.59 million in 2001 to 2.42 million in 2011 with increasing at the rate of 5.2% per year (CBS 2014a). The water demand in the valley was estimated to be 366 MLD in 2016 and is expected to reach about 482 MLD by 2021 (Udmale et al. 2016). The available surface water from the conservation zones is estimated to be about 133.88 MLD and 199.79 MLD during the dry and wet seasons, respectively (Udmale et al. 2016). The KUKL, an authorized agency supplying potable water in Kathmandu valley, is currently harnessing about 49% and 66% of the available surface water sources during the dry and wet seasons, respectively. This shows that there is scope to harness additional surface water for potable purpose in the valley. The estimated groundwater potential in the valley is about 1116 billion liters (KUKL 2014). However, the actual volumes of water supply during the wet and dry seasons are 115 and 69 MLD, respectively. The total water demand in service areas was approximately 361.6 MLD in 2016, with a supply deficit of 210 MLD (Fig. 17.9). The present deficit is currently met through private groundwater pumping, traditional water spouts and wells, private vendors (through surface, spring, and groundwater), and bottled water industries. This signifies an over-exploitation of groundwater storage, resulting in drawdown of the groundwater level and drying of wells.

According to Raina et al. (2018), 89% of the households are connected to the public piped water system in the valley, though only 70% use water from it. This is probably because the public water supply is unreliable, irregular, and inadequate. On average, the households receive water for only 1.5 h every five-days. Therefore, the households tend to diversify their sources of water for meeting water demand for different purposes, based on price and perceived quality.

(iii) *Urban wastes*

Urban wastes in Nepal include both solid and liquid wastes. Such wastes directly concern with sanitation and hygiene of the people living in the cities. Volume of wastes is directly related to the population size of cities. As the population size of

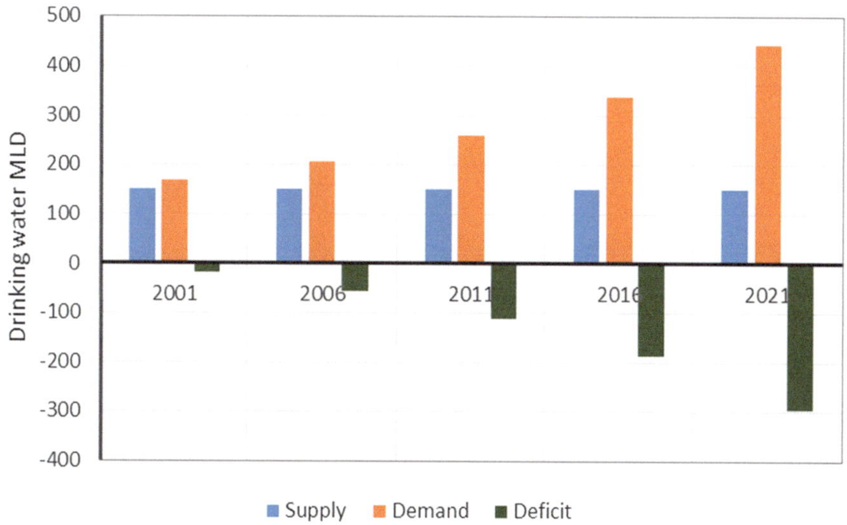

Fig. 17.9 Supply and demand condition of drinking water supply, Kathmandu Valley (*Source* Udmale et al. 2016)

cities increases, the volume of wastes also increases. In Nepal, the environmental pollution due to increasing generation of wastes is particularly a crucial problem of large cities, including the valley cities.

The urban Kathmandu valley produced 594 metric tonnes of wastes daily in 2013, grew from 62 metric tonnes in 2007. Of the total, Kathmandu city alone shared over three-fourths. This is the largest in the country's total urban waste generation (CBS 2014b). The organic waste constitutes about 70% of the valley's total urban wastes (Fig. 17.10).

Haphazard dumping of wastes is an acute problem in the valley, where one can see widespread heap of garbage littering the city streets or at dump sites on river banks, or at public places. Currently, 250 to 300 tonnes of wastes from Kathmandu valley are dumped, compacted, and covered with soil every day. However, the solid wastes being dumped into landfill sites are in most cases not environmentally safe, because the landfill sites do not have liners and leachate collection or treatment system.

In regard to liquid waste, majority of the buildings in the valley are not yet connected to the sanitary wastewater system (CBS 2014a). Domestic sewage generated by the valley cities' inhabitants makes up of about 93% of the pollution load and the per capita wastewater generation is estimated to be 90 L (DWSS 2016). In order to keep the valley cities clean, the valley contains wastewater treatment plants at seven locations, such as Dhobighat, Guheswori, Teku, Kodku, Hanumanghat, Madhyapurthimi, and Sallaghari. Of these, all six-treatment plants except the last one—the Sallaghari community based treatment plant, are now dysfunction and the cities' sewage mixes directly into the nearby rivers (Shrestha et al. 2015). Thus, all rivers flowing through the valley's urban areas are the repositories of untreated sewers.

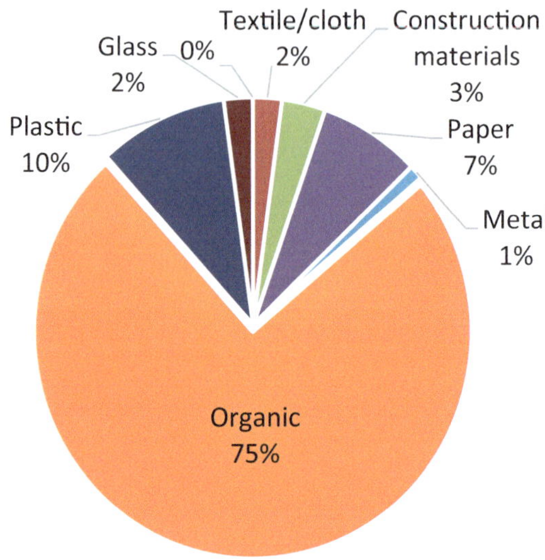

Fig. 17.10 Sharing of wastes by type, Kathmandu valley (CBS 2014)

(iv) *Urban sanitation*

Sanitation facility refers to toilet coverage in Nepal. In Kathmandu valley, the urban toilet coverage is almost 99% (Table 17.7). It means not all urban households do have access to toilets. It covers 946 households without toilets or 0.3% of the valley's total urban households (CBS 2014b).

Two important issues of sanitation coverage in the valley's cities are public toilets and households' toilet connection. The valley has one public toilet for 46,000 people, which is indeed fairly inadequate, and further, the condition of the public toilets in most cases is awfully poor in terms of cleanliness and hygiene (Ekantipur 2017). There are two connection types of urban household toilets such as toilets with septic

Table 17.7 Urban households' access to toilets, Kathmandu valley

Municipality	% Households (N = 366, 255 hhs)			
	No toilet	Flush toilet	Ordinary toilet	Not stated
Lalitpur	0.2	93.3	5.6	0.8
Bhaktapur	0.3	95.8	3.3	0.6
Madhyapurthimi	0.7	96.3	2.3	0.6
Kathmandu	0.2	92.1	6.8	0.9
Kirtipur	0.5	91.3	7.6	0.6
Total %	0.3	92.7	6.2	0.8

Source CBS (2014a)

tank and public sewerage connection. At the national level, they make up 48 and 30%, respectively (CBS 2014). Unfortunately, there is no data on both these types of connections of toilets in the valley. However, it can be estimated that the valley may have the same situation as with the national urban scenarios of toilet connections.

In Kathmandu valley, around 70% of the households dispose their excreta directly into the sewer line while remaining 30% households still depend on onsite systems such as pit latrines and septic tanks (HPCIDBC 2011). Fecal Sludge Management (FSM) is also poor and every year about 170,000 m^3 of fecal sludge in the valley was unsafely disposed into the nearby rivers (ENPHO 2015).

The Bagmati River as the main river system of Kathmandu valley, as well as the major source of water supply. The headwater area of the river is in pristine stage, but as this river together with all its tributaries flows downstream, it gets polluted in terms of biological and chemical parameters (Pradhan 1999). Further, when it passes through the urban core areas it is heavily polluted, as it also receives polluted water from its tributaries as well. The river leaves the valley without getting improvement in its quality (Pradhan et al. 2005a).

(v) *Health impacts*

Health of people is affected due to poor water, sanitation, and hygiene, causing water-related diseases such as diarrhea, dysentery and cholera, and skin diseases. Throughout Nepal, including Kathmandu valley, the incidence of diarrheal disease is found related to nature of precipitation and temperature. A highest occurrence of diarrhea cases or peak of diarrhea outbreaks was recorded during the monsoon season from June through August (characterized by peak precipitation and temperature) when heavy rainfall-runoffs transport terrestrial microbiological agents into the drinking water sources resulted in outbreaks and spreading of infectious waterborne diseases (NHRC 2009).

The analysis of average monthly incidence of diarrheal disease and rainfall pattern in the valley from 2002 to 2015 has shown a positive correlation with coefficient of correlation of 0.8, indicating high correlation between these dependent and independent variables (Fig. 17.11). Diarrheal disease often occurs with peak just after initial rain. Diarrheal epidemic and cholera outbreak also take place during this time (Dhimal et al. 2016).

There is a rising trend in incidence of diarrhea in Kathmandu valley. According to DOHS about 14% of the total patients' visits to the health care services in 2005 were related to the waterborne disease and that grew to 19% in 2015 (DOHS 2005; 2015). Figure 17.12 depicts that the incidence of diarrheal diseases among the under-five children has increased from around 90 per 1000 in 2003 to over 350 per 1000 in 2014 in Kathmandu valley (DoHS 2015).

Average burden of disease due to waterborne is increasing that accounted for 11% to the total burden of diseases in Kathmandu valley in 2017 (DOHS 2017). Cholera outbreak is also common in Kathmandu. The number of total laboratory-confirmed cases of cholera detected was 4, 80, 169 in 2013, 2015, and 2016, respectively (DOHS 2017).

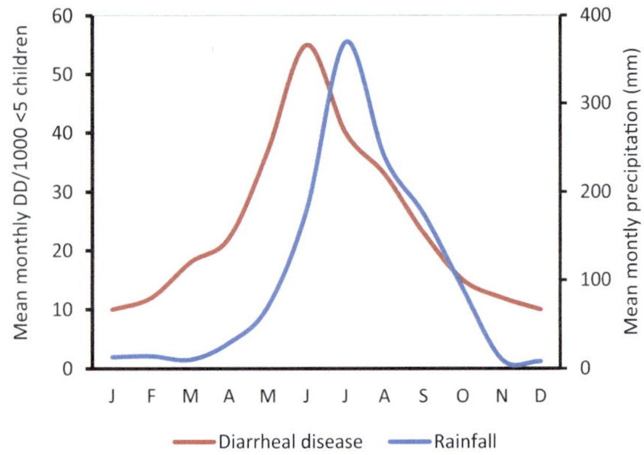

Fig. 17.11 Incidence of diarrheal diseases and rainfall pattern in < 5 children, Kathmandu valley (*Source* DHM 2017; Dhimal et al. 2016)

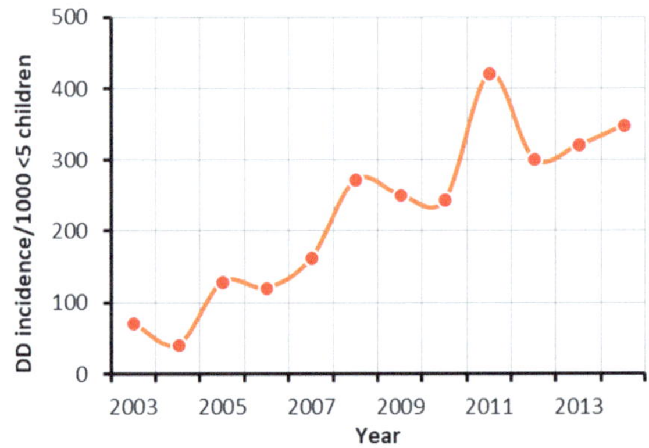

Fig. 17.12 Incidence of diarrheal disease (DD) under-five children, Kathmandu valley (DoHS 2005 and 2015)

17.4 Discussions

The town planning in Nepal has evolved eminently over the past four and half decades. The early town planning initiatives in the late 1960s focused on physical development plans (Pradhan 2014). Since the 1990s the integrated action plans remained as a major tool of town planning. In the 2000s and onward, the town planning constituted the strategic and periodic plans. Currently, the country has National Urban Development Strategy, which was initiated in 2017 (MOUD 2017). Throughout the planning period, the government has continued efforts to build urban infrastructure, but these have lagged behind the need for the rapid growth of urban population (Pradhan 2014).

In Nepal, the rapid growth of urban areas has taken place in a haphazard manner and has invited many unwanted environmental and health hazards. Yet, the urban areas have relatively better facilities than the rural areas and therefore the cities draw people towards them. The recent trend is that when people migrate, they choose larger cities like the valley's cities and others, instead of small towns. Thus, bigger cities continue to grow at faster rate than small towns, putting more pressure on available resources, infrastructure and facilities and above all, the urban environment (ADB 2010).

Rapid urbanization is intensifying the municipal infrastructure deficit. Urban sector initiatives as indicated by World Bank's study (2013) have taken 10.4% of the total budget, the third biggest share after roads (48.2%) and drinking water supply (13.4%). The latter two are somehow related to urban sector. Thus, the provision of infrastructure and facilities is less with respect to urban growth, meaning proliferation of unhealthy, poorly serviced, and infrastructure-deficient sites.

The government of Nepal has made several efforts to improving urban environment and life across the country through formulating several legislative arrangements such as Waste Water Management Policy 2006, Environment Protection Act 1996, Water Resources Act 1993, Industrial Enterprise Act 1992, Solid Waste Regulations 1989, Solid Waste Act 1987, and Aquatic Animal Protection Act 1960 (MoEST 2007). Of all, the municipal government, according to the Local Self-Governance Act-1999 of Nepal, is responsible for preparing urban plans, resource use and conservation, land use plan, etc. But yet, the municipality has virtually no legal instruments to control land use development effectively. There is no legislation that defines land use classes, land use planning, zoning planning, or similar. Above all, the Health Policy 2015 has described that each citizen shall have the right to access to clean water and hygiene.

The air pollution in the Kathmandu valley has a high level of particulate matter mainly due to the weak implementation of mass vehicle standard and not properly maintained vehicles running in the cities (MOPE 2012). Black smokes spewing vehicles are a common sight on the streets of the valley cities.

There is an increasing trend of public health expenditure in the recent years and prevention of waterborne diseases including diarrhea is a priority program of the Government of Nepal (MOHP 2009). But the public health expenditure as defined by the Department of Health Services does not include activities like emergency plans, environmental protection, water supply, sanitation, etc. though they certainly affect the quality of life (Shrestha et al. 2012). Huge expenditure is being made every year on the curative measures including diarrhea diseases. The curative care services accounted for almost half of the government's total health expenditure. For instance, the Government spending on curative care services increased by around 67% between 2007 and 2009 while on the other hand, preventive, and public health services consisted of 17% in 2009 (Shrestha et al. 2012), indicating less priority on preventative measures of diseases such as diarrhea. The incidence of diarrhea has in fact not reduced, rather constantly increased. Provision of safe water in urban areas has never got top priority. Top three diseases such as skin, diarrhea, and acute respiratory infection are among the top ten diseases (DoHS 2017). The polluted environment has also reduced the religious, recreational, and aesthetic values of the rivers.

Indeed, the Kathmandu valley was at pristine condition some decades ago. Most of the valley floor was used for agriculture and the surrounding hill slopes were covered with greenery. After the onset of political change in 2007–end of 104 years Rana regime, the valley became a focal point for investment in education, health, roads, water supply, sewerage, and other government activities and thus witnessed rapid in-migration. The arable land declined immensely at the cost of urban built ups and the inhabitants who were engaged in agriculture have shifted to other activities. All available water sources have been tapped for water supply, causing highly depletion of both surface (river) and groundwater sources. Other conspicuous instances include shrinking of greeneries, numerous sewerage lines mixing into the rivers, unpaved and poorly maintained roads and rapidly increasing motor vehicles, causing muddy streets in rainy summer season and dusts in dry winter season.

Of all cities in Nepal, the Kathmandu Valley's cities have received the most attention of the central government. The valley has the highest share of capital expenditure for municipal infrastructure (physical and social), where the infrastructure need is also greatest due to largest population size. But the per capita spending in the valley's cities is less than that in other cities and municipalities. According to World Bank (2013), the per capita capital expenditure on infrastructure in Kathmandu city in 2010 was US$ 6, the lowest among the municipalities in the country, but in Lalitpur city it was US$ 9.2, the highest among those spent in other large cities such as Biratnagar, Pokhara, and Birganj. In small municipalities, the per capita capital expenditures on infrastructure were US$ 11.0. Also the solid waste management expenditure is mostly concentrated in Kathmandu. Of total capital expenditure in the municipal infrastructure in 2008, solid waste accounted for 65% in Kathmandu, whereas it was only 2% in other municipalities. Yet, in terms of poverty level, Kathmandu city is next to Pokhara; the latter city has 1.3% below poverty line, the least poverty level in Nepal.

The Kathmandu Valley Development Authority Act 1988 has empowered its authorities to regulate the environmental situation of the Valley in a holistic manner. This gives due consideration to environmental pollution problems and recognizes the need to improve compliance with standards and to improve coordination among various stakeholders. More specifically, the Kathmandu Valley Development Authority has prepared a envision plan of the urban feature vision for the Kathmandu Valley comprising all fundamental and crucial aspects of urban planning and management by 2020.

17.5 Conclusions

The Kathmandu valley's urban areas have witnessed rapid growth by almost 280 times and the urban population by almost 500 times over the past five decades. Migration of population has been the main contributing factor, encroaching upon prime agricultural land, public land, riverbank, and forestland. The urban expansion in the valley has taken place in a haphazard manner, posing innumerable challenges. The fundamental problem now the valley facing is the lack of a modest urban land use zonation policy. Coordinated policy response to an integrated and economically vibrant and balanced urbanization process in the valley is lacking. Urban database on land, infrastructure, urban utility facilities, environment, etc. is poor and far from comprehensive and whatever available they lack in consistency.

The water supply in the valley is neither safe nor adequate. Yet there is lack of awareness among the majority of urban households about the polluted water supply, as a cause of diarrhea. The inhabitants are compelled to use or consume such polluted water, due to inadequacy of water supply. The community perception towards environmental sanitation and personal hygiene is very poor. There is an urgent concern that the government should give high priority on dealing with the diarrhea disease by intervening essential and proper preventative measures. The most hindrance with the government is the lack of seriousness for implementing whatever the policies and plans being formulated or designed.

Health of the people living in the urban areas is affected by the pollution of air, water, and sanitation, as well as by a huge loss of greenery and watershed coverage. Air and water pollution and poor sanitation are the most serious public health issues in the cities of Kathmandu valley. The recently adopted "Constitution of Nepal 2015" has mentioned about to ensuring citizens right to live in healthy environment through effective control environmental pollution for health protection and promotion. This is indeed a matter of exhilaration, but this will be fulfilled only when the said "matter" is achieved by the sincere and committed activities of the government in implementing the devised plans, acts, and regulations. Urbanization is essential for social and economic development in Nepal and so well-being of the people. Rapid urbanization, if well planned and the policies and acts so far existing implemented sincerely and dedicatedly, can be an encouraging trend and also a driving force for modernization, economic growth, and development, as well as for livable cities.

References

ADB (2010) Unleashing economic growth: region-based urban development strategy for Nepal. Asian Development Bank, Philippines

Axinn WG, Ghimire DJ (2011) Social organization, population, and land use. Am J Sociol 117(1):209–258

Baniya IC (2002) Problems. Prospects and policy measures in slums and squatters: a case study of Jagriti Tole and Hyumat Tole. Kathmandu. A Seminar Report. Department of Architecture and Urban Planning, Institute of Engineering, Kathmandu

Bishop RNW (1952) Unknown Nepal. Luzac and Company Ltd., London, p 101

CANN/CEN (2014) Air quality status and management in Kathmandu valley: Make The City Air Breathable. Kathmandu: MaYA (Manav-Kendrit Yatayat Abhiyan), Fact sheet 5, UN-Habitat/Clean Air Network Nepal/Clean Energy Nepal

CBS (2003) Population monograph-2003. Central Bureau of Statistics, Kathmandu

CBS (2012) National population and housing census 2011 (National Report). Central Bureau of Statistics, Government of Nepal, Kathmandu

CBS (2014a) Environmental statistics of Nepal 2013. Central Bureau of Statistics, Government of Nepal, Kathmandu

CBS (2014b). Population monograph of Nepal. vol 1–3. Central Bureau of Statistics (CBS), Government of Nepal, Kathmandu

CBS (2015) Nepal multiple indicator cluster survey 2014, Final report. Central Bureau of Statistics and UNICEF Nepal, Kathmandu

CEN/ENPHO (2003) Health impacts of Kathmandu's air pollution. Clean Energy Nepal/ Environment Public Health Organization, Kathmandu

Chen P, Kang S, Li C, Rupakheti M, Yan F, Li Q et al (2015) Characteristics and sources of polycyclic aromatic hydrocarbons in atmospheric aerosols in the Kathmandu valley. Nepal Sci Total Environ 538:86–92

Depuis J (1962) Les basins interiéurs du Kashmir et du Nepal'. Annals de Geographie. 71(384):156–166

Dhimal M, Karki K, Aryal K, Shrestha S, Pradhan B, Nepal et al. (2016) Effects of climate factors on diarrheal diseases at national and sub-national levels in Nepal. Nepal Health Research Council (NHRC)/ WHO Country Office, Nepal Kathmandu

DHM (2015) Study of climate and climatic variation over Nepal: final report. Department of Hydrology and Meteorology, Government of Nepal (GoN), Kathmandu

DHM (2017) Observed climate trend analysis in the districts and physiographic regions of Nepal (1971–2014). Department of Hydrology and Meteorology, GoN, Kathmandu

DOHS (2005) Annual report 2004/2005. Department of Health Services, Government of Nepal, Kathmandu

DOHS (2015) Annual report 2013/2014. Department of Health Services, Government of Nepal, Kathmandu

DOHS (2017) Annual report 2016/2017. Department of Health Services, Kathmandu, Government of Nepal, Kathmandu. https://doi.org/10.3126/nje.v6i1.14733

DOTM (2016) Vehicle registration record. Department of Transport Management, Kathmandu. http://www.dotm.gov.np/en/vehicle-registration-record/

DWIDP (2009) Disaster review 2009, Series XVII. Department of Water Induced Disaster Prevention (DWIDP), Ministry of Irrigation, Government of Nepal, Kathmandu

DWSS (2005) Nepal drinking water quality standard. Department of Water Supply and Sanitation, Kathmandu

DWSS/NMIP (2014) National coverage and functionality status of water supply and sanitation in Nepal. Department of Water Supply and Sewerage, Government of Nepal, Kathmandu

EKantipur (2017) http://kathmandupost.ekantipur.com/news/2017-06-22/kmc-to-build-300public-conveniences.html

ENPHO (2015) Annual report 2015. Environment and Public Health Organization, Kathmandu

ETG (2012) Kathmandu valley tourism cluster competitiveness assessment and action plan, background paper for this report. World Bank, Economic Transformations Group (ETG), Washington, DC

Gautam C (2006) Final report on action program on air quality management of Kathmandu valley, Ministry of Environment. Science and Technology, Government of Nepal, Kathmandu

Giri D, Murthy VK, Adhikary P (2008) The influence of meteorological conditions on PM 10 concentrations in Kathmandu valley. Int J Environ Res 2(1)

Giuseppe F (1799) Account of the kingdom of Nepal. Asiatick Researches. Vernor and Hood, London

Gurung A, Bell ML (2012) Exposure to airborne particulate matter in Kathmandu valley, Nepal. J Expo Sci Environ Epidemiol 22(3):235

Haack BN, Khatiwada G (2007) Rice and bricks: environmental issues and mapping of the unusual crop rotation pattern in the Kathmandu valley. Nepal Environ Manage 39:774–782

HMG/UN/UNESCO (1975) Kathmandu valley: the preservation of physical environment and cultural heritage protective inventory, vol 1. His Majesty Government of Nepal, Kathmandu

Hodgson BH (1874) Essays on the languages, literature and religion of Nepal and Tibet. Trübner and Co, London

HPCIDBC (2011) Status and strategy for faecal sludge management in the Kathmandu valley, High Powered Committee for Integrated Development of the Bagmati Civilization (HPCIDBC), Kathmandu, Nepal

IOM (2017) Annual report 2016/2017. Institute of Medicine, Tribhuvan University, Kathmandu
Kansakar VBS (1985) Land resettlement policy as a population distribution strategy in Nepal. In: Kosinski LA, Elahi KM (eds) Population redistribution and development in South Asia. D. Reidel Publishing, pp 111–122
Kansakar VBS (1974) Population change in Nepal: a study of population mobility during 1952/54–1961 Unpublished Ph. D. Thesis, Patna University, Patna, India
Karki KB, Dhakal P, Shrestha SL, Joshi HD, Aryal KK, Poudyal A, Puri S, Verma SC, Pokhrel A, Lohani GR, Dhimal M (2016) Situation analysis of ambient air pollution and respiratory effects in Kathmandu valley. Nepal Health Research Council, Kathmandu
Karki S, Fitzpatrick AL, Shrestha S (2014) Risk factors for pneumonia in children under 5 Years in a teaching hospital in Nepal. Kathmandu Univ Med J (KUMJ) 12(48):247–252
KUKL (2014) Kathmandu Upatyaka Khanepani Limited, Sixth Anniversary Report. Kathmandu, Kathmandu Upatyaka Khanepani Limited, Nepal
Kurmi OP, Semple S, Devereux GS, Gaihre S, Lam KB, Sadhra S, Steiner MF (2014) The effect of exposure to biomass smoke on respiratory symptoms in adult rural and urban Nepalese populations. Environ Health 13(1):92
LSGS (2008) Status of squatter communities along the Bagmati river and its tributaries in the Kathmandu valley. high-powered Bagmati area sewerage implementation and monitoring committee. Lumanti Support Group for Shelter, Kathmandu
MoEST (2007) Kathmandu valley environmental outlook. International Centre for Integrated Mountain Development (ICIMOD), Ministry of Environment, Science and Technology (MoEST) and United Nations Environment Programme (UNEP), Kathmandu
MOHP (2009) Public expenditure review on health sector 2003/04-2005/06. Ministry of Health and Population, Government of Nepal, Kathmandu
MoPE (1999) Environmental planning and management of the Kathmandu valley. GoN, Ministry of Population and Environment and IUCN, Kathmandu
MOPE (2012) National ambient air quality standard. Ministry of Population and Environment, GoN, Kathmandu
MOPE (2017) Air quality management action plan for Kathmandu valley. Ministry of Population and Environment/Department of Environment (DoEnv)/ Quest Forum Pvt. Ltd, Kathmandu
MOUD (2014) National water supply and sanitation sector policy 2014. Government of Nepal, Kathmandu
MoUD (2015) National urban development strategy 2015. Ministry of Urban Development, Government of Nepal, Kathmandu. http://www.moud.gov.np/pdf/NUDS-2015-final-draft.pdf
MOUD (2017) National urban development strategy 2017. Ministry of Urban Development, GoN, Kathmandu
NHRC (2004) Assessment of ambient air quality in selected urban areas of Nepal. Nepal Health Research Council, Kathmandu
NHRC (2009) Situation analysis of environmental health in Nepal. NHRC, Kathmandu
Pearce N, Blair A, Vineis P, Ahrens W, Andersen A, Anto JM et al (2015) IARC monographs: 40 years of evaluating carcinogenic hazards to humans. Environ Health Perspect 123(6):507–514
Pokhrel B, Gong P, Wang X, Wang C, Gao S (2018) Polycyclic aromatic hydrocarbons in the urban atmosphere of Nepal: Distribution, sources, seasonal trends, and cancer risk. Sci Total Environ 618:1583–1590. https://doi.org/10.1016/j.scitotenv.2017.09.329
Pradhan B (1999) Water quality analysis of the Bagmati river and its tributaries in Kathmandu valley using bacterial and saprobic measures. In: Proceedings of the 3rd national conference on science and technology. Royal Nepal Academy of Science and Technology, Kathmandu
Pradhan B (2004) Rural urban relations with particular reference to Nepal. RUPP/UNDP, Kathmandu
Pradhan B, Bajracharya R, Rajbhandari L (2005a) Water quality classification model in the hindu kush-himalayan region: the Bagmati river in Kathmandu valley, Nepal. Submitted to ICIMOD. Joint collaboration of MENRIS and WHEM programme. Department of community medicine and family health institute of medicine, Tribhuvan University Kathmandu, Nepal

Pradhan B, Gruendlinger R, Fuerhapper I, Pradhan P, Pradhanang S (2005b) Knowledge of water quality and water borne disease in rural Kathmandu Valley. Nepal. Nepal Aquat Ecosyst Health Manag 8(3):277–284

Pradhan PK (1998) Market centre and hinterland relation in Lalitpur district. Institute of geography and regional studies, Number 17, University of Klagenfurt, Austria, pp 1–212

Pradhan PK (2003) Population growth, migration and urbanisation, and environmental consequences in Kathmandu valley, Nepal. In: Unruh J, Krol M, Kliot N (eds) Environmental change and its implications for population migrations. Kluwer Academic Publishers B.V, Dordrecht (The Netherlands), pp 177–199

Pradhan PK (2008) Study on Status of unemployed slum youth and their possible linkages with job market in greater Kathmandu. CARE Nepal, Kathmandu

Pradhan PK (2013) Urbanization. In: Jha PK, Neupane FP, Shrestha ML, Khanal IP (eds) Environment and natural resources, Nepalpedia Series No.1. Nepal Academy of Science and Technology, Kathmandu, pp 477–484

Pradhan PK (2014) Urban infrastructure planning and investment issues in Nepal. In: Urban infrastructure investment: challenges and potentials. Special Issue-2070: 26th Anniversary.Town Development Fund, Kathmandu

Pradhan PK, Sharma P (2016) Nepal: urban environment analysis (Chapter 18). In: Singh RB, Prokop P (eds) Environmental geography of South Asia, advances in geographical and environmental sciences, pp 305–334. https://doi.org/10.1007/978-4-431-55741-8_18

Rabenau BV (1990) Issues in housing development. Kathmandu: Urban Development through Local Efforts (UDLE)

Raina A, Gurung Y, Suwal B (2018) Equity impacts of informal private water markets: case of Kathmandu valley. Water Policy

Sanjel S, Khana S, Thygerson SM, Khanal K, Pun Z, Tamang S, Joshi SK (2016) Airborne particulate matter and health condition in Brick Kiln workers in Kathmandu valley. Nepal Kathmandu Univ Med J (KUMJ) 14(54):159–166

Shrestha BR, Gauchan Y, Gautam GS, Baral P (2012) Nepal national health accounts, 2006/07–2008/09. Ministry of Health and Population, Government of Nepal, Kathmandu

Shrestha N, Lamsal A, Regmi RK, Mishra BK (2015) Current status of water environment in Kathmandu valley, Nepal. water and urban initiative working paper series (United Nations University), Number 03

Skeldon R (1985) Migration in South Asia. In: Kosinski LA, Elahi KM (eds) Population redistribution and development in South Asia. D. Reidel Publishing, pp 37–64

Tebbutt THY (1992) Principles of water quality control. Pergamon Press, Oxford

Thapa RB, Murayama Y (2009) Examining spatiotemporal urbanization patterns in Kathmandu valley, Nepal: remote sensing and spatial metrics approaches. Remote Sens 1:534–556. https://doi.org/10.3390/rs1030534:534-556

Thapa RB and Murayama Y (2010) Drivers of urban growth in the Kathmandu valley, Nepal: examining the efficacy of the analytic hierarchy process. Appl Geogr 30(1): 70–83

Udmale P, Ishidaira H, Thapa B, Shakya N (2016) The status of domestic water demand: supply deficit in the Kathmandu valley. Nepal Water 8(5):196

UN (1997) Glossary of environment statistics, studies in methods, Series F, No. 67. United Nations (UN), New York

WHO (1997) Guidelines for drinking-water quality Surveillance and control of community supplies. 2nd edn, vol 3. Geneva

World Bank (2013) Urban growth and spatial transition in Nepal: an initial assessment. International Bank for Reconstruction and Development, World Bank, Washington DC

Part III
Integrated Urban Risk Mitigation and Resilience Measures

Chapter 18
Impact of Urbanization on Megacities' Lakes Using Remote Sensing Technique: A Case Study of Water Quality Analysis in Ambattur Lake, Chennai, India

V. Chandrasekar, Shaik Mahamad and G. Latha

Abstract Urban lakes are the essential indicator factors, which not only preserve the water level in the cities, but also supply water to the urban areas which consist of Industrial sectors, Residential and Administrative sectors. With the help of emerging trends in remote sensing techniques and appropriate water quality analysis, one can access the impact of urbanization on the water bodies. Lakes are one of the significant water bodies that help in nourishing the natural resources in the world as they constitute 4% of the world's water bodies in land surface. This paper will study the overall Chennai lakes shrinkage, by using Landsat 8 Imagery of 2016. Land Cover and Land Use (LCLU) techniques will be used to analyze the shrinkage of lakes in Chennai. Ambattur Lake is a water body, which for the past few decades has suffered a drastic shrinkage due to human intervention. The quality of water has gradually depreciated due to the encroachments and also because of the disposal of large amounts of industrial effluents pertaining to factories in the neighborhood. This has resulted in complete compromise in the quality of water of the lake and has jeopardized its entire water ecosystem. To analyze this, water samples have been collected from the lake during the pre-monsoon season and water quality test has been conducted including selected heavy metals analysis. The total shrinkage of the lake is around 108.782 ha of land, which is approximately 40% of the lake area.

Keywords Urban lakes · Remote sensing · Shrinkage · Water quality · Chennai

18.1 Introduction

Once a lake forms, it develops a "Water Economy," a balance of some kind, in its inflow/out flow "budget". This involves several key factors, the first being the inflow. Lakes receive their water from precipitation, from ground flow (via runoff, rivers

V. Chandrasekar (✉) · S. Mahamad
Department of Geography, Presidency College, Chennai, India
e-mail: vchandrasekar1989@gmail.com

G. Latha
Department of Geography, Sri Venkateshwara University, Tirupathi, India

R. B. Singh et al. (eds.), *Urban Health Risk and Resilience in Asian Cities*, Advances in Geographical and Environmental Sciences,
https://doi.org/10.1007/978-981-15-1205-6_18

and the lakes and from groundwater seeping (Hanson 2007). Lakes are valuable resources for the people. Water bodies near developing cities are decreasing both by natural and anthropogenic activities, which deteriorate their water quality and push them to the point of extinction (Kistan et al. 2015; Ahuja 2015). In 1975, only four megacities existed in 2000, there were around 18 cities by the end of 2015 and UN has estimated that there will be 22 cities in future (Torrey 2004). Increase in population and extinction of urban limit have made people to occupy the wasteland which earlier used to be the lake area. The shrinkage of lakes is an alarming issue of major cities that demands immediate attention. Many lakes have vanished, with only their traces remaining in the form of names such as 'Lake view street,' 'Lake Area'. Healthy lakes and their shores provide us with a number of environmental benefits and also improve the quality of our life by strengthening our economy using the lake resources (Chandrashekar et al. 2014). The decline in the spatial coverage and recovery of the water body in many sectors are more frequently reported today (Dipson et al. 2015). The effluent from residential and industrial activities contain various inorganic and organic substances in different concentrations that affect the nature and quality of both lake and ground water (Pandey and Tiwari 2009; Nandhakumar et al. 2015). Pollutants such as herbicides, pesticides, fertilizers, and hazardous chemicals make their way into the lakes, rivers, and sea. When water supply is contaminated, it is a threat to human, animal, and plant health unless it undergoes an expensive purification procedure. The quality of surface water and groundwater is severely compromised in densely populated areas, due to the localization of industries. Almost 70% of the water in India has become polluted due to the discharge of domestic sewage and industrial effluents into natural water sources such as rivers, streams, and lakes (Sonune et al. 2015). The largest source of water pollution in India is untreated sewage and the other sources of pollution include agricultural runoff and unregulated small-scale industry. This paper analyzes the spatial shrinkage of the lake by Land Cover and Land Use method for preservation of the lake and the impact it has on the lake by collecting water samples from the water body (Gandhi et al. 2015). The LULC method will help to analyze the depletion and effect of urbanization on the lake water quality, thereby helping in creating protective measures for Ambattur Lake.

18.1.1 Study Area

Chennai District is located—northern latitudes of 12° 59′ 10″ and 13° 08′ 50″ and eastern longitudes of 80° 12′ 10″ and 80° 18′ 20″ as per the Survey of India Topographical Maps No. 66 C/4 and 66 D 1 & 5 (Shanmugam et al. 2006). The northeast monsoon which showers the city from October to December contributes to the rainfall in the city and heavy water source for the lakes and rivers. The average annual rainfall ranges from 1285.6 to 1232.7 mm. Ambattur Lake is an amalgamation of three lakes: Korattur Lake, Madhavaram Lake, and Ambattur Lake, which is spread over 990 acres at Ambattur, Chennai, Tamil Nadu, India (Priyanga et al. 2015). It is located in the north of the Chennai–Arakkonam railway line (13° 06′ 26.95″ N and 80° 8′

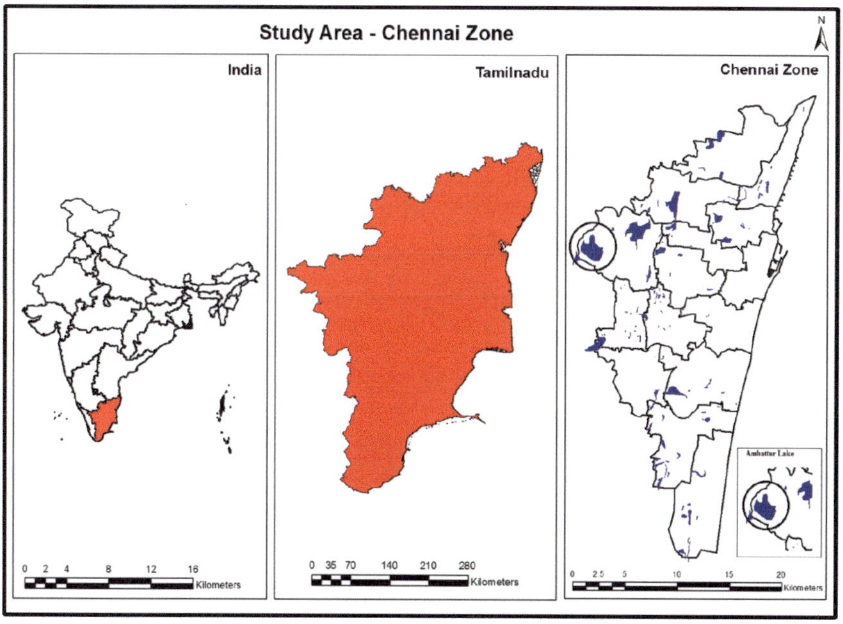

Fig. 18.1 Chennai Zone and Ambattur Lake

30.79″ E) and is one of the important lakes in the western part of Chennai (Fig. 18.1). The Ambattur Lake is located near Thiruverkadu road, Ayappakkam, which used to be an important water resource for the surrounding region, has now become an "open drain". The lake once used to be the main source of drinking water. It was also one of the alternatives to the Puzhal reservoir for Chennai Municipal Corporation. The lake is totally spread over 280 acres. Sadly, the lake has been encroached by the local residents. The remaining area of the lake is covered by water hyacinths. To make matters worse, sewerage is directly discharged from the neighboring housing board tenements; this has led to the deterioration of the lake.

18.1.2 Aim and Objectives

1. To analyze the shrinkage of overall Chennai akes using Land Use Land Cover Methods from 1972 to 2016
2. To examine the shrinkage and water quality of Ambattur Lake, by collecting water samples from various locations analyzing through Piper Diagram.

Table 18.1 Landsat 8 image used

S. No	Acquisition date of image	Website	Resolution in Mts	Path and row
1	25-09-2016	Earth explorer	30	142/51

Table 18.2 Toposheets used

Toposheet number	Scale	Year	Source
66c/4 & 5	1: 50,000	1976	Survey of India
66/c8	1: 50,000	1976	Survey of India

18.2 Methods

The Landsat Image (Table 18.1) and Toposheets, 1976, (Table 18.2) have been used for various years to denote the shrinkage of the lake using supervised classifications techniques. Google Earth Images have been used for locating the water samples. Water samples have been analyzed by Piper diagram and heavy metals have been analyzed using graphical representations.

The samples have been collected in five different locations and were analyzed for the following parameters: Electrical conductivity (EC), pH, total dissolved solids (TDS), chloride, calcium, magnesium (Table 18.3), using standard protocols. The parameters were studied based on World Health Organization, Bureau of Indian Standards and Indian Council of Medical Research (ICMR 1975; WHO 2004). The instruments used were of precise accuracy and the chemicals used were of analytical grade. Water quality has been tested in the Department of Geology, under the guidance of Dr. B Gowtham, Assistant Professor, Presidency College, Chennai. The sample will be also analyzed through Piper Diagram. Heavy metals have been tested under the guidance of Dr. Jaya Prakash, Dept. of Geology, University of Madras.

18.3 Analysis of Shrinkage of Chennai Zone Lakes

The shrinkage of Chennai lakes (Fig. 18.2) clearly shows the impact of urbanization. It has largely affected the water bodies in the city. Majority of small lakes have totally been vanished and have been converted to built-up lands. For the last few decades, real estates have become a trend in the cities and large number of apartments, complex and private colleges have been constructed. In Fig. 18.2, all the water bodies in Chennai zone have been affected in many ways; the lakes have been used as sewage dumps and many lakes have been affected by the industrial effluents which make the lakes worse and level of chemical content to increase, which also affect the groundwater of the surroundings. In the year 2015, flood has affected the places where lakes were converted into buildings. In the northern part of Chennai, many small to medium size lakes have vanished. Kodingayur was a lake which is now totally converted into

Table 18.3 Ambattur Lake chemical parameters

Sample No	Ca	Mg	$CacO_3$	HCO_3	SO_4	Cl	TDS	pH	EC
1	105.259	21.36	351	271	70.211	224.9	1004	7.73	1568.75
2	125.119	26.4	421	281	89.658	319.9	1240	7.3	1937.5
3	121.147	26.4	411	279	93.352	344.9	1203	7.5	1879.69
4	135.049	39.6	500	284	95.315	354.9	1215	7.7	1898.44
5	202.574	64.8	772	444	132.93	789.8	2120	7.97	3312.5

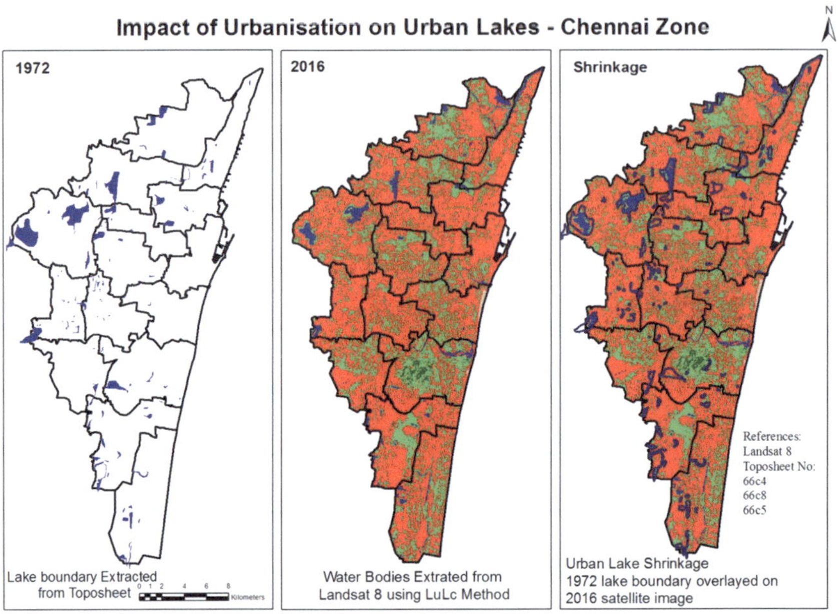

Fig. 18.2 Impact of Urbanization on Chennai Lakes. *Source* Landsat 8 Satellite Images (Open Source)

buildings. In the southern part of the Chennai zone, a large number of lakes have now been dumped with sand to construct new apartments. This paper also details about one particular lake, i.e., Ambattur Lake, which has seen its shrinkage and residents have used it as a sewage dump; to the right side there is a factory and the other side is covered with residential areas. In the southern zone, the lake area is affected largely. This alarming issue can be understood clearly from the above images.

18.3.1 Case Study of Ambattur Lake

The study was carried out by collecting water samples from Ambattur Lake (Fig. 18.3), located in the suburbs of Chennai city which is well surrounded by the largest Industrial Estate and residents. Ambattur Zone is the biggest industrial town in Chennai. The present investigation was to study the contaminant of water in Ambattur Lake.

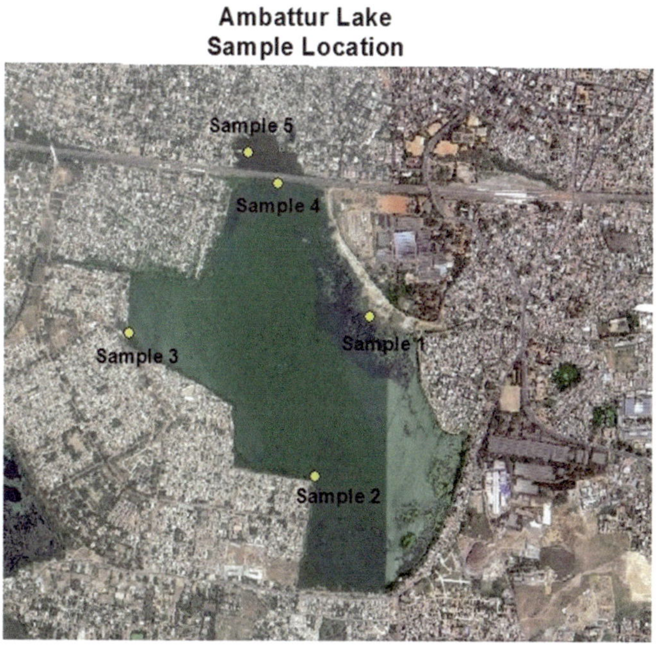

Fig. 18.3 Ambattur Lake—sample points

18.3.1.1 Analysis of Chemical Parameters in Lake Water of Ambattur, Chennai

From the (Fig. 18.3) abovementioned locations, the following parameters were collected and have been tested in the lab. Using Piper diagram, the overall chemical conditional of the lake is represented.

Total Dissolved Solids is a measure of the combined content of all inorganic and organic substances present in a liquid: in molecular, ionized or micro-granular (colloidal sol) suspended form. As per the BIS 2012, the desirable limit of TDS is 500 mg/l whereas the permissible limit is 2000 mg/l (BIS 2012). In the above-collected samples, the level of TDS is high in all the locations (Fig. 18.4). TDS values range from 1004 to 2120 mg/L (Table 18.3). Variations in TDS may be due to the inflow of industrial waste, dissolved minerals, iron and manganese, landfills, animal and agriculture wastes, and also evaporation and less rainfall (Deepa et al. 2016).

Sample 5 has the highest TDS level, which was collected from the northern part of the lake that has been separated by the railway line from the main lake area. This sample location has been surrounded by residential areas and the wastewaters from these regions directly flow into the water body. Even though the remaining four samples have high TDS values which are above the desirable limit, they were within the permissible limit.

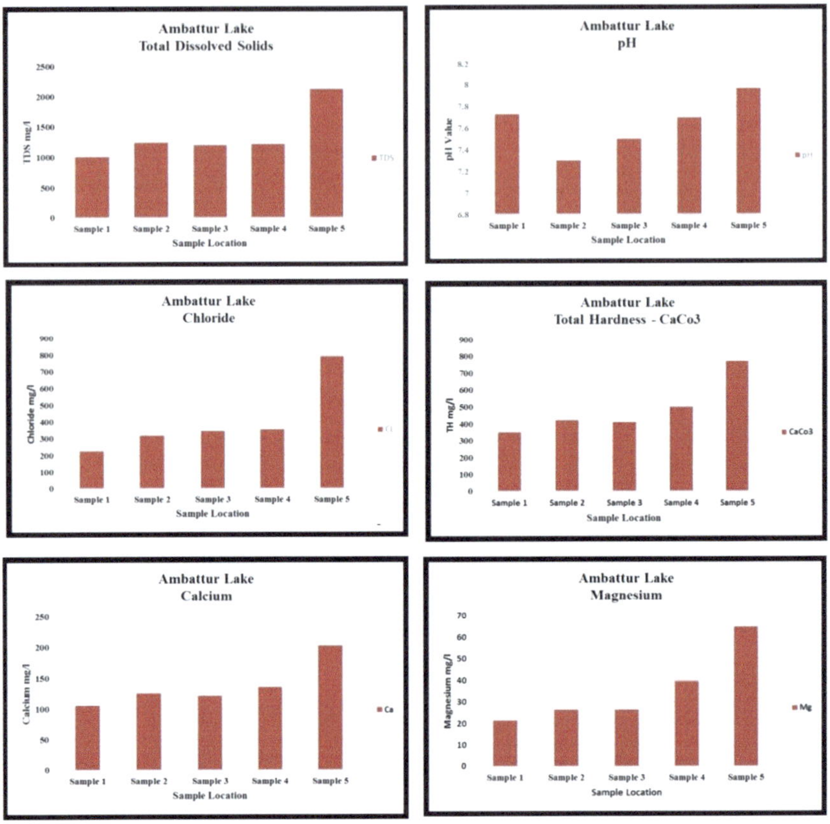

Fig. 18.4 Selected chemical parameters. *Source* Primary data

pH ranges from 7.3 to 7.97 in Ambattur Lake. The term pH is used to indicate the acidity or alkalinity of a substance. The desirable pH level ranges from 6.5 to 8.5 (Table 18.3) and there is strictly no relaxation for this limit (BIS 2012). The level of pH in the northern part of the lake is 7.97, which is the highest as it is surrounded by settlements (Sample 5) followed by Sample 1 (Industrial Area) with 7.73 of pH level. Increased level of pH appears to be related with increased use of alkaline detergents in residential areas and various alkaline materials from wastewater through industrial areas. Extremes in pH can produce conditions that are toxic to aquatic life. The pH of Ambattur Lake did not exceed the desirable limit of 8.5 (Fig. 18.4).

Chloride ranges from 224.93 to 789.75 mg/L (Table 18.3), the desirable limit of chloride is 250 mg/L and the permissible limit is up to 1000 mg/L in the absence of an alternate source (BIS 2012). Increase in chloride is due to fertilizers, industrial wastes, minerals, and seawater. In the Ambattur Lake, chloride is high in the upper part of the separated lake with 789.75 mg/l of chloride level. The level of chloride is high in the residential areas. Excess chloride reduces the DO content of water, which turns harmful to aquatic organisms. The high chloride concentration in water indicates the presence of a large amount of organic matter. Sample 1 which was taken from the eastern part of the lake has very low levels of chloride and falls under the desirable limit, but the remaining four Samples of Ambattur Lake water do not fall within the desirable limit but within the permissible limits (Fig. 18.4).

Total Hardness values range from 351 to 771 mg/l (Table 18.3) when water passes through or over deposits of limestone. The levels of Ca^{2+}, Mg^{2+}, and HCO^{3-} ions present in the water can greatly increase and cause the water to be classified as hard water. This phenomenon results from the fact that calcium and magnesium ions in water combine with soap molecules, making it 'hard' to froth. The Bureau of Indian Standards 2012 indicates that the ideal quality water should not contain more than 200 mg/L of total hardness, but the permissible level is up to 600 mg/l (BIS 2012). High levels of total hardness prove to be hazardous. The level of TH is more in the northern part of Ambattur Lake (Sample 5), but the remaining four Samples of Ambattur Lake water do not fall within the desirable limit but within permissible limits (Fig. 18.4).

Calcium values range from 105.25 to 202.57 mg/l (Table 18.3) in the Ambattur Lake; the desirable limit for calcium in water is 75 mg/L and the permissible limit in the absence of an alternate source is 200 mg/L (BIS 2012). In Ambattur Lake, the level of calcium is high in almost all the locations from where the samples were collected. Sample 5 has the highest calcium level 202.57 mg/l, which does not even fall under the permissible limit; remaining samples also have high levels of calcium but only fall within the permissible limits (Fig. 18.4).

Magnesium values range from 21.36 to 64.8 mg/l (Table 18.3) in the Ambattur Lake; the magnesium level is high in the northern part of the lake—sample 5 and sample 4, with 64.8–39.6 mg/l, which generally fall within the permissible limit. The remaining three samples have desirable limit of magnesium levels (Fig. 18.4).

Potassium ranges from 29 to 70 mg/l (Table 18.3) in the Ambattur Lake; high levels of potassium level are found in the upper (northern) part of the lake. Low levels of potassium level are found in the eastern part of the lake (Fig. 18.5).

Sulfate values range from 70.21 to 132.93 mg/l (Table 18.3) in Ambattur Lake. The desirable amount of sulfate in drinking water is 200 mg/L and can go up to 400 mg/L in the absence of an alternate source. Sulfate concentration in Ambattur Lake is below the desirable limit. Sulfate is present in fertilizers and contributes to water pollution by increasing sulfate concentration in the water body. Sulfate level is more in the northern part of the lake yet it is under the desirable limit (Fig. 18.5).

Electrical Conductivity values range from 1568.7 to 3312.5 μS/cm in Ambattur Lake (Table 18.3). EC of Ambattur Lake falls within the high conductivity indicating a rapid level of pollution in the lake. Low conductivity (0–200 μS/cm) is an indicator of pristine or background conditions. Mid-range conductivity (200–1000 μS/cm) is the normal background for most of the rivers and lakes. Conductivity outside this range indicates that the water is not suitable for certain species of animals. High conductivity (1000–10,000 μS/cm) (WHO 2004) is an indicator of saline conditions. All the samples collected in the Ambattur Lake fall under high conductivity and the

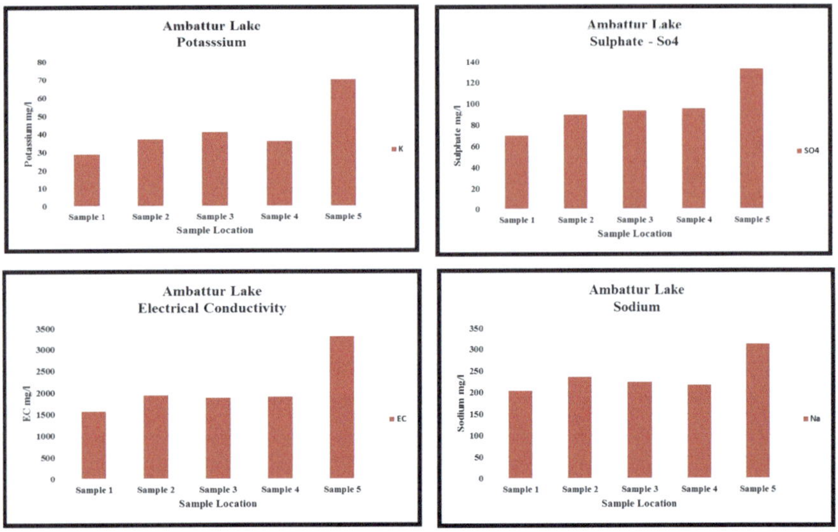

Fig. 18.5 Selected chemical parameters. *Source* Primary data

sample 5 has double the value of EC level when compared to the other locations (Fig. 18.5).

Sodium ranges from 204 to 312 mg/l (Table 18.3) in the Ambattur Lake; high levels of Sodium are found in the northern part of the lake. Low levels of Sodium are found in the eastern part of the lake (Fig. 18.5).

18.3.1.2 Analysis of Heavy Metals in Ambattur Lake

Zinc values are found to be within the range of 0.04–0.052 mg/l (Table 18.4) for lake water samples. All values are below the permissible limit of 5 mg/l. This shows that the content of zinc is low in the Ambattur Lake. However, some samples have recorded high values which are nearer to desirable limit, which is due to domestic waste and the industrial wastes discharged in the lake. Zinc is also one of the reasons

Table 18.4 Ambattur Lake chemical parameter

Heavy metals—Ambattur Lake						
S.no	Zinc	Iron	Manganese	Nickel	Copper	Chromium
1	0.033	0.0249	0.223	0.0152	0.0244	0.0304
2	0.052	0.0027	0.0023	0.025	0.0149	0.2499
3	0.041	0.0529	0.0189	0.0206	0.0037	0.2636
4	0.032	0.0102	0.0237	0.0209	0.0143	0.2831
5	0.04	0.108	0.129	0.0323	0.0122	0.3952

behind concentration in lake water, which get mixed into the runoff. Unsanitary and unhygienic practices may also increase the level of zinc. The low zinc concentration in lake water could be due to slightly alkaline pH of the water samples and the solubility of the pH being reduced. Hence, the samples from lake collected are below the maximum permissible limit as stated by the World Health Organization and Bureau of Indian Standards (2012) (Fig. 18.6).

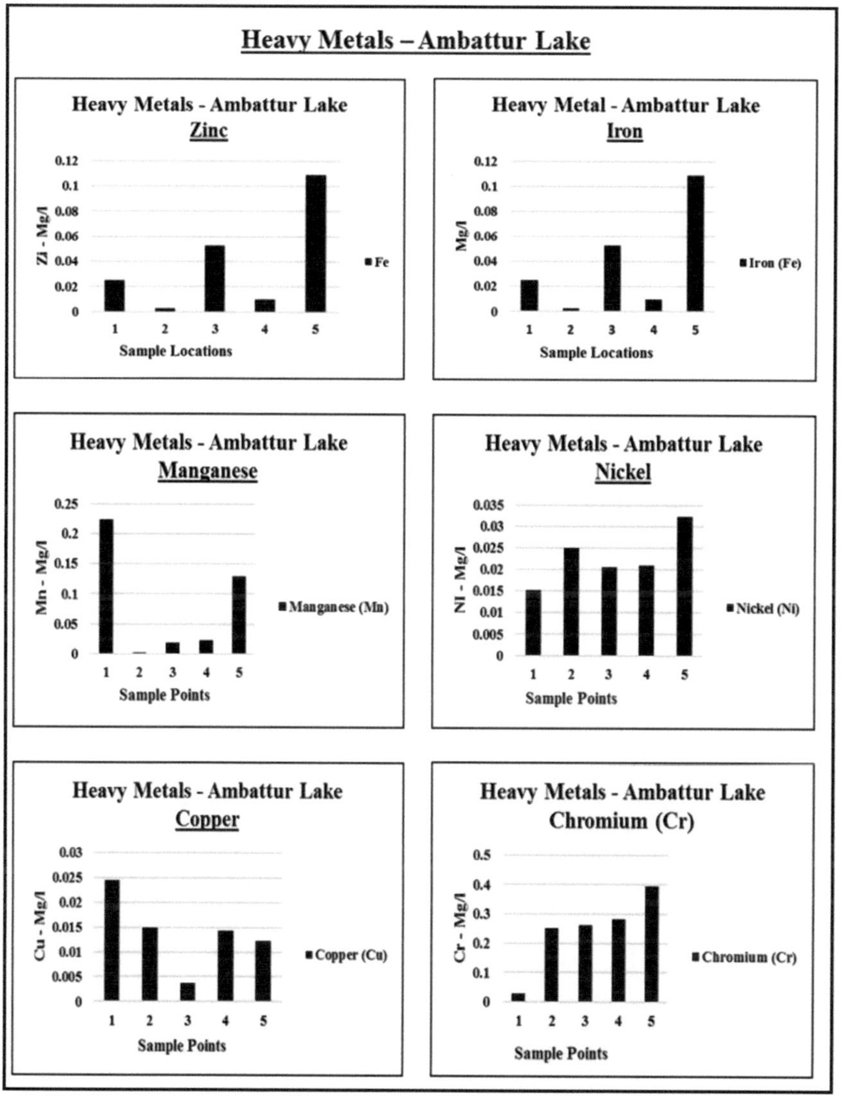

Fig. 18.6 Selected heavy metals in Ambattur Lake. *Source* Primary data

Iron

The Iron value in the lake ranges from 0.0027 to 0.1088 mg/l. All the values are lower than the permissible limit of 0.3 mg/l (Table 18.4). Higher concentration of iron content was analyzed in the fifth sample; this may be due to the discharge of waste and other effluent on surface that infiltrated into the water. Higher amounts seen particularly at few sample locations show that there were high domestic effluents. The content of iron in the water can also be due to the presence of organic substances in the sediments. Excessive iron also affects the presence of bacteria in lake water (Fig. 18.6).

Manganese

Manganese value ranges from 0.0023 to 0.223 mg/l. All collected samples fall under permissible limit of 0.1 mg/l (Table 18.4) suggested by BIS 2012. The samples that were analyzed show that the lake water is not contaminated by manganese. However, rise in its level will be due to the influence or discharge of sewage and other wastes (Fig. 18.6).

Nickel

Nickel values range from 0.0152 to 0.0323 mg/l (Table 18.4) from the collected lake water samples.

The values are higher than the prescribed limit of 0.02 mg/l. Nickel can be considered as a nontoxic element, but it affects physiological process at very high concentration (Fig. 18.6). The high **Ni** values are found in the sample location of 2, 3, 4, and 5. Higher levels of Nickel in the lake water are due to various wastes including automobile shops and sewages. Concentration of heavy metal in the lake is credited to runoff into the water body

Copper

In the Ambattur Lake, the Copper values (Table 18.4) are between 0.52 and 0.68 mg/l. The recorded values fall below the permissible limit of 0.05 mg/l. High content of copper reveals greater impact of contamination due to the discharge of domestic effluents. The low level may be due to the absorption process by the soil, which reduces the concentration of the heavy metals in water (Fig. 18.6). Higher copper levels may also be due to the presence of industrial and domestic wastes (Hussain and Sheriff 2013).

Chromium

Chromium values in the lake water ranges from 0.0304 to 0.3952 mg/l for lake water samples (Table 18.4). All the values recorded are below permissible limit of 0.5 mg/l. High content of Chromium can be due to various anthropogenic activities, industrial waste, and household discharges (Fig. 18.6).

Piper Diagram

The image (Fig. 18.7) explains the quality of groundwater locations through the piper plot. A piper plot is a way of visualizing the chemistry of a rock, soil, or

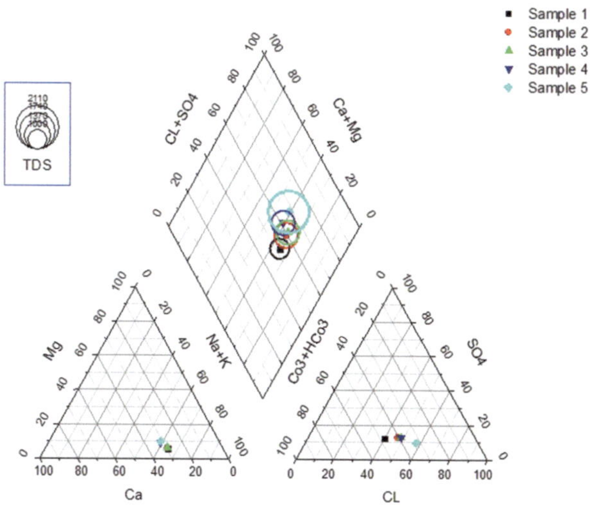

Fig. 18.7 Piper diagram—Ambattur Lake

water sample. It is included with three pieces: a ternary diagram in the lower left showing the cations, a ternary diagram in the lower right indicating the anions, and a diamond plot in the middle representing a combination of the two. The cations and anion fields are combined to explain in single point in a diamond-shaped field, from which inference is drawn on the given basis of hydro-geochemical facies concept. These trilinear diagrams are useful in bringing out chemical relationships among lake water samples in more definite terms rather than with other possible plotting methods (Piper 1953; Sadashivaiah et al. 2008).

In the above image the majority of the plot has fallen in the right quadrant diamond (Combination), which is sodium-chloride water (NaCl).

18.4 Conclusion

It has been clearly explained in the paper, that the water resources, especially in urban areas are "mis-managed" because of human interference in the name of development. Using the supervised classification in Chennai zone, it has been clearly seen that encroachment has been done by the residents in and around all the lakes. Extension of city limit and increase in population due to a lot of anthropogenic activities have been done near the lake which has worsened the lake condition. Lake boundary overlaid on the current LCLU map (Fig. 18.2, shrinkage) indicates, many major lakes have been converted to construction sites, due to extension of Chennai boundary and urbanization. Increase in population in the urban areas has led to all the open land

and lakes being utilized for the construction of apartments Remaining lakes have been used as sewage dumps (Ambattur Lake, Korattur Lake, Rettai Eri, etc.).

A few possible solutions are that the domestic and industrial wastes should be discarded in a controlled environment and not into the lake. Still if they are dumped into the water body, then they need to undergo elaborate purification procedures before being thrown into the lake. The residential areas having no proper facilities of sewer line must be encouraged to build septic tanks and must be motivated to cooperate with the government's "Swatch Bharat Campaign" that strives to make India a clean and disease-free country. If immediate actions are taken by the government and the public, the water bodies in the urban areas will be saved for the future generation.

The Presence of heavy metals is low in the lake, but is very near to the permissible limit, therefore they would affect the water, if the condition of the water remains the same. Also, these heavy metals can also contaminate the groundwater of Ambattur region. Due to climatic change, many bird sanctuaries are dried in summer seasons and these migrating birds finally depend on these types of lakes, which are highly polluted and are being affected.

References

Ahuja S (2015) Analysis of water quality from Salim Ali lake. Int J Adv Res Comput Sci Softw Eng 5(6):782–785

Bureau of Indian Standards (2012) Specification for drinking water. IS:10500, New Delhi, India

Chandrashekar H, Lokesh KV, Pushpa H, Ranganna G (2014) Studies on the assessment of tropic status of lakes. Int J Curr Microbiol Appl Sci 3(6):245–254

Deepa P, Rajasingh R, Venkatesan P, Subramanian A, Tennyson S (2016) Seasonalvariationsofphysicochemicalparameters of Korattur lake, Chennai, Tamil Nadu, India. Int J Chem Stud 4(3):116–123

Dipson P, Chithra S, Amarnath A, Smitha S, Nair MH, Shahin A (2015) Spatial changes of estuary in Ernakulam district, Southern India for last seven decades, using multi-temporal satellite data. J Environ Manag 148:134–142

Gandhi GM, Thummala N, Christy A (2015) Urban green cover assessment and site analysis in Chennai, Tamil Nadu—A remote sensing and GIS approach. ARPN J Eng Appl Sci 10(5):2239–2243

Hanson JK (2007) Lakes extreme earth. Chelsea House Publications, New York

Hussain AZ, Sheriff KM (2013) Status of heavy metal concentrations in groundwater samples situated in and around on the bank of Cooum river at Chennai City, Tamil Nadu. J Chem Pharm Res 5(3):73–77

ICMR (1975) Manual of standards of quality for drinking water supplies, 2nd edn. Indian Council of Medical Research, New Delhi

Kistan A, Kanchana V, Ansari AT (2015) Analysis of Ambattur Lake water quality with reference to physic-chemical aspects at Chennai, Tamil Nadu. Int J Sci Res 5(5):944–947

Nandhakumar S, Varun K, Sathyanarayanan N (2015) Interpretation of groundwater quality around Ambattur Lake, Chennai, Tamil Nadu. J Chem Pharm Res 4(7):1626–1633

Pandey SK, Tiwari S (2009) Physico-chemical analysis of ground water of selected area of Ghazipur city-A case study. Nat Sci 1(7):17–20

Piper AM (1953) A graphic procedure in the geochemical interpretation of water analysis. Geological Survey, Washington

Priyanga G, Abhirami B, Gowri NR (2015) Phyto-remediation of urban lakes in and around Chennai. J Chem Pharm Sci 8(2):288–291
Sadashivaiah C, Ramakrishnaiah CR, Ranganna G (2008) hydrochemical analysis and evaluation of groundwater quality in Tumkur Taluk, Karnataka State, India. Int J Environ Res Public Health 5(3):158–164
Shanmugam P, Neelamani S, Ahn Y, Philip L, Hong G (2006) Assessment of the levels of coastal marine pollution of Chennai city, Southern India. Water Resour Manag 21(7):1187–1206
Sonune NA, Mungal NA, Kamble SP (2015) Study of physico-chemical characteristics of domestic wastewater in Vishnupuri, Nanded, India. Int J Curr Microbiol Appl Sci 4(1):533–536
Torrey BB (2004) Urbanization: an environmental force to be reckoned with. Focus Area, Population Reference Bureau
World Health Organization (WHO) (2004) Guidelines for drinking water quality

Chapter 19
Evaluation of Mass Rapid Transit System (MRTS): A Case Study of Delhi

Menka Kumari and Anuradha Banerjee

Abstract Urban population in India is increasing at a very rapid pace. The number of metropolitan cities with a population of one million and above steadily increased from 5 in 1951 to 35 in 2001 and 53 in 2011. The urban areas work as engines of growth that create skills and wealth for the nation and generate employment for the migrants from the rural areas. Mobility is a very crucial factor for the economic growth of urban areas. Although the cities are growing at a great pace, the carrying capacity of the roads is very limited. The number of vehicles in urban areas are increasing, resulting in the slowing down of vehicular speed, unbearable pollution (air and noise) and increase in the number of road accidents. This has resulted in immense suffering for the commuters who face an ordeal travelling to and from their workplace. So, there is an urgent need to focus on the issue of providing an adequate transport system. Keeping this in mind, this paper aims to assess the importance of the role of the urban rapid mass transit system in context of sustainable urban development by analyzing data from the Annual Report of Delhi Metro Rail Corporation. The study also examines the performance of Delhi Metro and tries to explore its impact on the urban life of Delhi. Finally, the paper brings to the fore that Delhi Metro has a great impact in the reduction of road accidents, pollution level, vehicles taken out off-road and traffic congestion in Delhi.

Keywords Mobility · Pollution · Commuters · Mass-transit · Sustainable

19.1 Introduction

Metro rail is the most popular means of rapid mass transit in the modern world. The advent of the metro rail has been quite early in the developed nations. The history of metro in the world traces back to 1863 when the first underground metro had started

M. Kumari (✉) · A. Banerjee
Jawaharlal Nehru University, New Delhi, India
e-mail: menkachaudhary4@gmail.com

A. Banerjee
e-mail: banerjee.anb@gmail.com

R. B. Singh et al. (eds.), *Urban Health Risk and Resilience in Asian Cities*,
Advances in Geographical and Environmental Sciences,
https://doi.org/10.1007/978-981-15-1205-6_19

in London and then the technology spread to other cities of the world. For instance, New York City Subway was constructed, having the largest number of stations, i.e., 421 stations. Some of the other subways are Paris Metro (France), Seoul (South Korea), Tokyo and Beijing, etc. (Lin Luan 2014).

The history of the Indian Metro is a recent one when the first metro was introduced in 1971 in Kolkata with the help of the Soviet Union and Germany and started operating in 1984.[1] Subsequently, it was constructed in Chennai, Delhi, Bangalore, Mumbai, etc. At present, India has 8 operational metro systems with 324 km metro lines being operational and 520 km of lines are under construction.[2] The Delhi Metro opened its first corridor in 2002 between Shahdara and Tis Hazari. Presently, Delhi Metro consists about 213 km operational metro lines and 160 stations. The network of Delhi Metro is spread to cross boundaries also and connects Noida, Ghaziabad (Uttar Pradesh), Gurugram and Faridabad (Haryana) to Delhi.[3]

The metro rail has been working as an instrument by providing comfortable and eco-friendly services to the increasing population of urban centres like Delhi (see footnote 3). Urban population in India has been increasing in a very rapid pace. The number of cities with population of one million and above steadily increased from 5 in 1951 to 35 in 2001 and 53 in 2011.[4] The urban areas work as engines of growth and create skills and wealth for the nation and facilitate employment for the millions of migrants from the rural areas. Thus, mobility is a very significant determinant for the economic growth of urban areas. Economic activities like trade, commerce and industries will only flourish when accessibility is good with fast and placid mobility.[5] In this context, the transport infrastructure services and facilities in urban areas are thus amongst the most crucial factor for the growth of any urban economy (Nallathiga 2011). Urban development takes place only around such activity generators. When people and materials are transported at minimal investment and operating cost, only then the substantial contribution to city efficiency is possible. Therefore, an adequate, efficient and well-developed transport system permits cities and towns to become accelerators of social, economic, and industrial development (see footnote 5).

There has been accelerated population increase in towns and cities in contemporary times. With this growing population, the number of vehicles has also increased but the capacity of roads is limited which results in the slowing down of vehicular speed, high pollution (air and noise), and increase in the number of rail accidents (Advani and Tiwari 2005). All this lead to the considerable inconvenience among the commuters as well. More vulnerable groups like the poor people, suffer the most. They feel tired enough when they reach their workplace and that affects their productivity. In turn, the quality of life also suffers extremely (Advani and Tiwari

[1] Kolkata Metro Rail Corporation Limited.

[2] Business Standard India; (21 September 2016) "324 km of metro rail operational in India now, says Venkaiah Naidu".

[3] Delhi Metro Rail Corporation Limited.

[4] Census of India, 2011.

[5] www.populationfirst.org.

2005; Wright and Fjellstrom 2003). Thus, it is time to think in a proper perspective about the issue of providing an efficient and adequate transport system without losing any time in planning and implementing metro rail systems. If, however, looseness persists, the quality life of commuters and residents will deteriorate and further it will hinder the economic as well as social progress of the country (Breithupt et al. 2010; Mittal 2015). Therefore, there is a need of rail-based transport system in urban areas which will help in bringing the order in the city traffic. These would also bring the attitudinal and social changes among the citizens with a sense of discipline and cleanliness (Mittal 2015).

19.2 Objectives

Against such a backdrop, the aim of the study is to assess the importance of the role of urban mass rapid transit system in the context of sustainable urban development. An effort has been made to examine the performance of Delhi Metro and to find out its impact on the urban life of Delhi by studying the relationship between trends in ridership of metro train and patterns of road accidents and pollution level by considering various pollutants found in the ambient air of the city.

19.3 Database and Methodology

Data for the present study has been collected from various sources. Data on the trends in ridership and route length of metro has been collected from the Annual Report of Delhi Metro Rail Corporation. Data on fatal accidents in Delhi has been taken from the "Ministry of Home Affairs". The data on the trends in different pollutants has been compiled from the official site of Central Pollution Control Board. Other data used in the study has been collected from various reports and studies.

At first, an effort has been placed to compare the various options of Mass Rapid Transit System (MRTS) on the basis of literature and data. After establishing the fact that metro is the most viable option, though financially very expensive, the study focuses on the assessment of Delhi Metro Rail. The impact of Delhi Metro on road accidents, pollution level, vehicles taken out off-road and in reducing the congestion factor are presented through bar graphs.

19.4 Literature Review: Importance of MRTS

"Choices on public transit options are choices about the city's future".[6] The nature of a city centre is that it is the highest point of accessibility from both within and the peripheral areas of city. This high accessibility is very crucial for several functions, and in particular for those nodal functions that serves a large area and a wide labour market (Wright and Fjellstrom 2003; Phil 2013). Phil (Phil 2013) underlines the importance of MRTS in increasing accessibility. "The limited supply of road space feeding the city centre is exhausted, and the only possible relief would seem to be through the development of a mass transit system that makes the best use of the available road space" (Phil 2013). Mass transit system is the vein of metro cities. It makes the movement of urban population comfortable and saves time and reduces pollution and congestion on the road. "Mass Rapid Transit" (MRT) is used to elaborate the modes of urban transport which carry a large number of passengers quickly. They are considered to be well-defined corridors connecting suburbs to city centres (3–4, 7).

Halcrow (Halcrow 2000) asserts the importance of mass rapid transit in developing cities as it carries a large number of passengers rapidly. Also, Mass Rapid Transit (MRT) can play a significant role in improving overall condition of urban transport network. Clear efficiency and environmental benefits are derived from transferring urban travellers from their personal cars to public transport and non-motorized transport (Phil 2013; Acharya and Morichi 2007). Mass Rapid Transit System, as a technical improvement, is considered one of the most change-influencing transport modes. This is due to quality of influence it produces such as lessening pollution, providing orderliness of the flows of users and rising travelling speed of an accessibility increment (Javier 2004).

Among the means of mass rapid transits, metro rail is the best option. As Halcrow (Halcrow 2000) suggests that Metros are more viable option than any other means of MRT. Metro alone can carry very large flows of passengers, even several times more than other MRTS like Busway, Suburban Rails and the Light Rail Transit (LRT). All these Mass Rapid Transit Systems have a greater strategic impact on the city structures as they have the capacity to decongest the city corridor significantly. But the capital and operating cost is very high for them. At the same time they can be risky, so particular care is required in considering and developing them (Halcrow 2000; Acharya and Morichi 2007; Javier 2004).

Compared to other modes of mass transit networks, the metro is the best suited for a highly populated city like Delhi. A glance at Fig. 19.1 reveals that the metro has the highest capacity of carrying the passengers which is more than 80,000 people per hour per direction traffic (phpdt), whereas the other means like Bus Rapid Transport (BRT) has only 30,000 phpdt and buses (in mixed traffic) have only 10,000 capacity. In terms of Monorail and Light rail, the carrying capacity is up to 18,000 and 35,000 phpdt (Kochi Metro Project Report 2011). Monorails are best suited for narrow streets

[6]https://www.itdp.org/wp-content/uploads/2014/07/Sustainable-Transport-Mass-Transit-Options.pdf.

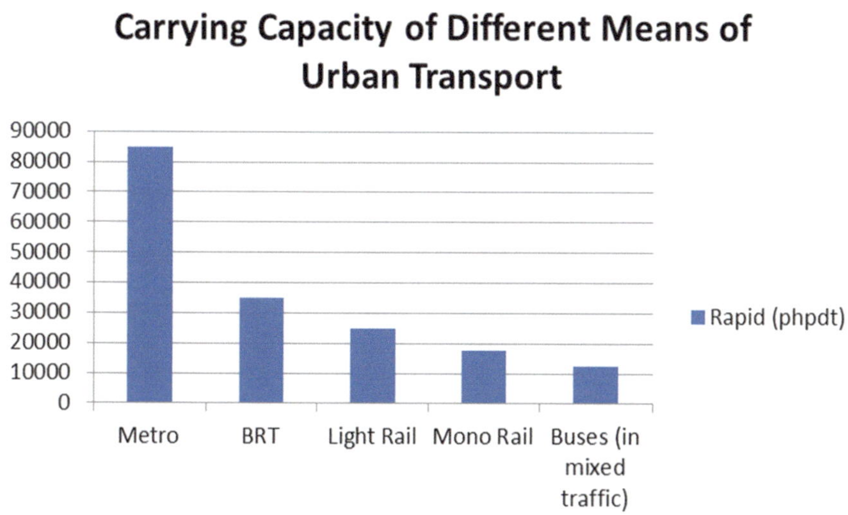

Fig. 19.1 Carrying Capacity of Different Means of Urban Transport in Kochi, 2011. (*Source* Kochi Metro Project Report 2011 prepared by DMRC)

and lLight rail trains are suited on the streets along with road traffic, but the metro needs a fully segregated/dedicated path (The Hindu (2015). In less dense urban areas, where commutation is low, the metro cannot be an economically viable option. This is because it needs a lot of resource and cost to construct underground or elevated metro network and also involves high operation cost. However, in cities like Delhi, which has a large population, and great number of daily commuters, metro rail can be the most viable option environmentally and economically as well (Advani and Tiwari 2005; The Hindu (2015).

19.5 Results and Discussions

Metro is one of the important means of mass rapid transportation in Delhi. Delhi Metro is still in the development stage and all parts of the city are not fully connected with metro network. Two phases of metro network has been completed and the third phase is in progress (Annual Report and Delhi Metro 2015). In 2009–10 (Fig. 19.2) there was only 95 km route length. In 2010–11, it rose to 161 km and in 2011–12 the route length increased to 190 km and remained the same for next two years. The total annual ridership is on rise since 2009–10. Annual ridership of metro increased by 50% in 2010–11 and 32% in 2011–12, for next two years, it increased by 15 and 13% in spite of no increase in length of metro route (Annual Report and Delhi Metro 2015).

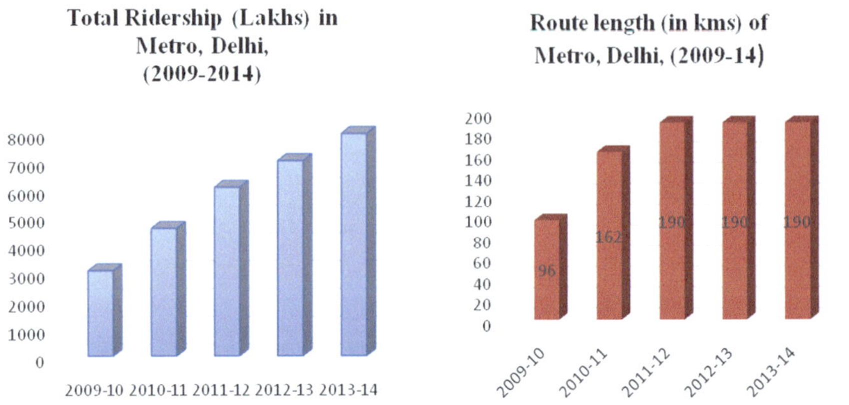

Fig. 19.2 Total Ridership and Route length of Metro in Delhi, 2009–2014. (*Source* Annual Report of Delhi Metro Rail Corporation, 2015–16)

19.5.1 Metro and Road Accidents

The increase in annual ridership of metro rail in Delhi coincides with the decrease in the total fatal road accidents in Delhi. This decline in the total fatal deaths has been attributed to the spread of metro rail as a means of public transportation as Badami and Haider (Badami and Haider 2007) also suggest that the public transport accounts for lower vehicular pollution and fatalities than private motorized modes. The improvement in the public transport system with the introduction of "Metro Rail" in Delhi since 2002 has been attributed by reducing number of road traffic accidents (Jha 2013). According to Murthy et al. (Murty and Dhavala 2006), "Delhi Metro provides an alternative and comfortable mode of transport for a large number of people in Delhi with reduction in travel time and atmospheric pollution". This relationship (Fig. 19.3) between increase in ridership and decline in road accidents (Annual Report and Delhi Metro 2013; Fatal Accidents in Delhi 2010) is explained by the fact that increase in the riders of metro lessens the number of private vehicles on the road. That will help in reduction in the road accidents due to reduction in traffic. As most of the accidents are caused by cars and motorcycles which are used by most of the commuters, reduction/withdrawal of the number of cars and motorcycles from roads will result in decline in road accidents (Advani and Tiwari 2005; Murty and Dhavala 2006).

Though the metro rail has contributed to the decline in fatal road accidents in Delhi, this cannot be solely responsible for the same as the reduction is not only because of decrease in the traffic on the road but because of other factors such as strict traffic rules, stringent law enforcement and other safety measures (The Hindu (2015; Jha 2013; Murty and Dhavala 2006). Data shows (Fig. 19.4) that there is decline in the total number of persons killed in road accidents per 10000 vehicles from 3.7 in 2009 to 3.2 in 2010 and 2.9 in 2011 (Total Number of Persons Killed in Road Accidents Per 10,000 Vehicles 2009). This declining trend in not only confined

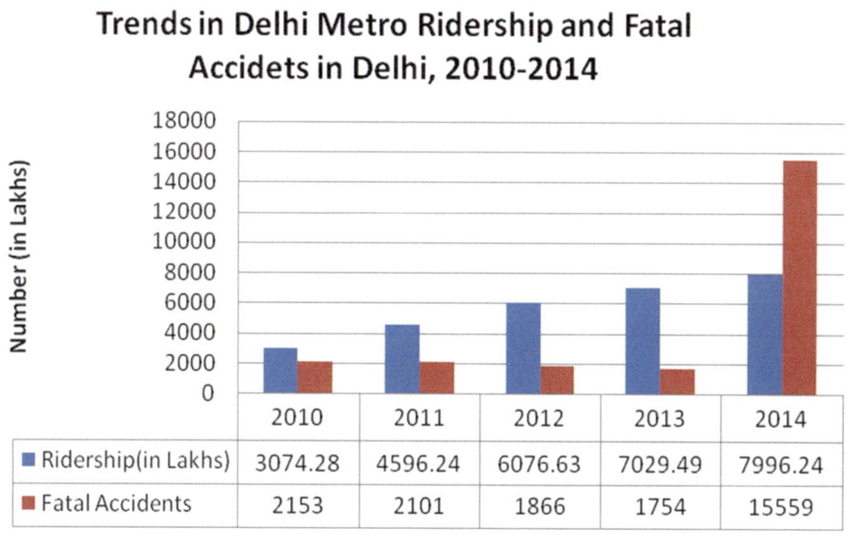

Fig. 19.3 Trends in Delhi Metro Ridership and Fatal Accidents in Delhi 2010–2014. (*Source* AnnualReport, Delhi Metro Rail Corporation, 2013–14; Ministry of Home Affairs, Government of India). (Accessed from Indiastat.com)

Fig. 19.4 Total Number of Persons Killed in Road Accidents Per 10,000 Vehicles 2009–2011. (*Source* Ministry of Road Transport and Highways, Government of India)

to Delhi but also very much conspicuous at the national level also and this might be the result of increased awareness about traffic rules and road safety.

19.5.2 Reduction of Road Congestion

Any public transport plays an important role in the reduction of congestion from roads. Due to increase in use of public transport, number of passengers using private

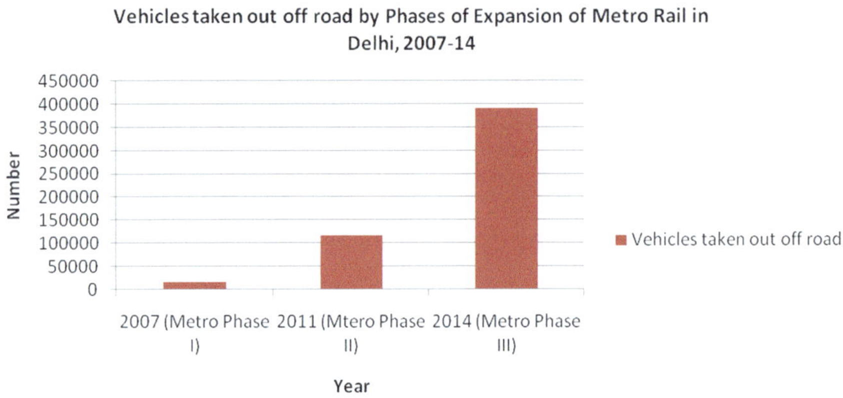

Fig. 19.5 Vehicles Taken Out off Road by Phases of Expansion of Metro Rail in Delhi, 2007–2014. (*Source* An assessment of DMRC by Central Road Research Institute, Indian Express, 1 January, 2015)

vehicles on the road declines and this helps in the reduction of traffic on the road. Less traffic results in reduction of jams and reduces the time taken to travel form one place to the other place within the city. Metro rail is very efficient in terms of carrying capacity and often results in the reduction of vehicular and commuters pressure on roads (Advani and Tiwari 2005; Halcrow 2000; Sabharwal and Goel 2011).

The Central Road Research Institute (Fig. 19.5) found that metro phases I and II, completed in 2006 and 2010, respectively, has taken off 16,895 vehicles in 2007; 1,17,249 vehicles in 2011; and 3,90,971 vehicles from roads in 2014 (Indian Express (1 January 2015). This is very significant, as it had helped in reducing not only congestion from the roads but also has helped in the reduction of pollution. Had the metro not been there, the level of pollution and congestion on the roads would have been even more worse than at present (Goel and Gupta 2015; Sharma et al. 2014).

19.5.3 Metro Rail and Air Pollution

The DMRC is being certified by the UN (United Nations) as the First Metro Rail and rail-based system in the world which will get Carbon Credits (worth 47 crore annually for next seven years and with increasing passengers this figure will also increase) for reducing Green House Gas Emissions (Goel and Gupta 2015; SocialCops (Feburary 4 2016). According to the data, the metro rail has helped to reduce the pollution levels by 6.3 tons every year which ultimately would help in reducing global warming as it is completely non-polluting and environment friendly.[7] Presently metro rails carry 31.7 lakhs (Maximum) passengers every day (Annual Report and Delhi Metro 2015; Goel and Gupta 2015; Sharma et al. 2014).

---

[7] http://www.delhimetrorail.com/press_reldetails.aspx?id=746xECETA6QIld.

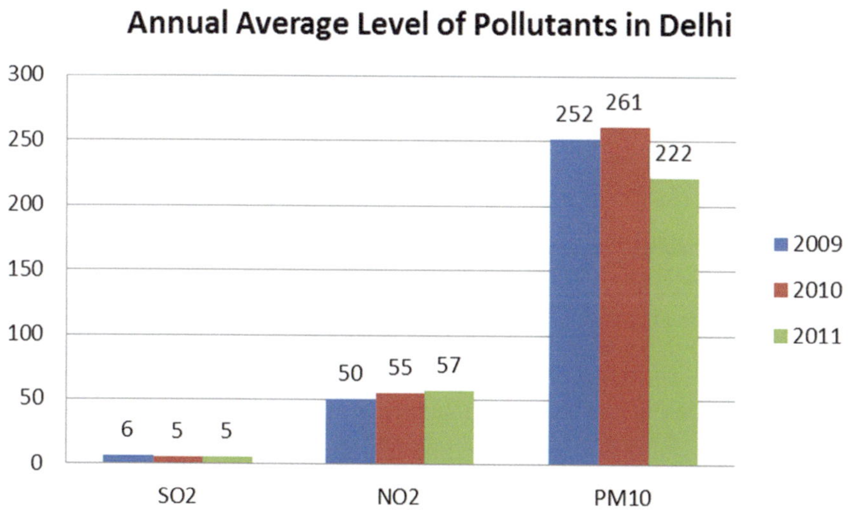

Fig. 19.6 Annual Average Level of Pollutants in Delhi, 2009–2011. (*Source* National Air Quality Measure Program (NAMP), Central Pollution Control Board, Accessed from Indiastat.com)

Though overall pollution level in Delhi has increased, the pollution level in case of absence of metro would have been even more severe. The annual average level of NO_2 concentration in the atmosphere of Delhi has been increasing continuously (Fig. 19.6) but with a decreasing rate between 2009 and 2011. PM10 has increased from 252 in 2009 to 261 in 2010 but declined in 2011 to 222. The concentration of SO_2 has declined from 6 in 2009 to 5 in 2010 and remained the same in 2011 (the numbers are in ug/m^3) (State-wise Annual Average Level of Pollutants 2009).

19.5.4 *Problems and Future Prospects of Delhi Metro*

Poor service of metro feeder buses is the main obstacle in the optimization of benefits from metro services. Feeder system are very time consuming and irregular in the services. Hence passengers often opt for cabs like (Uber pool and Ola share) at end points of metro route. There are some road transit means, which provide feeder services to metro, viz., Haryana Roadways, DTC, Feeder Buses, GrameenSewa, etc. Problem of noise pollution during the construction and operation of metro station is another very serious problem. The Government of NCT of Delhi has decided to launch a para-transit scheme where the high capacity three wheelers and similar other vehicles with seating capacity of six passengers will be given permit in rural areas, unauthorized, resettlement colonies and Jhuggi Jhopri (J.J) clusters.[8]

---

[8]http://www.delhi.gov.in.

DMRC is also planning for the integration of the ticket. If this happens, then the same ticket will be valid in metro train as well as in the buses (Annual Report and Delhi Metro 2015). However, it will only get success if the commuters will be willing to accept the added transfer time and transfer costs.

19.6 Conclusion

As metro carries a large number of passengers rapidly, it can play an important role in improving overall condition of urban transport. As the above discussion reveals, the high efficiency (in terms of carrying capacity and in reducing fatal accidents) and environmental benefits are derived by transferring urban travellers from personal cars to public transport and non-motorized transport, mainly the metro. This is due to the quality of influences it produces such as lessening pollution and providing orderliness in the flows of users and rising travelling speed. But construction cost, operational cost and maintenance cost of metro are so high that it is not financially viable for small cities. For big metropolitan cities where the number of daily commuters is very high, it is a boon. Delhi Metro had played great role in the reduction of traffic from road, reducing accidents and road congestion. It has also contributed in the reduction of emission of greenhouse gases and has helped in the control of environmental degradation. But there are certain obstacles in achieving cent percent efficiency of the metro and optimizing its utility, these being ineffective and unreliable feeder services. There should be frequent service of feeder buses and also integrated multi-modal ticketing system that will enhance the efficiency of metro and enhance comfort level to the commuters.

References

Acharya SR, Morichi S (2007) Motorization and role of mass rapid transit in East Asian megacities. IATSS Res 131(2)

Advani M, Tiwari G (2005) Evaluation of public transport system: case study of Delhi metro. Transport Research and Injury Prevention Programme, Indian Institute of Technology, Delhi

Annual Report, Delhi Metro (2013–14) Delhi Metro Rail Corporation

Annual Report, Delhi Metro (2015–16) Delhi Metro Rail Corporation

Badami MG, Haider M (2007) An analysis of public bus transit performance in Indian cities. Transp Res Part A: Policy Pract 41(10):961–981

Breithupt M, Wangner A et al (2010) Challenges of urban transport in developing countries- a summary. Transport Policy Advisory Services, GTZ

Fatal Accidents in Delhi (2010–2014) Ministry of Home Affairs, Government of India. Indiastat.com

Goel D, Gupta S (2015) Delhi Metro and air pollution. Department of Economics, Delhi School of Economics

Halcrow F (2000) World Bank urban transport strategy review- mass rapid transit in developing countries. Department for International Development, Halcrow Group Limited, London

Indian Express (2015) Delhi Metro keeps 3.9 lakhs vehicles off road in 2014

Javier P (2004) Mass rapid transit system and urban land spatial interaction and impacts: case study of LRT-1 metro Manila. Philippines, International Institute for Geo- Information Science and Earth Observation
Jha SK (2013) Road safety in Delhi: acritical analysis of the role of Public Institutions. International Institute of Social Sciences, The Hague, Netherlands
Kochi Metro Project Report (2011) Delhi Metro Rail Corporation
Lin Luan (2014) A short history of world metro systems-in pictures. The Rockefeller Foundation. https://www.theguardian.com/cities/gallery/2014/sep/10/-sp-history-metro-pictures-london-underground-new-york-beijing-seoul
Mittal T (2015) Smart urban mobility: road less travelled. Indian Institute of Management, Bangalore
Murty MN, Dhavala KK et al (2006) Social cost- benefit analysis of Delhi Metro. Institute of Economic Growth, Delhi, India. http://mpra.ub.uni-muenchen.de/1658/
Nallathiga R (2011) Mass urban transportation in Indian cities: features of three models and learnings. In: Asian urbanization conference, Hyderabad
Phil F (2013) Mass rapid transit systems for cities in the developing world. Transp Rev, 2003, 23(3):299–310
Sabharwal S, Goel M (2011) Delhi Metro: a solution to the problem of congestion and pollution in the modern city like Delhi. Int J Bus Econ Manag Res
Sharma N, Singh A et al (2014) Emission reduction from MRTS projects: a case study of Delhi Metro. Atmos Pollut Res 5:721–728
SocialCops (2016) How sustainable and successful is the Delhi Metro?
State-wise Annual Average Level of Pollutants, 2009–2011 under National Air Quality Measure Program (NAMP), Ministry of Statistics and Programme Implementation. Indiastat.com
The Hindu (2015) Impact of Public Transport on Delhi
Total Number of Persons Killed in Road Accidents Per 10,000 Vehicles (2009–2011) Ministry of Road Transport and Highways, Government of India. Indiastat.com
Wright L, Fjellstrom K (2003) Sustainable transport: a sourcebook for policy-makers in developing cities. Module 3a: Mass Transit Options, GTZ

Chapter 20
The First Survey of Flood Disaster Preparedness in Hanoi, Vietnam

Minh Dinh Nguyen

Abstract Typhoons and floods are annual threats to Vietnam due to its geographic location and socio-economic conditions. Therefore, it is necessary to establish flood disaster preparedness indices (FDPI) to improve flood disaster management by local governments and communities. This paper presents the first FDPI survey in Hanoi to assess local flood disaster preparedness levels. A total of 5 districts, 12 communes, and 1 ward were surveyed and 36 questionnaires were filled up by district, commune, and ward leaders and officers. The collected data were processed and analyzed to obtain useful FDPI. The survey results can help local communities and governments improve their plans to better cope with flood disasters. The survey contributes to capacity building for disaster management research in Vietnam.

Keywords Survey · Flood disaster · Preparedness indices · Vietnam

20.1 Introduction

Disaster management plays an important role in human survival and development. It is a challenging task due to the rise of disasters leading to heavy losses of human life and property in the context of rapid urbanization and climate change. This situation calls for stronger national and international efforts to increase disaster preparedness.

Understanding the role of flood disaster preparedness, The International Centre for Water Hazard and Risk Management (ICHARM) has developed a research project on Development and Application of Flood Disaster Preparedness Indices (FDPI). Within the FDPI project, different communities have been selected as survey targets in the Typhoon Committee (TC) member countries.

To date, in Vietnam, a TC member country, there have been some scientific and economic studies on natural disasters in general and flood disasters in particular (Benson 1997; GoV 2005; World Bank 2005, 2009, 2010; Huu et al. 2011; Chen

M. D. Nguyen (✉)
VNU University of Science, 334 Nguyen Trai, Thanh Xuan, Hanoi, Vietnam
e-mail: ndminhvnu@yahoo.com

R. B. Singh et al. (eds.), *Urban Health Risk and Resilience in Asian Cities*,
Advances in Geographical and Environmental Sciences,
https://doi.org/10.1007/978-981-15-1205-6_20

Chang et al. 2012). However, research to measure and monitor flood disaster preparedness in Vietnam is still lacking. Therefore, a direct FDPI survey in Vietnam is highly motivated and Hanoi was selected as a pilot city following the TC's consideration.

The objectives of the first FDPI field survey in Hanoi, Vietnam were to

(1) Introduce ICHARM, the FDPI project to local communities in Hanoi;
(2) Obtain FDPI survey data in collaboration with local partners;
(3) Learn from local experience, and
(4) Establish community network through FDPI activities.

20.2 Flood Disaster Preparedness Indices

Preparedness is usually seen as exercises to improve reactions and adaptabilities. Nevertheless, the emphasis is gradually being placed on recovery, i.e., not only in order to react adequately amid and after disaster but also in order to effectively explore difficulties related to short- and long-term recovery. Preparedness measurements comprise the different objectives or ends that preparedness is to achieve. Exercises are strong activities to achieve those objectives. Analysis units in disaster preparedness research include families, organizations, networks, and associations. A group of people is a social unit that could border on a nearby political competence. Networks range from small towns with limited administrative resources for open welfare and crisis to large regions with urban preparedness (Sutton and Tierney 2006).

To characterize, measure, and monitor flood disaster preparedness of a community, ICHARM has established eight flood disaster preparedness indices (FDPI) (Fig. 20.1), including hard countermeasure, flood disaster mitigation plans and standards, flood disaster mitigation systems, evacuation plans and systems, emergency and recovery plans and systems, leaderships and organizations' collaboration, information and education for local residents and community strength (Nakasu and Okazumi 2012).

The main applications of FDPI include

- Local community leaders can know what kinds of things should be done for coping with flood disasters and also can evaluate their preparedness level regularly, not only that, they can refer other communities experience and take specific actions for raising their preparedness level.
- FDPI can visualize the situations and progresses of their flood disaster preparedness level and encourage their activities.
- International societies and aid organizations can know communities' situations, needs, and other information for international cooperation.
- Utilizing FDPI platform, our experience and useful lessons can be shared.

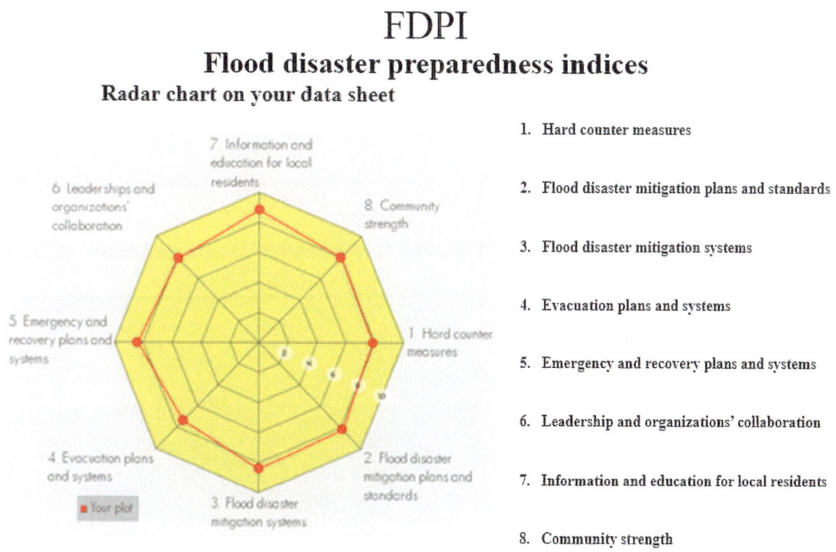

Fig. 20.1 Flood disaster preparedness indices

20.3 FDPI Survey in Hanoi

20.3.1 General Information of the Hanoi Area

The city of Hanoi is located on the floodplain of the Red River about 100 km from the sea. Geographically, the area is defined by latitudes 20°55′–21°23′ N and longitudes 105°44′–106°54′ E. It borders the provinces of Thai Nguyen, Vinh Phuc to the north, Bac Giang, Bac Ninh and Hung Yen to the east, Phu Tho, Hoa Binh to the west and Ha Nam to the south. Administratively, Hanoi consists of 10 urban districts, called quan (Tay Ho, Cau Giay, Dong Da, Ba Dinh, Long Bien, Hoan Kiem, Hai Ba Trung, Hoang Mai, Thanh Xuan and Ha Dong), Son Tay town and 18 suburban districts (Tu Liem, Gia Lam, Thanh Tri, etc.) (Fig. 20.2). Hanoi has a total land area of 3,344.6 km^2.

Hanoi's topography is gradually lower southward, eastward and from mountain areas to hilly and floodplain areas. The floodplain is generally flat and low lying with an elevation ranging from 3 to 13.6 m. A dike system was constructed along the Red River to protect the city from flooding. The Red River passes the central part of the study area carrying annually a large amount of sediments. Other smaller rivers in the area are the Da, the Duong, the Day, the Ca Lo, the Nhue, and the To Lich. Within the city there are several lakes, ponds and, canals which serve as drainage basin and other purposes.

Hanoi is characterized by a tropical monsoon climate. The year is roughly divided into two seasons: one is cold and dry, the other is hot and humid. The dry season starts

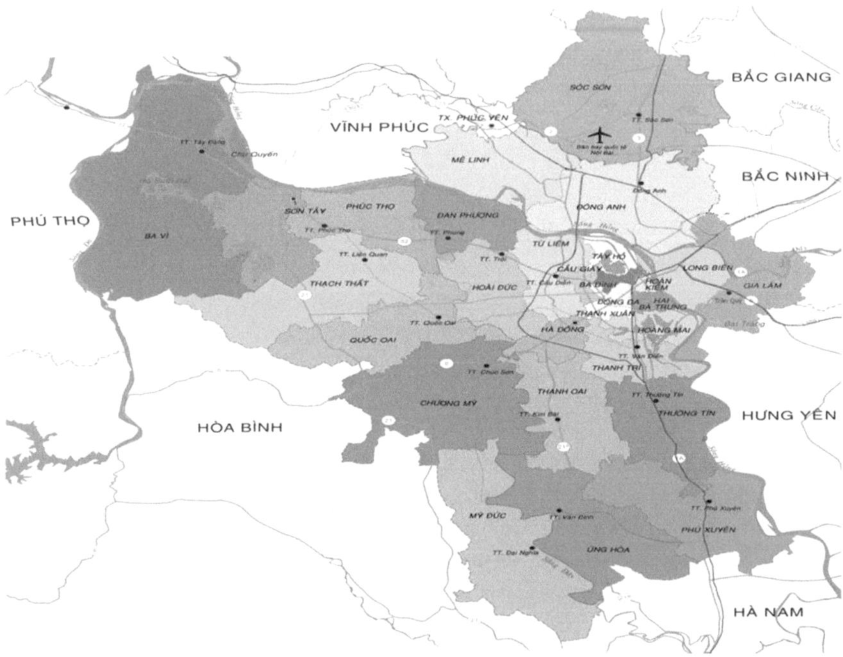

Fig. 20.2 Location of Hanoi and its administrative units. Reprinted from Ly 2008 licensed under CC BY 3.0

from November to April with the temperature ranging from 10° to 23 °C. Winters are cool with average temperature of 17 °C. The average annual temperature is 23 °C. The rainy season begins from May to October. The average rainfall is about 2000 mm for the whole year.

According to 2009 statistics, the urban population of Hanoi accounts for 2,632,087 in the total number of 6,451,909 inhabitants. The city population density is about 2,881 people/km^2, but it is not uniformly distributed among different districts. The population density reaches 19,163 people/km^2 in inner Hanoi and 1,721 people/km^2 in Hanoi suburbs. The most populated parts of Hanoi are the quarter of 36 old streets (Hoankiem district) and some wards in Haibatrung and Dongda districts where the population density can exceed 35,000 people/km^2. In rural districts such as Soc Son, Ba Vi, My Duc, the population density can be less than 1,000 people/km^2.

The urban population in Hanoi has continued to grow from 1274.9 in 1995 to 2046.1 in 2005 and 2804.0 thousand people in 2010. The year 2008 marked a sharp increase in Hanoi's urban population after widening the city administrative boundary to include the entire Ha Tay province and part of Hoa Binh and Vinh Phuc provinces.

20.3.2 The Need for FDPI Survey in Hanoi

Hanoi, the capital of Vietnam, was founded officially in the year 1010 by the Ly dynasty with the name of Thang Long, the soaring dragon. Historical floods with great damage for Thang Long occurred in the years 1078, 1121, 1236, 1238, 1243, 1270, 1445, 1467, 1491, 1506, 1630, 1713, 1728, 1806, 1809, 1821, 1827, 1844, 1893, and 1915). The dike system has never broken due to disasters since Independence Day in 1945 even during the biggest flood years of 1969, 1971, and 1996 (World Bank 2009).

Hanoi now is a fast-growing city. It faces many challenges in flood control and prevention due to its geographic location and socio-economic conditions in the context of rapid population growth, urbanization, industrialization, and climate change. As an unforgettable experience, the city suffered a historic flood in late October and early November 2008 (Fig. 20.3). This flood disaster caused heavy losses of life and property. However, the flood is not well documented yet and the knowledge of the flood is far from enough due to little and fragmented research. Therefore, more questions should be asked and answered by new and integrated research to better understand its root causes and impacts as a sound basis for improved management of the flood risk in the area for sustainable development. In this direction, the FDPI survey in Hanoi is highly motivated.

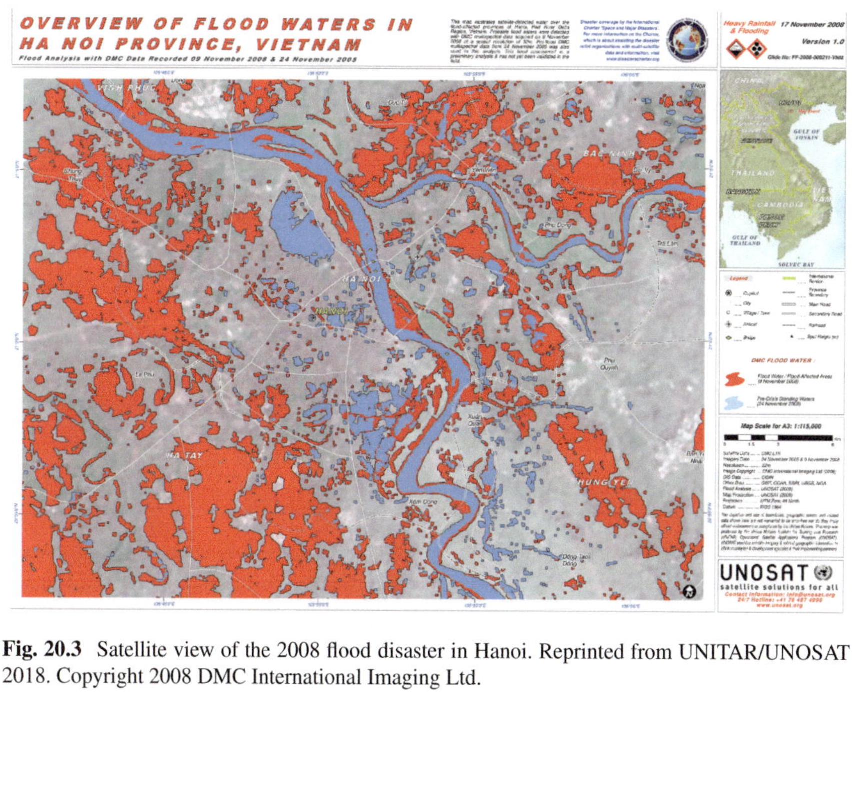

Fig. 20.3 Satellite view of the 2008 flood disaster in Hanoi. Reprinted from UNITAR/UNOSAT 2018. Copyright 2008 DMC International Imaging Ltd.

20.3.3 *Survey Methods*

20.3.3.1 Planning the Survey

Survey planning was done by ICHARM and VNU University of Science. The questionnaire was developed in Vietnamese. It is 18 pages long containing 79 questions. It is simplified and contains less questions than the version used in Japan. The questionnaire was photocopied for distribution and answering questions.

20.3.3.2 Selection of Survey Targets

The lowest or minimum administrative unit that has budget and staff in Vietnam is a *commune* in rural areas and a *ward* in urban areas. To localize the FDPI survey in Hanoi, this minimum administrative unit was selected as a target. Three types of communes in a district were selected for survey, including communes heavily affected, affected, and unaffected by flood.

20.3.3.3 Conducting the Survey

Following the meetings of the ICHARM researchers with VNU University of science and visiting Hanoi flood control facilities, the direct FDPI questionnaire survey was conducted at the commune and district levels.

The FDPI survey procedure at the commune and district levels consists of four main steps:

(1) Introducing ICHARM, the FDPI project and the main purpose of the survey;
(2) Doing the FDPI questionnaire survey (distributing questionnaires to participants, assisting participants in answering questions and collecting questionnaires filled up by participants);
(3) Getting the comments from the participants;
(4) Discussing the issue, how to share the flood disaster experience, local knowledge, etc. with participants.

20.3.3.4 Analyzing the Survey Data

In this second and important part of the survey, the data collected were reviewed, evaluated to establish its usefulness and appropriateness for inclusion in different formats to best fit the purpose outlined. The data were put in tables, diagrams, and maps. The scales and symbols were chosen so that they are suggestive and allow to extract as much information as possible. The locations of survey were mapped using Google Earth and Google Maps for visualization. The questions answered by participants in the questionnaire were statistically analyzed to present FDPI in the form of radar charts.

20.3.4 Survey Results

A visit was made to the flood control facilities in Hanoi such as the regulating lake of Dong Da, pumping station, JICA drainage project in Hanoi and West Lake, the largest oxbow lake in Hanoi.

The direct survey was conducted at 5 districts, 12 communes, and 1 ward. Figures 20.4 and 20.5 show the location of all communes and districts surveyed for FDPI development in Hanoi during August 2012.

The results of data processing and analysis in the form of radar charts with markers are shown as examples in Table 20.1.

These radar charts have a strong visual impact. Looking at the radar charts one can see the big picture of all eight FDPIs. The charts define full performance for each FDPI show the gaps between measured and full performance, and provide data to support priorities for improving performance. They are indeed very effective tools for comparing multiple communes, districts based on different FDPIs. The chart can be updated with new data to view progress over time.

Hence, these survey results help communities and local governments in Hanoi improve their plans to better deal with flood disaster.

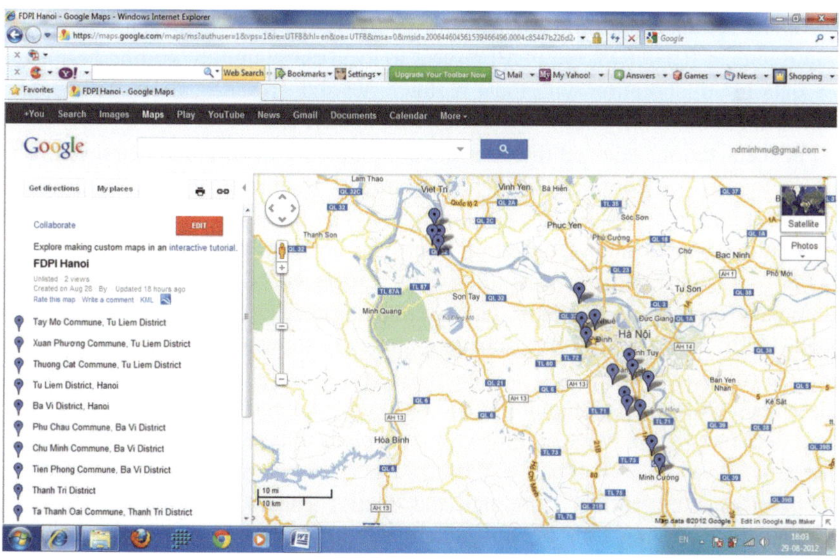

Fig. 20.4 Map view of surveyed communes and districts in Hanoi. Map data ©2012 Google-Edit in Google Map Maker

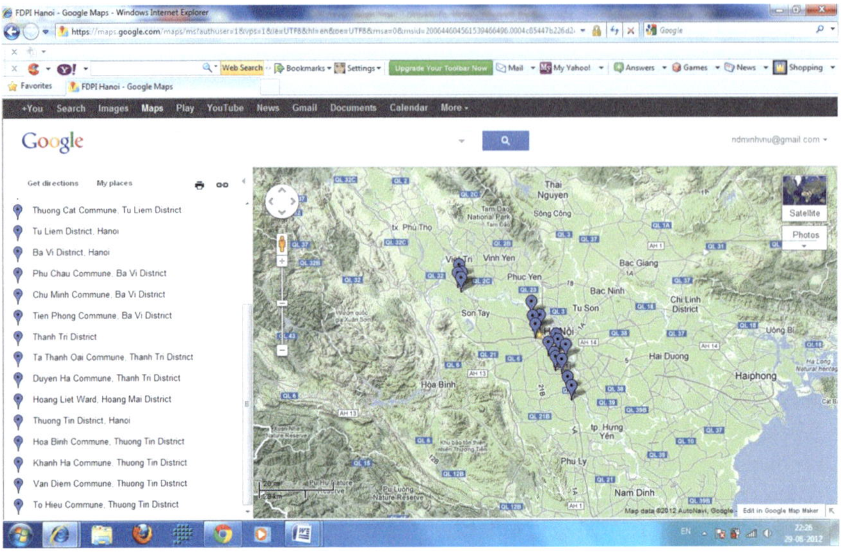

Fig. 20.5 Terrain view of surveyed communes and districts in Hanoi. Map data ©2012 Google-Edit in Google Map Maker

20.4 Conclusion and Recommendations

The first survey of FDPI in Hanoi was successfully completed by ICHARM and VNU University of Science during August–December, 2012. The survey team succeeded in introducing ICHARM, the FDPI project to local communities; getting FDPI survey data in collaboration with local partners; learning from local experience, and establishing community network through FDPI activities.

For localization, selected districts and communes in Hanoi were surveyed. The survey focused on commune leaders and officers in charge of flood control and prevention at the commune and district levels. The flood disaster preparedness at household, business, and organization levels was not surveyed.

Thirty-six questionnaires were filled up by leaders and officers from 5 districts, 12 communes, and 1 ward in Hanoi. They were later processed and analyzed to obtain useful FDPI. The resulting radar charts graphically show areas of relative strength and relative weakness, as well as depicting general overall performance of FDPIs. These results will help communities and local governments raise their awareness of flood disaster preparedness and improve their plans to better deal with future flood disasters.

For disaster mitigation, commune and district leaders in Hanoi are aware of the need for applying four in situ guidelines before, during, and after the disaster. The social capital is mobilized in flood control and prevention including mass organizations such as women's union, youth union, veteran association, etc.

Table 20.1 Radar charts showing the FDPI by commune and district in Hanoi, Vietnam

Commune	District	FDPI
Thuong Cat	Tu Liem	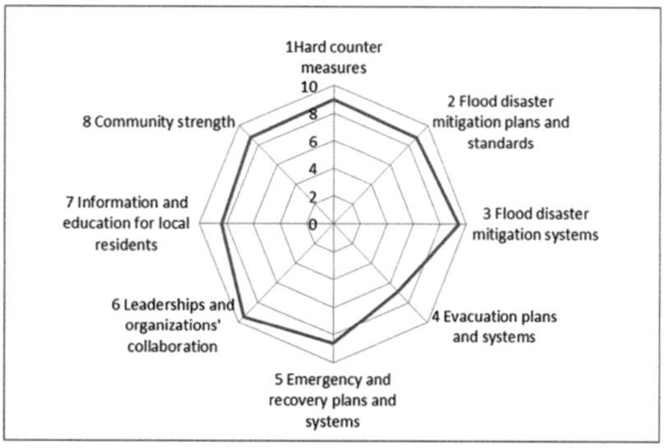
Phu Chau	Ba Vi	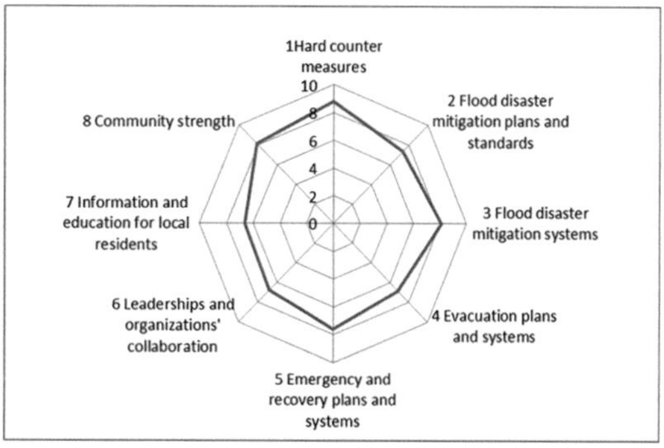
Duyen Ha	Thanh Tri	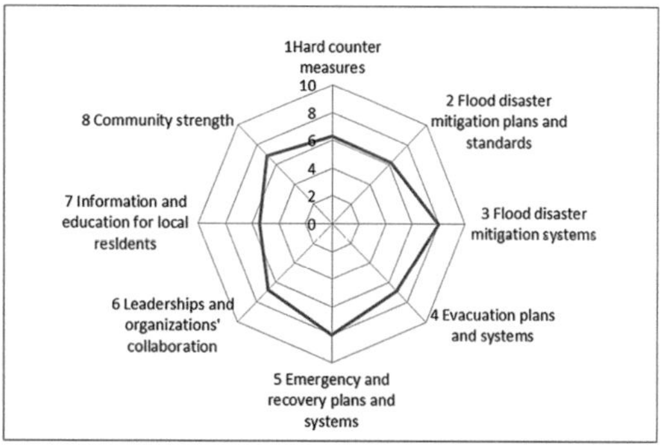

(continued)

Table 20.1 (continued)

Commune	District	FDPI
Hoang Liet Ward	Hoang Mai	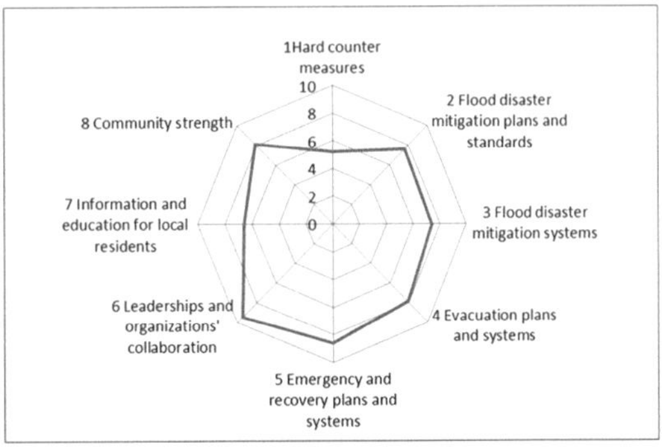
Khanh Ha	Thuong Tin	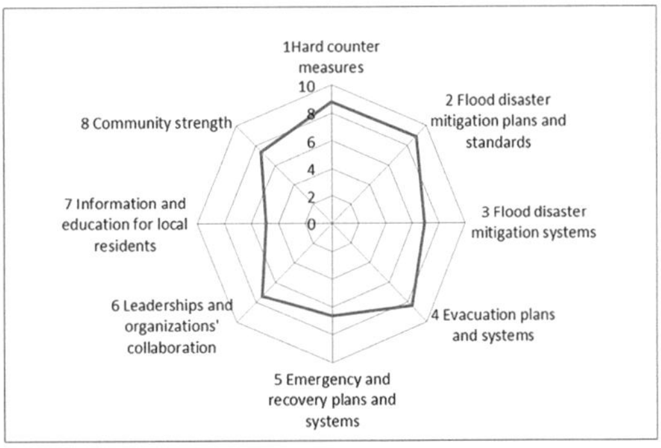

Natural, socio-economic conditions and flood situations vary from district to district, from commune to commune. All districts and communes surveyed have annual flood control plans with allocated budget and staff. However, all these administrative units still lack flood hazard maps prepared and published to guide proper land use to reduce flood disaster risks.

For successful planning and implementation of the FDPI survey at the commune or ward level in Hanoi, it is necessary to get permission and cooperation commitment from the city, district, and commune authorities in advance using different channels of communication. Questionnaires should be designed, tested, and improved to yield objective and useful results. They should be easy to read, understand, and fill up by target respondents.

It is necessary to construct and publish flood hazard maps for Hanoi at different scales to guide land use for flood control and prevention and future FDPI questionnaire surveys can and should be extended to household, business, and organization levels to better understand and raise flood preparedness levels for flood mitigation in the area.

The experience of the first FDPI survey in Hanoi can be applied to other districts, communes in Hanoi and other provinces in Vietnam to assess and improve flood disaster preparedness for disaster mitigation in the country in the years to come.

References

Benson C (1997) The Economic impact of natural disasters in Vietnam. In: Working Paper 98, Overseas Development Institute, London

Chen Chang T, Yen MC, Tsay JD, Nguyen Thanh NTT, Alpert J (2012) Synoptic Development of the Hanoi heavy rainfall event of 30–31 October 2008: multiple-scale processes. Mon Wea Rev 140:1219

GoV (2005) National report on disaster reduction in Vietnam, presented at the world conference on disaster reduction, Kobe-Hyogo, Japan, pp 18–22

Huu BT, Quang LN, Hal LTT, Hanh TTD, Guha-Sapir D (2011) Impacts of flood on health: epidemiologic evidence from Hanoi, Vietnam. Glob Health Action 4:6356

Ly L (2008) Map of Hanoi at the end of 2008. https://commons.wikimedia.org/wiki/File:B%E1%BA%A3n_%C4%91%E1%BB%93_H%C3%A0_N%E1%BB%99i.png

Nakasu T, Okazumi T (2012) Introduction to ICHARM and project on FDPI, ICHARM

Sutton J, Tierney K (2006) Disaster preparedness: concepts, guidance, and research. In: Report prepared for the Fritz Institute Assessing Disaster Preparedness Conference, Sebastopol, California

UNITAR/UNOSAT (2018) Overview of flood waters in Hanoi Province, Vietnam. https://disasterscharter.org/web/guest/activations/-/article/flood-in-vietn-2

World Bank (2005) Natural Disaster Hotspots: A Global Risk Analysis. Disaster Risk Management Series No. 5, Hazard Management Unit, The World Bank

World Bank (2009) City Profiles. A Primer on Reducing Vulnerabilities to Disasters, Climate Resilient Cities

World Bank (2010) Weathering the Storm: Options for Disaster Risk Financing in Vietnam

Chapter 21
Tradition Meets Innovation: Herbal Medicine as a Sustainable Source of Anticancer Agents

Shanoo Suroowan and Mohamad Fawzi Mahomoodally

Abstract Traditionally, herbs have been the immediate answer to human ailments in the pursuit of relief. The traditional claims surrounding the use of herbs to manage or treat various maladies have gained significant popularity throughout time. A common disease that requires urgent attention with regard to the development of novel therapeutic agents is cancer. In 2015, cancer accounted for more than 8.8 million deaths and was listed as the second top cause of mortality behind cardiovascular disorders. It is projected that by 2030, 21.7 million new cases of cancer will emerge resulting in around 13 million deaths due to an aging and growing population. Accordingly, a panoply of herbs have been attested to be used by various cultures around the globe in traditional systems of medicine against cancer. Even today, herbs remain an integral part of various cultures and traditions globally. A plethora of herbs are employed in the management of cancer and other ailments in traditional African, Ayurvedic, Arabic & Islamic, Chinese and Mexican medicine and have shown potent anticancer properties when assayed in vitro and in vivo. Hence, the validated therapeutic claims of herbs guarantee their innovative and sustainable use in an attempt to open avenues in anticancer extracts and or drug development.

Keywords Herbs · Traditional medicine · In vitro, in vivo · Anticancer agents

21.1 Introduction

Humans and herbs have shared a unique relationship even before recorded history (Rates 2001). Indeed, herbs have been a source of food, clothing, shelter, fuel, and medicine since time immemorial (Newman et al. 2000). Interestingly, the first reaction by primitive man in an attempt to assuage suffering has been the exploration of

S. Suroowan · M. F. Mahomoodally (✉)
Department of Health Sciences, Faculty of Science, University of Mauritius, 230 Réduit, Moka, Mauritius
e-mail: f.mahomoodally@uom.ac.mu; mohamadfawzimahomoodally@duytan.edu.vn

M. F. Mahomoodally
Institute of Research and Development, Duy Tan University, Da Nang 550000, Vietnam

R. B. Singh et al. (eds.), *Urban Health Risk and Resilience in Asian Cities*, Advances in Geographical and Environmental Sciences,
https://doi.org/10.1007/978-981-15-1205-6_21

the medicinal virtues of herbs (Panda et al. 2016). A plethora of herbs are used by various cultures around the globe such as in traditional Chinese, Egyptian, Indian, and Japanese systems of medicine. Even today, herbs remain an integral part of various cultures and traditions around the world (Boughrara and Belgacem 2016).

A common disease that requires urgent attention with regard to the development of novel therapeutic agents is cancer. It is the result of various genetic and epigenetic changes which lead to the initiation, progression, and propagation of carcinogenesis (Holmes et al. 2007). Moreover, it is the second leading cause of death worldwide after cardiovascular diseases. Accordingly, the International Agency for research on cancer projects that new cases from cancer will spike up to 14.1 million among which 8.2 million mortality figures will result (Cheung and Delfabbro 2016). Even in 2015, cancer accounted for 8.8 million deaths and was listed as the second top cause of mortality over the globe (Cheung and Delfabbro 2016). It is projected that by 2030, 21.7 million new cases of cancer will emerge resulting in around 13 million deaths due to an aging and growing population (Torre et al. 2015).

Various conventional medications have been developed for the management of cancer. Diverse classes of synthetic conventional anticancer medications available today include: alkylating drugs, anthracyclines and other cytotoxic antibiotics, antimetabolites, vinca alkaloids and etoposide and antineoplastic drugs among others. Nonetheless, the major toxicity experienced from conventional anticancer drugs coupled with their non-specificity in targeting cancer cells alongside their major side effects has drawn scientists toward the study of alternative forms of medicine, in this case, herbs (British National Formulary 2014).

Interestingly, the wide array of phytochemicals emanating from the plant kingdom offer an extraordinarily vast avenue for exploring new sources of template pharmacophores that can be harnessed as potential anticancer drug candidates. These pharmacological activities have resulted in the development of a plethora of conventional anticancer agents throughout time. For example, vinblastine; vincristine; the camptothecin derivatives; topotecan and irinotecan; etoposide derived from epipodophyllotoxin and paclitaxel (taxol) are all derived from plant origin (Cragg and Newman 2005; Rates 2001). Notably, of 50,000 plants known to withhold medicinal virtues worldwide, only around 5,000 have been screened for their active constituents while for a few others, only preliminary investigations have been conducted (Bibi et al. 2014).

For decades, the worldwide popularity and human fascination about herbs as a source of medicine in various cultures and traditions have prompted a major interest among scientists for research in this field. This has led to the emergence of a research area exclusively studying the medicinal uses of herbs in different cultures and traditions known as ethnobotany. Consequently, diverse herbs have been highlighted to be used in various traditional systems of medicine against cancer in ethnobotanical studies (Gurib-Fakim 2006; Traore et al. 2013). The goal of herb–drug research and development program is to discover single entity and multicomponent bioactive natural products that may serve as leads for the development of novel pharmaceuticals and address unmet therapeutic needs (Verpoorte 2000).

In this endeavor, this book chapter focuses to highlight herbs widely used since time immemorial against cancer in various traditional systems of medicine throughout the world. In addition, it is an attempt to validate their therapeutic claims evidenced from the results of in vivo and in vitro studies. Further study of the herbs underscored in this book chapter and of their corresponding phytochemicals will undeniably brighter avenues in cancer research. Consequently, to execute the search, major scientific databases such as Google Scholar, EBSCOhost, Science Direct and Pubmed were browsed for the keywords; "traditional medicine," "African traditional medicine," "Arabic Medicine," "Mexican Medicine," "Traditional Chinese Medicine," "Ayurvedic medicine" and "cancer," "in vitro" and "in vivo" studies among other terms. Altogether, 128 articles met the inclusion criteria and were incorporated to form part of this book chapter.

21.2 Herbal Medicine, Phytochemicals and Cancer

Traditionally, herbs have been the immediate answer to human ailments in the pursuit of relief. Even today, up to 80% of the world's population rely on herbs either as a source of food, shelter, or medicine (Boughrara and Belgacem 2016). The traditional claims surrounding the use of herbs to manage or treat various maladies has gained significant momentum recently, especially given the validated therapeutic potential of herbs through in vitro and in vivo studies (Mendonça-Filho 2006). Undoubtedly, studies on medicinal plants have proven that plants bear potentially low toxicity and exert minimal side effects (Ajay et al. 2009). These characteristics of medicinal plants are particularly of value in cancer therapy where patients are burdened with a panoply of side effects from allopathic drugs administration as well as their health is afflicted adversely due to the toxicity and non-specificity of modern drugs (Ramawat and Goyal 2008).

In furtherance, herbs are natural and are therefore the innate choice of various sufferers. They serve as potent anticancer agents and are also useful in the prophylaxis of cancer. Various grains, cereals, nuts, soy products, olives, beverages such as tea and coffee, and spices including turmeric, garlic, ginger, black pepper, cumin, and caraway confer a protective effect against cancer (Boon and Wong 2004; Chan et al. 2005; Clifford and Digiovanni 2010). The consumption of vegetables and fruits is another interesting avenue to explore in the prophylaxis of cancer as attested by several studies demonstrating the reduced risk of cancer following high consumption of cabbage, cauliflower, broccoli, Brussels sprout, tomatoes, apples and grapes (Chan et al. 2005; Vainio and Weiderpass 2006).

On the other hand, medicinal plants have been used safely by different cultures in various traditions for diverse purposes since antiquity and are by far more economical than synthetic agents. In contrast to allopathic medicine, herbal medicine bears the boon of addressing multiple diseases given the fact that their effects are concurrently extended to multiple systems of the body culminating into a broad spectrum of biological activities. Indeed, common biological activities of herbs include:

enhancement of the immune system, antibacterial, antiviral, anti-hepatotoxic, antiulcer, anti-inflammatory, antioxidant, anti-mutagenic, and anticancer effects among diverse other therapeutic effects (Rathee et al. 2009).

Over the years, a considerable number of herb-based secondary metabolites have been characterized and appraised for their role in cancer chemoprevention. In this regard, flavonoids are naturally occurring polyphenols and can be classified in these classes: flavonols, flavan-3-ols, anthocyanins, flavanones, flavones, isoflavones, and proanthocyanidins and are found in diverse plant sources. Interestingly, increased consumption of flavonoids is related to reduced cancer incidence. Flavonoids including apigenin, epigallocatechin-3-gallate (EGCG), luteolin, quercetin, chrysin, daizein and genistein, malvidin and citrus flavonoids have attracted particular interest in cancer research. Other secondary metabolites of interest in search of novel drug candidates against cancer include terpenoids, nitrogen, and sulfur containing compounds among others are illustrated in Table 21.1 (George et al. 2017).

21.3 African Traditional Medicine and Cancer

Africa, the largest continent of planet earth is the home of various ecosystems where dwells countless number of unique plant species (Light et al. 2005; Berrington and Lall 2012). Indeed, history has witnessed that the first phytochemical agents to be in widespread clinical use against cancer are vincristine and vinblastine derived from the Madagascar periwinkle plant *Catharantus roseus* (L.) G. Don which is traditionally employed to treat diabetes by various cultures (Mbele et al. 2017). Today, vincristine and vinblastine are still used in combination with other chemotherapeutic agents for the management and treatment of leukemias, lymphomas, advanced testicular cancer, breast and lung cancers, and Kaposi's sarcoma therapy (Cragg and Newman 2009).

Other success stories from herbs traditionally used for the management and treatment of warts and skin cancers include the discovery of the podophyllotoxin derived from the *Podophyllum* species (Podophyllaceae) such as *Podophyllum* peltatum L. and *Podophyllum emodi* Wall. ex Hook.f. & Thomson originating from the Indian subcontinent. Paclitaxel a common anticancer agent emanates from *Taxus brevifolia* Nutt., a plant widely used by native American tribes for the management of noncancerous conditions. Semisynthetic anticancer agents have also been derived from plant sources such as topotecan, irinotecan, etoposide, and teniposide (Kuete et al. 2014).

A profuse number of African traditional medicinal plants have shown promising activity against various cancer cell lines. Notably, *Salvia africana* L., *Zingiber officinale* Roscoe, *Vepris soyauxii* (Engl.) Mziray, *Ximenia americana* L., and *Ferula hermonis* Boiss, have demonstrated cytotoxic effects on cancer cell lines (Hussein et al. 2007; Kurapati et al. 2012; Kuete et al. 2013; Plengsuriyakarn et al. 2012; Voss et al. 2006; Kuete and Efferth 2015). In addition, *Anonidium mannii*, *Gladiolus quartinianus* A. Rich., and *Xylopia aethiopica* (Dunal) A. Rich. can induce apoptosis in

Table 21.1 Plant based metabolites and anticancer properties

Flavonoid	Plant species (Common English name)	Effect	Source
Apigenin (Flavone)	*Lycopodium clavatum* L. (club moss) and some vegetables such as *Petroselinum crispum* L. (parsley) and *Apium graveolens* L. (celery)	DNA protective effect against UV-B by removal of cyclobutene rings, inhibition of reactive oxygen species generation and down regulation of NF-kB in skin cells. Induction of cell cycle arrest and inhibition of cell proliferation in cancer cells. Dwindles the risk of ovarian cancer	Meyer et al. (2006), Das et al. (2013)
Epigallocatechin-3-gallate (Flavan-3-ol)	*Camellia sinensis* (L.) Kuntze (Green tea)	Traditional oxidative free radical scavenger Sets back phosphorylation of H2AX foci in HaCaT keratinocytes	Yin et al. (2008), Zhu et al. (2014)
Luteolin	*Salvia tomentosa* Mill. (Balsamic sage)	Induces apoptosis in human lung squamous carcinoma CH27 cells Activates the p53 gene in various cancer cells such as prostate and breast cancer cells and triggers apoptosis Induces apoptosis in multidrug-resistant cancer cells by promoting ROS generation, DNA damage activation of ATR/Chk2/p53 signaling, inhibit NF-kB signaling, activation of p38 and depletion of antiapoptotic proteins	Silva et al. (2008), Leung et al. (2005), Shi et al. (2007), Rao et al. (2012)

(continued)

Table 21.1 (continued)

Flavonoid	Plant species (Common English name)	Effect	Source
Quercetin (Flavonol)	*Pyrus malus* L. (apples), *Allium cepa* L. (onion) and *Allium sativum* L. (garlic)	Induce DNA damage in cancer cells Found to modulate DNA repair enzymes to withstand damage from hydrogen peroxide challenge in Caco-2 human epithelial colorectal adenocarcinoma cells The metabolite, quercetin-3-Oglucuronide inhibits the binding of noradrenaline significantly to α2-adrenergic receptor and suppresses DNA damage by challenge of 4-hydroxyestradiol and noradrenaline in MCF-10A normal human breast cancer cells	Yamashita et al. (1999), Min and Ebeler (2009), Yamazaki et al. (2014)
Daidzein and genistein	*Glycine max* L. (soybean) and *Pueraria mirifica* L. (Kwao Krua)	DNA protective effects against UV-B in skin fibroblasts Suppression of cyclooxygenase-2 and DNA-damage-inducible (Gadd45) genes effectively in treated skin samples When combined together daidzein and genistein have synergistic chemoprotective properties against UV-B	Iovine et al. (2011)
(Antocyanins) malvidin-3-O glucoside	Red wine	Promotes proapoptotic protein expression in endothelial cells	Paixao et al. (2012)

leukemia cells while *Tulbaghia violacea* Harv. triggers apoptosis in cervical cancer cells (Mthembu and Motadi 2014).

21.4 Ayurveda and Cancer

Ayurveda translated as the science of life is known to be the "mother of all healing" that brings well-being in diverse spheres of life such as the mind, body, spirit, and promotes social harmony. The roots of Ayurvedic medicine dates to more than 5000 years ago. For many years, Ayurvedic formulations have been the mainstream source of medicine for many Indian inhabitants. During the last decades, the quick efficacy of allopathic medicine especially in cancer therapy has easily overtaken the use of Ayurvedic formulations which tend to act slowly but definitely on the disease condition (Bhandari et al. 2015).

Nonetheless, the major toxicity, side effects, and inefficacy of allopathic medicine to treat diverse major mortality inducing chronic disorders have renewed interest in other natural sources of medicine such as Ayurveda (Bhandari et al. 2015). Indeed, cancer is explicitly defined as inflammatory or non-inflammatory swelling referred to as Granthi (minor neoplasm) Arbuda (major neoplasm) respectively in Ayurvedic scripts (Balachandran and Govindarajan 2005). Tumor growth can be reversed and destroyed by restoring metabolic defects and normal tissue function can be induced by improving lifestyle (Sumantran and Tillu 2012).

A plethora of plants mentioned in Ayurvedic scriptures deserves considerable attention in cancer research and treatment. Various leads derived from herbs mentioned in Ayurveda have been developed as anticancer agents and include: taxol, roscovitine, combretastatin A-4, betulinic acid, silvestrol, camptothecin, podophyllotoxin, topotecan, irinotecan, vincristine, vinblastin, and flavopiridol (Salim et al. 2008). Taxanes are a unique class of phytochemicals bearing significant cytotoxic potential. Its derivative taxol, the first taxane anticancer agent was initially isolated from the Pacific yew *Taxus brevifolia* Nutt. The use of several species of *Taxus* against cancer is well documented in Ayurveda (Wheeler et al. 1992). Available literature also suggests that they were employed for similar purposes by native American tribes. In 2015, 23 taxanes were in preclinical trial for cancer treatment (Bhandari et al. 2015).

Interestingly, *Dysoxylum binectariferum* (Roxb.) Hook.f. ex Bedd. is widely cited in Ayurveda for its therapeutic benefits (Mohanakumara et al. 2010). Rohitukine, a metabolite derived from *D. binectariferum* has received considerable scientific attention as its precursor flavopiridol (alvocidib) has shown significant anticancer potential in a number of in vitro investigations (Ali et al. 2009). Accordingly, flavopiridol can inhibit cyclin-dependent kinases 1, 2, 4 and 7 and thereby suppress lung, breast cancer, head, and neck squamous cell carcinomas (Mohanakumara et al. 2010; Ali et al. 2009). During the years 1998–2014, more than 60 phase I and II clinical trials have been performed on this metabolite (Phelps et al. 2009; Shindiapina et al. 2014).

Hence, the metabolite holds considerable promise in the hope of deciphering a novel anticancer agent.

Another commonly utilised culinary plant, *Curcuma longa* L. biosynthesizes the metabolite curcumin which has displayed potent anti-tumorigenic potential (Anand et al. 2007). It can target various types of cancer including: colon, skin, duodenum, and stomach (Shankar et al. 2008). An interesting feature of curcumin is that even at high doses, it does not display significant side effects (Marwat et al. 2009). In vitro and in vivo studies have demonstrated that curcumin possesses antiangiogenic and antitumor properties, respectively, in addition to phosphodiesterase 2 and 3 inhibition potential (Ahmad et al. 2009). All these properties are undeniably helpful in the inhibition of various types of cancer.

Notably, anticancer properties from several metabolites of *Andrographis paniculata* Nees commonly referred to as the king of bitters in Ayurveda have been reported by the scientific community. Major metabolites from the plant, andrographolide, neoandrographolide, and dehydroandrographolide have all displayed anticancer properties. Andrographolide has displayed anticancer activity against B16F0 melanoma syngeneic and HT-29 xenograft models (Rajagopal et al. 2003), 2-cell line panel containing MCF-7 (breast cancer cell line) and HCT-116 (colon cancer cell line) (Jada et al. 2008) and multiple myeloma (Matsuda et al. 1994; Gunn et al. 2011). A study on patients with late-stage cancer involved the ingestion of 500 mg of *A. paniculata* extract twice daily along with other nutraceuticals demonstrated an improvement in the health status of the patients (See et al. 2002). However, its poor oral bioavailability is a major issue to address.

Metabolites from the plant *Withania somnifera* (L.) Dunalare are also blessed with anticancer properties. An in vivo study has reported the antitumor properties of its roots against Sarcoma-180 solid tumor and its remarkably low toxicity. Withaferin A, its common metabolite has demonstrated antiproliferative activities against NCL-H460 (lung), HCT-116 (colon), SF-268 (central nervous system; CNS and MCF-7 (breast) human tumor cell lines (Jayaprakasam et al. 2003). Enthrallingly, when combined with oxaliplatin, it demonstrated synergistic antitumor properties.

21.5 Chinese Herbal Medicine and Cancer

The use of Chinese herbal medicine (CHM) in cancer therapy is recorded since 2000 years. Elaboration on the topic of cancer can be found in classical Chinese scriptures such as "The Yellow Emperor's Inner Canon" and "The Classic of medical Problems." Mention of the word tumor was found on an oracle bone dating more than 3500 years. Today, CHM is used either alone or in combination with chemotherapy in China. The efficacy of a CHM formulation is amplified through the combination of more than one species of herb (Liu et al. 2015).

The synergistic anticancer effects of a plethora of phytochemicals from the plant *Scutellaria baicalensis* Georgi has been demonstrated in vitro. The extract of the plant consists of various metabolites including: baicalin (80%), wogonoside

Table 21.2 Chinese herbs and anticancer potential

Plant specie/common English name	Cell line/cancer type	Mechanism	Source
Hydrolysable tannis From *Eugenia jambos* L.	HL-60 cell line	Apoptosis induction	Yang et al. (2000)
Anemarrhena asphodeloides Bunge	MNK45 and KATO-III cell lines/gastric cancer	Apoptosis induction	Takeda et al. (2001)
Scutellaria barbata D. Don	SKOV3, CAOV3 and OVCAR-3 cell lines/ovarian cancer	Apoptosis induction	Powell et al. (2003)

(16%), baicalein (2%), wogonin (1%) alongside other components in trace amounts which were shown to exhibit a synergistic anticancer effect. When assayed alone in vitro, the pure compound baicalein also demonstrated anticancer potential. The extract inhibited prostaglandin E2 while the pure compound did not (Zhang et al. 2003).

Laboratory investigations have concluded that activity of distinct plants on different cancer cell lines is dissimilar. For example, artesunate was found to be most active against leukemia and colon cancer cell lines. The butanol extract of *Mylabris phalerlata* displayed a specific cytotoxic activity on human monocytic leukemic U937 cells in comparison to peripheral blood mononuclear lymphocytes (Huh et al. 2003). On the other hand, there is also a theory hypothesizing that some plants might be more selective to certain cancer cells. This was epitomized when *Scutellaria baicalensis* Georgi was assayed against human head and neck squamous cell carcinoma (HNSCC) cell lines, which resulted in a strong growth inhibition versus normal growth of the non-tumorigenic cell line HaCaT cells (a non-tumorigenic cell line) (Zhang et al. 2003). In the management of lung cancer, five herbal components retain attention. These include: polyphyllin I, tanshinone IIA, isochaihulactone, 25-OCH3-PPD, and andrographolide which have shown potent anti-lung cancer activity both in cells and animal models (Chen et al. 2014). Other herbs used in CHM with anticancer potential are listed in Table 21.2.

21.6 Traditional Islamic and Arabic Plants in Cancer

Among the various synthetic drugs used in cancer therapy, around 73% are derivatives of natural products. Medicinal plants are still in profuse use in Arab and Islamic countries and their original uses kicked off at the time of Prophet Mohammad, Peace Be upon Him (PBUH). Spiritual healers and herbalists still practice in the Mediterranean part of the world and are actively involved in the treatment of various ailments such as cancer; skin, respiratory, digestive, and liver diseases; diabetes as well as other diseases (Ahmad et al. 2017). A number of these plants have been evaluated in various in vitro and in vivo cancer models (Table 21.3).

21.7 Mexican Traditional Medicine and Colorectal Cancer

The incidence of colorectal cancer is one of the highest among all the other forms of cancer. A multitude of risk factors are associated with the disease and involves aging, poor dietary habits such as the exclusion of fibers and consumption of large amounts of red meat, alcohol, tobacco, and obesity (González and Riboli 2010). All these factors combined accelerate the onset of the disease.

When detected in early stages surgical procedures involving local excision or resection and anastomosis for the carcinoma should prove sufficient. Nonetheless, in advanced stages, chemo and radio therapy may follow. Various conventional chemotherapeutic agents have proved successful among colorectal cancer patients. Notably, 5-fluorouracil, leucoverin, capecitabine, and oxaliplatin can either be employed singly or in combination (Binefa et al. 2014). Nonetheless, despite the availability of these synthetic agents, a high risk of recurrence and metastasis of tumors is current among colorectal cancer patients.

More than 3000 plants occurring around the world are acknowledged to play a role in cancer prophylaxis (Alonso-Castro 2011). Interestingly, Complementary and Alternative medicines (CAM) still holds a major role in the lives of Mexican cancer patients. It is known that around 30% of cancer sufferers still rely on the medicinal virtues of medicinal plants (Jacobo-Herrera 2016). The secondary metabolite synthesized by a plethora of traditionally used Mexican plants has demonstrated promising activity against a variety of cancer cell lines (Table 21.4).

21.8 Discussion

Herbs have been used by various cultures worldwide as part of various traditions and as a source of food, shelter, and clothing (Gurib-Fakim 2006). Even today, a plethora of herbs occurring throughout the world are used for medicinal purposes but have not been scientifically evaluated (Folashade et al. 2012). Indeed, herbs remain an armory of natural sources in the quest of deciphering novel anticancer agents. Concurrently, the failure of conventional medicines to address fully several forms of cancer has promoted interest in the field of complementary medicines which can either be used singly or as combination therapy with allopathic agents in an attempt to relieve cancer sufferers from their agony (Vincent et al. 1997).

Undoubtedly, herbs possess various advantages over conventional medicines. Firstly, being natural makes them a better choice over synthetic agents. Secondly, they can be more easily accessed and are cheaper than conventional medicines. Furthermore, in contrast to conventional drug agents, they have lower toxicity and side

Table 21.3 Mediterranean plants and anticancer assays

Cancer	Cell line	Plant specie/common English name	Compounds	Activity[Source]
Hepatocellular carcinoma	HepG2	*Acacia seyal* Delile/Acacia		Cytotoxic[a,b]
Breast adeno carcinoma	MCF-7	*Acacia seyal* Delile/Acacia		Cytotoxic[a,b]
Lung carcinoma	A549	*Acacia seyal* Delile/Acacia		Cytotoxic[a,b]
Colorectal carcinoma	HCT-116	*Acacia seyal* Delile/Acacia		Cytotoxic[a,b]
Human cancer cell line	HO-8910 & 7721	*Agaricus Campestris* L./Mushroom		Cytotoxic[c,d,e,f]
Cancer cell lines	Jurkat and K562	*Allium Ascalonicum* L./Leek		Anti-growth[g,h,i]
Human stomach cancer cell line	BGC-823	*Apium graveolens* L./Celery		Antiproliferative activity and apoptosis[j,k,l,m,n]
		Apium graveolens L./Celery	3-n-butyl phthalide and sedanolide	Tumor inhibition[j,k,l,m,n]
		Apium graveolens L./Celery seeds	Senkyunolide-N, Senkyunolide-J & 3-hydroxymethyl-6-methoxy-2, 3-dihydro-1H-indol-2-ol	Topoisomerase-I and II enzymes inhibitory compounds[j,k,l,m,n]
In vitro tumor cell lines		*Arum palaestinum* Boiss	A novel alkylated piperazine	Cytotoxic[o,p,q]
Human prostate cancer cells	PC-3	*Beta vulgaris* L./Beet-root		Potent anticancer activity[r,s,t,u]
Human breast cancer cells	MCF-7	*Beta vulgaris* L./Beet-root		Potent anticancer activity[r,s,t,u]
	HepG2 and HCT 116	*Boswellia carterii* Balf.f./Olibanum		Cytotoxic[v,w]

(continued)

Table 21.3 (continued)

Cancer	Cell line	Plant specie/common English name	Compounds	Activity[Source]
Human hepatocarcinoma cell line	7721	*Cinnamomum camphora* (L.) J. Presl/Camphor		Inhibition[x,y]
		Citrullus colocynthis (L.) Schrad./Colocynth	Cucurbitacin glucosides e	Pleiotropic effects on cells, causing both cell cycle arrest and apoptosis[z,aa]
	larynx HEp-2 cells	*Crataegus azarolus* L./Azarole Hawthorn		Cytotoxic[ab,ac]
	A549 and SPC-A1 and mice MCF-7 cell	*Crocus sativus* L./Saffron		Anti-Proliferation effects on human colorectal cancer cells; human lung adenocarcinoma cell lines lines[ad,ae]

[a]El-Hallouty et al. (2015), [b]Patel et al. (2014), [c]Shi-feng et al. (2005), [d]Fortes et al. (2009), [e]Ikekawa et al. (1969); [f]Elbatrawy et al. (2015), [g]Seyfi et al. (2010), [h]Mohammadi-Motlagh et al. (2011), [i]Hsu et al. (2009), [j]Zheng et al. (1992), [k]Zheng et al. (1993), [l]Momina and Naira (2002), [m]Subhadradevi et al. (2011), [n]Gao et al. (2011), [o]El-Desouky et al. (2007a), [p]El-Desouky et al. (2007b), [q]Ali-Shtayeh et al. (2008), [r]Kapadia et al. (1996), [s]Govind et al. (2011), [t]Georgie et al. (2010), [u]Nowacki et al. (2015), [v]Hui-Qing et al. (2008), [w]Frank et al. (2009), [x]Tannin-Spitz et al. (2007), [y]Ayyad et al. (2012), [z]Ling and Liu (1996), [aa]Lin et al. (2008), [ab]Zaid et al. (2010), [ac]Xiao et al. (2011), [ad]Aung et al. (2007), [ae]Abdullaev et al. (2003)

Table 21.4 Traditional Mexican plants potential against anticancer cell line

Plant species	Extract/metabolite	Cytotoxic activity (ED_{50} (μg/ml)/cell line	Source
Rolliniamucosa (Jacq.) Baill	Rollitacin (EtOH) Jimenezin Membranacin Desacetyluvaricin Laherradurin	4.6×10^{-3}/HT29 4.25/SW-480 3.04/SW-480 1.69/SW-480 IC50 = 0.015/SW-480	Shi et al. (1997), Chávez et al. (1998, 1999)
Annona diversifolia Saff	Cherimolin-2	IC50 = 0.5/SW-480	Schlie-Guzmán et al. (2009)
Avicularia purpurea Moc. &Sessé Exduna	CHCl3–MeOH (1:1) Purpurediolin Purpurenin Annoglaucin Annonacin A	1.47/HT29 <107/HT29 3.16×10^{-1}/HT29 <107/HT29 1.18/HT29	Chávez and Mata (1998)
Annona muricata L.	Annopentocinaa Annopentocinb Annopentocinc Cis- and trans-annomuricin-D-ones	1.63/HT29 1.64/HT29 1.24/HT29 $<10^{-2}$/HT29	Zeng et al. (1996), Moghadamtousi et al. (2014, 2015)
Viguieradecurrens (A. Gray) A. Gray	Hex; EtOAc; MeOHextractconstituitedby β-sitosteryl- 3-O-β-D-glucopyranoside; β-D-glucopyranosyl-oleano- late; β-sitosteryl-3-O-β-D-glucopyranosideandolea-nolic acidl-3-O-methy-β-D-glucuronopyranosiduronoate	3.6/colon carcinoma	Marquina et al. (2001)

(continued)

Table 21.4 (continued)

Plant species	Extract/metabolite	Cytotoxic activity (ED_{50} (μg/ml)/cell line	Source
Burserafagaroides (Kunth) Engl	Hydroalcoholicextract	$7.1 - 10^{-3}$/HF6	Rojas-Sepúlveda et al. (2012)
	Podophyllotoxin	$1.8 - 10^{-4}$/HF6	
	β-peltatin A. methylether	$3.8 - 10^{-2}$/HF6	
	5′-desmethoxy-β-peltatin-A-methylether	0.40/HF6	
	Desmethoxy-yatein	0.68/HF6	
	Desoxypodophyllotoxin	1.23/HF6	
	Burseranin	2.89/HF6	
	Acetylpodophyllotoin	2.41/HF6	
Hyptis pectinata (L.) Poit	Pectinolide A	1/Col2	Pereda-Mirandaet al. (1993)
	Pectinolide B	1.1/Col2	
	Pectinolide C	1.6/Col2	
	α-pyroneboronolide	4/Col2	
	Deacetylepiol-guine	3/Col2	
Hyptis verticillata Jacq	Dehydro-β-peltatin methylether	3.2	Novelo et al. (1993)
	Dibenzylbutyrolactone(-)-yatein	0.08	
	4′ demethyldeoxypodophyllotoxin	0.3	
	Deoxypicropodophyllin		
Perseaamericana Mill	1,2,4-trihydroxynonadecane	IC_{50}= 3/HT-29	Oberlies et al. (1998)
	1,2,4-trihydroxyhep-tadec-16-ene	IC_{50} = 2.6	
	1,2,4-trihydroxyheptadec-16-yne	IC_{50} = 8.9	
Linum scabrellum Planch	CHCl3 (roots)0.2HF6	0.2/HF6	Lautié et al. (2008), Alejandre-García et al. (2015)
	CHCl3 (aerial parts)2.3	2.3	
	BuOH (roots)0.5	0.5	
	DCM: MeOH	5.7×10^{-1}	
	6MPTOX	7.9×10^{-2}	
	PTOX	1.4×10^{-3}	

effects and have been used with confidence over several decades since the dawn of mankind. Consequently, the therapeutic benefits emanating from herbs remains time tested and evidence for this long-standing use guarantees their safety to some extent. Herbs are rich sources of secondary metabolites and hence possess diverse pharmacological properties such as antibacterial, antiviral, anti-hepatotoxic, antiulcer, anti-inflammatory, antioxidant, anti-mutagenic, and anticancer effects among diverse other therapeutic effects (Gurib-Fakim 2006).

A plethora of plants traditionally used to treat cancer symptoms as well as other ailments are the origin of various conventional drug agents. For example, *Catharanthus roseus* (L.) G. Don is the source of vincristine and vinblastine and *Taxus brevifolia* Nutt., the source of taxol (Kintzios 2006). The use of these conventional drug agents is still widespread presently. Plant-derived medicines have shown that they can be effectively combined with other chemotherapeutic agents and enable achieve desired therapeutic effects.

According to various traditional systems of medicine, cancer is the result of an unhealthy lifestyle with poor eating habits. Several fruits and vegetables have been postulated as prophylactic agents against cancer. Interesting classes of secondary metabolites in cancer research are: flavonols, flavan-3-ols, anthocyanins, flavanones, flavones, isoflavones, proanthocyanidins, terpenoids, nitrogen, and sulfur containing compounds according to the results generated from various scientific investigations (Shin et al. 2018).

Several barriers arise in the process of promoting the positive anticancer results from plant species obtained from in vitro studies. The most common being that rarely in vitro results are translated in vivo. Even if positive in vivo results are obtained, few clinical trials of the herbs against different forms of cancer are performed. Other issues involve the poor absorption of herbal extracts in the body particularly if the metabolite of interest is hydrophobic in nature as well as the low oral bioavailability of the active constituents of herbal products.

21.9 Conclusion

The use of herbs for the management and treatment of various disease conditions dates to the medieval ages in different cultures and traditions around the world. Several plant species have been assayed for their purported effects in in vitro and in vivo studies while a major proportion of plants known to be used for medicinal purposes have never been assayed for their therapeutic effects. While plants derived from different traditional systems mentioned in this chapter propose an extraordinary armory of anticancer agents, they also remain a sustainable source of inspiration for scientists in the quest of unveiling potent anticancer agents.

References

Abdullaev FI, Riveron-Negrete L, Caballero-Ortega H et al (2003) Use of *in vitro* assays to assess the potential antigenotoxic and cytotoxic effects of saffron (*Crocus sativus* L.). Toxicol *In Vitro*, 17(5–6):731–736

Ahmad M, Khan MA, Marwat SK et al (2009) Useful medicinal flora enlisted in Holy Quran and Ahadith. Am Eurasian J Agric Environ Sci 5(1):126–140

Ahmad R, Ahmad N, Naqvi AA et al (2017) Role of traditional Islamic and Arabic plants in cancer therapy. eJTCM 7(2):195–204

Ajay KM, Parveen B, Sanjiv K (2009) Plants-herbal wealth as a potential source of Ayurvedic drugs. Asian J Tradit Med 4(4):152–170

Alejandre-Garcia I, Alvarez L, Cardoso-Taketa A et al (2015) Cytotoxic activity and chemical composition of the root extract from the Mexican species *Linum scabrellum*: mechanism of action of the active compound 6-methox-y podophyllotoxin. Evid Based Compl Alt 298463

Ali S, Heathcote DA, Kroll SHB et al (2009) The development of a selective cyclin-dependent kinase inhibitor that shows antitumor activity. Can Res 69:6208–6215

Ali-Shtayeh MS, Jamous RM, Al-Shafie JH et al (2008) Traditional knowledge of wild edible plants used in Palestine (Northern West Bank): a comparative study. J Ethnobiol Ethnomed 4:13

Alonso-Castro AJ, Villarreal ML, Salazar-Olivo LA (2011) Mexican medicinal plants used for cancer treatment: pharmacological, phytochemical and ethnobotanical studies. J Ethnopharmacol 133(3):945–972

Anand P, Kunnumakkara AB, Newman RA et al (2007) Bioavailability of curcumin: problems and promises. Mol Pharm 4:807–818

Aung HH, Wang CZ, Ni M et al (2007) Crocin from Crocus sativus possesses significant anti-proliferation effects on human colorectal cancer cells. Exp Oncol 29(3):175

Ayyad SN, Abdel-Lateff A, Alarif WM et al (2012) *In vitro* and *in vivo* study of cucurbitacins-type triterpene glucoside from *Citrullus colocynthis* growing in Saudi Arabia against hepatocellular carcinoma. Environ Toxicol Pharmacol 33:245e251

Balachandran P, Govindarajan R (2005) Cancer-an ayurvedic perspective. Pharmacol Res 51:19–30

Berrington D, Lall N (2012) Anticancer activity of certain herbs and spices on the cervical epithelial carcinoma (HeLa) cell line. Evid Based Complement Altern Med 564927

Bhandari M, Ravipati AS, Reddy N et al (2015) Traditional Ayurvedic medicines: pathway to develop anti-cancer drugs. JMPOPR 3(3)

Bibi S, Sultana J, Sultana H et al (2014) Ethnobotanical uses of medicinal plants in the highlands of Soan Valley, Salt Range. Pakistan J Ethnopharmacol 155(1):352–361

Binefa G, Rodríguez-Moranta F, Teule, À et al (2014) Colorectal cancer: from prevention to personalized medicine. WJG 20(22):6786

Boon H, Wong J (2004) Botanical medicine and cancer: a review of the safety and efficacy. Expert Opin Pharm 5:2485–2501

Boughrara B, Belgacem L (2016) Ethnobotanical study close to the population of the extreme north east of Algeria: the municipalities of El Kala National Park (EKNP). Ind Crops Prod 88:2–7

British National Formulary (2014) Malignant disease and immunosuppression 8:553–630

Chan JM, Wang F, Holly EA (2005) Vegetable and fruit intake and pancreatic cancer in a population-based case-control study in the San Francisco bay area. Cancer Epidemiol Biomarkers Prev 14:2093–2097

Chávez D, Acevedo L, Mata R (1999) Tryptamine derived amides and acetogenins from the seeds of *Rollinia mucosa*. J Nat Prod 62:1119–1122

Chávez D, Acevedo LA, Mata R (1998) Jimenezin, a Novel Annonaceous Acetogenin from the Seeds of Rollinia mucosa containing Adjacent Tetrahydrofuran—Tetrahydropyran ring systems. J Nat Prod 61(4):419–421

Chávez D, Mata R (1998) Purpurediolin and purpurenin, two new cytotoxic adjacent bis-tetrahydrofuran annonaceous acetogenins from the seeds of Annona purpurea. J Nat Prod 61:580–584

Chen, Y, Zhu J, Zhang W (2014) Antitumor effect of traditional Chinese herbal medicines against lung cancer. Anti-Cancer Drugs 25(9):983–991
Cheung SY, Delfabbro P (2016) Are you a cancer survivor? A review on cancer identity. J Cancer Surviv 10(4):759–771
Clifford JL, Digiovanni J (2010) The promise of natural products for blocking early events in skin carcinogenesis. Cancer Prev Res (Phila) 3:132–135
Cragg GM, Newman DJ (2005) Plants as a source of anti-cancer agents. J Ethnopharmacol 100:72–79
Cragg GM, Newman DJ (2009) Nature: a vital source of leads for anticancer drug development. Phytochem Rev 8(2):313–331
Das S, Das J, Paul A et al (2013) Apigenin, a bioactive flavonoid from Lycopodium clavatum, stimulates nucleotide excision repair genes to protect skin keratinocytes from ultraviolet B-induced reactive oxygen species and DNA damage. J Acupunct Meridian Stud 6:252–262
Elbatrawy EN, Ghonimy EA, Alassar MM et al (2015) Medicinal mushroom extracts possess differential antioxidant activity and cytotoxicity to cancer cells. Int J Med Mushrooms 17:471–479
El-Desouky SK, Kim KH, Ryu SY et al (2007a) A new pyrrole alkaloid isolated from *Arum palaestinum* Boiss. and its biological activities. Article Drug Discov Arch Pharma Res 30:927–931
El-Desouky SK, Ryu SY, Young-Kyoon K (2007b) Piperazirum, a novel bioactive alkaloid from *Arum palaestinum* Boiss. Tetrahedron Lett 48:4015–4017
El-Hallouty SM, Fayad W, Meky NH (2015) *In vitro* anticancer activity of some Egyptian plant extracts against different human cancer cell lines. Int J Pharm Tech Res 8:267–272
Folashade O, Omoregie H, Ochogu P (2012) Standardization of herbal medicines-a review. Int J Biodivers Conserv 4(3):101–112
Fortes RC, Novaes MRCG, Recôva VL et al (2009) Immunological, hematological, and glycaemia effects of dietary supplementation with *Agaricus sylvaticus* on patients' colorectal cancer. Exp Biol Med (Maywood) 234:53–62
Frank FM, Yang Q, Osban J et al (2009) Frankincense oil derived from *Boswellia carteri* induces tumor cell specific cytotoxicity. BMC Complement Altern Med 9:6
Gao LL, Feng L, Yao ST et al (2011) Molecular mechanisms of celery seed extract induced apoptosis via s phase cell cycle arrest in the BGC-823 human stomach cancer cell line. Asian Pac J Cancer Prev 12:2601–2606
George VC, Dellaire G, Rupasinghe HV (2017) Plant flavonoids in cancer chemoprevention: role in genome stability. J Nutr Biochem 45:1–14
Georgiev V, Weber J, Kneschke E-M et al (2010) Antioxidant activity and phenolic content of betalain extracts from intact plants and hairy root cultures of the red beetroot *Beta vulgaris* cv. Detroit dark red. Plant Foods Hum Nutr 65:105–111
González CA, Riboli E (2010) Diet and cancer prevention: contributions from the european prospective investigation into cancer and nutrition (EPIC) study. Eur J Cancer 46(14):2555–2562
Govind JK, Magnus AA, Subba Rao G et al (2011) Cytotoxic effect of the red beetroot (*Beta vulgaris* L.) extract compared to doxorubicin (Adriamycin) in the human prostate (PC-3) and breast (MCF-7) cancer cell lines. Anticancer Agents Med Chem 11:280–284
Gunn EJ, Williams JT, Huynh DT et al (2011) The natural products parthenolide and andrographolide exhibit anti-cancer stem cell activity in multiple myeloma. Leuk Lymphoma 52:1085–1097
Gurib-Fakim A (2006) Medicinal plants: traditions of yesterday and drugs of tomorrow. Mol Aspects Med 27(1):1–93
Holmes K, Egan B, Swan N et al (2007) Genetic mechanisms and aberrant gene expression during the development of gastric intestinal metaplasia and adenocarcinoma. Curr Genom 8(6):379–397
Hsu YL, Chia CC, Chen PJ et al (2009) Shallot and licorice constituent isoliquiritigenin arrests cell cycle progression and induces apoptosis through the induction of ATM/p 53 and initiation of the mitochondrial system in human cervical carcinoma HeLa cells. Mol Nutr Food Res 53:826–835

Huh JE, Kang KS, Ahn KS et al (2003) *Mylabris phalerlata* induces apoptosis by caspase activation following cytochrome c release and Bid cleavage. Life Sci 73(17):2249–2262
Hui-Qing Y, Feng K, Xiao-Ling W et al (2008) Inhibitory effect of acetyl-11-keto-b-boswellic acid on androgen receptor by interference of Sp1 binding activity in prostate cancer cells. Biochem Pharmacol 5:2112–2121
Hussein AA, Meyer JJ, Jimeno ML et al (2007) Bioactive diterpenes from *Orthosiphon labiatus* and *Salvia Africana*-lutea. J Nat Prod 70(2):293–295
Ikekawa T, Uehara N, Maeda Y (1969) Antitumor activity of aqueous extracts of edible mushrooms. Cancer Res 29:734–735
Iovine B, Iannella ML, Gasparri F et al (2011) Synergic effect of genistein and daidzein on UVB-induced DNA damage: an effective photoprotective combination. J Biomed Biotechnol 2011:692846
Jacobo-Herrera NJ, Jacobo-Herrera FE, Zentella-Dehesa A et al (2016) Medicinal plants used in Mexican traditional medicine for the treatment of colorectal cancer. J Ethnopharmacol 179:391–402
Jada SR, Matthews C, Saad MS et al (2008) Benzylidene derivatives of andrographolide inhibit growth of breast and colon cancer cells in vitro by inducing G arrest and apoptosis. Br J Pharmacol 155:641–654
Jayaprakasam B, Zhang Y, Seeram NP et al (2003) Growth inhibition of human tumor cell lines by withanolides from *Withania somnifera* leaves. Life Sci 74:125–132
Kapadia GJ, Tokudab H, Konoshimac T et al (1996) Chemoprevention of lung and skin cancer by *Beta vulgaris* (beet) root extract. Cancer Lett 100:211–214
Kintzios SE (2006) Terrestrial plant-derived anticancer agents and plant species used in anticancer research. Crit Rev Plant Sci 25(2):79–113
Kuete V, Efferth T (2015) African flora has the potential to fight multidrug resistance of cancer. Biomed Res Int 914813
Kuete V, Fankam AG, Wiench B et al (2013) Cytotoxicity and modes of action of the methanol extracts of six Cameroonian medicinal plants against multidrug-resistant tumor cells. J Evid Based Complementary Altern Med 2013
Kuete V, Tankeo SB, Saeed ME et al (2014) Cytotoxicity and modes of action of five Cameroonian medicinal plants against multi-factorial drug resistance of tumor cells. J Ethnopharmacol 153(1):207–219
Kurapati KR, Samikkannu T, Kadiyala DB et al (2012) Combinatorial cytotoxic effects of Curcuma longa and Zingiber officinale on the PC-3M prostate cancer cell line. J Basic Clin Physiol Pharmacol 23(4):139–146
Lautié E, Quintero R, Fliniaux MA et al (2008) Selection methodology with scoring system: application to Mexican plants producing podophyllotoxin related lignans. J Ethnopharmacol 120:402–412
Leung HW, Wu CH, Lin CH et al (2005) Luteolin induced DNA damage leading to human lung squamous carcinoma CH27 cell apoptosis. Eur J Pharmacol 508:77–83
Light ME, Sparg SG, Stafford GI et al (2005) Riding the wave: South Africa's contribution to ethnopharmacological research over the last 25 years. J Ethnopharmacol 100(1–2):127–130
Lin RJ, Lo WL, Wang YD et al (2008) A novel cytotoxic monoterpenoid from the leaves of *Cinnamomum subavenium*. Nat Prod Res Former Nat Prod Lett 22:1055–1059
Ling J, Liu WY (1996) Cytotoxicity of two new ribosome-inactivating proteins, cinnamomin and camphorin, to carcinoma cells. Cell Biochem Funct 14:157–161
Liu J, Wang S, Zhang, Y et al (2015) Traditional Chinese medicine and cancer: history, present situation, and development. Thorac Cancer 6(5):561–569
Marquina S, Maldonado N, Garduño-Ramírez ML et al (2001) Bioactive oleanolic acid saponins and other constituents from the roots of *Viguiera decurrens*. Phytochemistry 56:93–97
Marwat SK, Khan MA, Khan MA et al (2009) Fruit plant species mentioned in the Holy Quran and Ahadith and their ethnomedicinal importance. Am-Eurasian J Agric Environ Sci 5:284–295

Matsuda T, Kuroyanagi M, Sugiyama S et al (1994) Cell differentiation-inducing diterpenes from *Andrographis paniculate* Nees. Chem Pharm Bull (Tokyo) 42:1216–1225

Mbele M, Hull R, Dlamini Z (2017) African medicinal plants and their derivatives: current efforts towards potential anti-cancer drugs. Exp Mol Pathol 103(2):121–134

Mendonça-Filho RR (2006) Bioactive phytocompounds: new approaches in the phytosciences. Modern phytomedicine: turning medicinal plants into drugs, pp 1–25

Meyer H, Bolarinwa A, Wolfram G et al (2006) Bioavailability of apigenin from apiin-rich parsley in humans. Ann Nutr Metab 50:167–172

Min K, Ebeler SE (2009) Quercetin inhibits hydrogen peroxide-induced DNA damage and enhances DNA repair in Caco-2 cells. Food Chem Toxicol 47:2716–2722

Moghadamtousi SZ, Karimian H, Rouhollahi E et al (2014) *Annona muricata* leaves induce G 1 cell cycle arrest and apoptosis through mitochondria-mediated pathway in human HCT-116 and HT-29 colon cancer cells. J Ethnopharmacol 156:277–289

Moghadamtousi SZ, Rouhollahi E, Karimian H et al (2015) The chemopotential effect of Annona muricata leaves against azoxymethane-induced colonic aberrant crypt foci in rats and the apoptotic effect of Acetogenin Annomuricin E in HT-29 cells: a bioassay-guided approach. PLoS ONE 10(4):e0122288

Mohammadi-Motlagh H-R, Mostafaie A, Mansouri K (2011) Anticancer and anti-inflammatory activities of shallot (*Allium ascalonicum*) extract. Arch Med Sci 7:38–44

Mohanakumara P, Sreejayan N, Priti V et al (2010) *Dysoxylum binectariferum* Hook.f (Meliaceae), a rich source of rohitukine. Fitoterapia 81:145–148

Momina RA, Naira MG (2002) Antioxidant, cyclooxygenase and topoisomerase inhibitory compounds from *Apium graveolens* Linn. seeds. Phytomedicine 9:312–318

Mthembu NN, Motadi LR (2014) Apoptotic potential role of *Agave palmeri* and *Tulbaghia violacea* extracts in cervical cancer cells. Mol Biol Rep 41(9):6143–6155

Newman DJ, Cragg GM, Snader, KM (2000) The influence of natural products upon drug discovery. Nat Prod Rep 17(3):215–234

Novelo M, Cruz JG, Hernández L et al (1993) Cytotoxic constituents from *Hyptis verticillata*. J Nat Prod 56(10):1728–1736

Nowacki L, Vigneron P, Rotellini L et al (2015) Betanin-enriched red beetroot (*Beta vulgaris* L.) extract induces apoptosis and autophagic cell death in MCF-7 cells. Phytother Res 29:1964–1973

Oberlies NH, Rogers LL, Martin JM et al (1998) Cytotoxic and insecticidal constituents of the unripe fruit of *Persea americana*. J Nat Prod 61:781–785

Paixao J, Dinis TC, Almeida LM (2012) Protective role of malvidin-3-glucoside on peroxynitrite-induced damage in endothelial cells by counteracting reactive species formation and apoptotic mitochondrial pathway. Oxid Med Cell Longev 2012:428538

Panda T, Mishra N, Pradhan BK (2016) Folk knowledge on medicinal plants used for the treatment of skin diseases in Bhadrak district of Odisha, India. Med Aromat Plants 5(4):1–7

Patel A, Hafez E, Elsaid F et al (2014) Anti-cancer action of a new recombinant lectin produced from *Acacia* species. Int J Med Sci 5:1

Pereda-Miranda R, Hernández L, Villavicencio MJ et al (1993) Structure and stereochemistry of pectinolides AC, novel antimicrobial and cytotoxic 5, 6-dihydro-α-pyrones from Hyptis pectinate. J Nat Prod 56(4):583–593

Phelps MA, Lin TS, Johnson AJ et al (2009) Clinical response and pharmacokinetics from a phase 1 study of an active dosing schedule of flavopiridol in relapsed chronic lymphocytic leukemia. Blood 113:2637–2645

Plengsuriyakarn T, Viyanant V, Eursitthichai V et al (2012) Cytotoxicity, toxicity, and anticancer activity of *Zingiber officinale* Roscoe against cholangiocarcinoma. Asian Pac J Cancer Prev 13(9):4597–4606

Powell CB, Fung P, Jackson J et al (2003) Aqueous extract of herba *Scutellariabarbatae*, a chinese herb used for ovarian cancer, induces apoptosis of ovarian cancer cell lines. Gynecol Oncol 91(2):332–340. https://doi.org/10.1016/j.ygyno.2003.07.004

Rajagopal S, Kumar RA, Deevi DS et al (2003) Andrographolide, a potential cancer therapeutic agent isolated from *Andrographis paniculata*. J Exp Ther Oncol 3:147–158
Ramawat KG, Goyal S (2008) The Indian herbal drugs scenario in global perspectives. Bioactive molecules and medicinal plants. Springer, pp 325–347
Rao PS, Satelli A, Moridani M et al (2012) Luteolin induces apoptosis in multidrug resistant cancer cells without affecting the drug transporter function: involvement of cell line-specific apoptotic mechanisms. Int J Cancer 130:2703–2714
Rates SMK (2001) Plants as source of drugs. Toxicon 39(5):603–613
Rathee P, Chaudhary H, Rathee S et al (2009) Mechanism of action of flavonoids as anti-inflammatory agents: a review. Inflamm Allergy Drug Targets 8(3):229–235
Rojas-Sepúlveda A, Mendieta-Serrano M, Antúnez Mojica M et al (2012) Cytotoxic podophyllotoxin type-lignans from the steam bark of *Bursera fagaroides* var. fagaroides. Molecules 17:9506–9519
Salim AA, Chin Y-W, Kinghorn AD (2008) Drug discovery from plants. Bioactive molecules and medicinal plants. Springer, pp 1–24
Schlie-Guzmán MA, García-Carrancá A, González-Esquinca AR (2009) In vitro and in vivo antiproliferative activity of laherradurin and cherimolin-2 of Annona diversifolia saff. Phytother Res 23:1128–1133
See D, Mason S, Roshan R (2002) Increased tumor necrosis factor alpha (TNF-alpha) and natural killer cell (NK) function using an integrative approach in late stage cancers. Immunol Invest 31:137–153
Seyfi P, Mostafaie A, Mansouri K et al (2010) *In vitro* and *in vivo* anti-angiogenesis effect of shallot (*Allium ascalonicum*): a heat-stable and flavonoid-rich fraction of shallot extract potently inhibits angiogenesis. Toxicol *In Vitro* 24:1655–1661
Shankar S, Ganapathy S, Chen Q et al (2008) Curcumin sensitizes TRAIL-resistant xenografts: molecular mechanisms of apoptosis, metastasis and angiogenesis. Mol Cancer 7:7–16
Shi G, MacDougal JM, McLaughlin JL (1997) Bioactive annonaceous acetogenins from *Rollinia mucosa*. Phytochemistry 45:719–723
Shi R, Huang Q, Zhu X et al (2007) Luteolin sensitizes the anticancer effect of cisplatin via c-Jun NH2-terminal kinase-mediated p53 phosphorylation and stabilization. Mol Cancer Ther 6:1338–1347
Shi-feng LI, Gui-Chen C, Yu-Rong BI (2005) Studies on antioxidative and antitumor activities for two wild edible fungi. Edible Fungi China 3
Shin SA, Moon S, Kim WY et al (2018) Structure-based classification and anti-cancer effects of plant metabolites. Int J Mol Sci 19(9):2651
Shindiapina P, Brown JR, Danilov AV (2014) A new hope: novel therapeutic approaches to treatment of chronic lymphocytic leukaemia with defects in TP53. Br J Haematol 167:149–161
Silva JP, Gomes AC, Coutinho OP (2008) Oxidative DNA damage protection and repair by polyphenolic compounds in PC12 cells. Eur J Pharmacol 601:50–60
Subhadradevi V, Khairunissa K, Asokkumar Ket al (2011) Induction of apoptosis and cytotoxic activities of *Apium graveolens* Linn. using in vitro models. Middle-East J Sci Res 9:90–94
Sumantran VN, Tillu G (2012) Cancer, inflammation, and insights from ayurveda. Evid Based Complement Altern Med 2012:306346
Takeda Y, Togashi H, Matsuo T et al (2001) Growth inhibition and apoptosis of gastric cancer cell lines by *Anemarrhena asphodeloides* Bunge. J Gastroenterol 36(2):79–90. https://doi.org/10.1007/s005350170135
Tannin-Spitz T, Grossman S, Dovrat S et al (2007) Growth inhibitory activity of cucurbitacin glucosides isolated from Citrullus colocynthis on human breast cancer cells. Biochem Pharmacol 73:56–67
Torre LA, Bray F, Siegel RL et al (2015) Global cancer statistics, 2012. CA Cancer J Clin 65(2): 87–108
Traore MS, Baldé MA, Diallo MST et al (2013) Ethnobotanical survey on medicinal plants used by Guinean traditional healers in the treatment of malaria. J Ethnopharmacol 150(3):1145–1153

Vainio H, Weiderpass E (2006) Fruit and vegetables in cancer prevention. Nutr Cancer. 54:111–142

Verpoorte R (2000) Pharmacognosy in the new millennium: lead finding and biotechnology. JPP 52(3):253–262

Vincent C, Furnham A, Richardson P et al (1997) Complementary medicine: a research perspective. Wiley

Voss C, Eyol E, Berger MR (2006) Identification of potent anticancer activity in *Ximenia americana* aqueous extracts used by African traditional medicine. Toxicol Appl Pharmacol 211(3):177–187

Wheeler NC, Jech K, Masters S et al (1992) Effects of genetic, epigenetic, and environmental factors on taxol content in *Taxus brevifolia* and related species. J Nat Prod 55:432–440

Xiao D, Zeng Y, Prakash L et al (2011) Reactive oxygen species-dependent apoptosis by gugulipid extract of Ayurvedic medicine plant Commiphora mukul in human prostate cancer cells is regulated by c-Jun N-terminal kinase. Mol Pharmacol 79:499–507

Yamashita N, Tanemura H, Kawanishi S et al (1999) Mechanism of oxidative DNA damage induced by quercetin in the presence of Cu(II). Mutat Res 425:107–115

Yamazaki S, Sakakibara H, Takemura H et al (2014) Quercetin-3-Oglucronide inhibits noradrenaline binding to alpha (2)-adrenergic receptor, thus suppressing DNA damage induced by treatment with 4-hydroxyestradiol and noradrenaline in MCF-10A cells. J Steroid Biochem Mol Biol 143:122–129

Yang LL, Lee CY, Yen KY (2000) Induction of apoptosis by hydrolyzable tannins from *Eugenia jambos* L. on human leukemia cells. Cancer Lett 157(1):65–75. https://doi.org/10.1016/s0304-3835(00)00477-8

Yin ST, Tang ML, Su L et al (2008) Effects of epigallocatechin-3- gallate on lead-induced oxidative damage. Toxicology 249:45–54

Zaid H, Said O, Saad B (2010) Cancer treatment in the Arab-Islamic medicine: integration of tradition with modern experimental trails. Jamia (Jāmi'ah) 14:13–40

Zeng L, Wu FE, Oberlies NH et al (1996) Five new monotetrahydrofuran ring acetogenins from the leaves of *Annona muricata*. J Nat Prod 59(11):1035–1042

Zhang DY, Wu J, Ye F et al (2003) Inhibition of cancer cell proliferation and prostaglandin E2 synthesis by *Scutellaria baicalensis*. Cancer Res 63(14):4037–4043

Zheng G, Kenney PM, Zhang J et al (1993) Chemoprevention of benzo[a]pyreneinduced forestomach cancer in mice by natural phthalides from celery seed oil. Nutr Cancer 19

Zheng GQ, Kenney PM, Lam LK (1992) Anethofuran, carvone, and limonene: potential cancer chemoprotective agents from dill weed oil and caraway oil. Planta Med 58(04):338–341

Zhu W, Xu J, Ge YY et al (2014) Epigallocatechin-3-gallate (EGCG) protects skin cells from ionizing radiation via heme oxygenase-1 (HO-1) overexpression. J Radiat Res (Tokyo) 55:1056–1065

Chapter 22
Urban Health Infrastructure in Small Cities: Is It Availability or Accessibility?

Krishna Udnoor, Priya Narayanan and Ramesh Anguluri

Abstract Infrastructure is the basic lifeline on which any city thrives. The mere presence of infrastructures does not guarantee the services until the geographic accessibility ensures availability of infrastructure at proximity. Therefore, infrastructure has to be a cross viewed with spatial distribution and spatial accessibility. The research work attempts to exemplify the application of Geoinformatics in analysing the distribution and space-time accessibility of selected infrastructure related to health in Gulbarga city, India. The outcome of the study also exhibits the application of Geoinformatics for designing the placing of infrastructure to achieve the spatial balance of distribution. People to service analysis are a unique technique through which GIS helps to visualise distribution aspects of urban infrastructure facilities.

Keywords Urban · Infrastructure · Health · Gulbarga

22.1 Introduction

The dictionary meaning of the word infrastructure is "the underlying foundation or basic framework." The term infrastructure typically refers to the technical structures that support a society, such as roads, water supply, sewers, power grids, telecommunications, schools and hospitals. To satisfy the basic needs such as food, water, clothing and shelter and other requirements every human being requires access to certain facilities like water supply, electricity, housing and communication (Al-hader and Al-hader 2016). These infrastructures form the essential part of our life. Access to critical infrastructure is the most important aspect of the quality of urbanisation.

K. Udnoor
Adithi GeoInfo Pvt Ltd, Gulbarga, India

P. Narayanan (✉)
NAAC, Bangalore, India
e-mail: Priya.srihari@gmail.com

R. Anguluri
Ministry of Environment, Forest and Climate Change, New Delhi, India

R. B. Singh et al. (eds.), *Urban Health Risk and Resilience in Asian Cities*, Advances in Geographical and Environmental Sciences,
https://doi.org/10.1007/978-981-15-1205-6_22

The amenities like electricity, water, sanitation and clean fuel are the essential determinants of living conditions and health of the urban people. As reported as Census of India 2011, there are 48 urban agglomerations in India in the cities in Megacities that have a population of more than one million. Earlier studies suggest that metro cities act as urban magnets population. These unprepared growths of population with the lack of proper urban planning have created wide gaps between infrastructure demands and supply. Some visible effects of urbanisation result in the lack of basic amenities, congestion due to inadequate infrastructure, higher amount of pollutants in the environment, contamination of water, scarcity of water and electricity, increasing crime rates, etc. (Sudhira et al. 2003).

There is a rapid shift in human activities from primary to non-primary mainly due to changing resource base and environmental conditions. With the advent of science, technology and management facilitate the study of the distribution of infrastructures to plan for equal distribution, even in small cities. Such a shift naturally enhances human affordability and capabilities to achieve improved standard of living through better access to infrastructure and services (State and Sastry 2006). The varying levels of different infrastructural services that are available to inmates are because of regional disparities in economic development. Improvement in these infrastructural services is necessary for enhancing the efficiency of the productive process and for raising the productivity of any economic being.

Infrastructure distribution has two parameters service and customer. Between the service and customer there exist space and time. Hence the distribution of services and its access by the customers can be studied in spatial and temporal dimension. The temporal dimension of dispersed facilities plays a significant role in accessing the infrastructure (Miller 1999). Measuring the accessibility of Basic Infrastructure gives an understanding of the performance of the system in the city.

22.2 Urban Infrastructures

Urban infrastructure is a framework of system usually envisaged in the engineering perspective. They can have several subsystems of a city. Urban infrastructure consists of education system, sanitation, sewage systems, electricity and gas distribution, potable water, urban transport and primary health services (Pandit et al. 2015). Infrastructure can be naturally given and/or artificially located. The artificially created basic infrastructure through investments plays a crucial role in the location of services (Wenban-smith 2006). As Indian cities should be prepared to host the rapid growth and to provide services for the entire society, it is important that they should have a plan developed for urban infrastructure. Cities not only need basic infrastructure but also need to put a system in place to check the delivery of services of standards to all inmates and also should be affordable to underprivileged. There is a range of technologies that help as Decision Support System to assist the urban planners and administrators to organise, analyse, modify and re-evaluate the distribution of urban infrastructures (Coutinho-rodrigues et al. 2011). Urban infrastructure facilities are strategic, and the investments have a long-standing impact on the region. Hence the cost-benefits of these investments have to be guided spatially. While assessing the range of Urban Infrastructure, it is inherent and imperative to analyse the distribution and delivery of basic services like safe potable water, sanitation, waste management, health and emergency services, management of open spaces have a long-standing impact on the quality of life in cities. The functional capability of these services is just not merely guided by its spatial location. There exist invisible inter-linkages that adversely impact the delivery of these services. Knowledge infrastructure governs such inter-linkages. Geoinformatics acts as a tool to guide the knowledge infrastructure for planning, regulating and strategic delivery of these infrastructures.

22.3 Geoinformatics in Infrastructure Distribution Analysis

There is an increase in urban population and maintaining information about the infrastructure available, and their distribution is crucial. Traditional tool and techniques used to design cities prove to be inadequate while designing Tier I city and metro cities. Advancement in technology has created a robust system wherein Geoinformatics can integrate multilayer spatial information with statistical attributes to help urban planners to make decision in the creation and distribution of the basic infrastructures. The use of Geoinformatics in the planning of infrastructure and its distribution has expanded considerably over the years. The fact cannot be ruled out that unplanned urbanisation with less or no holistic approach leads to lack of infrastructure and basic amenities (Ramachandra et al. 2012). But unbalanced distribution of basic infrastructure leads to overuse or underuse of infrastructure facilities (Moss 2008). Geoinformatics guides infrastructure distribution assessments through land suitability analysis (Khahroa et al. 2014). Apart from infrastructure distribution, Geoinformatics can be used as an efficient tool to plan the structural procedure of project management that can include several variables like cost estimation, planning, scheduling and control (Palve 2013). The advancement of information technology has supported the generic activity of information analysis for infrastructure development in urban areas through an array of tools and techniques that are specially designed for specific application (Bhailume and Das 2013). Physical accessibility of people to urban services and facilities is a key component of quality of life in urban areas, and Geoinformatics helps to achieve equality in accessibility of the same (Pacione 2008). Geographic Information System (GIS) are effective tools to foster community need-based urban planning (Al-rasheed and El-gamily 2013). GIS also assists the decision-making in site suitability for several developmental activities involving multiple criteria evaluation at municipal level planning (Coutinho-rodrigues et al. 2011). The capabilities of GIS in transport network analysis are plenty. Regression models, structural equation models (SEM), logit models, artificial neural network models are the models that are frequently used in utility analysis and modelling (Rahman et al. 2016). Transport offers opportunity of mobility to the inmates in a city and it acts as the possibility to connect the spaces of demand with that of services (El-geneidy et al. 2016).

22.4 Accessibility and Availability

Accessibility is the comfort by which any infrastructure that is accessed from a particular location utilising available transport facility. Development of human life is the amount to which man can satisfy the basic needs such as food, clothing and shelter. To fulfil these, man requires access to these necessary facilities such as market, housing, water supply, electricity and adequate transportation (Aderamo 2011). Accessibility is an essential concept of interpretative studies of public service delivery. Accessibility is understood as the ease with which individuals can contribute to desired activities given the available transportation system and land use pattern (Pirie 1979; Pooler 1987; Urban et al. 2016). Accessibility is fundamental and much neglected in infrastructure planning. Social accessibility to infrastructure facilities more specifically to the low income and the socially disadvantaged population who are more like to use public transit should be of more concern to planning (Ibes 2015). Accessibility has not only been an important explanatory factor in a host of geographic phenomena (Huff 1964; Lakshmanan and Hansen 2007), it has also been functional in the past for various analytical and evaluative purposes. The disparities have led to series of codified standards by which differences can be measured (Golub et al. 2013). Improvement in geographic information system (GIS) has opened up several ways to measure accessibility (Yoshida and Deichmann 2009). GIS also acts as an enhanced tool for decision support system (DSS) (Naude et al. 1999). Space-time prism concept in GIS has been extensively applied to study accessibility of services (Taylor et al. 2007). Optimal planning and development of the city not only depends on the landscape and its limitation. Nevertheless, the inhabitants need should govern the distribution of infrastructure. Geographic information system (GIS) environs facilitate with array of techniques to map and analyse the same (Jaroslav 2008). While Production and consumption of goods play a significant role in economics, the space between the two plays strategic role in determining the demand, supply and friction (Boarnet et al. 2017). This space between availability and accessibility determines the attractiveness of availability in geographic space. Understand individual travel behaviour on a day-to-day basis is also an important parameter for service delivery (Cherchi et al. 2017). Over repeated observations, the same can be modelled and simulated. Supply chain management practices and freight management from warehouse to customers also need to use accessibility (Sakai et al. 2017). Accessibility is also considered and studied in time and money perspective (Mouter and Chorus 2016). Apart from quantitative methods of studying transport, there exist many qualitative aspects of travel-related studies and researches (Sadhukhan et al. 2016).

22.5 Infrastructure and Basic Amenities in Urban India

The population of urban towns in India has been ever growing, but the lack of matching investments in urban development has created an acute deficiency in the availability of basic infrastructure in the towns and cities of urban India. Census data and NSS data provide us bare minimal input to analyse some of the basic amenities like sanitation and drinking water, educational facilities. However, mere extracts of numbers do not explain their spatial accessibility or distribution. While analysing these data over a period, one can come across increasing trends; however, the question service to the sparsely served areas remain unanswered. The Habitat II Agenda (1996) refers basic amenities of cities as safe drinking water, sanitation, waste management, social welfare like schools, transport and communication facilities, energy health and emergency facilities, public safety and management of open spaces (Annan 1997). The 2001 Census of India expresses that only 43% of Indian population have access to piped drinking water supply, 67% of the Indian population have access to electricity, only 44% of the Indian urban drains are closed drainage system, only 47% of the households have the basic household sanitation latrines. 67% of Indian household still use firewood for their domestic fuel, but in contradiction, 63.2% of household use mobile or landline phones.

22.6 Study Area

Gulbarga city is one of the largest cities in North Karnataka. The geographic coordinates of the city are 17.330 latitude North and 76.8340 longitude East. The city is characterised with hot climate ranging from 22 °C to 40 °C in summers, and cool and dry winters with temperature ranging from 15 °C to 33 °C. The annual average rainfall of the city is 777 mm (IMD). The average number of rainy days is 46, (1971–2000) (http://www.imd.gov.in). The city gets more rainfall in the Southwest monsoon ranging up to 186.6 mm. Convectional rainfall and thunder storms are quite usual phenomena in the in the month of April–May. According to Census 2011, the population of the city is 543147 with 102830 numbers of households. The male and female population is 276552 and 266595, respectively, with sex ratio of 964. The average literacy rate is 82.20%. The city is governed by the Gulbarga Mahanagar Palike (GMP), which is divided into 55 Administrative Wards and 3 Outgrowths. Gulbarga city covers a spatial spared of 172 km^2 (Fig. 22.1). The urban area is about 82 km^2.

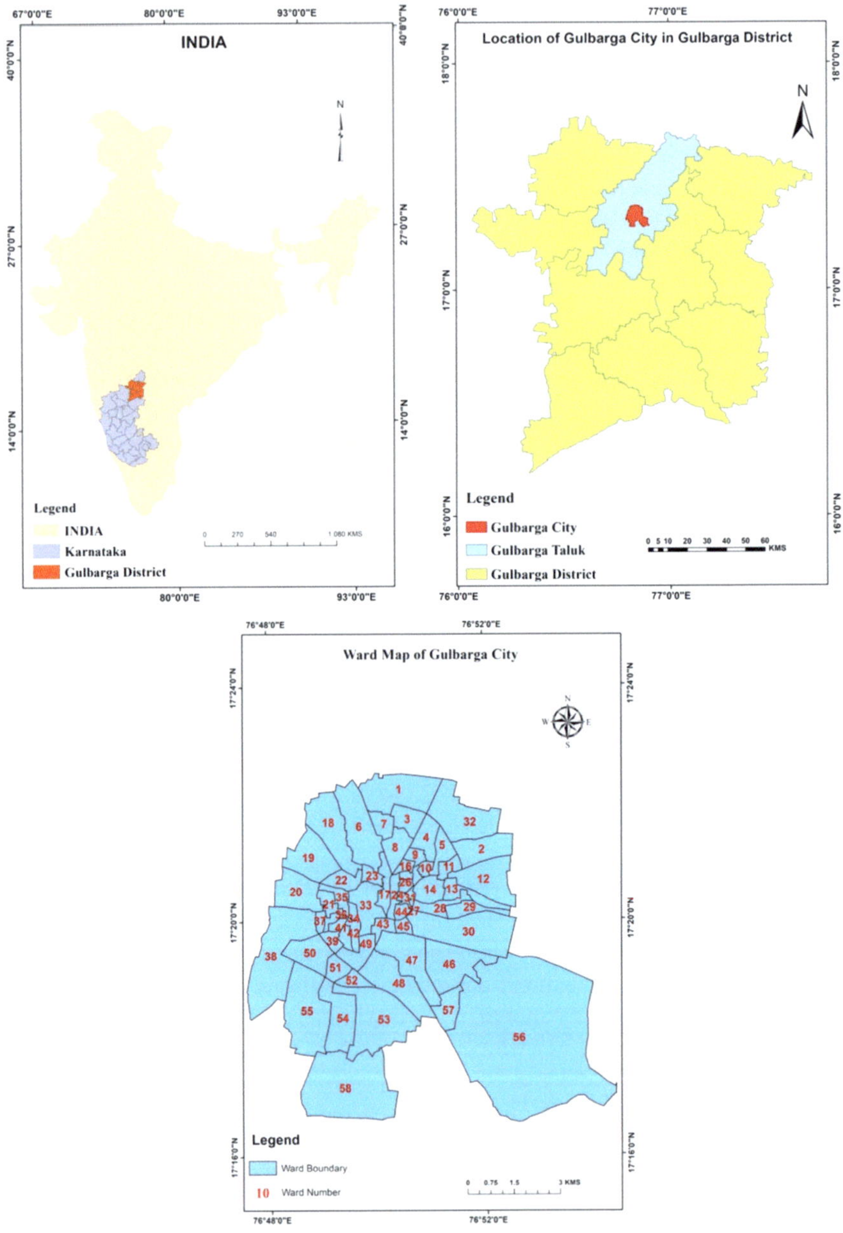

Fig. 22.1 Location of study area

22.7 Methodology

The analysis comprises the component of distribution of health and emergency facilities and its accessibility and its reach. These components were studied in two-dimension space. Immediate accessibility within a ward using accessibility index and overall accessibility using service area analysis based on the location of infrastructure. Solid waste bins and hospitals were the infrastructures considered for analysis. The attractiveness index provided an accurate quantitative result of the attractiveness of facilities that is distributed within every ward about the distance nearest wards. Hence the results thus obtained are more realistic that if a service is not available within a ward, people usually rely on its neighbouring ward. Considering the distance of the nearby ward being higher, the attractiveness of the ward taken for analysis will be lower as depicted in Fig. 22.2.

The index can be quantified by applying the formula

$$Ai = j = \ln SijDij \quad (1)$$

where, Ai is the accessibility in Zone I, Sij = Size of the attractiveness of facilities in Zone I, Dij is the distance between zone I and j. Dij can further augment by assigning an exponential K which is known as friction. Friction can represent connectivity, traffic jam, the number of traffic signals or any other parameter that could hamper travel time. However, friction was not considered in this research due to lack of pertinent data. The infrastructures considered for analysis in this study are educational institutions, hospitals and solid waste disposal bins. Through infrastructure do not restrict only to the three, these three layers had the ease of being represented as points. The attractiveness of facilities in the zone I represented as Sij is just the number of hospitals/bins in a ward. However, the wards are not standardised in their size. Hence, the number of facility available was divided by ward area to get the Sij, which represents the density of service available in a ward. In real word context, regarding service usage, the ward boundaries do not exist. Hence, it is not essential that a consumer or user, located in the corners of ward I will use the facilities of ward which s located in its centre. The user considers would also look out for the facilities in terms of distance that is surrounding him, and hence the distance of other wards plays crucial role here. Hence centroids of each ward were obtained, and the nearest ward and the distance between them as ij were worked out. The numerator showing the attractiveness of one ward, the denominator describes the chances of the neighbouring ward to attract users. As the distance increases between two administrative unit increases, the attractiveness decreases.

The aspect of Ai describes the availability of infrastructure. The accessibility is measured conditioning with the ability to reach that facility through some means of transportation. In Gulbarga, city roads are the primary mode of transportation. There exists no metro or rail connectivity within the city. Footprints of building and roads were the results of image extraction from quick bird satellite image. The accessibility analysis focuses on service area in terms of time taken to travel. It determines the

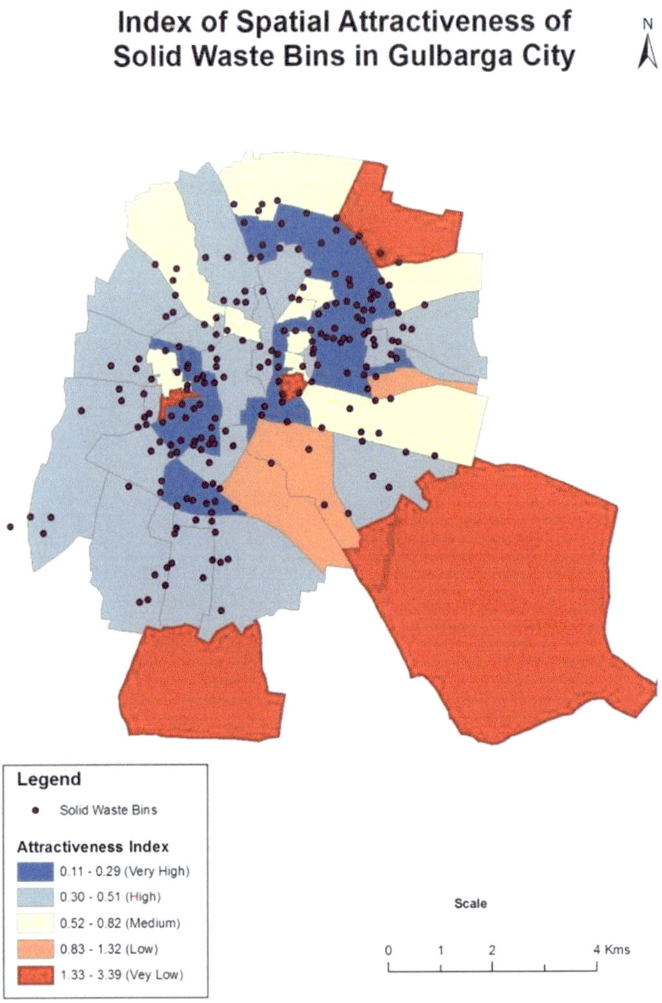

Fig. 22.2 Distribution of solid waste collection bins and its spatial service index

service area that one hospital, or one bin and its ability service area in given travel time. There different time slots of five, ten and fifteen minutes are analysed and results obtained has been discussed.

22.8 Results

Figure 22.3 represents the distribution of solid waste collection bins. One aspect of smart city is the cleanliness. How adequate is the service are of solid collection binds determines the efforts by the local government to keep the city clean. It is evident through the analysis that the distribution is unequal serving only the CBD the periphery of the city is inadequately serviced. Few wards represented in Blue enjoy the privilege of adequate service of solid waste collection and hence littering is prominent in the other ward represented in colours other than blue. Figure 22.4 the

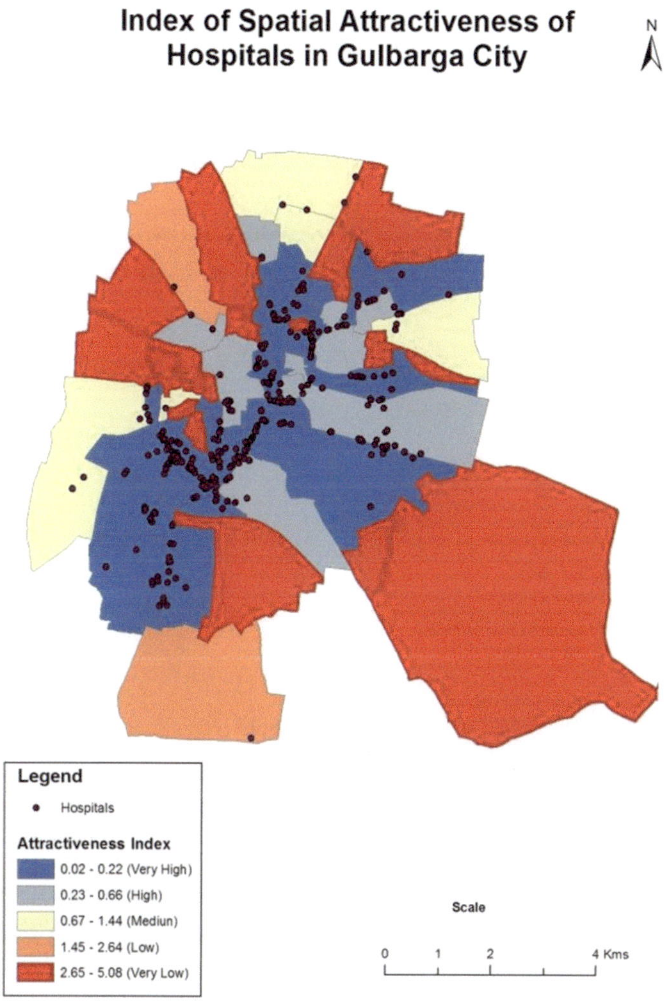

Fig. 22.3 Index of spatial attractiveness of hospitals in Gulbarga city

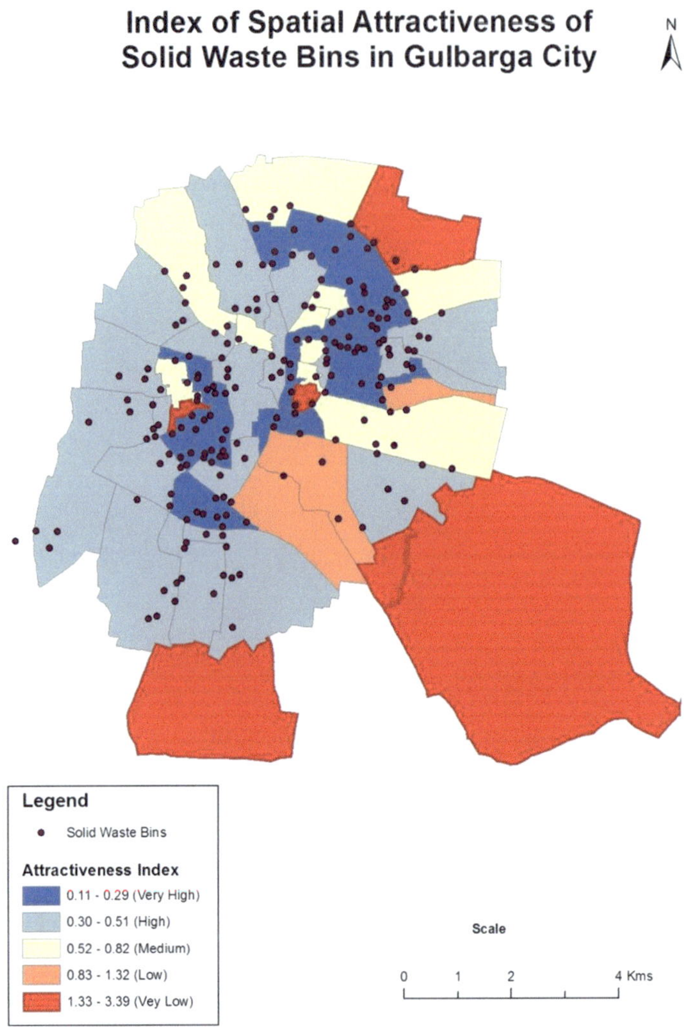

Fig. 22.4 Index of spatial attractiveness for household waste disposal

distribution of emergency infrastructure such as hospitals and clinics in Gulbarga city. The hospitals and clinics also are clustered. In the real-world situations, hospitals and clinics have several complicated patients as user pattern. Specialisation, the experience of the physicians, government or private, cost, quality of health services, governs the user in the health sector in India. However, mere location gives access to health services during emergencies. The current research focuses on the place of health services situated and the user accessibility to it in spatial terms.

The index of the attractiveness of hospitals shows a clear north–south divide in the city. The northern part of the city has minimal clinics, and the outgrowth in the south also has minimal health facilities. As revealed by Figs. 22.3 and 22.4 also orient to the fact that health amenities in minorities dominant area are minimal. Such spaces give scope for further probe on health awareness and health schemes offered by the government of India.

The index of attractiveness of solid waste collection bins again shows clustered pattern and a north–south divide.

22.8.1 Emergency Service Infrastructure

The Gulbarga City Fire Department serves the entire municipal area and provides aid response into neighbouring jurisdictions also. To extend timely firefighting cover in any urban areas during an emergency to the remotely located areas is hardly possible with limited fire stations. The initial 5–10 min of any disaster or emergency are very crucial and to reach the incident spot within that stipulated time and save life and property makes the infrastructure more efficient. The location of the fire station plays a significant role in the emergency service to the nearby locality. Fire stations located in the core area are not much suitable as it causes the bottleneck of roads during busy hours.

Gulbarga has a population of 540,000, as per the URDPFI Guidelines the city should have a minimum of two fire stations to serve the town in any major outbreaks. The below images simulates 5, 10, 15 min drive time from the fire station to the neighbourhood locations. There are 102830 households in the city.

The five, ten and fifteen minutes time of drive from the fire station serves only 8.5%, 59% and 93%, respectively, of the total households. Hence the service is not uniformly available to all the families. Figure 22.5 explains the reach of fire exhaust trucks to the different parts of the city through roads. While applying the isochrones analysis, it is evident that the station is at the centre of the city can serve 8.5% of the city in its first 5 min. The reason being the traffic and narrow roads in the core of the city. According to the objectives of the Karnataka State Fire and Emergency Service Department, and the URDPFI norms mentioned in Table 22.1, aims to reach the fire spot, with a minimum response time of three minutes. With regard to the scenario at Gulbarga, to the number of fire stations and its spatial location, this service is seldom possible. Hence one more new outpost of the fire station on a spatially suitable location is to be established for the city.

22.8.2 Healthcare Infrastructure

Sufficient and unbiased access to healthcare facilities for local populations in urban areas is an important issue for the urban planner and policymakers. Disparities in the spatial distribution of population and health care facilities lead to poor accessibility of the infrastructure (Wang and Luo 2005). The city has four medical colleges (two

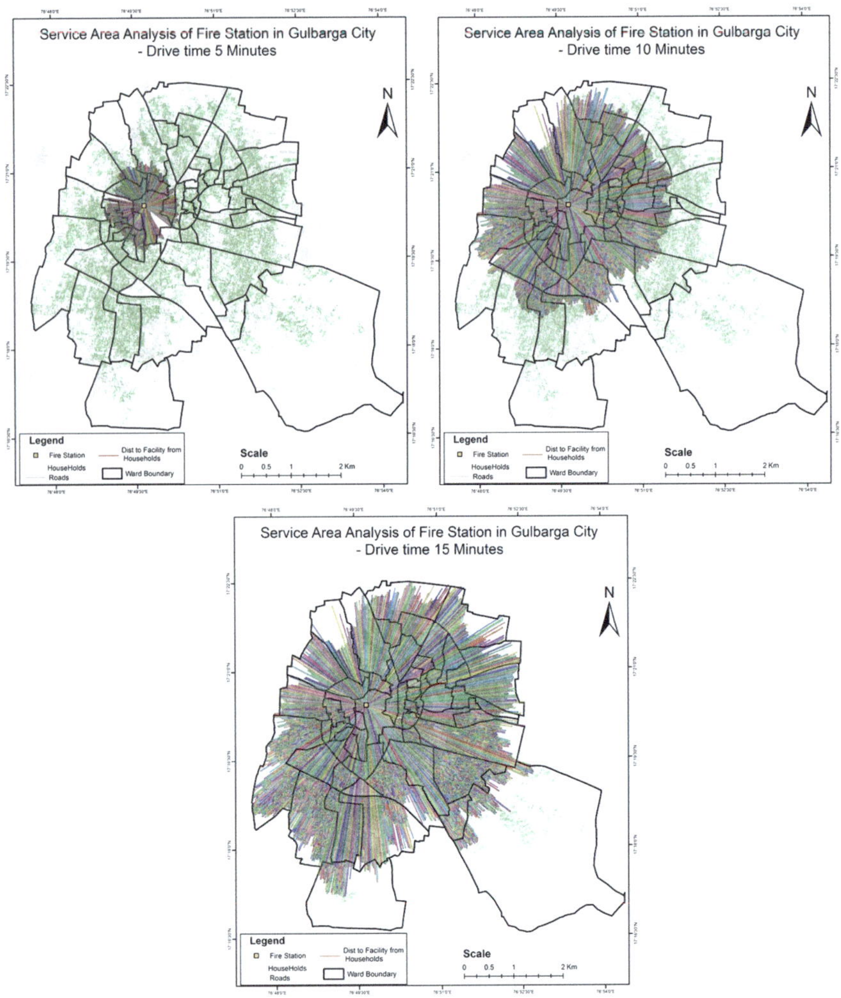

Fig. 22.5 Isochrones of emergency fire service infrastructure

Table 22.1 URDPFI norms for fire station

Sl. no	Category	Distribution or population served per unit	Area requirement
1.	Sub fire station/Fire Post	Within 3–4 km radius	0.6 Ha (with essential residential accommodation)
2.	Fire station	2 lakh population or 5–7 km radius	1 Ha with residential accommodation

Source Urban and Regional Development Plansformulation and Implementation (URDPFI) Guidelines, 2014. Ministry of Urban Development and MPD 2020

Table 22.2 Healthcare centres of Gulbarga city

Sl. no.	Type of hospital	No of hospitals	Total No of beds
1	Government hospital	1	400
2	Children hospital	8	91
3	Maternity hospital	58	743
4	General hospital	9	1987
5	Cardiac hospital	6	60
6	Ortho hospital	7	118
7	Multispecialty hospital	28	579
8	Diagnostic centre	33	–
9	Eye hospital	11	86

Government Colleges and two Private Colleges). Table 22.2 shows the details of hospital types and beds accommodated. The spatial distribution of these hospitals is a major concern to the inmates of the city.

There are 161 healthcare centres in Gulbarga city. But their spatial distribution is clustered in such a way that these facilities are not uniformly available to all the inmates of the city. While analysing, the geographical service are of these healthcare centres, Fig. 22.6 portrays the spatial reach of every hospital in five, ten and fifteen minutes drive time.

While analysing the isochrones of the healthcare centre in Gulbarga city, it is well evident from Fig. 22.6, that there are many blank spots unserved with healthcare facilities.

22.8.3 *Solid Waste Infrastructure*

Total waste generated in the city is estimated to be about 197 tonnes per day (TPD). The door–to-door collection of waste and segregation of waste is available at sources in few wards of the city (Table 22.3). There are 201 container bins spread across the city, out of which there are about 145 trash bins of size 3 m^3 and 56 trash bins of size 4.5 m^3. A landfill area located at village Udnoor, having 28.19 acres of land at a distance of 13 kms from the city serves as dumping yard. Each day residential, commercial and institutional waste accounts around 197 tonnes.

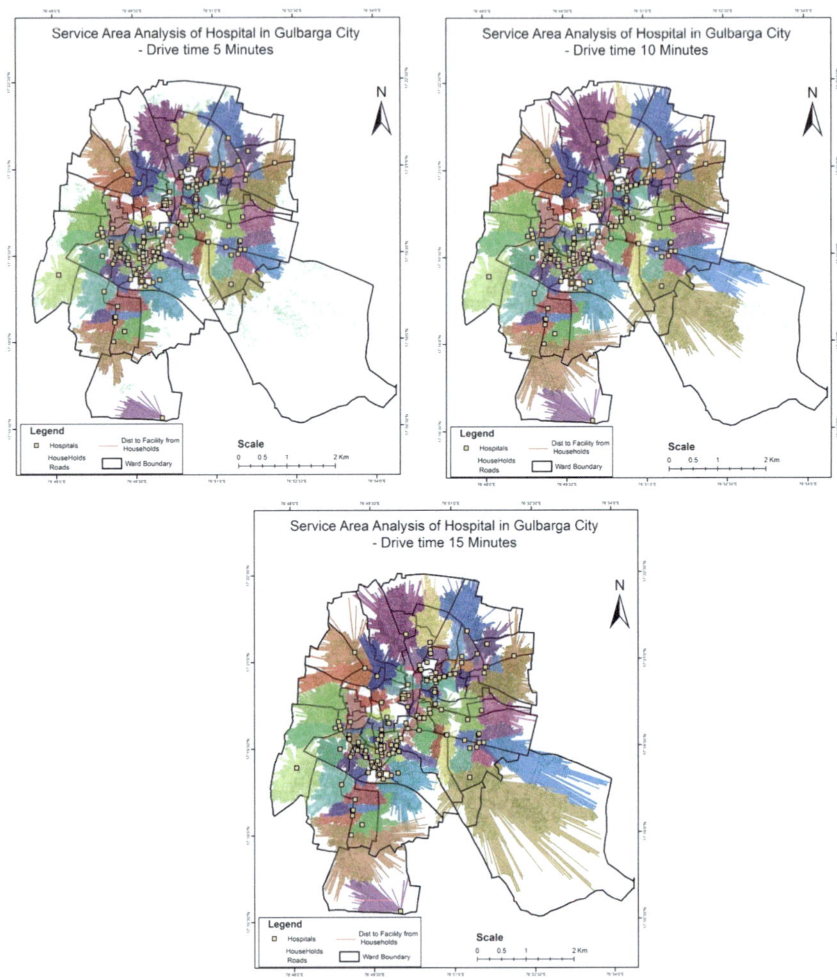

Fig. 22.6 Isochrones of emergency hospital service

Figure 22.7 explicitly shows the bin service areas as the collection point of solid waste. Though the number of trash bins is adequate in the centre, there is a vacuum and an unserved spot in the core of the city and the peripheries are unserved.

Table 22.3 Details of solid waste generation of Gulbarga city

Sl. no.	Waste generators	No of waste generators	Avg waste per source in kgs	Total waste in Kg	Total waste in tonnes
1	Household	102,830	1.2	123,396	123.4
2	Commercial shops	14,861	1.6	23,777.6	23.78
3	Hospitals (MSW)	514	5.1	2621.4	2.62
4	Small hotels	530	4.1	2173	2.17
5	Large hotels	53	11	583	0.58
6	Markets (major)	15	1230	18,450	18.45
7	Street sweepings				
	Type A	202.47	29	5871.63	5.87
	Type B	645.31	9.1	5872.321	5.87
	Type C	138.78	4.1	568.998	0.57
8	Educational institutes	682	6.4	4364.8	4.36
9	Miscellaneous waste	5%		9383.93	9.38
	Total			197,062.69	197.06

22.9 Discussions

Landscape planning is instrumental in allocation of land use and proposal for landscape maintenance or redevelopment that can be implemented by different regulations (Haaren et al. 2014). The blank spots of infrastructure lacunae can be addressed only through landscape planning of the cities that are realised through city master plans. Further such data can be updated through volunteered or user-generated spatially referenced data that can be collected through ubiquitous technologies (Cope 2015). Land use models and statistical packages and models can be used to understand transportation decisions (Hankey et al. 2012). Infrastructure developments and expansions would need land and hence per-urbanisation is an unavoidable process whose pattern and process will affect urban landscape (Hersperger et al. 2012). Hence prior planning can be at place taking into account of existing infrastructure. "Strong sustainability" framework is an integrated approach that uses multiple indicators for sustainability assessment (Huang et al. 2016). Sustainable urbanisation can be a holistic framework of integrated infrastructure assessment as the one carried out in this research paper. Establishment of appropriate infrastructure for number of targeted people has ever been a challenge (Oh et al. 2005). There is always an accumulated deficit with the demand and availability of infrastructure. It is apparent to understand the carrying capacity of infrastructure and projection of population in

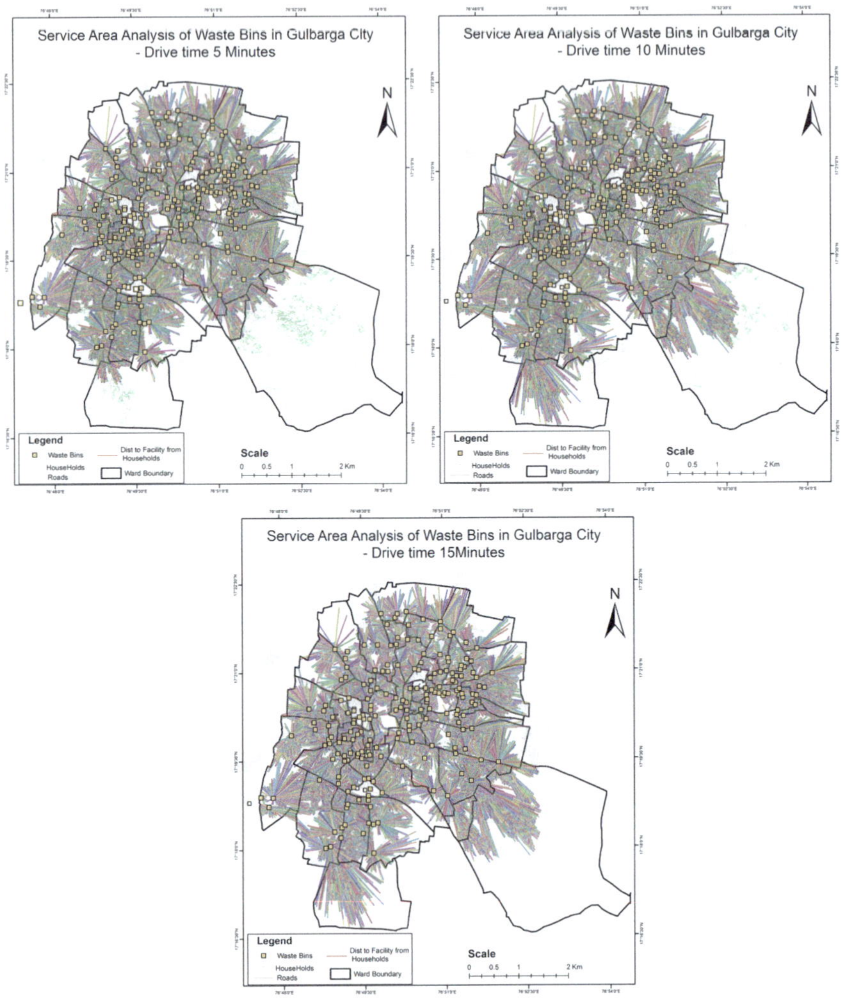

Fig. 22.7 Isochrones of solid waste collection bins

order to understand the future needs. Accessibility is a spatial phenomenon. If the availability of access can be varied in different parts of an area, this pattern might affect location pattern of different urban activities (Schrijnen 2000).

22.10 Conclusion

In the consolidation of all the findings, the blank spots of health services, emergency services and solid waste are showing common areas as some pockets in the centre of the city, and some of the south and peripherals of the city. The efficiency of services can be improved by spatially planning the new services to achieve spatial efficiency. Analysis of this kind will help the landscape planners to place the services to achieve efficiency and spatial coverage.

References

Aderamo AJAO (2011) Spatial inequalities in accessibility to social amenities in developing countries: A Case from Nigeria. Aust J Basic Appl Sci 5(6):316–322

Al-hader M, Al-hader M (2016) The smart city infrastructure development & monitoring THE SMART CITY INFRASTRUCTURE Theoretical and Empirical Researches in Urban Management. Theoretical and Empherical Research in Urban Management 2(11):86–94

Al-rasheed K, El-gamily HI (2013) GIS as an Efficient Tool to Manage Educational Services and Infrastructure in Kuwait. Journal of Geographic Information System 2013(February):75–86. https://doi.org/10.4236/jgis.2013.51008

Annan K (1997) Cities and Homes for All: The Habitat Agenda. Retrieved from http://www.chs.ubc.ca/archives/files/Cities and Homes for All.pdf

Bhailume S, Das S (2013) Geoinformatics Application for Urban Utilities Information System: A Case Study of Pune City. International Journal of Computer Applications 65(22):33–39

Boarnet MG, Hong A, Santiago-bartolomei R (2017) Urban spatial structure, employment subcenters, and freight travel. J Transp Geogr 60:267–276. https://doi.org/10.1016/j.jtrangeo.2017.03.007

Cherchi E, Cirillo C, Dios JD (2017) Modelling correlation patterns in mode choice models estimated on multiday travel data. Transp Res Part A 96:146–153. https://doi.org/10.1016/j.tra.2016.11.021

Cope M (2015) Geographies of digital lives: Trajectories in the production of knowledge with user-generated content. Landscape and Urban Planning 142:212–214. https://doi.org/10.1016/j.landurbplan.2015.08.009

Coutinho-rodrigues J, Simão A, Henggeler C (2011) A GIS-based multicriteria spatial decision support system for planning urban infrastructures. Decis Support Syst 51(3):720–726. https://doi.org/10.1016/j.dss.2011.02.010

El-geneidy A, Levinson D, Diab E, Boisjoly G, Verbich D, Loong C (2016) The cost of equity: Assessing transit accessibility and social disparity using total travel cost. Transp Res Part A 91:302–316. https://doi.org/10.1016/j.tra.2016.07.003

Golub A, Robinson G, Nee B (2013) Making accessibility analyses accessible: A tool to facilitate the public review of the effects of regional transportation plans on accessibility. The Journal of Transport and Lan Use 6(3):17–28. https://doi.org/10.5198/jtlu.v6i3.352

Haaren CV, Warren-kretzschmar B, Milos C, Werthmann C (2014) Landscape and Urban Planning Opportunities for design approaches in landscape planning. Landscape and Urban Planning 130:159–170. https://doi.org/10.1016/j.landurbplan.2014.06.012

Hankey S, Lindsey G, Wang X, Borah J, Hoff K, Utecht B, Xu Z (2012) Estimating use of nonmotorized infrastructure: Models of bicycle and pedestrian traffic in Minneapolis. MN. Landscape and Urban Planning 107(3):307–316. https://doi.org/10.1016/j.landurbplan.2012.06.005

Hersperger AM, Langhamer D, Dalang T (2012) Inventorying human-made objects: A step towards better understanding land use for multifunctional planning in a periurban Swiss landscape. Landscape and Urban Planning 105(3):307–314. https://doi.org/10.1016/j.landurbplan.2012.01.008

Huang L, Yan L, Wu J (2016) Assessing urban sustainability of Chinese megacities: 35 years after the economic reform and open-door policy. Landscape and Urban Planning 145:57–70. https://doi.org/10.1016/j.landurbplan.2015.09.005

Huff DL (1964) Defining and Estimating a Trading Area. Journal of Marketing 28(3):34–38. https://doi.org/10.2307/1249154
Ibes DC (2015) A multi-dimensional classification and equity analysis of an urban park system: A novel methodology and case study application. Landscape and Urban Planning 137:122–137. https://doi.org/10.1016/j.landurbplan.2014.12.014
Jaroslav B (2008) GIS Analytical Tools for Planning and Management of Urban Processes. GIS Ostrava 27:1–12
Khahroa SH, Matorib AN, Chandiob IA, Talpurb MAH (2014) Land Suitability Analysis for Installing New Petrol Filling Stations Using GIS. Procedia Engineering 77(November):28–36. https://doi.org/10.1016/j.proeng.2014.07.024
Lakshmanan JR, Hansen WG (2007) Journal of the American Institute of Planners, (July 2014), 37–41. http://doi.org/10.1080/01944366508978155
Miller HJ (1999) Transportation Networks: Basic Theory and Computational Procedures 31(1):1–26
Moss T (2008) " Cold spots" of Urban Infrastructure : " Shrinking" Processes in Eastern Germany and the Modern Infrastructural Ideal. International Journal of Urban and Regional Research, 32(June 2005), 436–451. http://doi.org/10.1111/j.1468–2427.2008.00790.x
Mouter N, Chorus C (2016) Value of time – A citizen perspective. Transp Res Part A 91:317–329. https://doi.org/10.1016/j.tra.2016.02.014
Naude A, Jong TD, Teeffelen PV (1999) Measuring Accessibility with GIS-Tools: A Case Study of the Wild Coast of South Africa. Trans GIS 3(4):381–395
Oh K, Jeong Y, Lee D, Lee W, Choi J (2005) Determining development density using the Urban Carrying Capacity Assessment. System 73:1–15. https://doi.org/10.1016/j.landurbplan.2004.06.002
Pacione M (2008) Access to urban services — the case of secondary schools in Glasgow. Scott Geogr Mag 105(1):37–41
Palve SN (2013) Application of gis in infrastructure project management. International Journal of Structural and Civil Engineering Research 2(4):110–122
Pandit A, Minné EA, Li F, Brown H, Jeong H, Newell JP, Crittenden JC (2015) Infrastructure Ecology: An Evolving Paradigm for Sustainable Urban Development. J Clean Prod. https://doi.org/10.1016/j.jclepro.2015.09.010
Pirie GH (1979) Measuring accessibility: a review and proposal 11:299–312
Pooler J (1987) Measuring Geographical Accessibility: Review of Current Approaches and Problems in the Use of Population. Potentials 18(3):269–289
Sadhukhan S, Banerjee UK, Maitra B (2016) Commuters' willingness-to-pay for improvement of transfer facilities in and around metro stations – A case study in Kolkata. Transp Res Part A 92:43–58. https://doi.org/10.1016/j.tra.2016.07.004
Sakai T, Kawamura K, Hyodo T (2017) Spatial reorganization of urban logistics system and its impacts: Case of Tokyo. J Transp Geogr 60:110–118. https://doi.org/10.1016/j.jtrangeo.2017.03.001
Schrijnen PM (2000) Infrastructure networks and red ± green patterns in city regions. Landscape and Urban Planning 48:191–204
State FL, Sastry GS (2006) Working paper 174
Sudhira HS, Ramachandra TV, Raj KS, Jagadish KS (2003) Urban growth analysis using spatial and temporal data. Journal of the Indian Society of Remote Sensing 31(4):299–311. https://doi.org/10.1007/BF03007350
Taylor P, Miller HJ, Miller HJ (2007) Modelling accessibility using space-time prism concepts within geographical information systems within geographical information systems. International Journal of Geographic Information Systems 5(3):37–41
Ramachandra TV, Aithal BH, Sanna DD (2012) Insights to urban dynamics through landscape spatial pattern analysis. Int J Appl Earth Obs Geoinf 18:329–343. https://doi.org/10.1016/j.jag.2012.03.005

Wang F, Luo W (2005) Assessing spatial and nonspatial factors for healthcare access: Towards an integrated approach to defining health professional shortage areas. Health and Place 11(2):131–146. https://doi.org/10.1016/j.healthplace.2004.02.003

Wenban-smith H (2006) Urban infrastructure: Density matters, not just size. Research Papers in Environmental & Spatial Analysis 104:1–16

Further Readings

Rahman F, Das T, Hossain S (2016) Perceived service quality of paratransit in developing countries: A structural equation approach. Transp Res Part A 93:23–38. https://doi.org/10.1016/j.tra.2016.08.008

Rodney E. Hero Urban T, Delivery S, Questions S, Author C, Source REH, Press C, Linked UTCR (2016) The Urban Service Delivery Literature : Some Questions & Considerations Author (s): Published by : University of Chicago Press Stable URL : https://www.jstor.org/stable/3234887 References Linked references are available on JSTOR for this, 18(4), 659–677

Yoshida N, Deichmann U (2009) Measurement of Accessibility and Its Applications. Journal of Infrastructure Development 1(1):1–16. https://doi.org/10.1177/097493060900100102

Chapter 23
An Approach to Social Sustainability in Chennai—Understanding the Dynamics of Public Places

Reshmi Panicker

Abstract Urbanization is the most significant trend in the twenty-first century. It is no longer a demographic phenomenon involving migration, but has evolved into a transformative process in the core of global development. It has also brought about its own issues and challenges. We have many guidelines to bring about growth in a wholesome manner. Urban sustainability is another key phenomenon which is being explored and practised in the current world. Though enough is said and done about economic sustainability and environmental sustainability, the third leg of the triad, social sustainability is often forgotten or ignored. In this age of social unrest and inception of exclusive spaces the very fabric that sustains a city, that of the web of social connectivity within its denizens is not strong. Even though the factors for it may be multifold, the design of our public places is in a large way responsible. This study deals with a cross-cultural city like Chennai and its public spaces. There is a dearth of planned public places in Chennai. The way in which existing places behave gives an understanding of the varied dynamics that are at play within these spaces. An attempt is made to characterize the factors which make a place more desired and vibrant. Hierarchy of spaces, layers of activity, controlled or often-uncontrolled chaos and temporality all define the public places in Chennai. The paper argues that in the rapid urbanization process, not only provisions for public places have to be effected, but also equal importance has to be given to maintain these places as vibrant hubs linking the cultural, economic, social and ecological sensitivities of the people. An approach to social sustainability through certain broad guidelines can be defined with these factors.

Keywords Urbanization · Social sustainability · Public places · Vibrant hubs

---

R. Panicker (✉)
School of Architecture, Hindustan Institute of Technology & Science, Chennai, India
e-mail: resh.rp@gmail.com

R. B. Singh et al. (eds.), *Urban Health Risk and Resilience in Asian Cities*, Advances in Geographical and Environmental Sciences,
https://doi.org/10.1007/978-981-15-1205-6_23

23.1 Introduction

Urbanization is the most significant trend in the twenty-first century. It is no longer a demographic phenomenon involving migration, but has evolved into a transformative process in the core of global development. It has also brought about its own issues and challenges. We have many guidelines to bring about growth in a wholesome manner. Urban sustainability is another key phenomenon which is being explored and practised in the current world. Though enough is said and done about economic sustainability and environmental sustainability, the third leg of the triad, social sustainability is often forgotten or ignored. Social sustainability thrives and sustains in public places. It is in these hubs that the web which connects and holds the social fabric of a city exists. These spaces need to be understood and analysed for better sustainability.

Space is not an inert physical entity. It changes hues and colours with respect to the usage of the people. In turn, a space is capable of changing the dynamics of the people occupying it. In this context, interplay between public spaces and its neighbourhood is very important. Public space has always been the centre and heart of any locale. The strength and vitality of this place define the bond within the community. Here, architecture plays a decisive role. It can create, change or stimulate the vibrancy of a gathering space, making it more meaningful.

In this age of social unrest and inception of exclusive spaces the very fabric that sustains a city, that of the web of social connectivity within its denizens is not strong. Even though the factors for it may be multifold, the design of our public places is in a large way responsible.

23.1.1 Methodology and Hypothesis

The study deals with Chennai, a city in South India. The usage pattern of public spaces has undergone a transformation over the years. India's globalized economy over time has propagated a mass transnational culture to the ultimate exclusion of local identity. So the public behaviour, their choice and usage of public spaces have undergone a drastic change. The paper argues that more than designing a public space, efforts for social sustainability create interwoven patterns linking the cultural, economic, social and ecological sensitivities of the people. In this context, this study attempts to arrive at some guidelines defining the social sustainability concepts which can be adopted in a progressive, cross-cultural city like Chennai.

23.2 Chennai City—The Study Area

Chennai, capital to the state of Tamil Nadu is the fourth largest metropolitan city in India. Located on the Coromandel Coast of the Bay of Bengal, it is the biggest industrial and commercial centre in South India, and a major cultural, economic and educational centre. The city of Chennai has a rich heritage of being the port of the Pallava kingdom in the southern region of India. Before the accession to the British, the area now known as Chennai was a group of small villages existing rather independently under different local Chieftains. Portuguese came and settled in the sixteenth century followed by British in the seventeenth century. It became the first major British settlement in India (CMDA 2016).

Currently, Chennai is one of the fastest-growing cities in the world. The changing economic policies have given Chennai's IT sector a major boom attracting investors from all over the globe. This in turn brought in a workforce from across the country turning it into a bustling metro city.

23.2.1 Public Spaces in Chennai: Current Scenario

Chennai is a city which is at crossroads between traditionalism and globalization. It is a complex relationship. It is complex because globalization changes the existing social fabric with new ideas of space, usage, inclusiveness/exclusiveness, consumption and mobility. However, the existing traditional framework mediates this change into globalization and in some way arrests it. This can be understood from the popular public spaces in the city. The advent of a multitude of malls has not diminished the importance of traditional public spaces. In spite of an array into the modern world, the interrelationship between the people and the urban form is not severing its ties with traditional sensibilities.

A design or social sustainability in public space in Chennai cannot be approached without a specific understanding of the existing public spaces and their vibrancy. The case studies will help in understanding the parameters which contribute a sense of belonging and feeling of proximity as a community. These case studies will be analysed in terms of 6 dimensions of Urban design (Carmona et al. 2003).

- Morphological dimension
- Social dimension
- Perceptual dimension
- Visual dimension
- Functional dimension
- Temporal dimension.

23.2.2 *Kapaleeshwara Temple, Mylapore*

Mylapore is a neighbourhood in the southern part of Chennai. It is a major commercial centre as well as a residential neighbourhood. It is one of the oldest neighbourhoods in Chennai and its origin can be traced as a temple town from first century AD (Muthiah 2008). Mylapore rose as the commercial and intellectual hub of Madras city, and home to a vast array of British-educated lawyers and statesmen. In recent times, increasing commercialization has changed its look. Today, Mylapore is known for its residential colonies, temples, shopping malls, kutcheri halls and sabhas and old nineteenth century residential buildings and houses.

The streets around the Kapaleeswarar temple and the tank area have been delineated as the study area in Mylapore.

23.2.2.1 Morphological Dimension

The buildings along North Mada Street, South Mada Street and East Mada Street are commercial on the ground floor and residences on the top. The spaces contained within these streets are old residential buildings. The residences adjacent to the temple are said to be more than 100 years old. The busiest stretch is the North Mada Street with its row of temporary shops along the tank boundary.

The temple tank is huge and dominant. But access to the tank is restricted and used only for temple purposes. Examining the street pattern around the temple gives a glimpse of the earlier temple town settlement with grid pattern. The building heights vary from single floor to four floors. The relative low heights of the buildings retain the dominance of the temple gopuram.

23.2.2.2 Perceptual Dimension

Every citizen has had long associations with some part of this city, and his image is soaked in memories and meanings (Lynch 1960). Kapaleeshwara temple and the area surrounding it being one of the oldest neighbourhoods are associated with many images. The temple obviously is an important icon for Hindu religion in the city. The commercial space around is known for the availability of religious, music and dance artefacts. Also giving it a different colour is the legendary eateries especially at East Mada Street.

23.2.2.3 Social Dimension

The temple being the focal point makes it more popular to one religious group. However, the commercial establishments are frequented by all.

The social activities are mostly temple related. The processions move generally around the madaveethis and also on the R K Mutt Road. Also, Navaratri time is associated with cultural events within the temple. Sometimes competitions are held even on the Mada Streets. The residential associations and Ramakrishna mutt nearby organize events around this area.

The roads are choked with immovable vehicular traffic and high pedestrian movement. The space is not able to cater to these multiple functions with vehicles, people and hawkers struggling for space. However, there are a lot of memory associations with these events which bring people in spite of hardships.

23.2.2.4 Visual Dimension

The temple gopuram is the predominant structure in the skyline. The commercial establishments in pucca buildings are 3–4 floors high. The hawkers have permanent places which are ground-level structures.

On a pedestrian level, the vast expanse of the temple tank is hidden partially by the trees from the R K Mutt Road. The hawkers along North Mada\Street completely block it.

In earlier days, the approach through a narrow path and a large open space beyond would have been there, but now the vehicles and hawkers have made the place congested. But the array of flower shops along Vinayakar Street marks the visual entry to the temple.

23.2.2.5 Functional Dimension

The R K Mutt Road stretches from Adyar to Royapettah. It is an important traffic corridor and hence plays a vital role in the prominence of this public place.

The MRTS station is also within walking distance of this area. The junction of North Mada Street and R K Mutt Road is the most congested because of the vehicular and pedestrian traffic. The South Mada Street is predominantly vehicular. The access to the temple is congested. It is either through Vinayakar Street or East Mada Street.

23.2.2.6 Temporal Dimension

The entire vibrancy of the study area is because of its temporal dimension. The temple offers an array of festivals in different months. The East Mada Street houses the temple chariot right in front of the entrance to the temple. It is taken around the temple annually during April as a part of a festival. During festival seasons like Diwali and Navaratri extra hawkers and increased pedestrian inflow in all the three Mada Streets can be observed.

The hawkers along temple tank are there throughout the day. Extra shopping stalls come up during evenings. The South Mada Street is notable for the evening market where several vendors come and sell vegetables on the street.

Another aspect of the temporal dimension is the transformation along the years. The area around Kapaleeshwara temple was a quiet residential suburb during British times. One of the major factors which brought about the transformation of this place is its access and linkage. The R K Mutt Road literally burst opened a profusion of commercial activity along its stretch. The temple is somewhat drowned by the influx of buildings and activities around it. But still it manages the image of being a focus of attention. The activities are slowly taking over.

23.2.3 Marina Beach

Marina Beach is a natural beach along Bay of Bengal and is a popular leisure place for the people of Chennai. It is claimed to be the second-longest urban beach in the world. On an average, a lakh people visit the beach during weekends and 30,000 during weekdays (Times of India 2011). There are a number of historic buildings constructed on the opposite side of the beach, including the Senate House of the University of Madras, Chepauk Palace (now the Public Works Department office), the nineteenth century Presidency College building, Ezhilagam and the office of the Director General of Police.

23.2.3.1 Morphological Dimension

This space is a linear stretch with perpendicular roads like Wallajah Road, Radhakrishna Salai and Lloyd's Road which connects Mount Road, Dr. Besant Road and Pycrofts Road to Triplicane.

The buildings facing the marina promenade, though only 3–4 floors are big in scale and proportions. They are interspersed with a large expanse of open land further accentuating the massive form of these heritage structures.

23.2.3.2 Perceptual Dimension

Irrespective of where they are, accessible waterfronts are always popular public places. For the residents of Chennai, Marina Beach is their pride and glory. In spite of the development of other neighbourhood beaches, the images and memories associated with Marina are strong. Also contributing to the factor are the presence of heritage structures. The hand-cranked merry-go-rounds, horse rides and make shift eateries are all part of Marina's image. However, the lack of sufficient civic facilities and unhygienic conditions has made it a little less popular in recent years.

23.2.3.3 Social Dimension

The social space in Marina Beach is at three levels. The recreational space bordering the sea, a parking area parallel and the third level constitutes of a landscaped area with various amenities. The activities within the public space include playing or relaxing along sea, eateries, merry-go-rounds, shooting, horse rides, etc.

The parking areas which are in between are bordered again with a long lane of eateries and shops. The peripheral edge along Kamaraj Salai has a pedestrian walkway wide enough to be used as a jogging track.

The landscaped stretch has multiple activity zones like seating and spaces for social interaction, Skating rink and Swimming pool.

These social activities are haphazard and do not follow any pattern. Because of the vast expanse on beach there are areas which are sparsely populated and hence becoming unsafe.

Another important social aspect of Marina are the two memorials of Anna and MGR, which have become part of the whole visit to Marina.

23.2.3.4 Visual Dimension

The long stretch of Kamaraj Salai with the unhindered sandy beach on one side and Indo-Sarcenic stately buildings on the other side is a sight to behold. The landscaped area is designed in contemporary style and has features to improve social interactions.

However, the inter-spaced eateries, shops and kids' playthings do not have any visual or spatial character.

23.2.3.5 Functional Dimension

Kamarajar Salai is a six-lane road and one of the major arterial roads of Chennai City. Three bus stops are located at different points on the beach. Railway stations alongside the beach include the Chepauk, the Tiruvallikeni and the Lighthouse MRTS railway stations. The Chennai central station is also close by. The location of Marina Beach is important in that it is very near to old Chennai areas like Royapuram, Parrys and newer suburbs like Adyar. This contributes very largely to its popularity.

In terms of area, Marina Beach is a vast expanse. But the spaces or activity zones are not clearly defined. This combined with the unhygienic condition of the sandy area is slowly dwindling the number of visitors to this scenic spot.

The landscaped area is clearly defined and has many interesting features. But the things which work against it are

- Too exposed to the busy Kamaraj Salai. No sense of enclosure
- The parking bays and vehicular traffic separates the landscape from the actual beach. The activities get bifurcated resulting in a sparsely used designed space.

There is increased activity near parking and near the sea. Some spaces in between gets very dark and isolated even with thousands of people around, making it unsafe. There is a long lane of shops at one end which gives a definition to the space.

23.2.3.6 Temporal Dimension

Marina Beach being an expansive open space come alive mostly by evening. But one can see visitors at all time of the day. Morning time is very popular for joggers, who make use of the wide pedestrian way along Kamaraj Salai. Towards the light house side, a fish market caters to people looking for fresh catch from sea. Seasonal activities include kite flying and beach cricket. An assembly at Marina Beach is deeply symbolic. Denizens of Chennai have many a time gathered here to let their views know. Another important event associated with this place is the Chennai marathon.

The design of the Marina, connecting Fort St George to San Thome Church, changed the scale of the city in the nineteenth century. Over time, this coastal promenade was lined with a number of significant institutional buildings, a magnificent façade to welcome ships from Europe. Due to the heritage nature of the buildings facing Marina Beach, no built structures were allowed so that this view line is maintained. All the changes and physical developments are curtailed to that narrow strip of land between the sand and promenade. The vibrancy of Marina is dwindling.

23.2.4 Thyagaraya Nagar

Thyagaraya Nagar (T. Nagar) is an important commercial and business centre of Chennai City. It attracts many people from different parts of the city and its suburbs, due to its location and transportation links. T. Nagar has the highest commercial catchment population when compared with other commercial centres of the city (Rakesh 2007). There are a number of apparel, jewellery and utensil stores based in Thyagaraya Nagar. It is the most important gold market in South India. The road is full of big name establishments side-by-side smaller and petty shops that sell all sorts of household goods and garments.

23.2.4.1 Morphological Dimension

During 1923–25, the township of "Thyagaraya Nagar" named after Sir Pitti Thyagaraya Chetty was carved out of the southern part of the erstwhile Mambalam zamindari, with a park developed at the centre (Muthiah 2008). The planned nature of the area is visible from the map. Main roads diverge from the Panagal Park and secondary roads are interconnected in patterns.

Commercial activity has spread over the years on all the major roads, in a linear pattern. Now it is slowly encroaching on the by lanes of the major commercial streets. Single storey shops have given way to multi-storey shopping malls.

The proportion of built space to open space is very high, making it a very choked and congested urban space. It has transformed itself into a centre of intense commercial activity from its original residential nature over a period of time. As a result of this, there is a remarkable change in the overall land use pattern and subsequent development of commercial establishments without any comprehensive planning, which has resulted in the overburdening on sectors like land use, services and traffic and transportation network.

23.2.4.2 Perceptual Dimension

T. Nagar has a strong image of being the ultimate shopping paradise not only for the residents of Chennai, but also for entire South India. Anything worth buying will have a shop in T. Nagar. Also the very chaos, which is criticized becomes an entreating factor to many a shopper. This area has got shops, eateries, hawkers, etc. This mayhem is the vibrancy of T. Nagar.

23.2.4.3 Social Dimension

The residential areas still clinging for space in this teeming commercial district, try to organize some cultural events. The lack of space and intrusion of parking vehicles into any space which is free, jade the efforts. The predominant activities are shopping. Eateries of different levels coexist with these. For shoppers being able to get their hands on anything they want to buy or eat by just walking a few metres is the difference that T. Nagar makes from malls and shopping outlets.

The nature of this shopping paradise is such that people across all ages from kids to elderly people throng this place in search of their favourite goods.

23.2.4.4 Visual Dimension

T. Nagar is characterized by its streets. The long array of shops, hawkers along the road, the vehicles, people… all create a visual chaos. There are no visible architectural characteristics. The major shops have raised their facade to uncharacteristic proportions, defying corporation regulations. It is only visible from the flyover and does not interact well with the pedestrian down below.

23.2.4.5 Functional Dimension

The study area is bounded by Kodambakkam High Road on the north, Mount Road on the east, Mount Road on the south and the suburban railway line on the west, comprising an area of 6.94 km^2.

The bus stand nearby is an advantage and blight. It takes the common man right to the heart of T. Nagar. But the bus traffic creates a lot of issues for vehicular as well as pedestrian traffic.

In T. Nagar, pedestrian amenities are virtually non-existent. There are no resting places, no coherent signage system, no unified design language adopted for street furniture, no play equipment, no landscaping or works of art. The sidewalks—the only area allocated for the pedestrian—are dilapidated and encroached upon by official and illegal structures.

The hawkers are along South Usman Road and Pondy Bazaar Road. The sidewalk at Pondy Bazaar Road barely manages to cater to pedestrians. South Usman Road with its pedestrians, hawkers and narrow road with a flyover make it very congested. Vehicles and people clamour for space here.

The parking facilities along Venkataraman Road and Usman Road are virtually non-existing. The Pondy Bazaar Road is wider and offers parking facilities along the road. Still they are grossly insufficient.

23.2.4.6 Temporal Dimension

The morning starts with all major shops opening for business. Late morning some hawkers set up their wares. By late afternoon moving peddlers slowly start emerging. The evenings see the full burst of activities. Observing T. Nagar throughout the day is like adding layer after layer of activity in the space of time. Weekends and festival seasons see a boost in the pedestrian and vehicular traffic. There will also be a rise in the number of vendors.

The down factor of this teeming public space is its lack of civic amenities.

23.3 Conclusion

Quality of life is most influenced by the neighbourhood you live in. The social interaction with people in your city and its rich cultural experience can transform and affect the way in which an individual functions. Social sustainability—an urban planning trend that focuses on making better places to live in is an idea one must pay attention to. Social sustainability derives from crafting places as destinations of the neighbourhood which are lively and vibrant. It also benefits from tactically crafting places as destinations that people are attracted to. This kind of approach invigorates public and private open spaces, animates built environments, stimulates local businesses and ensures public safety. Qualities such as rich cultural experience as well as presence of social events and activity are keys to successful social sustainability.

The main crux on which the idea of an ideal public place stands is irrespective of the geographic location of the place, and they are

- Sociability,
- users and activity,
- access and linkage,
- comfort and image.

It is in the interpretation of these elements that the approach to the design of public places enhancing social sustainability varies.

23.3.1 Approach to Social Sustainability in Chennai

A successful approach to social sustainability in the city of Chennai should involve many dimensions:

- Functional needs of the people using the public space,
- Culture and sensibilities,
- Memories and images associated to a public place,
- The lines between the four elements will be blurred as they affect and change the dimension of others.

The parameters of social sustainability to be followed in the city of Chennai can be concluded as

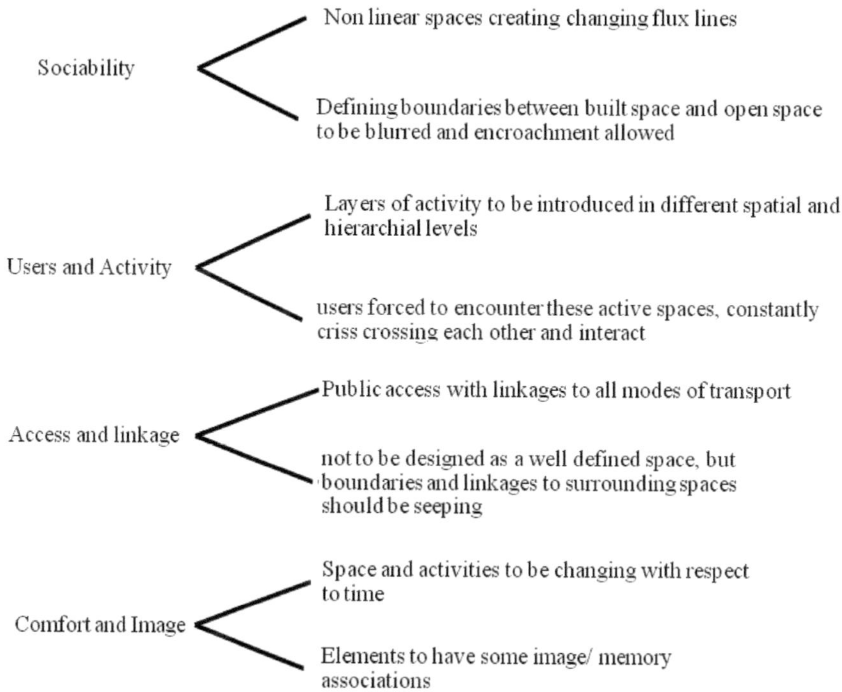

Parameters of social sustainability. Source Compiled by author

Hierarchy of spaces, layers of activity, controlled or often-uncontrolled chaos and temporality all define the public places in Chennai. An approach to social sustainability through certain broad guidelines can be defined with these factors which would positively affect the cultural, economic, social and ecological sensitivities of the people, defining social sustainability.

References

Carmona M, Heath T, Oc T, Tiesdell H (2003) Public places, urban spaces: the dimensions of urban design. Architectural Press, Oxford, England

Chennai Metropolitan Development Authority (CMDA) (2016) Structure of Chennai, 15th June 2016. http://www.cmdachennai.gov.in/Volume3_English_PDF/Vol3_Chapter01_Structure%20of%20Chennai.pdf

Lynch K (1960) Image of a city. The Technology Press and Harvard university Press

Muthiah S (2008) Madras rediscovered. Westland Ltd., Chennai

Rakesh KS (2007) Evaluation of the Pedestrian environment—A Qualitative approach—Case study of T Nagar—the Commercial hub of Chennai city. In: Prashad, Deeependra, New Architecture and Urbanism Development of Indian Traditions, Cambridge Scholars publishing

Times of India (2011) Corpn seizes filthy food and beverages from Marina, Chennai edition, 18th Octo 2014. http://timesofindia.indiatimes.com/city/chennai/Corpn-seizes-filthy-food-and-beverages-from-Marina/articleshow/8229665.cms

Chapter 24
Water Scarcity in Delhi: Mapping for Solutions and the Way Forward

Rituparna Bhattacharyya and Sanjay Prasad

Abstract Almost the whole of India is currently witnessing a massive drought due to the relentless increase in water demand to cater its more than one billion people alongside the growing demand for water for agricultural activities, industries and other allied activities. However, there is a lack of research focusing on water scarcity in India. Central Water Commission data shows that water levels in 91 major reservoirs have reached staggeringly low. This chapter, however, aims to assess the water supply system in Delhi, the capital city of India and the world's third largest conurbation after Tokyo and Mumbai. Currently, there are nine major Water Treatment Plants (WTP) in Delhi, National Capital Territory responsible for catering water to its 16.8 million people. Using Geographical Information System (GIS) alongside the 2011 Census data of Delhi and taking into account the water supply norm of Delhi Jal Board (DJB), which is 60 Gallon Per Capita per Day (GPCD), a simple metric is developed to calculate the freshwater demand of its people residing within the command area of each WTP. For this, the 2011 Census population size of each WTP is multiplied by 60 GPCD to retrieve the approximate water demand of the people residing in each WTP, which in turn allows us to seek the amount of water scarcity/surplus for each of the WTP command area. Based on the findings and the current understanding of the guiding indicators of water scarcity, we map for possible solutions.

Keywords Water scarcity · Water demand · Geographic information system · Delhi jal board · Delhi · India

R. Bhattacharyya (✉)
Independent Researcher; Editor-in-Chief (Joint), Journal Space and Culture, India and in Charge of Training and Development, Alliance for Community Capacity Building in North East India, London, UK
e-mail: rituparna.bhattacharyya@accb.org.uk

S. Prasad
SAMNE Associates Private Limited Delhi, Provision of GIS Consulting Services under the Institutional Development Program of the JICA Assisted Ganga Action Plan Project, Varanasi, India Providing Consulting services for Greater Yangon Water Supply Improvement Project Ph-2, in Yangon Myanmar, New Delhi, India
e-mail: sanjayprasad.gis@gmail.com

R. B. Singh et al. (eds.), *Urban Health Risk and Resilience in Asian Cities*, Advances in Geographical and Environmental Sciences,
https://doi.org/10.1007/978-981-15-1205-6_24

24.1 Introduction

Water as a resource and the lifeline for every individual (both human and animal) increasingly suffer from the risk of scarcity. Approximately, 80% of the population of the world (in 2011, 5.6 billion people) continue to face a threat to water security (Gilbert 2010; Pulla et al. 2018); out of which 4 billion people suffer from severe water scarcity (Mekonnen and Hoekstra 2016; Pulla et al. 2018). While water scarcity is defined as 'the shortage in the availability of freshwater relative to water demand' (Gain and Giupponi 2015: 120), water security is defined 'as the availability of an acceptable quantity and quality of water for health, livelihoods, ecosystems and production, coupled with an acceptable level of water-related risks to people, environments and economies' (David and Sadoff 2007: 545).

There is a large literature on water scarcity risk, emerging crises, sustainable allocation and privatization of water across the living planet (Arnold (Tony) 2009; Dinar et al. 1997; Eckstein 2009; Gain et al. 2011; Gain and Giupponi 2015; Glennon 2005; Hoekstra et al. 2012; Mekonnen and Hoekstra 2016; Pulla et al. 2018; World Water Development Report 2015). The notion 'water scarcity' also embraces the notions of water stress, water shortage (or deficits) and water crisis. Studies suggest that alongside limited water resources, a number of other intersecting geographical factors contribute to water scarcity—these are—'flawed water planning and management approaches, institutional incapability to provide water services, unsustainable economic policies, unequal power relationships, inequality and poverty' (Gain and Giupponi 2015: 120; Human Development Report 2006; World Water Development Report 2015).

In the context of India, there is a lack of geographical research focusing on water scarcity. Few studies suggest that the issues of water scarcity could be tackled through practices of rainwater harvesting and wastewater recycling (Suresh 2001; Srinivasulu 2008). Very little, however, is known about the prevailing conditions and the constraints of the freshwater supply system of Delhi, the capital city of India and the world's third largest conurbation after Tokyo and Mumbai. Considering the acute water shortage that Delhi continues to face apace with its rapidly growing population pressure, we use Geographical Information System (GIS) based on a simple water demand calculation developed by us to examine the water supply and the growing demand of the city while teasing out the deficits and mapping for possible and sustainable solutions.

24.2 Background

Based on a resolution taken by the United Nations General Assembly in 1992, every year the 22nd of March is celebrated as a World Water Day. The key aim of this day is to generate awareness about different issues connected to water. As far as urban water supply in India is concerned, more than 90% of its population has access to drinking water, although, no cities of India have access to piped water supply 24 × 7 and '[w]ater quality has deteriorated in most receiving bodies and shallow groundwater as a result of uncontrolled discharge of raw domestic and industrial waste-water'[1] (Bhattacharya and Borah 2014; Mandal and Sanyal 2019). This crisis, however, is not unique to India only—indeed, as stated above, the growing scarcity of water and depletion of groundwater and drying up of rivers are phenomena of many regions of the globe (Mekonnen and Hoekstra 2016; Hoekstra et al. 2012; World Water Development Report 2015). Nonetheless, the water scarcity problem was confronted openly by honourable Prime Minister of India, Mr. Narendra Damodar Modi when he officially launched the *Swachh Bharat Abhiyan* (Clean India Mission) on Gandhi Jayanti on 2 October 2014 at Rajghat, New Delhi emphasizing on building toilets for 600 million, who defecate in open space; and develop infrastructure (water security projects via *Jal Kranti Abhiyan*)[2,3] for 130 million households that fail to access drinking water (Bhattacharyya 2014: 2, 2015).[4] Statistics retrieved from the Ministry of Water Resource reveals that the whole country is facing an acute crisis of water shortage arising from the lowering of potable water table. That said, in 91 major reservoirs, the live storage[5] capacity stand at just 22% capacity signalling that 34.082 Billion Cubic Meter (BCM) of water was available in these reservoirs by the end of April 2016 against their total capacity of 157.799 BCM, which is approximately 35% lower when compared to the previous year available stock of 53.5 BCM and 24% lower when compared to the 10-year average storage levels of 46.5 BCM for the same period.[6] Figure 24.1 illustrates the region-wise differences in 'live storage'. Indeed, in 2015 out of a total of 676 districts, 302 suffered from drought hit (Mohan 2015).

The magnitude of water scarcity has been so grave that currently, 540 million people are struggling for everyday survival where rural women and young girls in most parts of India {to name a few—Jammu, Bundelkhand region of Uttar Pradesh (UP), Uttarakhand, Odisha, Maharashtra, Karnataka and Telangana} trudge the rugged landscape for miles to collect drinking water (Dyson 2014; Singh 2014, 2015a, b; also, Bhattacharyya 2015), where they remain at high risk of going missing/kidnapped or being sexually assaulted (Bhattacharyya 2017, 2019). Alongside water scarcity, these regions have been suffering from crippling food and fodder shortages. Ostensibly, these problems have been transformed into perennial phenomena in most areas

[1]Urban Water Supply in India, The World Bank, Retrieved from: http://www.worldbank.org/en/news/feature/2011/09/22/urban-water-supply-india.

[2]Water Security Pilot Projects, Ministry of Drinking Water and Sanitation. Retrieved from: http://www.mdws.gov.in/water-security-pilot-projects.

[3]Jal Kranti Abhiyan, Step by step implementation Guide, Ministry of Water Resources, River Development and Ganga Rejuvenation, Retrieved from: http://wrmin.nic.in/writereaddata/JalKrantiAbhiyan_StepByStepGuide.pdf.

[4]Ministry of Drinking Water and Sanitation, retrieved from:http://tsc.gov.in/tsc/nba/nbahome.aspx.

[5]Live storage capacity refers to that part of the reservior which could be used for production of power, flood control, navigation and downstream releases.

[6]Ministry of Water Resources, River Development & Ganga Rejuvenation, Retrieved from: http://wrmin.nic.in/Default.aspx.

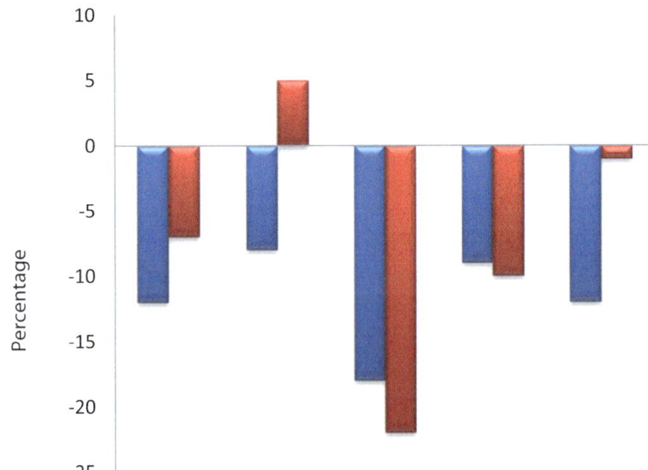

	Northern	Central	Western	Southern	Eastern
Difference from last year (in %)	-12	-8	-18	-9	-12
Difference of average of last 10 years (in %)	-7	5	-22	-10	-1

Fig. 24.1 Difference in live storage, region wise, India, April 2016 (*Source* Reservior storage, central water commission; Courtesy: Authors)

(Bathla and Kannan 2016). Unsurprisingly, water scarcity had sparked row over T20 cricket tournament of Indian Premier League (IPL) for 20 matches that started in Maharashtra from 09 April 2016 where approximately 60,000 L of water was required daily to maintain the pitches. In the wake of the extreme drought in Maharashtra as a whole (and Latur district in particular, the epicenter of drought), on 27 April 2016, the honourable Apex Court quashed the joint Special Leave Petition of Maharashtra and Mumbai Cricket Associations to upheld the Bombay High Court's order to shift out all IPL matches post-01 May 2016.[7] Sadly, it is worth mentioning here that between 2012 and 2014, approximately 3000 farmers claimed their own lives,[8] where drought is perceived to be one of the central causes alongside other causes like debts, alcohol and drugs.[9] The National Water Policy (2012), which stressed on treating water as an 'economic good', also recommends drafting a National Water Framework Law,

[7]Supreme Court dismisses MCA's plea against moving matches out of Maharashtra (2016, 27 April). *The Times of India, Sports*. Retrieved from: http://timesofindia.indiatimes.com/sports/ipl/news/Supreme-Court-dismisses-MCAs-plea-against-moving-matches-out-of-Maharashtra/articleshow/52007508.cms.

[8]Over 3,000 farmer suicides in the last 3 years (2015, 22 April). *The Hindu*. Retrieved from: http://www.thehindu.com/data/over-3000-farmers-committed-suicide-in-the-last-3-years/article7130686.ece.

[9]NCRB DATA: 2,568 farmers ended lives—another highest (2015, 20 July). *The Indian Express*. Retrieved from: http://indianexpress.com/article/cities/mumbai/ncrb-data-2568-farmers-ended-lives-another-highest/.

wherein *inter-alia*, it suggests that 'every individual should have a right to a minimum quantity of potable water (not less than 25 L per capita per day) for essential health and hygiene and within easy reach of the household, which may be provided free of cost to eligible households, being part of pre-emptive need' (Annual Report 1: 15). Against these backdrops, and on consideration, that to date, apart from a plethora of news reports, there is no systematic study on the issues of water scarcity in Delhi. The next section discusses the geographical location and methodological issues deployed by this study. This is followed by a section that paints the existing freshwater supply scenario of Delhi.

24.3 Location and Methodological Issues

Geographically, Delhi is located in between 28.61 °N latitude and 77.23 °E longitude (Fig. 24.2). The state of Haryana borders in its three sides—north, south and west while UP is located in its east. The area of Delhi, the National Capital Territory (NCT) is 1483 km^2 (573 sq. mile) and is home to 16.8 million (urban—16.4 million or 97.5% and rural—0.4 million) (Census of India, 2011). The population distributions of Delhi (Census of India 2011) are mapped in Fig. 24.3. In 2016, the population was estimated to be 19.9 million and projected to reach 23 million by 2021 by Delhi Development Authority (DDA). The 2016 population of NCT was equivalent to the total population of Romania (19.4 million) but more than the population of Netherlands (16.9 million), Zambia (16.7 million), Guatemala (16.6 million) and Ecuador (16.3 million).[10]

Since the 1990s, NCT has been expanding rapidly towards its periphery, thereby, decreasing its rural area and subsequently, increasing its urban area reproducing a 'conurbation'. In 1991, the rural area of NCT was 797.66 km^2, while its urban area was 685. 34 km^2. In 2014–2015, while NCT's rural area reduced to 369.35 km^2, its urban area, increased to 1113.65 km^2, signalling that NCT is a region of the ever-growing urban population, which also indicates, inter-alia its pressure on its water supply and other basic amenities.

Delhi is located in the Alpide belt or Alpine-Himalayan orogenic belt, a high-risk seismic zone prone to earthquakes ranging between the intensity of 8–9 on the Richter scale. Topographically, Delhi is flat, approximately, 210 MSL except for certain bits of small rocky (ridges) areas rolling out from north-northeast to south-southwest in central and south Delhi, respectively. The climate of NCT is humid subtropical with an average temperature above 36 °C (97 °F) during April–July; its average winter temperature between December and February is below 18 °C (64 °F). NCT depends mainly on monsoon between July and August, with its mean annual rainfall being 28.1 inches (714 mm). According to the World Health Organization (WHO), the air pollution level of NCT has triggered a lethal hazard and is one of the ninth worst

[10]Countries in the world by population (2016). *Worldometers*. Retrieved from: http://www.worldometers.info/world-population/population-by-country/.

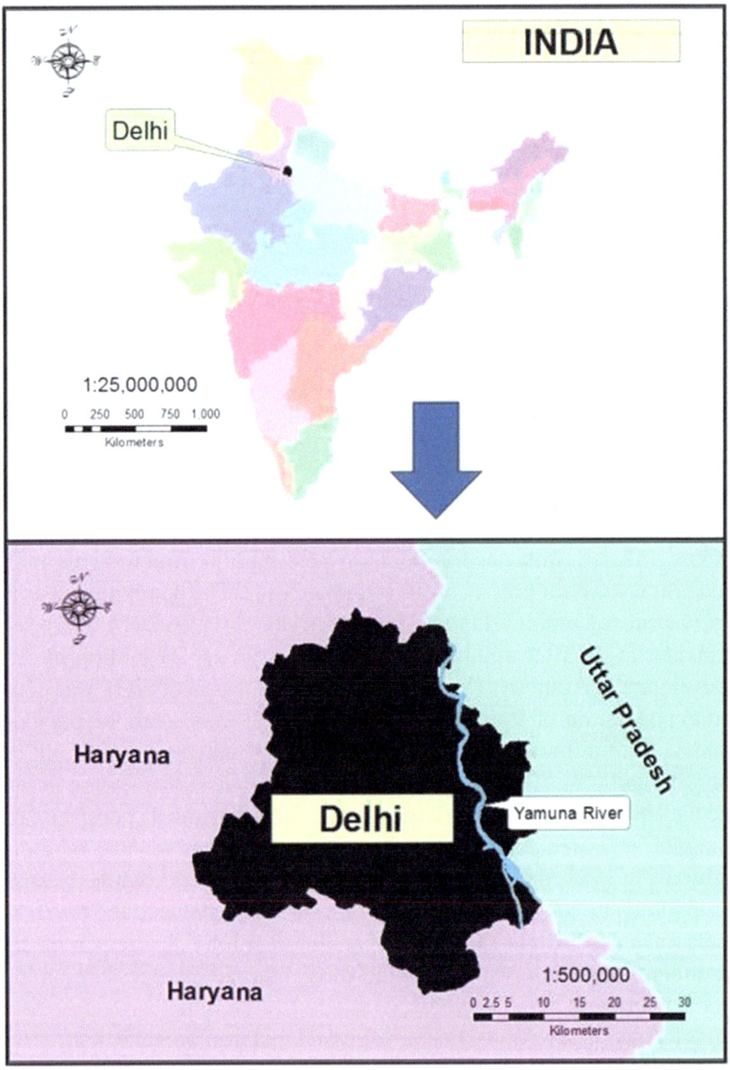

Fig. 24.2 Location map of Delhi (*Source* Produced by Authors)

polluted cities in the world,[11] which has far-reaching implications on its increasing population and of course, on its water quality and demand.

---

[11]Vienna Tops Latest Quality of Living Rankings (2015, 4 March). *Mercer.* Retrieved from: http://www.uk.mercer.com/newsroom/2015-quality-of-living-survey.html and Four out of world's five most polluted cities in India: WHO report (2016, 12 May). *Hindusthan Times*. Retrieved from: http://www.hindustantimes.com/delhi/four-out-of-top-five-polluted-cities-are-in-india-delhi-not-among-them/story-Gn2htcLbESB3BpeYJ4mY8K.html?utm_source = browser& utm_medium = push_notification&utm_campaign = PushCrew_notification_1463019233&_p_c.

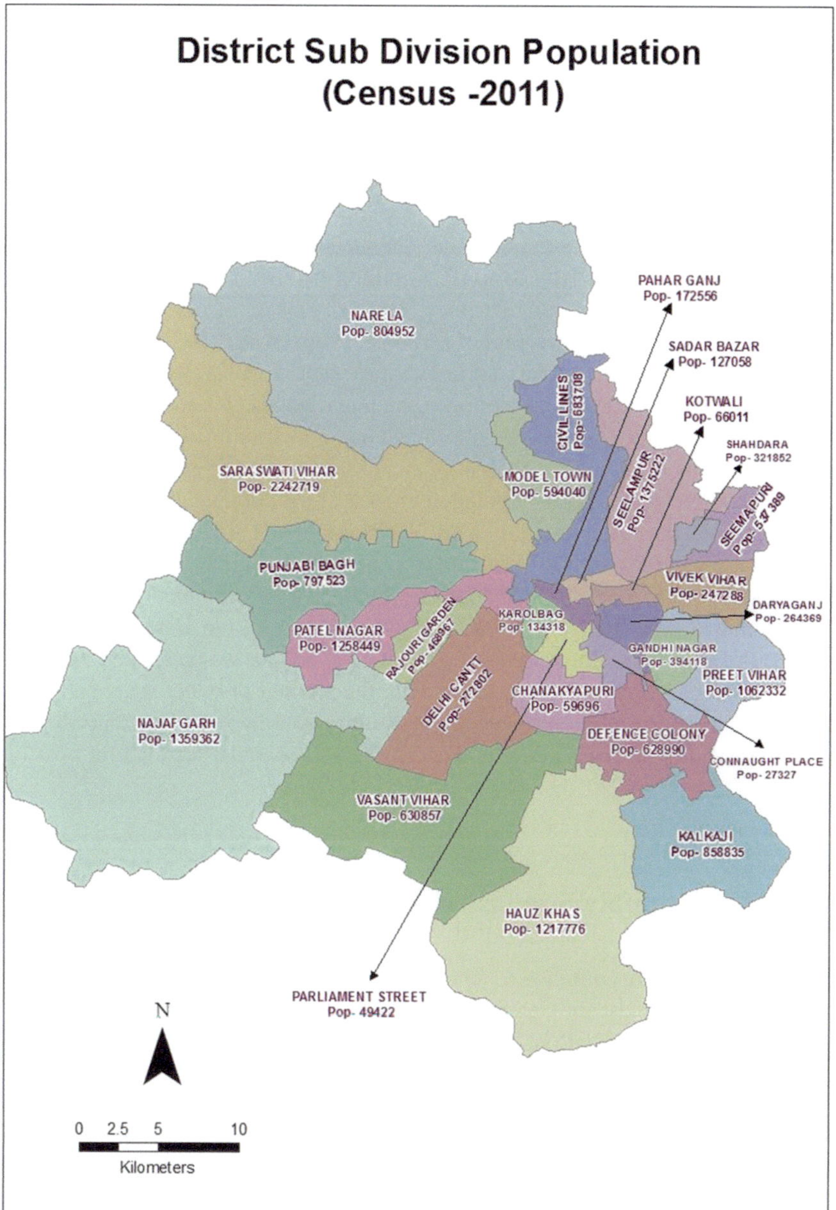

Fig. 24.3 Distribution of population (*Source* Census of India, 2011; Courtesey: Authors)

This research aims to assess the freshwater supply vulnerability for the population of NCT using Arc GIS. For this, it considers the studies titled (Study on Improvement of Water Supply System in Delhi in the Republic of India 2011) developed by Japan International Cooperation Agency (JICA) and Ministry of Urban Development National Capital Territory of Delhi, Delhi Jal Board (DJB), an autonomous body of the Government of India (GOI), and Economic Survey of Delhi (2014)–(2015) and Delhi Statistical Handbook 2015.

There are currently nine major Water Treatment Plants (WTP) in Delhi, NCT responsible for catering water to its 16.8 million people. The water supply norm of DJB is 60 Gallon Per Capita per Day (GPCD). Hence, as per 2011 Population Census, the total water requirement of Delhi is estimated at 1020 MGD, which also includes industrial, commercial and community requirements, fire protection and floating population and special uses in restaurants, hotels and embassies. In order to calculate the water demand, we developed a simple metric to retrieve the water scarcity/surplus of Delhi, NCT (Table 24.4). For this, using GIS and Census of India (2011) data, we calculated the population of each of the command area of the major WTPs. To calculate the population of the WTP command areas, the population density was calculated for each sub divisions (please refer to Fig. 24.3), which was then extrapolated to the WTP command areas, which gave an estimated population for each command area (Fig. 24.4). Following this, based on the water supply norm of DJB, and together with the help of Figs. 24.3 and 24.4, we calculated the water demand, which in turn helped us to estimate the amount of water scarcity/surplus of each WTP command area (Table 24.4). Building on these findings and the current understanding of the guiding indicators of water scarcity, we map for possible solutions.

24.4 Water Supply Scenario of Delhi

As stated above, Delhi continues to witness water scarcity across its ever growing and thickly populated landscapes. Insufficient raw water remains a key driver for inadequate water supply distribution. The problem is indeed deeply embedded (Bathla and Kannan 2016). According to the Census of India 2011, Delhi houses 3.34 million households, out of which 2.72 million households (81.30%) have access to piped water supply—from this, 75.20% households receive supply from treated sources whereas 6.10% from untreated sources. While 0.461 million households (13.8%) are depended on tube wells/deep bore hand pumps/public hydrants, the rest 0.164 million (4.90%) live on other sources such as river, canal, ponds, tank, spring, etc. (Fig. 24.5). As far as the availability of drinking water is concerned, 78.40% of households have water facilities within their premises, 15.40% near the premises and 6.20% of households have away from their premises. DJB is at the helm and accountable for bulk water acquisition and its treatment in Delhi; it is responsible for supplying drinking water not only in the three Municipal Corporations of Delhi but also accountable for supplying treated water to Delhi Cantonment Board and New

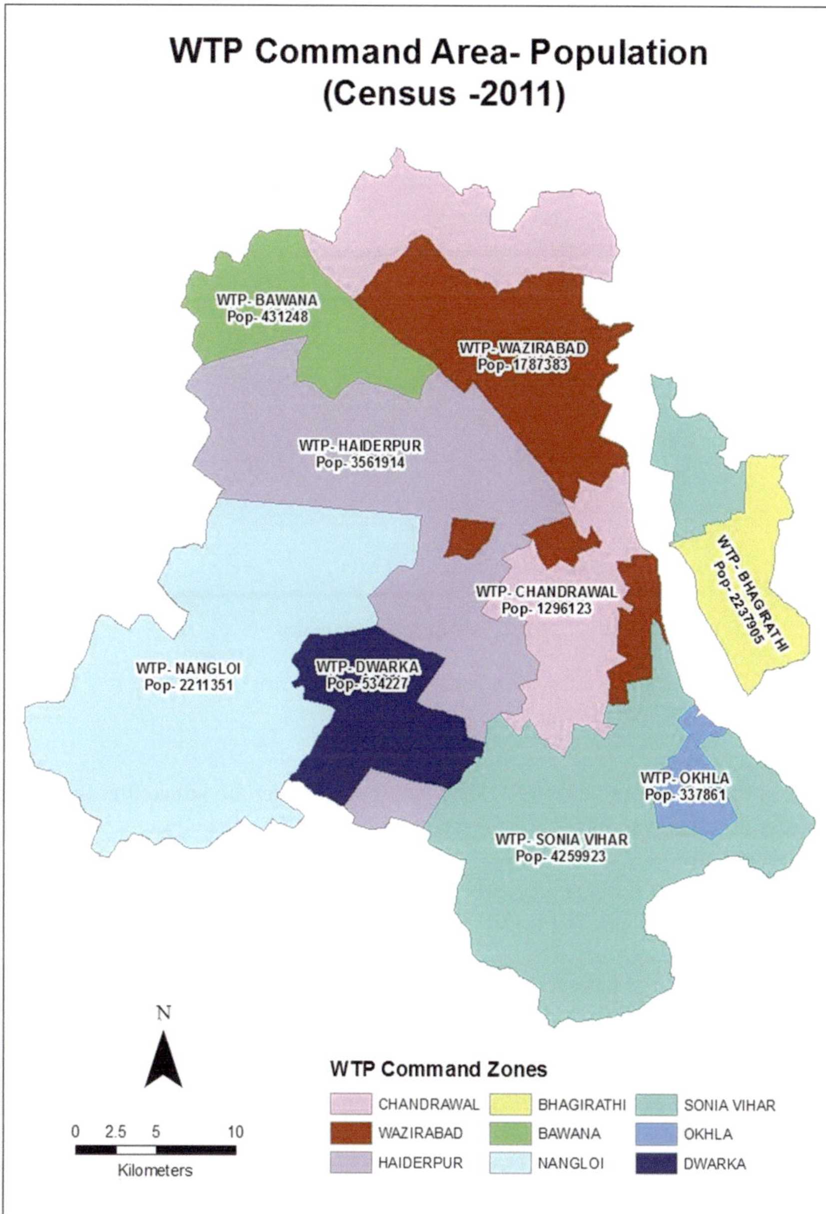

Fig. 24.4 Population catered by nine major water treatment plants (*Sources* Census of India 2011 and study on improvement of water supply system in Delhi in the Republic of India (2011); Courtsey: Authors)

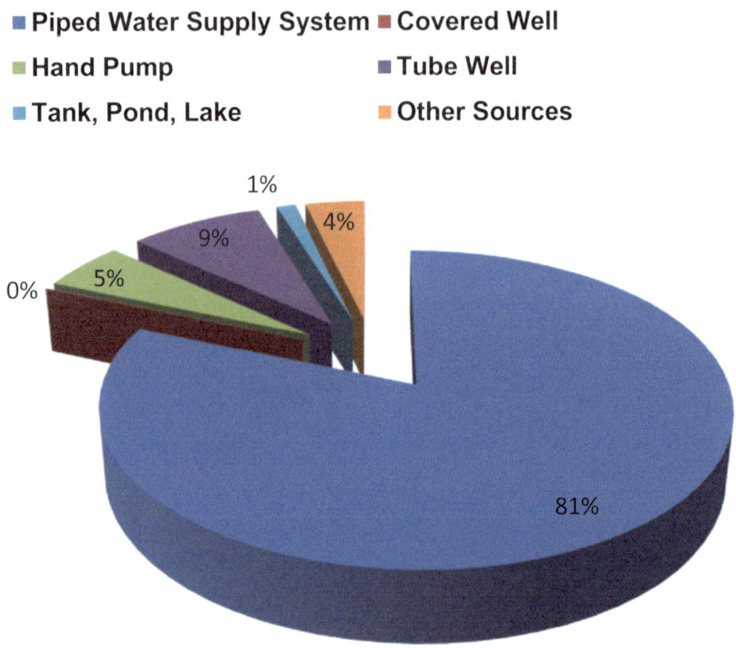

Fig. 24.5 Distribution of households by Sources of drinking water in Delhi (*Source* Delhi Jal Board; Courtesy: Authors)

Delhi Municipal Corporation (NDMC). Sources of water are situated at the lower levels of the river Yamuna; here, all water requires to be pumped out to supply the treatment plants.

The breakdown of the sources of raw water treated by DJB WTPs is outlined in Table 24.1. As of March 2014, DJB WTPs treated, respectively, 755 MGD surface water and 80 MGD groundwater.

Clearly, Table 24.1 indicates that the sources of raw water for DJB mainly feed on rivers Yamuna and Ganges, Bhakra Storage and Groundwater/Ranney Wells/Tube wells. These sources, however, pass through a number of states. For example, the river

Table 24.1 Sources of raw water treated by Delhi Jal board water treatment plants (March 2014)

Sources	Quantity (MGD)
Yamuna river	330
Ganges river	207
Bhakra storage	218
Ground water/Ranney well/Tube well	80
Total	835

Source Delhi Jal Board and Chap. 13: Water Supply and Sewerage, Economic Survey of Delhi, 2014–2015

Yamuna which is 1,376 km (855 miles) in length is the second largest tributary to the river Ganges and flows through Uttarakhand, UP, Haryana, then it crosses Himachal Pradesh (HP) and then Delhi. Similarly, the river Ganges, which is the largest river in India, is 2,510 km (1,560 miles) flows through Uttarakhand, UP, Bihar, Jharkhand and West Bengal before entering Bangladesh. And the Bhakra Storage built on the river Sutlej in HP supplies power and water to its partner states HP, Rajasthan, Punjab, Haryana, Delhi and Chandigarh. Although, currently, Delhi has not been connected with the Interlinking of Rivers (ILR) Programme of the Ministry of Water Resources for equitable distribution of water; given the current water scarcity scenario across the country, it is not difficult to predict that maintaining even the current levels of availability from these sources (Table 24.1) would be difficult in the foreseeable future; water sharing could soon become a site of contestation triggering a tumultuous development process.

Nonetheless, DJB is currently responsible for supplying water to approximately 16.8 million people of NCT using 14,000 km long pipelines and approximately 107 Underground Reservoirs (UGRs) covering both its planned and unplanned areas. The UGRs fitted with booster pumping stations have been provided in the water transmission system for onward delivery via the transmission distribution systems. However, there remain certain areas where the water is tapped directly from the transmission systems. Nevertheless, DJB supplies water to 903 unauthorized colonies. In addition, 50 unauthorized colonies that embrace a population of 0.2 million, benefit from the supply of filtered water by DJB. DJB also supplies water through tanker system. For efficient water tanker supply delivery system, DJB has introduced GPS fitted 407 stainless steel water tanker. Currently, DJB controls 3961 functional tube wells and 14 Ranney wells. However, water supply service remains highly uneven and the volume of supply ranges from 2 to 73 m^3/month. It is distressing to note that the Yamuna river water suffers from high seasonal turbidity of 5000 Nephelometric Turbidity Unit (NTU). Moreover, depletion of groundwater is a grave concern—in few locations of North-West, South, South-West Delhi—here, the level of groundwater has attained 20–30 m below the ground level containing a concentration of high salt and fluorine, surpassing the prescribed limit of 1.5 mg/l in 30% of the wells. In Shahdara and Kanjhawala locations, the content of nitrate has exceeded more than 1000 mg/l. Unquestionably, uncontrolled exploitation and reduced recharge of groundwater due to massive concretization aggravate the debilitating crisis. Besides, the Ranney wells located in the Yamuna flood plain suffer from increasing coliform pollution and high concentration of iron and ammonia. Evidently, the discharges of effluent waste also cause serious damage to the availability of water for use. Apparent consequence results in shutting of the WTPs for days and sometimes for weeks. Ostensibly, like the sources of raw water, the major canals carrying water to Delhi, NCT also cross the neighbouring states—Haryana, UP, Uttarakhand and HP. All these states have their water crises and other social problems too, which could sporadically trigger public and political unrest. For instance, in February 2016, the newly built 102 km parallel *pucca* (permanent) Munak canal from Munak to Haiderpurat at an estimated cost of INR5.2 billion to plug seepage, and which feeds the WTPs of Delhi, NCTwas sabotaged by the *Jat* community protesters of Haryana over an issue of reservations in education and jobs. This triggered unprecedented water crisis among the residents of Dwarka, Munirka, Janakpuri, Rajouri Garden, Palam, Vasant Kunj, Punjabi Bagh, Green Park, Saket and Lodhi Colony when the angry mobs coerced in shutting of several WTPs by breaking its control apparatus in Haryana.[12] The insufficient water back up plan of DJB to manage this crisis further tantalized the problem. This had caused heavy losses to the Government exchequer. Arguably, the ever-changing political scenarios and an increasing tussle between the states might further embroil situations. With the plummeting level of water sources and groundwater, it remains paramount that the existing WTPs (the majority of which were built before 2000 and manually operated) and pumping stations are maintained efficiently and economically to sustain the water demand for its ever-growing population. The following section discusses the installed treatment capacities of the WTPs.

24.4.1 Water Treatment Plants

Currently, there are nine WTPs responsible for distribution of water to the people of Delhi. Conventional treatment methods by sedimentation/coagulation/flocculation and rapid sand filtration are used in six WTPs. Two new WTPS have been built at Dwarka (50 MGD) and Okhla (20 MGD). These two WTPs access raw water from the Munak canal. In addition, Bawana WTP (20 MGD) is likely to be commissioned after the availability of raw water. To treat the water of Ranney wells, DJB applies a special treatment by aeration; for the newly built Okhla Treatment Plant, biological filtrations are applied. However, the water quality transmitted from all these WTPs complies with the WHO standards after meeting the Indian Standards requirements. The trends of the installed capacity of the WTPs are illustrated in Fig. 24.6. Following this, in Table 24.2 water supply capacity trend of the individual WTPs is shown. This is followed by a discussion of water demand.

24.4.2 Water Demand

Water demand is calculated by 'multiplying planned population by per capita demand' (Summary Part I Master Plan 2011: 4–5). Studies unfold that basic water requirement for an individual for *drinking purpose* is 5 L per day; for sanitation, an individual requires 20 L per day; it is 15 L per day for bathing; and for food preparation, it is 10 L per individual per day. Adding these requirement norms, the total water requirement is calculated at 50 L per day per individual (Brown and Matlock 2011; Gleick 1996). As stated above, the water supply norm of DJB is 60 GPCD

[12]Top priority to open supply today: Haryana cop on Delhi water crisis (2016, 21 February). *Hindusthantimes*. Retrieved from:http://www.hindustantimes.com/cities/jat-stir-in-haryana-may-dry-up-delhi-aap-govt-moves-sc-over-water-crisis/story-nPdIGg1yKq9cGB3LjsdkZI.html.

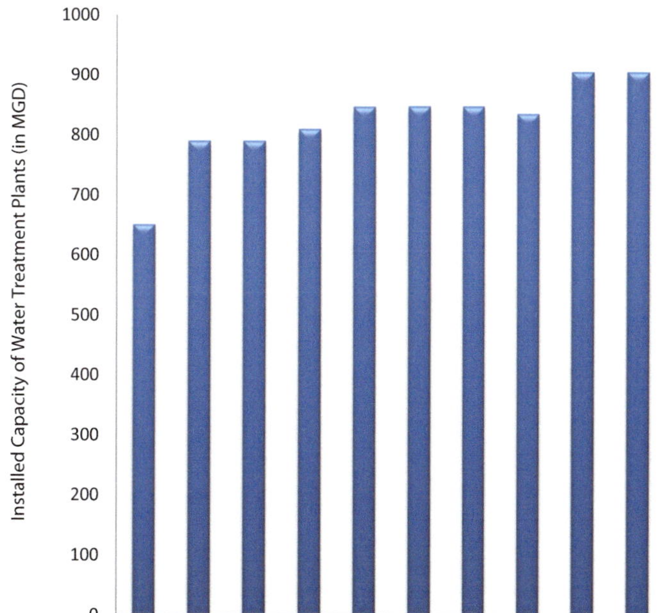

	06	07	08	09	10	11	12	13	14	15
Treatment Capacity of Water Treatment Plants (in MGD), 2006-March 2015	650	790	790	810	847	848	848	836	906	906

Fig. 24.6 Installed treatment capacity of water treatment plants (in MGD), Delhi, 2006–March 2015 (*Source* Economic survey of Delhi, 2014–2015)

but according to the Master Plan of Delhi (MPD), 2021 prepared by DDA, it is 80 GPCD. Of this, domestic requirement entails 50 GPCD, while for non-domestic requirement, it is 30 GPCD. However, 50 GPCD of domestic requirement consists of 20 GPCD for non-potable water needs and 30 GPCD for potable usage as illustrated in Table 24.3.

It is apparent from Table 24.4 that except for Wazirabad, Chandrawal and Dwarka, all other WTPs suffer from water scarcity, albeit Ranney wells and tube wells, recycling of water at Bhagirathi, Haiderpur and Wazirabad and Commonwealth games village supplies an additional 146 MGD (Table 24.2). Nevertheless, the amount of water supplied is far less than the water demand—999 MGD are required per day for a population of 16.8 million (Census of India 2011). This, of course, excludes other miscellaneous requirements and the floating population where the total estimated requirement is 1020 MGD; in addition, it should be noted that the current 2016 population is estimated to be 19.9 million. This observation of water scarcity reinforces the previous literature on water scarcity problems linked to drinking water in the cities of Bengaluru, Chennai and Guwahati (Bhattacharya and Borah 2014; Suresh 2001; Srinivasulu 2008).

Table 24.2 Water supply capacity of individual water treatment plants, 2006–March, 2015

Name of plants	Capacity (MGD)									
	2006	2007	2008	2009	2010	2011	2012	2013	2014	2015
Chandrawal Water house No. I and II	90	90	90	90	90	90	90	90	90	90
Wazirabad I, II and III	120	120	120	120	120	120	120	120	120	120
Haiderpur	200	200	200	200	200	200	200	200	200	200
North Shahdara (Bhagirathi)	100	100	100	100	100	100	100	100	100	100
Bawana				20	20	20	20	20	20	20
Nangloi	40	40	40	40	40	40	40	40	40	40
Sonia Vihar		140	140	140	140	140	140	140	140	140
Ranney wells and Tube wells	100	100	100	100	100	100	100	100	100	100
Recycling of water at Bhagirathi, Haiderpur and Wazirabad					37	37	37	45	45	45
Commonwealth games village						1	1	1	1	1
Okhla									20	20
Dwarka									50	50
Total	650	790	790	810	847	848	848	836	906	906

Source Delhi Jal Board and Economic Survey of Delhi, 2014–2015

Table 24.3 Water requirement norms as per DJB and MPD-2021

As per DJB water requirement norms		As per MPD-2021 water requirement norms			
Details	Requirement of water	Norms	Quantum(GPCD)		Sources of non-potable water
			Potable	Non-potable	
Domestic	172 LPCD[a]	**Domestic @50 GPCD**	30	20	
Industrial, commercial and community requirement based on 45000 L per hectare per day	47 LPCD	Residential	30	20	Recycling and permissible ground Water extraction at community Level
Fire protection based on 1% of the total demand	3 LPCD	**Non-domestic @30 GPCD**	5	25	
Floating population and special uses like Hotels and Embassies	52 LPCD	a. Irrigation, horticulture, Recreational, construction, fire @6.65LPCD		10	Recycling from Sewerage Treatment Plants (STPs) and Permissible Ground Water Extraction
		b. Public, semi-public, industrial and commercial	5	15	Recycling from Common Effluent Treatment Plants (CETPs)
Total	274 LPCD **(60 GPCD)**	**Total@80GPCD**	**35**	**45**	

[a]LPCD = liter per capita per day

Source Delhi Jal Board and Chap. 13: water supply and sewerage, economic survey of Delhi, 2014–2015

Table 24.4 Estimated water Deficit/surplus among the population of the water treatment plants command area

Water treatment plants	Population (Census of India, 2011)	Water demand (in Gallons) = PopulationX 60 GPCD	Water demand (in MGD)	WTP installed capacity (in MGD)	Deficit/Surplus (in MGD)	Deficit/Surplus
Bawana	431249	25874935	26	20	6	Deficit
Bhagirathi	2237906	134274356	134	100	34	Deficit
Chandrawal	1296123	77767406	78	90	−12	Surplus
Dwarka	534228	32053659	32	50	−18	Surplus
Haiderpur	3561914	213714846	214	200	14	Deficit
Nangloi	2211351	132681091	133	40	93	Deficit
Okhla	337861	20271672	20	20	0	Balanced
Sonia Vihar	4259923	255595395	256	140	116	Deficit
Wazirabad	1787383	107242985	107	120	−13	Surplus
		Total	999	780	219	Deficit

Source Calculated by the Authors

The rate of water demand notwithstanding is higher in the summer; DJB is committed to maintaining a consistent production of 835 MGD/day and at times, even allocate water tankers when demand supersedes the amount of supply. Evidence suggests that in July 2014, DJB recorded its production reaching 843 MGD. Reconnaissance observations, however, suggest that the WTPs like Dwarka and Nangloi fail to receive enough raw water to match the installed capacities. Arguably, the inadequate supply of raw water has resulted in irregular water supply. Nonetheless, based on the 60 GPCD norms, the estimated requirement of water in 2021 for 26 million (3 million more than as estimated by DDA) would be approximately 1,560 MGD [which also includes 15% of Non-RevenueWater (NRW) (JICA 2011). NRW is that amount of water which has been produced and transmitted but cannot be invoiced to its customers because it is the water that is lost either through leakage or overflow or both or even used by people who are not billed for the same]. Hence, to tackle the water scarcity, it remains paramount for efficient management. In the concluding section, we map for possible solutions.

24.5 Mapping for Possible Solutions

MPD-2021 aims for equitable and continuous water supply vis-à-vis extension of water supply to its peripheries [covering 15.3 million people of the urban areas and 7.7 million people outer (rural) areas of Delhi]; demand management via curtailment of water loss and reviewing of tariff to dissuade water wastage; energy management aims at preventing excessive pumping through proper rationalization of the system; lastly, to accomplish 24 × 7 water supply.

In 2002, the quantity of water treated by DJB was 650 MGD, which increased to 906 MGD by 2015 (Table 24.2). Evidently, in 2004, 40% of water supply was lost due to leakage in the pipelines, and therefore, only 60% could reach its residents, which was estimated to be 368 MGD against the demand of 600 MGD in the said year. Regrettably, in the water carrier system of Western Yamuna Canal System and river Yamuna, 30–50% raw water is wasted in its discharge from Tajewala Headworks. To address these leakages, various measures have been adopted. For instance, a Leakage Detection and Investigation (LDI) cell has been launched by DJB; and approximately, 1200 km of old, damaged and leaky pipelines have been replaced, bringing down the loss to approximately 20%, albeit the whole water supply system, as mentioned above, is entangled with approximately, 14,000 km of supply mains, of which, majority of the pipelines are approximately 40–50 years old. Notwithstanding, the newly built Munak canal, which became dysfunctional due to the *Jat* protesters for days was built by the Haryana Government (on behalf of the Delhi Government), to control water loss and anticipated that water availability would increase to about 95 MGD within the prevalent phenomenon at Munak. There remains a consensus nevertheless that water supply system of Delhi suffers from mismanagement, poor governance, lack of accountability alongside poor infrastructure, low levels of technical and managerial skills. This observation bears resonance with earlier literature (Gain and Giupponi 2015; Human Development Report 2006; World Water Development Report 2015). The following are the possible measures to improve the freshwater supply system of Delhi.

24.5.1 *Water Tariff, Willingness to Pay and Affordability to Pay*

'Use more pay more' was the principle of DJB until the new government of Aam Admi Party (AAP) took an oath to rule Delhi in February 2015. Before the AAP rule, statistics unfold that the total tariff collection as of March 2015 was INR 10922.5 million as against the estimated revenue collection of INR 12686.9 million, a collection efficiency estimated to be 86% (Economic Survey of Delhi 18). However, up to 20 kl per month (700 L a day) of free-water-scheme promises as made in the AAP election manifesto was transformed into reality with effect from 14 August 2015 for those households having functional water meter (Table 24.5). Evidently, within one and a half years of AAP Government, 1.05 million consumers including co-operative group housing societies' residents became the beneficiaries of this scheme. In addition, the AAP Government has reduced the water development charges and sewer development charges to the consumers of unauthorized colonies, respectively, from

Table 24.5 Water tariff of Delhi Jal Board 2015

Category- I (Domestic consumers):		
Monthly Consumption (Kiloliter)	Service Charge (INR)	Volumetric charge (Per Kiloliter)
Up to 20	146.41	4.39
20 to 30	219.62	21.97
Above 30	292.82	36.61
Category- II (Commercial/Industrial)		
00 to 06	146.41	14.64
06 to 15	292.82	21.96
15 to 25	585.64	29.28
25 to 50	1024.87	73.21
50 to 100	1171.28	117.13
Above 100	1317.69	146.41
Plus Sewer maintenance charge: 60% of volumetric water charge (for both domestic and commercial/industrial)		
Category- II A (Rainwater harvesting or wastewater recycling)		
Category- C consumers owning 2000 sq. Yards or larger size plot area will receive 10% refund from the total bill amount if rainwater harvesting or wastewater recycling exists in operational forms and 15% rebate from the total bill amount if both these services exist.		
Sewerage Maintenance Charge	60 % of Water Consumption Charge	
Water Cess Charge	@ 2 Paise Per Kl	
Late Payment Surcharge 5% of the total bill amount		

Source Delhi Jal Board

INR 440/sq. m to INR100/sq. m and from INR 494/sq. m to INR 100/sq. m.[13] While we agree with and praise the AAP government in helping reduce the bills of the residents of the unauthorized colonies, we also acknowledge that supplying 20 kl per month of water might motivate people to spend less water in order to receive water for free, but there are no incentives for those households (comprising of one or two people), who can manage their monthly requirements even for less than 10 kl. Table 24.5 illustrates a detailed picture of water tariff of DJB as applicable from 14 August 2015. Notwithstanding, Delhi's domestic consumer water tariff is one of the least when compared to some countries of the world (Table 24.6). Arguably, NCT is home to 545 Ultra-High-Net Worth Individuals (UHNWI) with net assets of over $30 million after Mumbai (with 1,094 UHNWI).[14] Moreover, as evidenced

[13]100-day AAP govt: Free 20 kl water, low charges Delhi Jal Board's feats (2015, 24 May). *The Financial Express*. Retrieved from: http://www.financialexpress.com/article/economy/100-day-aap-govt-free-20-kl-water-low-charges-delhi-jal-boards-feats/75339/.

[14]AAP govt: Free 20 kl water, low charges Delhi Jal Board's feats (2015, 24 May). *The Financial Express*. Retrieved from: http://www.financialexpress.com/article/economy/100-day-aap-govt-free-20-kl-water-low-charges-delhi-jal-boards-feats/75339/.

Table 24.6 Water tariff in some countries

Country/City	Water tariff		Average monthly payment for a family of 4 consuming 20 KL	
	Local currency	INR	Local currency	INR[a]
India/Delhi	INR 4.39/21.97 (20/Above/KL)	4.39	0	0
Japan/Tokyo	22/128 Yen/m3 (10/20 KL)–Average75 yen/KL	46	1500 yen	920
Brazil/Sao Paulo	1.35 $/m3	90	US $ 17	1130
Ukraine	0.14 $/m3	9.5	US $ 2.8	186
China/Beijing	RMB1.98/m3	22	US $ 6.6	440
South Africa/Durban	Rand 9.5/11.25/m3 (9/Above)	48	US $ 14.52	965

[a]Conversion rate as on 28 April 2016@1 US$ = INR66

Sources Delhi Jal Board; Family Income and Expenditure Survey in 2014 by Ministry of Internal Affairs and Communications; Las Tarifas De Agua Potable Y Alcantarillado En América Latina, The World Bank Group, Retrieved from: http://www-wds.worldbank.org/servlet/WDSContentServer/WDSP/IB/2005/07/06/000011823_20050706103726/Rendered/PDF/Las0tarifas0de1llado0Borrador0Final.pdf; Ukrainian Upgrades. Retrieved from: http://www.waterworld.com/articles/wwi/print/volume-25/issue-4/regional-spotlight/easter-europe/ukrainian-upgrades.html; China Water Risk. Retrieved from: http://chinawaterrisk.org/opinions/pricing-water/; Hartley, Wyndham (2010, 5 May). South Africa: Water Tariffs Set to Rise to Reduce Need for Borrowing. Business Day, allAfrica. Retrieved from: http://allafrica.com/stories/201005050084.html

from the Delhi (Statistical Handbook 2015), in 2014–2015, among all the states, Delhi recorded the highest per capita income of INR 2,40,849; a rise of INR 28,630 when compared to the financial year 2013–2014. That said, in a survey conducted by JICA in the WTP command areas of Chandrawal, Wazirabad, Haiderpur, Bhagirathi and Sonia Vihar, the 'willingness to pay' water tariff was found to be 'high' in the Chandrawal WTP command area followed by 'medium' willingness to pay in the command areas of Wazirabad, Haiderpur and Sonia Vihar, while in the Bhagirathi WTP command area, 'low' level of willingness to pay was found. Although, water is a basic human need, but due to its rising demand and as per the National Water Policy (2012), we urge the DJB to treat water as an 'economic good'and re-revise its water tariff and implement the methods of 'willingness to pay' and 'affordability to pay' as and where it applies to its consumers.

24.5.2 Non-revenue Water

There are no adequate statistics on NRW. Nevertheless, in the financial year 2008–2009, it was estimated that the loss was as high as 66% (including leakage). The

World Bank recommends an NRW loss of less than 25%. Evidently, cities like Melbourne (3%), Singapore (4%), Amsterdam (6%), Osaka (7%), New York (10%), Toronto (10%) and Tokyo (2.7%) (Bardies 2012) have reduced their NRW benchmarks rigorously.[15,16] Diligent low-level NRW benchmarks for NCT, Delhi can be accomplished by the following measures:

- Installation of 100% metering, and stringent meter reading, along with replacement of faulty meters.
- Corruption at all functionary levels—meter readers, law enforcement agencies and planning levels should be checked regularly and plug the loopholes in these functionaries for better coordination.
- Detection of illegal connection plus proper and regular maintenance of the facilities like pipelines, valves including detection of underground leakages, etc. This will not only reduce the distribution/transmission losses and leakages but also recover the expenses invested in the measures for reduction of NRW.
- Successful NRW programmes of the cities above, perhaps, Tokyo (that supplies water to its 12 million people and its NRW level is one of the lowest in the world) must be learnt and implemented according to the local needs of Delhi supply system (see, Bardies 2012).

24.5.3 Mandatory Rainwater Harvesting

As stated above, the average annual rainfall of Delhi is 714 mm (28.1 inches) and depends mainly on monsoon. The breakdown of average annual rainfall in NCT, Delhi is shown in Table 24.7.

DJB has already initiated mechanisms for mandatory harvesting of rainwater aimed at recharging of groundwater or water storage in reservoirs through 151 installations and imparted financial assistance of INR 8.2 million for 172 cases in the institutional category. In addition, DJB has launched a 'rebate scheme' to attract rainwater harvesting (Table 24.5: Category-II A) to its consumers. Given the current crisis, more needs to be done for storage of rainwater; perhaps, more reservoirs need to be installed to capture water to its fullest potentials (in colonies, temples, hotels, educational institutions and in industrial estates). But unlike North East India, Kerala, Goa, the pattern of Delhi's rainfall is mainly from June to September. Hence, for better management of water demand, larger tanks are perhaps required as the water crisis is more likely to be never-ending. Apart from the colonies, institutions and industries, very few individual homeowners of East of Kailash, Greater Kailash, Saket, Nizamuddin West, Kirti Nagar and Najafgarh have started harvesting rainwater

[15]Figures in the parentheses indicate the values of NRW levels.

[16]Stated NRW (Non-Revenue Water) Rates in Urban Networks (2011). Smart Water Networks Forum (SWAN). Retrieved from: http://www.swan-forum.com/uploads/5/7/4/3/5743901/stated_nrw_rates_in_urban_networks_-_swan_research_-_august_2011.pdf.

Table 24.7 Rainfall in national capital territory, Delhi

Stations	Rainfall (mm)
Chandrawal	886.5
Safdarjung	712.2
Delhi university	887.6
Palam	793.9
Okhla	792.4
Mehrauli	499.0
Delhi sadar	647.6
Nangloi	337.2
Shahdra	451.9
Najafgarh	398.9
Badli	516.1
Alipur	448.9

Source www.rainwaterharvesting.org

through the installation of rainwater harvesting structure within their own premises. Depending on their sizes, the prices of these rainwater harvesting structures ranges from INR 1500 to INR 6000. These structured systems have been making positive impacts not only in yielding tube well improvements but also in the enhancement of groundwater quality.[17] The success stories of these few homeowners (Suresh 2001; Srinivasulu 2008) should perhaps be aired in the media, which could encourage other homeowners to follow suits. Also, to attract homeowners for installation of rainwater harvesting structure, the Delhi government should perhaps launch a new irresistible incentive policy. Concomitantly, strict measures should be taken to stop illegal groundwater extraction.

24.5.4 Reuse of Wastewater

There is evidence to suggest that out of the total water consumed per person only 20% is used for drinking or potable purposes. The rest 80% goes back to the drain contributing to the sewerage flow. So, it is advisable not to be using potable water for toilets, washing clothes, cars, gardening, etc. Instead, sewage water should be recycled and treated to encourage consumers for non-potable uses. Dual pipe system should be encouraged and made mandatory in upcoming and existing planned colonies. JICA (2011) argues that although the dual piping system sounds impressive theoretically, however, there is a high risk associated with it from connecting drinking/potable water pipes with pipes of other usages, until and unless the engineers

[17] Database, New Delhi. Retrieved from: http://www.rainwaterharvesting.org/people/People-urban_database_delhi_individ.htm?#yudhavir.

and other humanpower delegated for these jobs carry extra vigilance and carefulness with all the nuanced issues linked to fitting pipes. Nonetheless, we argue that the dual pipe system would reduce water demand considerably. Therefore, we urge DJB to implement a policy for execution of the dual pipe system, where highly skilled dedicated workforce should be entrusted to implement this system. At the same time, the residents must be made aware of its positive impacts and emphasized that the system could only be executed with full cooperation from every resident. In addition, while it remains paramount to creating public awareness in making sustainable use of water, it is also important to increase 'will power' at the policy levels.

24.5.4.1 Crisis Management Including Earthquake

As mentioned above, Delhi is located at a high-risk intensity seismic zone capable of yielding high-frequency earthquakes. Hence, for natural emergencies and other fabricated crises (as in the case of *Jat* protestors), there needs to be preparedness. The existing water storage facilities must be evaluated against possible crises (including earthquake) and strengthened its capacities in addition to strengthening a backup plan.

24.6 Conclusion

A novel water scarcity scenario of NCT, Delhi has been presented in this chapter. As far as we are concerned, this is one of the first comprehensive studies on water scarcity in Delhi. In doing so, we tried to address three pertinent issues—using GIS, we calculated the water deficit/surplus among the population residing within the WTP command areas; that water scarcity in Delhi is both a natural and a human-made problem, and we have addressed few solutions to improve the effectiveness of the water supply system. Though the solutions presented herein are not going to solve the water scarcity problem of Delhi completely, nonetheless, if rationally followed and implemented, it would improve the sustainable efficacy of the system and build people's lives more comfortable. Nevertheless, if DJB fails to achieve these solutions, then the degree of water scarcity would be further widened, which would make it difficult to achieve an equitable distribution.

While this study is an addition to water scarcity literature, it is a useful research to proselytize public awareness, which can also be replicated to examine the water scarcity scenarios of other parts of the country.

Acknowledgements We express our deep sense of gratitude to Professor (Dr.) R. B. Singh and Dr. Subhash Anand, IGU Commission on Geoheritage, Department of Geography, Delhi School of Economics, University of Delhi, Delhi-110007 for giving us the opportunity to contribute this article on water scarcity in Delhi. And a 'Big Thank You' to Dhritman Biswa Sarma for copy-editing the manuscript.

References

Annual Report (2014–2015) Ministry of water resources, river development and Ganga Rejuvenation, Accessed from: http://wrmin.nic.in/forms/list.aspx?lid=451

Arnold CA (Tony) (2009) Water privatization trends in the United States: human rights, national security, and public stewardship. William Mary Environ Law Policy Rev 33:785–849

Bardies G (2012) Well-managed Tokyo Waterworks Reach New NRW Benchmark, Country Focus. Accessed from: http://www.shpmedia.com/images/AW_MayJune12_Country%20Focus-Tokyo_V3.pdf

Bathla S, Kannan E (2016) Water will continue to be scarce. The Hindu. Accessed from: http://www.thehindu.com/opinion/op-ed/comment-on-water-drought-by-seema-bathla-elumalai-kannan-water-will-continue-to-be-scarce/article8520173.ece

Bhattacharya P, Borah R (2014) Drinking water in Guwahati city: its past, present status and associated problems. Space Cult, India 1(3):65–78. http://dx.doi.org/10.20896/saci.v1i3.46

Bhattacharyya R (2015) Working childhoods: youth, agency and the environment in India. Prog Dev Stud 15(3):292–294. https://doi.org/10.1177/1464993415578570

Bhattacharyya R (2017) Sociologies of India's missing children. Asian Soc Work Policy Rev 11(1):90–101. https://doi.org/10.1111/aswp.12116

Bhattacharyya R (2019) Child sponsorship: exploring pathways to a brighter future. Prog Dev Stud 19(1):83–84. https://doi.org/10.1177/1464993418799895

Bhattacharyya R (2014) Good governance and development mandate. J Space Cult, India 2(1):1–4. https://doi.org/10.20896/saci.v2i1.65

Brown A, Matlock MD (2011) A Review of Water Scarcity Indices and Methologies, Food, Beverage and Agriculture, White paper No. 106, © 2011 The Sustainability Consortium, Accessed from: https://www.sustainabilityconsortium.org/wp-content/themes/sustainability/assets/pdf/whitepapers/2011_Brown_Matlock_Water-Availability-Assessment-Indices-and-Methodologies-Lit-Review.pdf

Census of India (2011) Primary Census Data Highlights - NCT of Delhi. Accessed from: http://www.censusindia.gov.in/2011census/PCA/PCA_Highlights/pca_highlights_file/Delhi/DATA_SHEET_PCA_DISTRICTS_NCT_OF_DELHI.pdf

Central Water Commission, Ministry of Water Resources, River Development & Ganga Rejuvenation

David G, Sadoff CW (2007) Sink or swim? Water Secur Growth Dev, Water Policy 9:545–571. https://doi.org/10.2166/wp.2007.021

Delhi Jal Board, Government of NCT of Delhi

Dinar A, Rosegrant MW, Meinzen-Dick RS (1997) Water allocation mechanisms: principles and examples. World Bank Policy Research, Working Paper No. 1779

Dyson J (2008) Harvesting Identities: youth, work, and gender in the Indian Himalayas. Ann Assoc Am Geogr 98(1):160–179. https://doi.org/10.1080/00045600701734554

Eckstein G (2009) Water scarcity, Conflict, and Security in a Climate Change World: Challenges and Opportunities for International Law and Policy. Wis Int Law J 27(3):409–461

Economic Survey of Delhi (2014–2015) Government of National Capital Territory of Delhi. Accessed from: http://delhi.gov.in/wps/wcm/connect/DoIT_Planning/planning/economic+survey+of+dehli/economic+survey+of+delhi+2014+-+2015

Gain AK, Giupponi C (2015) A dynamic assessment of water scarcity risk in the lower Brahmaputra river basin: an integrated approach. Ecol Ind 48:120–131. https://doi.org/10.1016/j.ecolind.2014.07.034

Gain AK, Immerzeel WW, SpernaWeiland FC, Bierkens MFP (2011) Impact of climate change on the stream flow of the lower Brahmaputra: trends in high and low flows based on discharge-weighted ensemble modelling. Hydrol Earth Syst Sci 15:1537–1545. https://doi.org/10.5194/hess-15-1537-2011

Gilbert N (2010) Balancing water supply and wildlife. Nature. https://doi.org/10.1038/news.2010.505

Gleick P (1996) Human population and water: To the limits in the 21st Century. Human Population and Water, Fisheries, and Coastal Areas. American Association for the Advancement of Science Symposium, 1995, Washington, D.C
Glennon R (2005) Water scarcity, marketing, and privatization. Tex Law Rev 83(7):1873–1902
Hoekstra AY, Mekonnen MM, Chapagain AK, Mathews RE, Richter BD (2012) Global monthly water scarcity: blue water footprints versus blue water availability. PLoS ONE 7(2):e32688. https://doi.org/10.1371/journal.pone.0032688
Human Development Report (2006) Beyond Scarcity: Power, Poverty and the Global Water Crisis, United Nations Development Programme, Palgrave Macmillan, New York
Mandal J, Sanyal S (2019) Geospatial analysis of fluoride concentration in groundwater in Puruliya District, West Bengal. J Space Cult, India 6(5):1–16. https://doi.org/10.20896/saci.v6i5.369
Mekonnen Mesfin M, Hoekstra Arjen Y (2016) Four billion people facing Severe water scarcity. Sustain, Adv Sci 2(e150032):1–7. https://doi.org/10.1126/sciadv.1500323
Mohan V (2015) Half of India is Drought-hit, but States Yet to Seek Central Help. The Times of India. Accessed from: http://timesofindia.indiatimes.com/india/Half-of-India-is-drought-hit-but-states-yet-to-seek-central-help/articleshow/49838565.cms
National Water Policy (2012) Ministry of water resources, Government of India. Accessed from: http://wrmin.nic.in/writereaddata/NationalWaterPolicy/NWP2012Eng6495132651.pdf
Pulla V, Ahmed Z, Pawar M (2018) Water and communities in South Asia: a case for regional cooperation. Space Cult, India, 6(3):23–40. https://doi.org/10.20896/saci.v6i3.392
Rainwaterharvesting.org. Accessed 14 May 2015 from, http://rainwaterharvesting.org/
Singh S (2015) Increased Burden on women and male out-migration: an analysis of Khul gad micro watershed of the Kumoun Himalaya. Int J Soc Work Hum Serv Pract 3(3):118–123. https://doi.org/10.13189/ijrh.2015.030303
Singh SB (2015) Women as milieu managers in integrated watershed management: perspectives from the hilly areas of Uttarakhand. J Space Cult, India 2(4):71–79. http://dx.doi.org/10.20896/saci.v2i4.130
Singh S (2014) Women, environment and sustainable development: a case study of Khul Gad Micro Watershed of Kumoun Himalaya. Space Cult, India, 1(3):53–64. http://dx.doi.org//10.20896/saci.v1i3.45
Srinivasulu Rajendran (2008) Water Scarcity Chennai: Financ-Econ Apprais Rain Water Harvest. https://doi.org/10.2139/ssrn.1152239
Statistical Handbook (2015) Government of National Capital Territory of Delhi. Accessed from: http://www.delhi.gov.in/wps/wcm/connect/doit_des/DES/Our+Services/Statistical+Hand+Book/
Study on improvement of water supply system in Delhi in the Republic of India (2011) Japan International Cooperation Agency (JICA) and Ministry of Urban Development National Capital Territory of Delhi, Delhi Jal Board
Suresh TS (2001) An Urban Water Scenario: A Case Study of the Bangalore Metropolis, Karnataka, India, Regional Management of Water Resources (Proceedings of a symposium held during the Sixth IAHS Scientific Assembly at Maastricht, The Netherlands, July 2001). IAHS Publ. no. 268
World Water Development Report (2015) Water in a sustainable world, UN water. Accessed from: http://www.unwater.org/publications/publications-detail/en/c/281166/

Chapter 25
Assessing the Relationship Between Neighbourhood Socio-economic Status and Environmental Quality

Saleha Jamal and Uzma Ajmal

Abstract There is a growing awareness that purely individual-based explanations of causes of ill health are insufficient and fail to capture important disease determinants. Most studies concluded that socio-economic status of an individual is strongly associated with risk of disease and mortality but a neighbourhood can also adversely affect health of the residents by its poor environmental quality like poor water quality, air quality, noise pollution, litter on the streets, lack of access to green spaces and crime/violence in the neighbourhood, etc. It has been found that not only individual socio-economic status but neighbourhood socio-economic status too can influence the health of the residents. Thus, in order to understand the distribution of health and diseases, there is a need to consider not only individual characteristics but also characteristics of the neighbourhood in which people live. In the present study, an attempt has been made to study the association between neighbourhood socio-economic status and neighbourhood environmental quality. Azamgarh, a small-sized city located in the Indian state of U.P. has been chosen as a study area, of which, administrative wards of the city are taken as neighbourhood units. The study is based on the primary source of data collected through questionnaire interviews from respondents belonging to different wards of Azamgarh City. Neighbourhood socio-economic status has been measured by socio-economic and housing condition of households of different wards. Environmental quality of the neighbourhood is measured by waste collection and disposal, green spaces, presence of stagnant water, quality of drinking water, air and noise pollution and crime/violence in the neighbourhood. The study finds that environmental quality in a neighbourhood varies with its socio-economic status and the poor environmental quality is found in deprived neighbourhoods which makes the residents more susceptible to health damages.

Keywords Neighbourhood · Socio-economic Status (SES) · Environmental quality · Health

---

S. Jamal (✉) · U. Ajmal
Department of Geography, Aligarh Muslim University, Aligarh, India
e-mail: salehajm@gmail.com

U. Ajmal
e-mail: uzmakhan667@gmail.com

R. B. Singh et al. (eds.), *Urban Health Risk and Resilience in Asian Cities*, Advances in Geographical and Environmental Sciences,
https://doi.org/10.1007/978-981-15-1205-6_25

25.1 Introduction

Neighbourhood is simply the vicinity in which people live, like the area around the house and neighbourhood socio-economic status refers to the socio-economic status of the residential area. There is a growing awareness that purely individual-based explanations of causes of ill health are insufficient. Hence, there is a need to consider not only individual characteristics but also characteristics of the neighbourhood in which people live. Neighbourhood level effects on human health have gradually become acknowledged as potentially significant factors of health disparities (Messer et al. 2006). Neighbourhoods or localities have appeared as a very important context as they clearly hold both physical and social features which can feasibly disturb the health of the residents (Diez-Roux 2007). Evidence is accumulating that the place where a man or woman lives may affect their health, even after controlling for individual risk factors (Stafford and Marmot 2003). It has been found that it is not only individual socio-economic status (SES) but neighbourhood socio-economic status too, which can influence the health of the residents. Various recent studies have suggested that neighbourhood socio-economic status (SES) can independently affect an individual's health above and beyond individual socio-economic status (Lemstra et al. 2006). A number of studies have found that residents of low-income neighbourhoods suffer from poor health than that of high-income neighbourhoods. Major serious environmental problems related to water supply, sanitation, air pollution, solid waste and housing, etc. are the problems found in all neighbourhoods but particularly abundant in poor neighbourhoods (Songsore and Macgranahan 1993) and impact of major environmental problems on the health of the residents is now a well-known fact. Other studies have also concluded that households in lower income neighbourhoods are more likely to be vulnerable to local environmental issues as compared to those in middle- and higher income residential areas (Rahman 2008). Therefore in the present study, an attempt has been made to examine the socio-economic status and environmental quality of different neighbourhoods in Azamgarh City to find out the association between neighbourhood socio-economic status and neighbourhood environmental quality and to understand the distribution of health and diseases among the residents of different neighbourhoods.

25.2 Study Area

Azamgarh city is a small-sized city located in the eastern part of Uttar Pradesh in the fertile Gangetic plain (Fig. 25.1). The city lies between 26° 4′ north latitude and 83° 1′ east longitude. It comprises 25 wards. The total population of Azamgarh City is 110,892, out of which 52.3% are males and 47.7% are females. Azamgarh City has 12.91% SC population and 918 sex ratio (District Census Handbook and Azamgarh 2011) (Fig. 25.1).

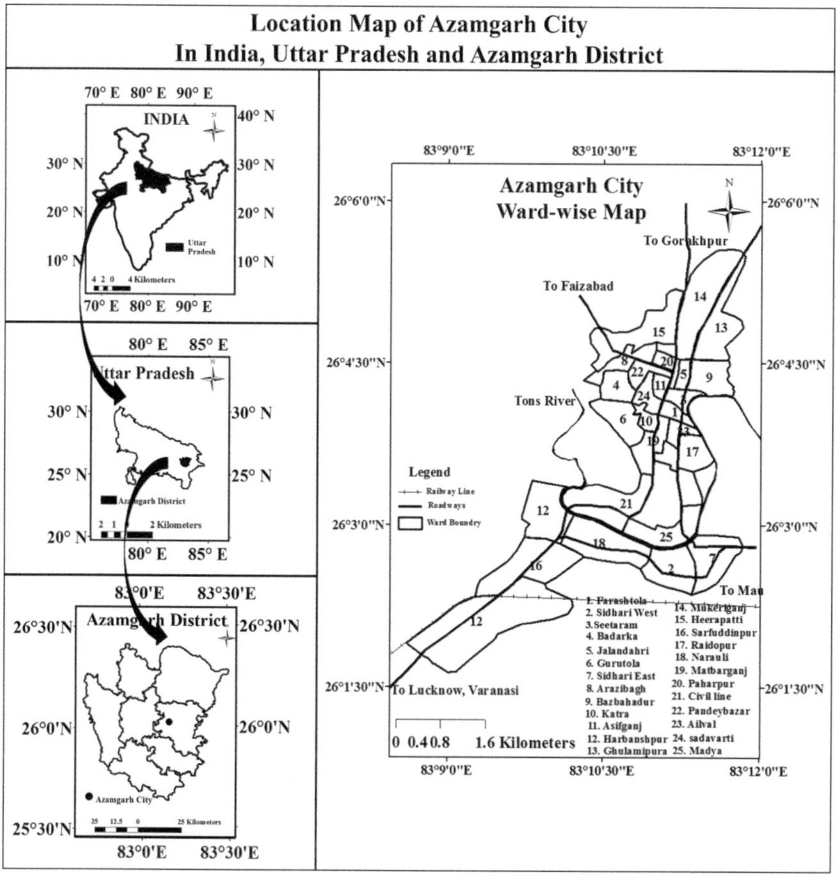

Fig. 25.1 Location map of Azamgarh City. *Source* (Municipal Office, Azamgarh City 2017)

25.3 Database and Methodology

- This study is based on primary as well as a secondary source of data which was collected with the help of questionnaire and interviews. The survey was conducted in the year 2017–2018.

The following methodology has been employed.

- Administrative wards of the city are taken as a neighbourhood unit for micro-level analysis. 10% households from each ward were taken for the study. The total sample size consisted of 1629 households.
- Neighbourhood socio-economic status (SES) has been measured by socio-economic and housing condition of households of different wards.
- Household's socio-economic status is measured by Kuppuswamy's scale of socio-economic status (SES), based on three variables, i.e. education and occupation of

the head of the households and income of the family (Raj et al. 2015). Housing condition is measured by type and material of houses, i.e. *pukka*, *semi-pukka* or mixed and *Kuccha* houses.

- Environmental quality of the neighbourhood is measured by perception of air quality, noise and traffic, crowding and congestion, water supply and quality, perception of solid waste and water stagnation, green and open spaces and perception of crime in the neighbourhood (Anderson et al. 2009). Problems that are prevalent in the city were selected for analysis.
- The data obtained has been standardized into standard scores based on the Z score technique for identifying high, medium and low socio-economic status (SES) of the neighbourhood as well as their environmental quality.

25.4 Results and Discussion

25.4.1 Level of Neighbourhood Socio-economic Status in Azamgarh City

According to Kuppuswamy's scale of SES, on the basis of education and occupation of the head of the households and income of the family, households of Azamgarh City have been divided into the following five categories.

1. Upper
2. Upper Middle
3. Lower Middle
4. Upper Lower
5. Lower

Neighbourhood SES in Azamgarh City has been measured by

(i) Ward-wise percentage of low SES households (includes lower and upper lower category of Kuppuswamy's SES scale).
(ii) Ward-wise percentage of substandard housing (percentage of *kuccha* and *semi-pukka* houses in the ward).

Household SES data and percentage of substandard housing are converted into Z score to obtain standard scores, and further composite Z score for each ward has been calculated to obtain neighbourhood SES of each ward (Table 25.1).

Table 25.1 Ward-wise percentage and Z score of household SES and substandard housing in Azamgarh City

Ward no.	Percentage low SES households	Percentage substandard housing	Z low SES households	Z substandard housing	Z composite
1.00	62.96	44.94	1.64	1.37	1.5
2.00	7.14	8.33	−1.73	−1.41	−1.57
3.00	53.33	10.00	1.05	−1.28	−0.11
4.00	25.71	34.99	−0.61	0.61	0
5.00	65.00	43.00	1.76	1.22	1.49
6.00	30.00	25.00	−0.35	−0.14	−0.25
7.00	37.04	24.95	0.07	−0.15	−0.04
8.00	25.93	36.00	−0.60	0.69	0.04
9.00	43.33	22.80	0.45	−0.31	0.07
10.00	45.00	24.00	0.55	−0.22	0.17
11.00	27.78	22.50	−0.49	−0.33	−0.41
12.00	69.7	27.40	2.04	0.04	1.04
13.00	26.47	26.67	−0.57	−0.02	−0.29
14.00	13.34	14.00	−1.36	−0.98	−1.17
15.00	12.00	8.33	−1.44	−1.41	−1.42
16.00	30.00	40.67	−0.35	1.04	0.34
17.00	25.00	35.00	−0.66	0.61	−0.02
18.00	50.00	42.33	0.85	1.17	1.01
19.00	25.00	5.00	−0.66	−1.66	−1.16
20.00	36.85	24.00	0.06	−0.22	−0.08
21.00	24.00	5.00	−0.72	−1.66	−1.19
22.00	54.17	49.44	1.11	1.71	1.41
23.00	45.00	43.64	0.55	1.27	0.91
24.00	35.28	29.00	−0.04	0.16	0.06
25.00	26.66	25.60	−0.56	−0.10	−0.33

Source Based on the field survey (2017–2018)

25.4.2 *Identification of Neighbourhoods According to Their Socio-economic Status*

With the help of data collected from field survey and use of composite Z score technique (Table 25.1), an attempt has been made to identify the socio-economic status of different wards in Azamgarh City. The neighbourhoods of Azamgarh City have been categorized into three levels (Table 25.2 and Fig. 25.2).

- Most deprived neighbourhoods

Table 25.2 Level of neighbourhood socio-economic status in Azamgarh City

Neighbourhood socioeconomic status	Composite Z score range	No, of wards	Percentage of the total wards	No, of wards	Name of wards
Most deprived	> 0.88	6	24	1, 5, 12, 18, 22, 23	Farashtola, Jalandhari, Harbanshpur, Narauli, Pandeybazaar, Ailval
Less deprived	0.88 to –0.88	14	56	3, 4, 6, 7, 8, 9, 10, 11, 13, 16, 17, 20, 24, 25	Seetaram, Badarka, Gurutola, Sidhari East, Arazibagh, Bazbahadur, Katra, Asifganj, Ghulami ka pura, Sarfuddinpur, Raidopur, Paharpur, Sadavarti, Madya
Least deprived	< –0.88	5	20	2, 14, 15, 19, 21	Sidhari West, Mukeriganj, Heerapatti, Matbarganj, Civil Line

Source Based on the field survey (2017–2018)

- Less deprived neighbourhoods
- Least deprived neighbourhoods

High SES neighbourhoods are considered as least deprived neighbourhood, medium SES neighbourhoods as less deprived neighbourhoods and low SES neighbourhoods as least deprived neighbourhoods (Lee et al. 2007).

Most deprived neighbourhoods include 6 wards namely Farashtola, Jalandhari, Harbanshpur, Narauli, Pandeybazaar and Ailval. These are basically low-income wards, primarily occupied by poor population. Mostly these wards lie in a very congested central old part of the city, along the river Tons and dominated by poor population compromising both Hindus and Muslims. As far as housing condition is concerned, there is a high density of low-income households living in slums and dilapidated housing. Majority of the residents in these wards are less educated and engaged in menial jobs like vendors, drivers, tailors, etc.

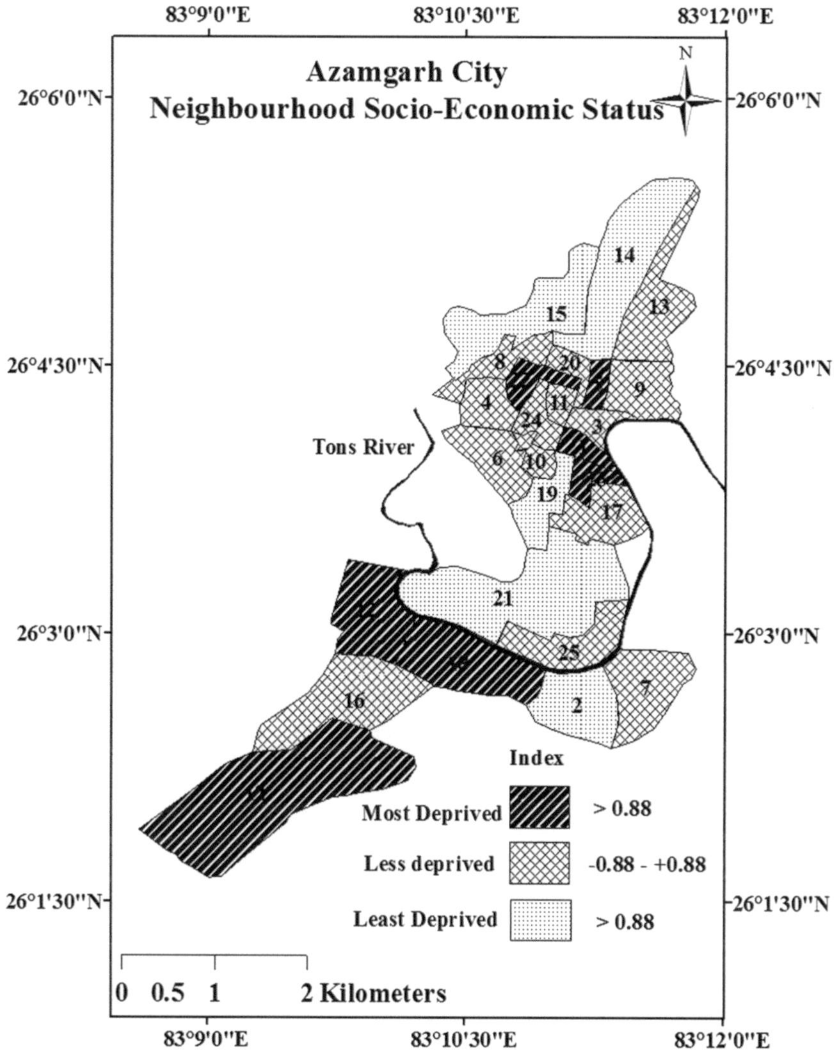

Fig. 25.2 Identification of neighbourhoods according to their socio-economic status

Second category includes less-deprived wards namely Seetaram, Badarka, Gurutola, Sidhari East, Arazibagh, Bazbahadur, Katra, Asifganj, Ghulami ka pura, Sarfuddinpur, Raidopur, Paharpur, Sadavarti and Madya. This category covers a major part of the city. These wards are having a mixed type of population, compromising high-, medium- as well as low-income households. A variety of housing type is observed in these wards ranging from well-furnished *pukka* houses to *semi-pukka* and *kuccha* houses.

The third category includes least-deprived wards namely Sidhari West, Mukeriganj, Heerapatti, Matbarganj and Civil Line. The major part of these neighbourhoods is located in the outer part of the city and inhabited by high-income counterparts of the city mainly business and service class residents (Fig. 25.2).

25.4.3 Level of Neighbourhood Environmental Quality in Azamgarh City

Environmental quality around immediate housing has an important influence on the lives of the residents. People's health is strongly affected by the robustness of their physical environment. The effect of pollutants and hazardous materials on people's health is substantial. Environmental quality also matters innately because most people cherish the beauty and health of the place where they reside and care about the degradation of environmental quality (Streimikiene 2015). In those areas where environmental quality deteriorates, i.e. proper supply of water, adequate drainage system, sewers, facilities like garbage collection and electricity are lacking, air and water quality is worsened, people are most susceptible to health damages. In this part, an attempt is made to present an overview of environmental quality in different neighbourhoods of Azamgarh City. Environmental quality in Azamgarh City is measured by the existence of major environmental problems like air pollution, noise pollution, incidence of crime, litter on the streets, crowding and congestion, stagnant water in the neighbourhood, problems with water supply and quality and absence of open green spaces, based on resident's perception.

Data collected from 1629 households living in 25 wards of Azamgarh City has been presented in Table 25.3. This table shows the distribution of major environmental quality indicators in Azamgarh City. It is observed that air and noise pollution is a major problem in wards Sarfuddinpur (100%), Asifganj (66.67%), Civil Lines (64.28%), Bazbahadur (50%), Narauli (50%) and Farashtola (35 and 70%) where most of the residents have reported these problems. Traffic is the main source of worsening air quality in these wards due to the presence of central market as well as their location in such parts of the city, which are inevitable for transport uses. Similarly, incidence of crime is also a major environmental issue faced by the residents. Crimes like theft, robbery, street fighting, rowdy group activity and drug dealing are abundant in wards mainly in Badarka (80%), Seetaram (76.45%), Bazbahadur (71.45%), Narauli (68.56%), Katra (61.93%), Farashtola (57.77%) and Pandeybazaar (55.97%). Litter on the streets is a major issue reported by 65–80% households, in almost all the wards due to poor waste collection and disposal services in the city. Crowding and congestion were also reported high by almost all the wards except some less dense wards. However, it was reported by more than 70% households in wards like Katra (100%), Jalandhari (95.87%), Badarka (90%), Bazbahadur (89.90%), Seetaram (85.30%), Paharpur (79.33%) and Matbarganj (73.56%). As far as waterlogging is concerned, stagnant water in the neighbourhood was reported high in Farashtola

Table 25.3 Ward-wise neighbourhood environmental quality and composite Z score

Ward no.	Air pollution*	Noise pollution*	Incidence of crime*	Litter on the streets*	Crowding and congestion*	Stagnant water*	Problem with water supply and quality*	Absence of open green spaces*	Composite Z Score
1	35.00	70.00	57.77	94.66	64.33	69.99	58.57	77.00	1.02
2	25.00	23.00	16.82	59.00	18.23	37.78	38.62	43.64	−0.77
3	0.00	0.00	76.45	89.70	85.30	55.65	41.43	77.00	0.28
4	30.00	30.00	53.93	80.00	90.00	40.11	40.00	100.00	0.24
5	0.00	0.00	45.45	80.31	95.87	48.25	46.00	100.00	0.18
6	0.00	0.00	49.77	84.66	54.65	65	53.75	100.00	0.38
7	0.00	0.00	24.42	70.00	50.00	47.67	44.29	33.34	−0.52
8	0.00	0.00	24.21	50.00	40.00	45.67	44.00	57.00	−0.65
9	50.00	50.00	71.45	79.00	89.90	64.88	54.00	100.00	1.02
10	0.00	0.00	61.93	60.00	100.00	63.85	47.00	100.00	0.31
11	66.67	66.67	10.17	66.67	66.66	33.33	39.53	100.00	−0.01
12	85.00	100.00	9.22	100.00	30.00	52.38	40.40	15.00	0.16
13	0.00	0.00	45.70	74.66	74.01	68.54	46.00	33.33	−0.02
14	25.00	38.65	23.75	39.85	45.86	38.78	41.56	27.00	−0.69
15	0.00	0.00	17.53	50.00	0.00	32.86	34.94	30.00	−1.28
16	100.00	100.00	9.22	75.00	0.00	53.46	53.33	50.00	0.32
17	0.00	0.00	30.51	50.00	43.67	61.89	38.56	17.86	−0.7
18	50.00	50.00	68.56	90.00	59.41	63.54	50.00	38.77	0.63
19	30.00	56.67	18.00	63.52	73.56	46.67	43.23	77.00	−0.03

(continued)

Table 25.3 (continued)

Ward no.	Air pollution*	Noise pollution*	Incidence of crime*	Litter on the streets*	Crowding and congestion*	Stagnant water*	Problem with water supply and quality*	Absence of open green spaces*	Composite Z Score
20	0.00	0.00	40.45	96.71	79.33	68.25	40.00	90.00	0.24
21	64.28	64.28	31.44	53.21	0.00	45.78	33.75	8.21	−0.64
22	50.00	30.00	61.84	85.32	64.33	66.33	47.40	67.00	0.58
23	15.30	17.83	55.97	84	61.21	73.12	39.22	43.00	0.14
24	0.00	0.00	51.57	84.66	67.45	61.33	40.00	100.00	0.13
25	33.33	33.33	33.73	75.00	33.33	39.57	42.20	40.00	−0.33

Source Based on the field survey (2017–2018) *Based on resident's perception

(69.99%), Ailval (73.12%), Ghulami ka pura (68.54%), Paharpur (68.25%), Gurutola (65%), Bazbahadur (64.88%), Katra (63.85%), and Narauli (63.54%). Problem with water supply and quality was found higher in wards Farashtola (58.57%), Gurutola (53.75%), Sarfuddinpur (53.33), and Narauli (50%). Water problems were mainly linked with municipal water due to erratic supply and most of the times it comes with dirt, colour and odour due to inadequate cleaning of water tanks and pipes. The table also revealed that there is lack of open green spaces in the city especially in wards Badarka, Jalandhari, Gurutola, Bazbahadur, Katra, Asifganj and Sadavarti where 100% residents have claimed absence of green spaces.

25.4.4 Identification of Neighbourhoods According to Their Environmental Quality

An attempt is made to categorize wards according to their environmental quality. For this, the data obtained is standardized into standard scores based on the Z score technique (Table 25.3). The neighbourhoods of Azamgarh City have been categorized into three levels (Table 25.4 and Fig. 25.3).

- Most degraded neighbourhoods
- Less degraded neighbourhoods
- Least degraded neighbourhoods

Most degraded neighbourhoods include 4 wards namely Farashtola, Narauli, Pandeybazaar and Bazbahadur. Environmental quality in these wards are worst reporting majority of the problems like air and noise pollution, litter on the streets, crowding and congestion, stagnant water and absence of open green spaces (Table 25.3). These wards cover 16% of the total wards of the city.

Less degraded neighbourhoods include 15 wards namely Seetaram, Badarka, Jalandhari, Gurutola, Sidhari East, Katra, Asifganj, Ghulami ka pura, Sarfuddinpur, Matbarganj, Paharpur, Ailval, Sadavarti and Madya. These wards cover 60% of the total area of the city inhabiting more than half of the population. These wards are not as degraded as wards of the first category, however, environmental problems are reported far from satisfactory in them.

Six wards come under the category of least degraded neighbourhoods namely Sidhari West, Mukeriganj, Heerapatti, Raidopur, Civil Line and Arazibagh. These wards cover 24% of the total area of the city and represent the highest level of environmental quality in the city.

The relationship between neighbourhood SES and environmental quality among the wards of Azamgarh City is dimensionally shown in Fig. 25.4. The abscissa represents the environmental quality and ordinate shows the neighbourhood SES. Three wards, Farashtola, Narauli and Pandeybazaar, represent the most deprived neighbourhood socio-economic status as well as most degraded environmental quality. Eleven wards, Sarfuddinpur, Sidhari East, Madya, Gurutola, Katra, Badarka, Asifganj, Sadavarti, Jalandhari, Seetaram and Ghulami ka pura, fall into the category of

Table 25.4 Level of neighbourhood environmental quality in Azamgarh City

Category	Composite Z score range	No. of wards	Percentage of the total wards	No, of wards	Name of wards
Most degraded	> 0.57	4	16	1, 9, 18, 22	Farashtola, Narauli, Pandeybazaar, Bazbahadur
Less degraded	0.57 to –0.57	15	60	3, 4, 5, 6, 7, 10, 11, 12, 13, 16, 19, 20, 23, 24, 25	Seetaram, Badarka, Jalandhari, Gurutola, Sidhari East, Katra, Asifganj, Ghulami ka pura, Sarfuddinpur, Matbarganj, Paharpur, Ailval, Sadavarti, Madya
Least degraded	< –0.57	6	24	2, 14, 15, 21, 17, 8	Sidhari West, Mukeriganj, Heerapatti, Raidopur, Civil Line, Arazibagh

Source Based on the field survey (2017–2018)

less deprived for neighbourhood socio-economic status as well as less degraded in case of environmental quality. Four wards, Mukeriganj, Heerapatti, Civil Lines and Sidhari West, fall into the category of least deprived neighbourhood socio-economic status having least degraded environmental quality. However, two of the less deprived socio-economic status wards namely Arazibagh and Raidopur show least degraded environmental quality because on an average they come under the category of less deprived, but most parts of these are occupied by high-class residents and environmental services are performed satisfactorily under their influence. Three most deprived wards, i.e. Jalandhari, Harbanshpur and Ailval show less degraded environment quality which in case of Jalandhari and Ailval is due to minimal interference of traffic and consequently lesser air and noise pollution and in case of Harbanshpur due to of low incidence of crime and negligible crowding. One least deprived ward Matbarganj shows much less degraded environmental quality due to the presence of the market leading to crowding and traffic flow which is responsible for air and noise. However, except these 6 wards which represent 24% of the city area, in the remaining 76% wards neighbourhood environmental quality follows the same distribution as neighbourhood socio-economic status.

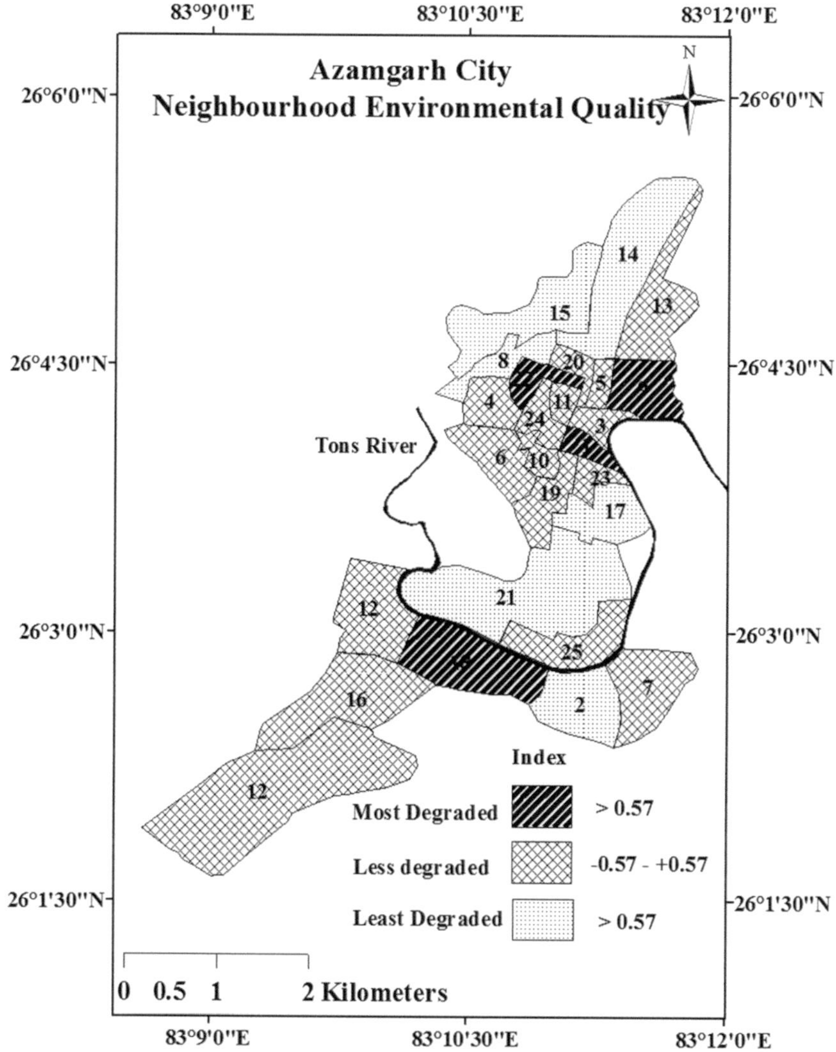

Fig. 25.3 Identification of neighbourhoods according to their environmental quality

25.5 Conclusion

The foregoing analysis revealed that environmental qualities of neighbourhoods vary with their socio-economic status. Most deprived neighbourhoods are associated with most degraded environmental quality and least deprived neighbourhoods show the least degraded quality of the environment. Most and less deprived neighbourhoods show deteriorating environmental quality in the form of various environmental issues like air pollution, noise pollution, incidence of crime, litter on the streets, crowding

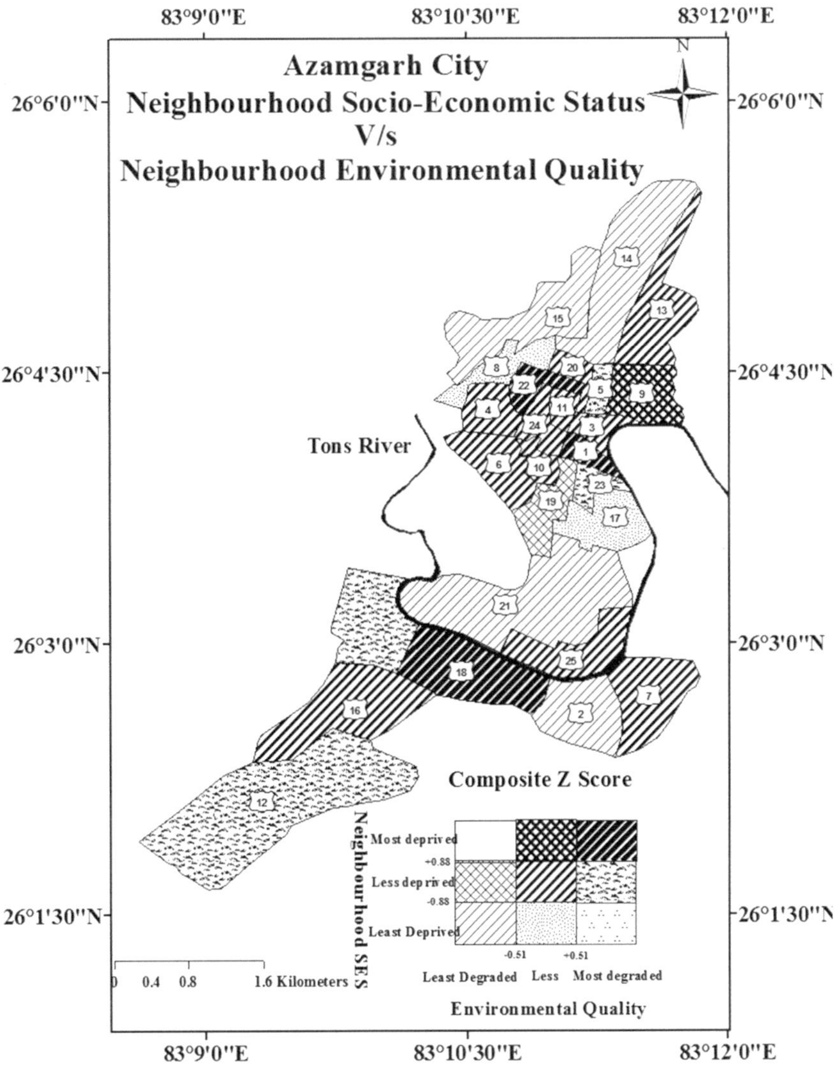

Fig. 25.4 Neighbourhood socio-economic status versus environmental quality

and congestion, stagnant water in the neighbourhood, problems with water supply and quality and absence of open green spaces. Therefore, residents living in deprived neighbourhoods are more susceptible to health damages due to poor environmental quality, which is deteriorating their health. The study thus concludes that, to improve the health of the residents, improvements in the environmental conditions at the local neighbourhood level should be done with proper planning to offer good environmental quality to the residents which in turn will lead to decreased vulnerability to diseases and poor health.

References

Anderson R, Mikuliç B, VermeylenG, Lyly-Yrjanainen M, Zigante V (2009) Second European quality of life survey-overview

Diez-Roux AV (2007) Neighborhoods and health: where are we and where do we go from here? Revue d'epidemiologie et de santepublique 55(1):13–21

District Census Handbook, Azamgarh (2011)

Lee RE, Cubbin C, Winkleby M (2007) Contribution of neighbourhood socioeconomic status and physical activity resources to physical activity among women. J Epidemiol Community Health 61(10):882–890

Lemstra M, Neudorf C, Opondo J (2006) Health disparity by neighbourhood income. Can J Public Health/Rev Canadienne de Sante'ePublique, pp 435–439

Messer LC, Laraia BA, Kaufman JS, Eyster J, Holzman C, Culhane J, O'campo P et al (2006) The development of a standardized neighborhood deprivation index. J Urban Health 83(6):1041–1062

Municiple Office, Azamgarh City (2017–2018)

Rahman M (2008) Urban environmental problems in Bangladesh: a case study of Chittagong city. Doctoral dissertation, University of Hull

Raj G, Shilpa S, Maheshwaran R (2015) Revised socio-economic status scale for urban and Rural india–revision For 2015. Socioecon: Sci J Theory Pract Socio-Econ Dev 4(7):167–174

Songsore J, Macgranahan G (1993) Environment, wealth and health; towards an analysis of intra urban differentials within greater accra metropolitan area. Environ Urban 5:2

Stafford M, Marmot M (2003) Neighbourhood deprivation and health: does it affect us all equally? Int J Epidemiol 32(3):357–366

Streimikiene D (2015) Environmental indicators for the assessment of quality of life. Intellect Econ 9(1):67–79

MIX
Papier aus verantwortungsvollen Quellen
Paper from responsible sources
FSC® C105338

If you have any concerns about our products,
you can contact us on
ProductSafety@springernature.com

In case Publisher is established outside the EU,
the EU authorized representative is:
Springer Nature Customer Service Center GmbH
Europaplatz 3, 69115 Heidelberg, Germany

Printed by Libri Plureos GmbH
in Hamburg, Germany